Manufacturing African Studies and Crises

Paul Tiyambe Zeleza

CODESRIA BOOK SERIES

Manufacturing African Studies and Crises

First published in 1997 by CODESRIA
Reprinted in 2003

CODESRIA is the Council for the Development of Social Science Research in Africa, headquartered in Senegal. It is an independent organisation whose principal objectives are facilitating research, promoting research-based publishing and creating multiple fora geared towards the exchange of views and information among African scholars. Its correspondence address is:

B.P. 3304, Dakar - Senegal

ISBN: 2-86978-066-4 (soft back)
       2-86978-067-2 (hard back)

Cover designed by Alla Kleekpo
Typeset by Sériane Camara, CODESRIA
Printed by Antony Rowe Ltd.
Distributor : The African Books Collective

CODESRIA would like to express its gratitude to the Swedish Agency for Research Cooperation (SAREC), the International Development Research Centre (IDRC), the Ford Foundation, the Norwegian Ministry of Foreign Affairs and the Danish Agency for International Development (DANIDA) for support of its research and publication activities.

CONTENTS

PREFACE

In November 1995, I attended for the first time, within the United States, the Annual Meeting of the African Studies Association held in Orlando, Florida. It was the largest academic conference I had ever been to. I met a lot of old colleagues and friends that I had not seen for many years, put faces to names that I had read in print, and made acquaintance with people who have since become friends. It was an impressive gathering certainly to someone recently moved from Canada where our African studies conference earlier in the year, which I and a colleague had organized at Trent University, was a small, intimate affair of a couple of hundred people or so.

What left an indelible impression on my mind, however, were not the seminars I attended, or the animated conversations with colleagues late into the night, let alone the fabled tourist attractions of Orlando, but one particular session on 'Ghettoizing African Studies?: The Question of Representation in the Study of Africa'. It was prompted by Philip D. Curtin's (1995) piece 'Ghettoizing African History', which appeared in the *Chronicle of Higher Education* in early March, 1995, in which he decried the growing numbers of Africans and African Americans teaching African history in American universities and the consequent 'lowering' of scholarly standards.

The piece ignited furore. I was still in Canada when it appeared, but I recall receiving frantic faxes from friends in the United States and burning the telephone wires with outraged conversations. Its notoriety soon crossed the Atlantic to Africa itself where it was reprinted in the Harare-based *Southern African Political Economy Monthly* and the Dakar-based *Codesria Bulletin* together with the vigorous rebuttal from dozens of African and African American historians (Black Historians Response, 1995). The hall where the session was held was packed to capacity. There must have been several hundred people in attendance. The place was palpable with anger, anguish, and anxiety. The speeches, interventions, and exchanges were sharp, sardonic, spiteful. The idea for this book began taking shape.

A few years before I had published a short story, which opens this book, on the growing antagonisms and gulf between Africans and Africanists in the study of Africa. There were some uncanny similarities between the story, which had been provoked by unpleasant personal and African colleagues' encounters with some Canadian Africanists, and what was unfolding before my eyes in that heated hall. In the days and weeks that followed the ASA conference the 'Curtin debate' continued on the Internet and doubtlessly in myriad phone conversations and personal meetings. I decided to revisit and revise some of the essays I had written on the paradigms and practices in African studies, to unravel the discursive practices and political processes behind the widening chasm. Several friends, especially Thandika Mkandawire and Mamood Mamdani had been urging me to do this for some time. Over the years, first as a graduate student in England and Canada in the late 1970s and early 1980s, and later as a teacher in Jamaica, Kenya, and Canada in the 1980s and 1990s I had become struck by a number of disconcerting tendencies in Africanist scholarship.

Much of the Africanist literature was self-referential: few paid attention to works by African scholars, at least going by their citations and analytical preoccupations which swung with every fad passing through the western academies. Many seemed anxious to play out Claude Mackay's poetic indictment of Gertrude Stein: 'eternal faddists who exist like vampires on new phenomena' (Echeruo 1996:175). Each generation produced its Livingstones who rediscovered Africa through the prevailing epistemological fad. Thus, Africa always appeared as nothing more than a testing site for theories manufactured in the western academies. In the 1960s, modernization was all the rage, then it was dependency theory at the turn of the 1970s, soon followed by modes of production, and since the mid-1980s Africa has been analyzed through the unrelieved gloom of 'Afropessimism' or the depoliticized posturings of post-structuralism, post-modernism, post-coloniality, and other post-prisons. Except for a brief moment in the late 1960s and early 1970s when 'radical' Africanists were looking for revolutions abroad following the failures of 1968 and read Fanon, Cabral and Nyerere, none of the subsequent theoretical fads were derived from an engagement with African social thought[1]. Indeed, from the 1980s as the Right gained ascendancy in one western capital after another, and following the end of the Cold War in the 1990s, discussions of African leaders, states, and societies were increasingly saturated with epithets, and there seemed to be a reputational lottery for those who could coin the most demeaning defamations of Africa and its peoples. This was one source of the growing rupture between Africans and Africanists in the study of Africa. Africans did not have the moral luxury of cursing their mothers, or dismissing their children's future.

A close reading of western writings on Africa clearly demonstrates that the regimes of representation are suffused with Africa's implacable

'otherness' and pathology. As we come to the end of the twentieth
century Africa remains a deeply contested intellectual and ideological
terrain; a continent that is perhaps still as misrepresented and
misunderstood as it was at the beginning of the century. Ever since
Africa's tragic encounter with Europe in modern times, each generation's
social imaginary of Africa, especially in the outside world, but
sometimes within Africa as well, has been dominated by powerful
metaphors and images through which Africa is constructed and
consumed, its histories and futures confiscated and condemned. To be
sure, the stereotypes that structure discourses about Africa mutate, but
each mutation carries with it past discursive genes, and the prevailing
social rhetoric always sets Africa up against the current conceptions of
western modernity. In the days of the slave trade, African 'paganism'
and 'primitivity' condemned millions of Africans to slavery. Later
African 'backwardness' and 'laziness' rationalized colonial conquest and
exploitation. After independence 'development' became the conduit for
neo-colonial interventions. Most recently, 'democratization' has been
added to the ideological repertoire, with the West presenting itself as
Prospero to Africa's Caliban. Almost invariably, then, Africa is
constructed or reconstructed as a representation of the West's negative
image, a discourse that, simultaneously, valorizes and affirms Western
superiority and absolves it from its existential and epistemological
violence against Africa[2]. The arrogant mobilization and deployment of
this discursive power in recent years in Africanist scholarship has
widened the rupture between Africans and Africanists.

This book, then, is an interrogation of African studies, its formulations
and fetishes, theories and trends, possibilities and pitfalls. As a discursive
formation, African studies is, of course, immersed in the contexts and
configurations of the western epistemological order. The state of flux,
some would say crisis, in African Studies in North America and
apparently in Britain[3], reflects changing cultural politics as a result of the
shifting ethnic and gender composition of classrooms, transformations in
the global positions of these countries, and the crisis of liberal values,
which manifests itself in the academy in the savage wars over curriculum
and canon, 'multi-culturalism' and 'political correctness', and in the
wider society in battles over the moral and fiscal boundaries of the
welfare state and the politics of identity and entitlement[4]. This explains
why Curtin's diatribe was received with dismay and indignation by
African and African American scholars: here was a renowned and senior
Africanist academic joining hands, whether wittingly or unwittingly, with
rightwing zealots who argued that affirmative action had gone too far,

that it had turned into 'reverse discrimination' and should therefore be scrapped. And he was launching his attack on their competence precisely at the moment that Murray and Hernstein's (1994) book, *The Bell Curve,* was proclaiming from all the mainstream best-seller lists that Blacks were genetically inferior to the other races. On the continent itself African scholars were perturbed by the analytical and prescriptive collusion between the international financial institutions and many Africanists over structural adjustment programmes, which the Africans attacked for undermining their countries' economies and their own intellectual production and reproduction. The parallels were too close for comfort.

These issues must be faced squarely if the study of Africa within the continent and abroad is not to diverge further and if the cultural antagonisms are to be transformed into creative agonisms. The challenges also lie on the African side. As argued in several chapters in this book, African institutions and scholars have not always been tolerant of each other, or welcoming to other African scholars, let alone African Americans. The need for academic democratization and decency, productivity and panAfricanism is compelling, indeed. Above all, there is an imperative case for public intellectual involvement, tethering theoretical paradigms and scholarly activity to actual social forces and struggles, for engagement with the burning questions of the day. African scholars cannot afford the disengaged academic recreations of faddish theorizing others seem to be able to indulge in. Their countries and communities cry out for clear and committed analyses, not the superficial travelogues they often get from foreign fly-by-night academic tourists.

The book consists of five sections, each a cluster of a set of the epistemological issues that have dominated African studies in the last few decades as filtered through my personal and generational encounters with the field. The analysis is conducted at two levels sometimes in the same chapter or in different chapters within each section. First, it seeks to deconstruct the discursive architecture of African studies in general, and African history, development economics, political science, and literature in particular. Second, it tries to reconstruct alternative narratives, especially of the processes and crises of development and democracy. This double analytical agenda is based on a conviction that deconstruction without reconstruction is a futile intellectual exercise, that if we aspire to be organic intellectuals we must seek not only to expose what is wrong, but also the social forces that could make it right. It also underscores the several meanings in which African studies and crises are manufactured: both are concretely constituted historical processes and socially constructed discursive regimes. And just as crisis in Africa, indeed, in the world political and cultural economies, creates discourses of crisis and crisis of discourses, the latter arouse perceptions and authorize policies that, as with structural adjustment programmes, reinforce the actual crisis. This is to suggest that Africanist discourse is in crisis and sees nothing but crisis crippling its object of study. By Africanist I mean the entire intellectual enterprise of producing knowledge based on a western epistemological order in which both educated Africans and non-Africans are engaged.

The five chapters in Part One examine the processes and practices of knowledge production in African studies within and outside the continent, while those in Part Two focus on the development of African studies paradigms. Part Three opens up with an examination of the discourse on development, tracing its inauguration and implementation in Africa. This is followed by critiques of neo-classical analyses of the continent's agrarian and general economic crises of the 1970s and 1980s, and attempts to reconstruct the making of these crises and the emergence of the social forces that sought to unmake them. Part Four focuses on democracy, beginning with a critical assessment of conventional Africanist perspectives, and proceeds to unravel the complex, and sometimes contradictory, visions of freedom and democracy in post-independence African social thought through a reading of African fiction. The last section, Part Five, explores the contributions of African studies to the explosion and disintegration of western 'metanarratives' and 'metascripts', but argues that discarding Eurocentric systemic perspectives and theories should not take us into the political netherworld of post-modernism with its glib cultural hybridities and rootless subjectivities, and celebrations of ambivalences, uncertainties, and contingencies, for the structural claws of imperialism remain real; indeed, they are more powerful than ever - grabbing and penetrating all available global spaces. Struggles against imperialism and neo-colonialism, therefore, cannot be waged by atomized individuals preoccupied with reinventing their personal identities everyday. The section, and the book, ends up stating the case for re-imagining panAfricanism for the twenty-first century, for constructing transformative panAfricanist paradigms and politics.

Some of the chapters first appeared in journals or as book chapters. Except for the short stories and a couple of the chapters, the original essays have been extensively rewritten and I am grateful to the editors of the various journals and the publishers for allowing me the privilege to revisit and revise my thoughts. Among the unpublished essays is the current Chapter 23, which has been retained in the form it was first presented at Columbia University in April 1996 as a 'commentary' on readings from my fiction. The book begins with a short-story and ends with excerpts from two short-stories, for I think that in the end all we try to do as intellectuals is to tell stories. That is what makes us human, the capacity to imagine. I believe we can imagine an Africa that is different from the one we encounter in many Africanist writings and the western mass media, an Africa that is not a disembodied caricature of dismal economic statistics and distressing political eruptions, but an organic world of human beings in all their bewildering complexity.

I have shared the ideas and images contained in this book with many friends and colleagues over the years in different countries and under diverse circumstances. In Jamaica I thank Patrick Bryan for welcoming me to my first 'real' academic appointment, Swithin Wilmot for being a friend, and the late George Beckford for being an inspiration; in Kenya I will always be grateful to my three 'mentors' William Ochieng, Bethwell Ogot, and Kwesi Darkoh for their encouragement when I needed it so badly, and to Tabitha Kanogo for being a thoughtful colleague. In Canada I remember Jane Parpart's generosity, and at Trent University I cherish the support of Doug McCalla and Joan Sangster, both caring and productive History Department chairs, and to Joan also for the privilege of writing a joint article on academic freedom in Canada in the face of rising rightwing intellectual intolerance; in Champaign to Ezekiel Kalipeni for bringing the sunshine of Malawi to the cornfields of the Mid-West; and to the network of old and new friends and intellectual companions scattered all over the panAfrican map: Ayesha Imam and Amina Mama for deepening my understanding of gender analysis and much else besides; Dickson Eyoh and Alamin Mazrui for being there for political and personal conversations; my Codesria 'family' - Thandika Mkandawire, Mamood Mamdani, and Tade Aina for exemplary commitment to African intellectual liberation; Lynette Jackson and Pearl Robinson for shared devotion to panAfricanism; Jack Mapanje in admiration of his fine poetry and for sacrificing so much; Yvonne Vera for the love of literature and her evocative writing; and Tiyanjana Maluwa for nearly thirty years of friendship. And, as always, to my daughter, Natasha Thandile, whose exuberance gives me hope for tomorrow. It is to her and her friends that this book is dedicated: may they create a more humane world for themselves and the future.

Paul Tiyambe Zeleza Champaign, Illinois.

Notes

1. I am using the terms 'social thought' in the sense that Amin (1994) uses it to encompass a much broader, liberatory and humanistic system of knowledge formation than implied in the objectivist and scientific pretensions of the term 'social science.'

2. For recent studies on western images of Africa and Africans and constructions of Africa, see Miller, (1985, 1990), Mudimbe (1988, 1994), Hawk (1992), and Pieterse (1995).

3. The literature on African studies is growing. For earlier comparative studies see Fyfe (1976) and Jewsiewicki and Newbury (1986). For more recent studies of African studies in Britain, see Fage (1989, 1993) and Fyfe (1994); for Canada see Ray (1991); and for the United States see Guyer (1995), *Issue: A Journal of Opinion* (1995), and West and Martin (forthcoming).

4. For a succinct summary on cultural politics and international studies in the US academy, see Lee (1995); and see Marable and Mullings (1994) on the effects of changes in the US political economy on African Americans.

About the Author

Born in Harare, Zimbabwe, in 1955, and raised in Malawi, Paul Tiyambe Zeleza did graduate work in England and Canada, and has taught at universities in Malawi, Kenya, Jamaica, and Canada where he was Professor of History and Principal of Eaton College at Trent University. He is currently Professor of History and African Studies and Director of the Center for African Studies at the University of Illinois at Urbana-Champaign. He has published numerous books, including three works of fiction, and dozens of scholarly essays. He is the recipient of many fellowships, research grants, and literary awards, including the prestigious 1994 Noma Award for his book, *A Modern Economic History of Africa. Volume 1: The Nineteenth Century.* He has served on the executive boards of scholarly and arts organizations, as well as the editorial boards of several journals including currently, *Afrika Zamani* and *The Toronto Review*, and in 1974 was a founder member of the *Malawian Writers' Series.*

PART ONE

THE PERILS OF
ACADEMIC TOURISM

Chapter One

TRIAL OF AN ACADEMIC TOURIST

Professor Clegg was a renowned scholar. He was widely published and extensively travelled. He was revered, envied, and sometimes criticized, but never ignored, dismissed, or pilloried. At conferences he was swarmed by young, aspiring scholars with freshly minted Ph.D.s looking for his acquaintance and future recommendations for jobs and publications. His peers mobbed him, too, for a little of his intellectual glow, so that they could later tell their mesmerized students that they had personally debated Professor Clegg on his latest theory at a recent conference. He never wore a name tag, like everyone else. He was genuinely baffled if someone accidentally asked him his name.

Professor Clegg did not attend the African studies conferences regularly. He came once every three years. It was an occasion to behold. His papers were always delivered in the plenary session, where everybody would attend, and not in the stuffy, small, and half-empty conference rooms. He would shuffle into the hall, a short, stocky man, with ruffled white hair, rumpled clothes and glasses hanging on his forehead, and pronounce in his deep baritone voice his latest thoughts on Africa, and like acolytes at some religious ritual, the participants would hang on every word he uttered, observe every gesture he made, savouring them like revelations from an oracle. His reflections would be debated in the corridors for the next few days, and in journals and books for years, until his next appearance and divination.

Professor Clegg was a remarkable scholar, indeed. On his retirement, the Association for the Study of Africa decided to honour him. This was just one of a chain of tributes he received from his department, college, and university. Some of his former students, who were now distinguished scholars in their own right and scattered all over North America, had already presented him with a collection of essays. Over three decades Professor Clegg had also trained hundreds of African students at all levels. Some of them had become prominent scholars, university administrators, top government officials and business executives. It was one of these former students who came to haunt Professor Clegg at his farewell conference.

It was the association's best attended conference ever. The theme of the conference was 'African Studies: The Last Thirty Years', the span of Professor Clegg's career. When he started teaching and researching on Africa, one could count the number of Africanists in North America. It was a small band of dedicated men — for they were mostly men — excited by the implications of the emergence of new nations on global politics. Some were attracted by the thrill of cultivating an almost virgin field, others sought to test their pet theories. There were also those whose interest in African studies developed because of the growing civil rights movements, or a chance encounter with an African student, or curiosity triggered by school textbooks, films, and anecdotes by that eccentric uncle or grandfather who had been to Africa as a missionary, colonial official, or soldier, with their stories of an exotic, primitive, and faraway continent that needed salvation, modernization and development.

Professor Clegg was among the best of his generation, a pioneer who went where others feared to tread, and created and followed academic fashions as effortlessly as he discarded them. His mind had the agility of a leopard, the vision of a giraffe, and the slipperiness of a snake, so that he hopped from one region to another, and embraced and abandoned theories with consummate ease. He got his doctorate in political science in 1960, that heady year of African independence. For the next decade his turf was West Africa, and he wrote profusely on the politics of modernization and nation-building. When the wave of military coups started he lauded the military as the only cohesive force in the fragile post-colonial states that could ensure and promote development and national unity.

But the agony of the Nigerian Civil War disabused him of that, so that he deserted West Africa and the politics of modernization, and found a new sanctuary in East Africa and the politics of dependency, a theory he borrowed from a casual reading of Latin American history and society. Africa, now incorporated into the Third World, could not develop, he proclaimed, so long as it was integrated into the exploitative world capitalist system, which had drained the continent of its resources for centuries. Disengagement and self-reliance were the only road to salvation, to development. And so he fell in love with Tanzania. He spent a sabbatical year there, and published a big book extolling African socialism.

But his restless mind got bored with a socialism that looked decidedly timid when compared to the resolute anger and revolutionary slogans of the guerrillas routing Portuguese fascism just across the border in

Mozambique. So he packed his academic baggage and descended upon the new African Marxist paradise. He vowed passionately in another fat book that at last he had found Africa's future, one that worked. Every summer he took a safari to Mozambique, and occasionally to Angola, to marvel at, and help build, scientific socialism in the African bush. He read Mao and Fanon and realized that Marx was wrong to call peasants rural idiots; they were natural rebels, uncorrupted by the materialism and pretensions of the neo-colonial ruling elite. So he extolled them, celebrated their harsh lives, and dreamt about the nobility of their poverty from the comfort of his hotel room in Maputo and his study in Toronto.

Unfortunately, civil wars, sponsored by imperialist forces and the Rhodesians and South Africans, increasingly made life in Maputo and Luanda dreary, and Professor Clegg could not make his brief pilgrimages to his beloved peasants, now cut off by land mines in the ravaged hinterlands. In 1980 he trekked across the border into newly liberated Zimbabwe, and to the delight of his palate but the consternation of his mind he found all the comforts he was used to back in Toronto. The peasants were not as enchanting, the rhetoric of the new rulers not as radical, and so a year later he published a bitter tome denouncing the betrayal of Zimbabwe's revolution.

Professor Clegg began despairing about Africa. Development remained as elusive as ever. He had run out of hope, countries, and theories. Only South Africa remained, but it was an artificial European transplant, destined to survive in its splendid isolation for the foreseeable future. In Central Africa there was Zaire with its banal, kleptocratic regime. North Africa was not Africa, for it sat above the sands of the Sahara, and its people were not Black. Africa was Black. It was Black Africa, the 'Dark Continent', the ultimate other of Europe, White Europe, of White Western Civilization. He had never been to North Africa, or to the Middle East, as he thought of it, except one time when he stopped in transit at Cairo airport.

Thus, two decades after receiving his doctorate Professor Clegg had mastered Africa, traversed it from West to East to South and Central, and witnessed it squander its opportunities. He had weighty books and articles and promotions to show for it, and a smattering of words from a few African languages. Not learning an African language had not been a handicap. The Africans he dealt with spoke English, his mother tongue, or Portuguese, which he had studied when he discovered Mozambique and Angola, and for his cherished peasants he had interpreters.

He decided it was time to evaluate everything. Perhaps he had been wrong to assume that these countries could discover a new path to development. He had let guilt warp his judgement, guilt over Europe's enslavement of Africans for four centuries and an extra century of colonization. Africans should not be patronized by being absolved of their failings. Perhaps there was something in Reaganomics and Thatcherism that state intervention stifled growth, that social welfare killed enterprise. Africa had too much of both: the hand of the state was too long, and the grip of the economy of affection was too tight. There were no shortcuts to progress, except through age-old, exploitative capitalism, upon which the affluence of the West and Japan rested.

So it was that Professor Clegg stopped taking his academic safari trips to Africa during summer. He became immersed in administrative work, as Director of the African Studies Center, later Departmental Chair, and finally Dean. And he discovered the lucrative world of consultancy. He did consultancy work for Canadian government development agencies and NGOs, as well as the World Bank, and a host of United Nations and international business organizations. He finished paying his mortgage, bought a cosy cottage by Lake Ontario, and was able to pay for his three children through college with relative ease.

Consultancy was more than profitable. It bestowed the power of prescription, not just the impotent titillation of analysis. And it was certainly more congenial than trudging in search of petty politicians in some nondescript African capital and peasants in inaccessible regions. All he did was sit in his office, make appropriate phone calls, and reread previous reports on the subject, for what was required was the environmentally friendly art of recycling, of confirming conclusions already made. Only very occasionally was there need to actually go to Africa. And when he did he would spend a week at most, tanning and hiding in the local Hilton or Sheraton, away from his former friends and admirers at the local university. After a decent interval he would issue his report, with all the appropriate phrases, code words, charts, and recommendations.

But the academic in him was dissatisfied with seeing his name buried in reports that gathered dust in development aid offices. He craved for the intellectual distinction of a real book. Besides, his CV needed updating. So he published a massive volume on the need for African countries to undertake Structural Adjustment Programmes. He went beyond the normal recommendations of the mighty IMF and World Bank for currency devaluation, cuts in government expenditure, removal of protective tariffs, and privatization, and added the importance of 'good governance', a phrase once commonly used by British colonial governors to characterize their 'civilizing mission' in Africa. He stumbled upon it in a history book. That was a major intellectual breakthrough, this linkage of development with 'good governance', and Professor Clegg got more consultancies and was even asked to join a United Nations team of 'Wise Men' to divine Africa's future.

Soon 'good governance' became 'democratization', and Professor Clegg coined that term, too. Academic minds work in wondrous ways. He discovered the new term after Gorbachev came to power in the Soviet Union, and started talking of *perestroika* and *glasnost*. Professor Clegg was intrigued by Gorbachev's reforms. In fact, he decided to join the crowded bandwagon of Sovietologists. He felt there was nothing more one could say about Africa. His African experience had given him a special understanding of how reforms should work in multi-ethnic, corrupt and statist societies. And it was time to touch base with his roots, for his late mother was a Russian immigrant. The fall of the Berlin Wall in 1989 confirmed the wisdom of switching his research interest to Eastern Europe.

He published a book that was, by his standards, rather slim and tentative, on the Second Russian Revolution. It was ignored by the experts in the field. That came as a shock to Professor Clegg, long used to accolades in African studies. He found that his new field was dominated by recent Eastern European émigrés or intelligence experts, unlike African studies which was virtually an exclusive preserve, not of people of African descent, but of European-American and European-Canadian scholars like him. He decided to go back to the easier glories of African studies, to taste once again, that awe and reverence reserved for founders of secret societies.

Professor Clegg's return to African studies was prompted by the release of Nelson Mandela on 11 February, 1990. As he watched the picture of Mandela on television gingerly walking out of Victor Vester prison in Cape Town, dignified and unbowed, with his right hand raised in a fist of victory and defiance, beside a triumphantly smiling Winnie, whose hand was also raised, his love and fascination for Africa welled up inside him, and he decided right there and then to make South Africa his new area of study. He convened a conference on 'The Construction of a Post-Apartheid Democratic South Africa', to which he invited his former students who had become experts on Southern Africa, current students whom he had convinced to change their dissertation topics to South Africa, the exhilarating new frontier in African studies, and renowned South African scholars themselves, none of whom was African. It was a successful conference. It resulted in an impressive collection which he edited, and was quickly published to set the new agenda for South African studies.

Professor Clegg was a truly great scholar. He was visibly moved as he listened to the long citation from Professor Montgomery, President of the Association, and the tributes from Professor Willoughby, the President of the university, Professor McDonald, representing his former students from North America, and Mr. Davidson, his lifelong publisher. They praised him as a pioneering Africanist, one who had brought great honour to the university, an inspiring teacher who had moulded the young minds of a whole generation of North American Africanists, a tireless writer whose numerous publications were indispensable reading for an understanding of that most fascinating and turbulent of continents, Africa.

There was only one more speaker before the presentation of the gift. The great auditorium was filled to capacity. Some in the audience had their luggage with them, for this was the last day of the conference. Some were getting a little impatient, for they had already been there for almost two hours. They hoped the new speaker would be brief. They were gratified when he began by saying he would not read the speech he had prepared.

'My grandfather', Professor Muwonalero chuckled, 'used to tell us that words on paper do not breathe, so one cannot tell whether or not they are lying, that's why I won't read my prepared speech. I will speak from my heart'.

The gathering laughed, amused by this infusion of African folk wisdom.

'And there is also a saying in my language', he continued after the laughter had died down, 'that it is bad to lie to a dying man, for the lie haunts you after the man is dead. Retirement is a form of death'. Only a few laughed this time, many just grinned.

Some people could be seen whispering, wondering who this man was. Professor Muwonalero was representing Professor Clegg's former African students. He was apparently the latter's first African Ph.D. student. He had gone on to become a distinguished political scientist, a Cabinet minister, and was now back in academe as vice-chancellor of one of Nigeria's top universities.

'I am also not going to read my speech because of what happened last night'. There were a few audible sighs of both impatience and curiosity.

'Some students, African students, of Professor Clegg, came to see me at my hotel'. The men on the dais stirred a little uneasily. The President of the Association who was chairing scribbled something on a piece of paper and passed it on to Professor Muwonalero.

'I am told I only have three more minutes', he smiled weakly. 'So let me come to the point. Like all of you, I had not come here to criticize Professor Clegg, but to praise him. Many of us believe that only good things need to be said about a colleague and friend on retirement. I thought so, too, until the students who came to see me last night reminded me of the saying I told you earlier'.

The feeling of unease was spreading to the rest of the audience. It was at that time that the doors of the auditorium swung open and a group of African students rushed in carrying placards whose message shocked everybody. They called Professor Clegg a charlatan, an intellectual thief, or an academic tourist. Murmurs of shock shot through the audience.

'These students told me that you have been taking their work and publishing it as if it was your own. I believe they call that plagiarism'. Professor Muwonalero's voice was firm, with a touch of sarcasm. Gasps of dismay reverberated throughout the auditorium. Professor Montgomery stood up and whispered something to him.

'I will not be silenced!' Professor Muwonalero exclaimed, 'like I was silenced twenty-five years ago and like these students were silenced. They wrote to the President of this university and the President of the Association, but they were ignored. Professor Clegg', he paused and turned in his direction. Professor Clegg's face had become ashen white, as if blood had stopped flowing in his veins. 'Twenty-five years ago you published a book on Nigerian politics. Two of the chapters in that book were shoplifted word for word from draft chapters of my dissertation. Then you asked me to change my topic a little. I was afraid of confronting you then because I needed my certificate, that piece of paper called Ph.D., and I was afraid you would close all doors to me if I protested, just as the students who came to see me are afraid', he paused. There was deafening silence.

'Well, I am no longer afraid of you now, and these students need no longer be afraid of you. Those days are gone, long gone. We now control our own universities, and when these students apply for jobs, your recommendation will not count in Nigeria where you stopped coming after I and others threatened to expose you. Perhaps that is why you kept hopping from one African region and country to another. Now, I hear you have become an expert on South Africa. God bless them!'

Then Professor Muwonalero turned to face the stunned audience. 'I did not come here to condemn Professor Clegg, but to remember him. We have another proverb in my language, which says that, like an onion, a person has many layers. So does our esteemed Professor Clegg, and in appraising him we must not only see the outer layer of success, but also the inner layer which may have some rot. My grandfather used to tell me that it is more rewarding to tell people in power one small truth than a basketful of lies. Professor Clegg is a powerful man, indeed, otherwise none of us would have been here this afternoon praising him. He built a remarkable career, one, however, based on lies'.

The atmosphere in the auditorium became brittle with tension. Professor Muwonalero scanned the audience, mocking, taunting it. The students carrying the placards at the back chanted their approval.

'It's not just the lie of plagiarizing his students' work', he resumed in a raspy voice, 'or of coming to African campuses and collecting research papers that are later used without acknowledgement, but also the lie of irresponsibility, of not owning up to his advice when that advice goes wrong. You all sit here comfortably condemning African governments for all sorts of policy failures when it is you who devised many of those policies, and with the help of your governments, corporations, and agencies imposed them on those governments. Orwell's Thought Police would be proud of your Newspeak and Doublethink. Every five years or so, you concoct new policies, and dress them up as new theories and models. Is Africa for you nothing more than a lab, and its people disposable rats?'

The question hang over the auditorium like a heavy cloud before a storm. Some in the audience stood up, muttered to themselves, and angrily walked out. The chanting students jeered at them. Both presidents of the association and the university called for calm, and stared at Professor Muwonalero with utter disbelief. Who was this ingrate? Why hadn't he been carefully checked out? Was this what the so-called 'political correctness' movement was coming to, a vengeful crusade against innocent white males, white people?

'I can see my tribute to Professor Clegg makes some of you uncomfortable. You are so used to describing other people that you think it is impudent of me to describe you, the way I see you, the way I and my colleagues describe you in the security of our privacy? That's why you can come and jump across Africa without learning a single African language, conduct convoluted discourses among yourselves and in your journals and books about Africa, without the voice of Africans. That's the arrogance of power, the power of definition, of seeing and judging others on your own terms, which Europeans have been doing since the fifteenth century. A few centuries ago you were interested in appropriating our bodies, then with colonialism you appropriated our countries, and now you are trying to appropriate our souls'.

Professor Muwonalero wiped his brow as he watched more people leave, among them Professor Clegg. Professor Montgomery tried to hold him back, pointing to the unopened gift.

'How come thirty years later', Professor Muwonalero bawled above the din and the chants, 'if you think you have done such a great job of explaining Africa, your people still display the same level of ignorance and racism against Africans? And, where are the local African-Americans and African-Canadians in your association? How many of you have a true friend or a colleague from these communities? My best wishes to Professor Clegg. Let us follow his shadow, but avoid his footsteps', he ended cryptically.

Professor Muwonalero descended from the dais and walked out of the auditorium. There was pandemonium. Nobody could locate Professor Clegg. The older participants were reminded of a similar disturbance in 1969 in Montreal when Black activists broke into the association's annual conference denouncing the participants as bleeding heart liberals, at best, or irredeemable racists, who thought they had a divine right to define Africa and Africans. So much had changed, so little had changed.

Chapter Two

THE LIGHTNESS OF BEING AN AFRICAN EXPATRIATE SCHOLAR

Introduction

The lack of academic freedom in Africa is often blamed on the state. Although the role of the state cannot be doubted, the institutions dominated by the intellectuals themselves are also quite authoritarian and tend to undermine the practices and pursuit of academic freedom. Thus, the intellectual communities in Africa and abroad, cannot be entirely absolved from responsibility for generating many of the restrictive practices and processes that presently characterize the social production of knowledge in, and on, Africa[1]. In many instances they have internalized the coercive anti-intellectualist norms of the state, be it those of the developmentalist state in the South or the imperialist state in the North, and they articulate the chauvinisms and tyrannies of civil society, whether of ethnicity, class, gender or race.

This chapter focuses on the often forgotten phenomenon of African expatriate scholars working in African and overseas universities, of whom I have been one for many years. The experiences of these scholars throw into sharp relief the intellectual conditions, contradictions, and constraints prevalent in universities within and outside Africa today. The chapter attempts, first, to examine the factors that motivate these scholars to migrate from their respective countries to other African countries. The argument is made that intellectual considerations, such as research interest in the countries to which they migrate, tend to play an insignificant role. This has far-reaching consequences on the way these scholars are perceived and treated. Next, the restraints that these scholars face in the host countries compared to the indigenous and Western expatriate scholars are examined. Finally, the chapter contends that the environment for expatriate African scholars in many African countries has, for various reasons, deteriorated, resulting in the acceleration of the 'brain drain' to the western countries. The challenges that confront the African expatriate scholars in these countries are briefly explored. The conclusion is made that from these debilitating exilic conditions vibrant panAfrican intellectual networks and traditions can, should, and indeed, have been, and are being, constructed.

The Pushes and Pulls of Academic Labour Migration

In the last two decades or so, the African scholar has discovered labour migration, long the preserve of ordinary workers. The migration of academic workers across national boundaries has, in fact, grown, while that of the labourer has declined. The labour migration waves of the colonial and early post-independence periods have largely been replaced by the flow of refugees. General labour migration in Africa has declined in proportion to the consolidation of the exclusionary fictions of the nation-state, each jealously guarding its 'sovereignty' and 'national economy'. This has meant that only highly trained personnel have found it relatively easy to migrate freely. But the African academic migrants are, structurally, not all that different from their less fortunate compatriots who flock to the refugee camps. For the most part, the academics are themselves refugees.

Most African academic migrants flee from their countries for a number of political and economic reasons. At independence the universities most of which were created by the newly independent states as factories to churn out the personnel for the Africanization of state apparatuses and as emblems of cultural modernity, enjoyed cosy relations with the new nationalist rulers, many of whom were themselves intellectuals, or at least aspiring ones. But it did not take long for the honeymoon to turn sour. As liberal institutions, in inspiration and aspiration, the universities initially resisted turning into glorified party schools. They subjected the political rhetoric and dreams of nation-building and development to critical analysis. This did not enamour them to the new impatient and insecure ruling elites, who were unduly concerned by the trappings and realities of power. The drive for centralization and control that this led to pitted the universities, as vibrant mediators of civil society, against the state, which was increasingly flexing its authoritarian reflexes as the triumphs of nationalism were eclipsed by the challenges of independence. The universities came to be seen as potential saboteurs of the national mission, defined narrowly according to the shifting ideological, religious, ethnic, regional and class predilections of the incumbent regime.

The universities and critical academics were increasingly accused of being purveyors of 'foreign ideology', a charge that was as damning as it was hypocritical coming from leaders who themselves worshipped at the altars of modernization made in the capitalist West or the communist

East. It was damning because it reflected the contradictory mandate of the African university as a vehicle of modernization and the transmission of western cultural capital, on the one hand, and as crucibles through which national cultures could be forged out of the splendid or secessionist diversities simmering in the belly of the inherited colonial leviathan. As reservoirs of concentrated critical consciousness, despite their internal confusions and contradictions, the universities, almost by their very existence, frustrated and mocked the drive for political monopoly. And so their space was watched constantly, carefully, capriciously. Every attempt was made to tame them through concessions and closures, donations of fiscal support and detentions of fearless staff and students. In many countries once branded a 'subversive', through arrest or detention, it becomes virtually impossible to make it back to the university. In short, as relations between the universities and the state deteriorated, particularly as the euphoria of independence evaporated into the thick clouds of rumbling economic recessions and authoritarian structural adjustment programmes, many academics opted for migration for reasons of personal security and intellectual integrity.

The university is of course no ivory tower, living in splendid isolation from the rest of society, wrecked as is so often the case by political strife and social dislocation. Like the majority of refugees, the African academic migrants are usually displaced persons who vote with their feet for the uncertain safety and security of other lands. The conflicts in the wider society are condensed and reproduced in the university. Societal polarizations, combined with schisms within the university, create stubborn structures that generate and thrive on the persecution of those who do not stay within the prescribed intellectual and ideological boundaries. The university, as a haven of petty-bourgeois ambitions, aspirations and fantasies, engenders a culture of careerism and fierce competition, fertile breeding grounds for the transmission of political repression and intellectual persecution. Periods of civil conflict and social strife often reinforce the university community's fratricidal struggles for the limited spoils of academia. Political and intellectual witch-hunts rear their ugly heads. The victors, those who support the successful forces, are rewarded with the transient privileges of promotion and corporate appointments, until they, too, slip and fall in Africa's treacherous political quicksands. The victims, mostly those who belong to the 'wrong' ethnic or religious groups, or subscribe to 'subversive' political or ideological tendencies, pay with their jobs, freedoms, and, occasionally, lives. Many of them end up joining the ranks of refugees into the wilderness of intellectual exile.

Unlike the other labour migrants, however, who are pushed by the lack of employment opportunities at home, the academic migrants belong to the privileged end of the labour market in their home countries. After independence there was an enormous expansion in university education throughout Africa. Consequently, university employment grew rapidly. But this growth was accompanied by important structural and institutional changes in the politics of control of the university labour market. Virtually all the universities established were national in character, unlike the regional universities belatedly set up in the twilight years of colonial rule. Indeed, the regional universities were mostly dismembered into their constituent national components soon after independence. The national universities were more amenable to control than the regional ones, easier to subordinate to the national will as dictated by the state. And like so many other sectors of civil society, the universities enjoyed diminishing autonomy from the state. Increased state intervention in university affairs meant that, regardless of the number of universities in a country, running foul of the authorities in one practically foreclosed employment opportunities in all. In such cases, the only realistic alternative lies in looking for a job outside the country. Thus the nationalization and politicization of the university labour market helped to generate and sustain academic labour migration.

The economic crisis that hit many African countries from the mid-1970s further compounded the problems of intellectual production and reproduction. It led to the deeper incorporation of these countries into the world capitalist system, the decomposition of the social and economic advances they had made, and the demobilization of the ideologies and strategies of self-reliant development. As part of the social sector targeted for retrenchment by the structural adjustment programmes ostensibly adopted to rectify the crisis, the universities, and education in general, were forced to bear some of the awful costs of these programmes: overcrowded classrooms, under-equipped laboratories, empty library shelves, dwindling research grants, falling real wages, and plummeting morale. Faced with deteriorating material conditions, African academics responded in several ways. Many resorted to moonlighting, or sought to exercise their entrepreneurial skills in the nebulous world of the 'informal' sector, or they tried to endear themselves to the state for lucrative appointments in parastatal boardrooms. Others found refuge in consultancy services and hawked their talents to the ubiquitous donor agencies. For the rest, they could always migrate to the relatively greener pastures overseas or in a neighbouring country. And many did.

It would seem that these political and economic factors have played an overwhelming role in generating academic labour migration in, and from, Africa. Research considerations are of marginal importance. Not many Africans do research on other African countries. This is partly because few Africans study in other African countries. All academic roads lead to the North. And the exchange of scholars, curricula, and research findings between countries is quite limited. Academics in one country are often hardly aware of what their colleagues in the same discipline are doing in the neighbouring country. Northern universities are usually the first places of choice for sabbaticals. PanAfricanism has never been emptier of substance as it is among the academics themselves, among whom are to be found some of its most vocal proponents. The dismantling of the regional universities after independence, it can be argued, reinforced the underdevelopment of panAfricanist research practices. The earlier generation of African academics, trained in the regional universities, were far more panAfricanist in their orientations and expectations than their successors. This is one instance of nationalism, the project of nation-building, devouring its more beneficent progenitor, panAfricanism.

The process began as the post-colonial state sought to set the spatial and epistemological boundaries of research in pursuit of developmentalism and national integration. Academics, especially in the Social Sciences and Humanities, were implored to show commitment to the problems of their country, to study and find solutions to the great challenges of development and nation-building, to design a workable future, as prescribed, of course, by the political class. In short, social science research was firmly locked into an instrumentalist, prescriptive mode. Many African academics cheerfully responded to the call, keen to play their part in the epic drama of turning the dreams of the nationalist struggle into the fruits of independence. But in so doing, they became narrowly nationalistic in their concerns, analyses and expertise, and left research on regional and continental problems and processes to European and North American based Africanists, who would eagerly collect and synthesize the local and national studies and package them appropriately in whatever theory was in fashion, thereby setting the terms of debate in African studies.

The typical African academic migrating to another African country differs from his western counterpart, who often goes to a country of his primary research interest. This has seriously affected the nature of the relationships between the expatriate African academics and their local colleagues. It may have led to a situation whereby they are seen primarily as refugees with little to offer to the national research programme. In contrast, expatriate western scholars tend to be more readily accepted, and even respected, for their potential contributions. The few African scholars who actually have research interest in the host country are too anomalous to change these perceptions. Accordingly, their work receives little support and is often ignored.

Days in the Life of the Migrant Academic

The academic migrant arrives in the new country relieved and apprehensive. The label refugee is written all over him. He is often received in a manner befitting a refugee. There is little of the fanfare that accompanies the arrival of the western expatriate scholar, who is usually given a better office, housing and furnishings, and the small favours that help one to settle down. The distinction is clear. The western academic is not a refugee, but an expert, who is making a sacrifice in coming to the 'Third World', so special efforts need to be made to accommodate him, make the 'sacrifice' more bearable, the transition smoother. He is given a generous package that includes 'inducement', leave and educational allowances, all of which turn his actual salary, nominally pegged to local levels, into an allowance. The African expatriate scholar is sometimes given the same privileges, but grudgingly, indeed, resentfully. After all, a refugee is, in popular perception, a liability not an asset, a miserable destitute who should be grateful just for being let in. Occasionally, he is denied these privileges, on the pretext that he is as good as the locals. Thus, the African expatriate scholar soon discovers that he is marginal, valued only as a source of cheap, docile labour.

The reasons for this are not far to seek. He lacks the material and ideological assets of the western expatriate and the indigenous academics. To the latter, the western expatriate academic can be a useful link to universities in the North for sabbaticals, conferences, visits and study, and to western research foundations and consultancies. So he is carefully courted. The western expatriate academic, who is normally of European rather than African descent, is also seen as less of a threat, because he lacks the potential anonymity of the African academic migrant, who is usually indistinguishable from the local population, and whose very status as a refugee makes his stay indeterminate. For his part, the indigenous academic can manipulate various constituencies, whether ethnic, regional, political, or religious that determine and mediate the discretionary allocations of resources within the university system. The African expatriate academic has little lucrative contacts to offer his indigenous colleagues, and neither has he any of the latter's protective mantles. The alienation of his exile deepens.

He is confronted by authoritarian attitudes and chauvinistic contempt at every turn. He is discriminated against when it comes to promotions, research grants, and leave. The problem starts with the recruitment process itself. Often he is a recruit of last resort. Rarely is he recruited to the high ranks, even if he may have all the right qualifications. He is more likely to be slotted at the bottom of the scale, regardless of his previous position. It is all justified in the lofty name of standards. As a refugee, the academic migrant often has little choice, and the recruiters gleefully prey on that, and so he takes up the job. With that the migrant's marginality and powerlessness are established, to be re-enacted and re-enforced each time there is a vacancy for promotion. The migrant is sometimes prevailed upon, subtly and not so subtly, not to apply for promotions and research grants. His contract, which has to be renewed every two or three years, hangs like a sword over his head, controlling him, always reminding him of his temporality and transience.

The migrant academic is exceptionally vulnerable because of the discretionary decision-making processes of universities. Patron-client networks determine recruitment and promotion procedures, and the allocation of work loads and resources. In such a system, the migrant academic is disadvantaged because he cannot be a good or useful client. Clients have to offer their patrons some social returns for the reputational capital bestowed on them, which the migrant cannot because he has no base in the local society's competing constituencies. Migrants are only good for scapegoating, especially in times of crisis.

Many African universities do not have ample resources for research, partly because they have insufficient funds, for they mostly depend on governments which have other pressing needs to meet. Governments also tend to view universities primarily as teaching institutions, factories to churn out skilled personnel. In addition, the available funds are often channelled through the corrupting patron-client networks to which the migrant does not belong, or they are frittered away in conspicuous consumption by the university administrative elite, including the vice-chancellors, their deputies, the registrars, deans, and sometimes even heads of departments, who are provided with official cars and residences and expense accounts. In the meantime, teaching facilities deteriorate. The misplaced priorities of many African universities can be seen in the institutionalization of the power imbalance between the administrative and teaching staffs. The latter are often expected to execute the former's unimaginative decisions.

Corruption and patron-client networks reinforce each other to undermine research. Research grants are often made on a discretionary basis by whoever is in charge of the research grants committees. External sources are likely to be hoarded by heads of departments for themselves or for redistribution to favourites. The same happens to invitations for conferences, or book collections. Once again, the African expatriate academic is left out in the cold. He is rarely entrusted with positions of power within the university hierarchy to be part of any network. Efforts to carve out an independent research path are frustrated by the lack of institutional support. Indeed, being too actively involved in research often provokes animosity and may endanger one's job. A migrant is supposed to work hard, but not too hard to threaten or show up the locals. In a situation where many people are trying to make ends meet through extra-curricular entrepreneurship, being actively involved in research is not always an asset for the migrant academic. On the contrary, that may expedite his day of departure. In any case, a good research record does not guarantee career mobility since the patron-client networks often determine such matters.

The research of the African expatriate scholar suffers not only because of the scholar's marginality, but also the marginality of his research interests. Funding, whether by the university itself or by external agencies, concentrates on national topics in which he is not likely to have much expertise. The chances of getting funding for a research project on his area of specialization, usually his home country, that is if he can even afford or dare to go back, are remote. So he gradually slides into intellectual inertia. Of course, he can develop a research interest on the country of his residence, a daunting task and a gamble to someone whose job security hangs by the thread of periodic contract renewals. Doing so might, in fact, raise the eyebrows of the indigenous academics, for that subverts the transience of his position upon which their tolerance of him is based.

The African expatriate academic who actually does research on the country of his exile is an oddity. He is initially treated with bemused surprise, for he is trespassing the exclusive domain of the indigenous and the western expatriate scholars. The former base their intellectual proprietary rights on birthright, and the latter on the pitiless privileges of colonial conquest and neo-colonial domination, for which they are both resented and followed, but never ignored. A product of colonialism and neo-colonialism himself, the African academic who chooses to study countries other than his own, defies the mould, thus reinforcing his image as a 'radical', a potential 'troublemaker'. He is particularly unwelcome because, unlike the western academic who has his own sources of research funds, he competes with the indigenous scholars for the limited funds available, especially from external agencies that may seek to exploit his 'objectivity' both as a foreigner and a non-western scholar. His interventions in controversial local debates are dreaded, derided, and dismissed. The accumulated grievances against western scholars metamorphose into a crusade against him, an African, a refugee, who dared to be an expert on their country. They seek to put him in his place, and if he proves recalcitrant, there is always the contract.

One of the most important resources for research is time. In this the African migrant scholar also receives the raw end of the stick. To begin with, as a worker on short-term contracts, he is often not entitled to sabbaticals, and should he take an unpaid leave of absence he is not assured of his job upon return. So year in, year out, he toils, while the indigenous members of staff come and go from their sabbaticals and western expatriates make their periodic academic safari trips. He is also the first to be asked to shoulder more teaching responsibilities, although the last to be rewarded. In a sense he becomes the academic housekeeper, assuming the chores of senior professors who are too busy being directors of parastatal corporations, or going to every available conference recycling the same old paper written long back when they still had a few fresh ideas. The African migrant academic shares his misfortune with the hapless junior staff. But while for the latter this is a rite of passage to the segregated spoils of academia, for him it is an endless Sisyphean ordeal. Denied time, the migrant's services are fully exploited, but his potential is dulled, thus ensuring and reproducing his lowly status on the academic totem pole.

He comes to discover that his survival increasingly depends on the art of self-effacement and self-censorship. He embodies most poignantly the anti-intellectualism of modern universities, in Africa and elsewhere. Everyday he lives the glaring contradiction between the myth of the university as a centre of critical intellectual enquiry and the realities of scholastic conformism in pursuit of job security. Universities of course do not pin the blame of intellectual censorship on their enlightened shoulders, but on the state and other misguided outsiders. Intolerant state policies and external political pressures, real and imagined, give the academics, individually and collectively, an alibi when they abscond, as they often do, from their intellectual calling in pursuit of mindless careerism and mutual cheering.

The African expatriate academic is soon left in no doubt that exile is no haven, that many of the forces and pressures that pushed him out of his homeland are no less powerful in his new country of residence. Indeed, he witnesses the flight of the indigenous scholars into exile to other countries. Many, in fact, wonder why he came, a question that gnaws at him and makes him miss home, idealizing it at times, and occasionally he even wishes to return. Exile becomes an existential nightmare. The material benefits, imagined and acquired, suddenly become token, and the security of survival becomes less tangible. Despair and apathy often set in. The search begins for a new country of exile. To many the seductive lure of the West proves irresistible. The bags are packed once again. Innocence and hope are shattered one more time.

Seasons of Migration to the North

The decision to migrate to the North demonstrates, despite its despair, the relative privilege of the African expatriate scholar, for he is saved from the fate of millions of hapless people who trek to eke out incomplete lives in teeming refugee camps. In the' lost decade' of the 1980s the multitudes of refugees and migrants within Africa increased, while tolerance for them decreased. Labour migrants in the continent faced mass expulsions and harassment. Universities were not innocent bystanders in this sad saga. They created conditions which helped push some of their own faculty into involuntary exile, or they watched meekly as some of them were arrested and detained, and refused to rehire them upon their release, fearful of political contamination from their alleged 'subversion'. Some universities also dutifully employed and dismissed exiled scholars from neighbouring countries, and built and broke inter-university relations, according to the vagaries of inter-state relations and the whims of political leaders. Academics have been known to be refused permission by their universities to visit other universities for conferences or as external examiners because the presidents of the two countries are not on speaking terms! Many university authorities have not hesitated to cooperate with the security services to weaken or break students' movements, and they have brazenly expelled student leaders following demonstrations and unrest. To students, university administrations often act and appear as coercive extensions of the state, rather than as bastions of liberty and critical thought. Once expelled, the student activists either languish in jail or 'tarmac' the streets in search of jobs. The lucky few manage to go abroad to finish their studies, where they sometimes decide to stay. The expatriate academics make easy scapegoats for administrations anxious to exonerate themselves in the eyes of intolerant state authorities. And so they are usually blamed and deported in times of student unrest, thus saving both the state and the university the embarrassment of introspection and reform. The deportees rarely go back to their home countries but head to the North as well.

In the 1980s, the brain drain spared few countries in Africa, including those that were once net recipients of academic migrants, such as Nigeria during the oil boom years of the 1970s and early 1980s. Not only did more African professionals migrate to the North, but fewer of those trained abroad returned. Predictably, African leaders condemned the migrants for their lack of patriotism. Similar charges were levelled, more honourably, in progressive African intellectual circles. In fact, many African academic migrants themselves exhibited deep ambivalences, anxieties, and anger about their displacement, notwithstanding Mazrui's (1978:314-18) heroic claims about 'counter-penetration' and Said's (1993:326-36) fantasies of 'liberatory voyages in[to]' the belly of the imperial beast where they are destined to lead the liberation of both the western and post-colonial worlds[2]. They were only too painfully aware that the brain drain to the North, unlike migrations within Africa, represented a net loss to the continent as a whole, that it reflected and reinforced Africa's intellectual underdevelopment and dependence, which, in turn, further undermined academic freedom, if academic freedom is not seen solely in relation to internal institutional conditions, but also in terms of subordination to other societies' ideological systems and intellectual traditions. Far from loosening the tentacles of repression, the brain drain provided both the authoritarian states and universities a safety valve through which the boiling steam of internal resistance could be filtered and defused. This might explain their ambivalent attitude towards the brain drain: they deplored it, but did little to stem its flow by rectifying the policies and practices that generated and reproduced it in the first place.

The brain drain has sapped Africa of its intellectual resources and increased the continent's dependence on western expatriates, who are far more expensive to recruit and maintain than their African counterparts. According to one source, 'at any given moment, sub-Saharan Africa has at least 80,000 expatriates working for public agencies under official aid programmes. More than half of the $7-8 billion spent yearly by donors goes to finance these people' (Timberlake 1988:3). Needless to say, many of these expatriates and their policy prescriptions are half-baked. The tragedy is that many African governments have been importing them while at the same time inadvertently exporting their own experts, whose understanding of African problems and commitment to their solution is far greater. But regardless of how qualified and well-meaning the western expatriates might be, they are a poor substitute for local expertise, and cannot help much in the development of vibrant local intellectual traditions.

The African expatriate migrant in the North finds that the political and intellectual insecurities of his African exile are multiplied by new racial and cultural marginalities. If he should be so unlucky as to go to the former imperial metropole, where he is reborn as a colonial native, the chances of university employment are slim, even if he earned his impeccable degrees there. Should he cross the Atlantic and choose the 'Great White North', he is recast as a 'visible minority', squeezed between Canada's warring solitudes of the so-called two founding peoples, and after paying his dues as a taxi driver, sessional lecturer, or research assistant to some aspiring Africanist, he may get a job in a department or a university as the lone 'employment equity' candidate, and suffer the indignities of always being mistaken for a student, always trying to prove himself to cocky students and patronizing colleagues, always being interviewed at cocktail parties about the latest calamity, flashed on last night's television news, to befall the 'Dark Continent'. And he finds his African research is as peripheral as his presence, seen as a polite concession to political correctness. And so he hardly gets funding to go back to do field research, and he begins to dabble in the solipstic cults of post-something scholarship, or the angry discourses of minority nationalisms. He watches his Africanist colleagues, who resent the potential of his authenticity in the competition for African studies students, go and come from their beloved African countries with weary tales and jokes of collapsing economies, vampire states, and incurable epidemics, which they recount in smoke-filled senior common rooms, oblivious to his frowning presence. The nightmare of his exile deepens, as he struggles for tenure and promotion, for recognition and relevance, for a sense of belonging.

Should he move to 'God's Own Country' in the south his colour and cultural configuration are accented and affirmed in America's much bigger African diaspora population, with its relatively large professional middle class and political presence. But his blackness also becomes an exclusionary marker in America's eternal racial war, a slate on which the historic white pathologization of both Africa and Africa-America is scripted. He negotiates, awkwardly, unsteadily, tripping sometimes, between the intolerant demands of colour and career. Often he ends up in the under-funded historically black colleges and universities where the comforts of racial solidarity are not matched by the reputational possibilities of the large historically white research universities, whose ivy citadels remain resolutely white both in faculty and student composition. That, too, seems to be the constitution of the Africanist establishment, of the large African studies programmes, the prestigious Africanist journals and book series, the foundations and agencies that fund African research, of the voices that are called upon in the popular media to explain Africa's peculiar genius for biblical afflictions. The sense and reality of his marginality deepens, only periodically broken by drunken conversations and exaggerated laughter with fellow African expatriate scholars at the occasional conferences.

Conclusion: The Necessity of Becoming Public PanAfrican Intellectuals

The constraints and challenges facing African expatriate academics are serious and sobering. But they should not be allowed to petrify us into intellectual paralysis. To beginning with, we need to become more than academics, to become intellectuals, insurgent intellectuals, as West (1991) implores us to do. There is a difference between an academic and an intellectual. Academics tend to engage in narrow scholarly work, in analyses confined to internal disciplinary issues and debates, often writing for each other in constipated, convoluted jargon, while intellectuals transgress and transcend discursive frontiers and critically and visibly engage the large and burning public issues (hooks 1991). Academics are like technicians who tinker with the bits and pieces of broken machinery, while intellectuals are craftspeople who conceive and create and trade the machinery of ideas in the public marketplace. Becoming public intellectuals entails conscious and critical immersion in community movements and popular politics, embarking on a vocation of creative commitment to collective insurgency against oppression and exploitation, the principled pursuit of humane values and a more egalitarian and generous global social order. Such involvement provides a powerful shield against the alienating pretensions and pettiness of academic careerism, and the reactionary antihumanisms that dominate much of contemporary academic writing, with their depoliticized, defeatist, and disempowering analyses. It offers protection from the seductions of wealthy western universities and their superficial multi-cultural cravings for native ventriloquists, for complicit authentic 'others'[3]. Specifically, for Africanist scholarship, it insures against the almost gleeful pathologization of African societies and realities, against the crippling concessions to Afropessimism.

From the pain and pathos of our migratory experiences as African expatriate intellectuals lie the possibilities of building emancipatory panAfricanist bridges and organic intellectual traditions. In contemporary Africa one can distinguish three forms of panAfricanism: that of the presidents, of the professionals, and of the popular masses. Presidential panAfricanism, often louder in political rhetoric than in practical resolve, is periodically consummated at OAU and other regional summits; the panAfricanism of the professional classes, including scholars, finds expression in concerns for the developmental issues and over the negative images facing the continent; the panAfricanism of the popular masses is realized most concretely in the derelict camps of migrant workers and refugees. The migrant African intellectuals are as well poised as any group to promote more meaningful, broad-based, and enduring forms of panAfricanism not only through their analyses, but also by building infrastructures of intellectual discourse and dialogue. It is, indeed, revealing that some of the most dynamic independent continental and sub-regional research institutions in Africa, such as the Council for the Development of Social Science Research in Africa (CODESRIA) and the Southern African Political and Economic Series (SAPES), just to mention two, were formed, and continue to be nourished by the constant infusion of critical talents forged in the crucibles of voluntary and involuntary intellectual labour migration. These networks need to be extended to link up with those in the African diaspora, so that out of the tragedies of our various dispersals we can weave strong political and intellectual garments to protect us from the elements of racism and imperialism, and fortify us in our struggles against them. The case for creating serious and sustainable panAfrican networks of intellectual exchange is overwhelming. The basis for such a project lies in our very conditions as expatriate African intellectuals.

Notes

1. This subject is discussed in greater detail in the subsequent chapters. On academic freedom in Africa and general overviews of the development of university education in Africa see Diouf and Mamdani (1994), Ajayi et al. (1996), Ngara (1994). For a brief analysis of the issue of academic freedom in the Canadian context see Sangster and Zeleza (1994).

2. For a trenchant critique of Said, see Ahmad (1992, chapter 5); also see Chapter 21 below.

3. Term ventriloquist is borrowed from Spivak (1995:175).

Chapter Three

AFRICAN SOCIAL SCIENTISTS AND THE STRUGGLE FOR ACADEMIC FREEDOM

Introduction

Africa has been undergoing tumultuous political change in recent years, as struggles rage for democracy and human rights, forcing dictatorships to tumble one after the other. These struggles are multi-dimensional. This chapter examines the struggles for academic freedom. It seeks to identify the forces that have shaped and conditioned the social production of African intellectuals in general, and social scientists in particular, which include the state, civil society, institutions of higher learning in Africa and abroad, and international donor agencies and research foundations. It is shown that the relationship between African intellectuals and each of these forces has been complex and contradictory and, collectively, they have all undermined the production of critical social knowledge. The current drive for democracy appears to offer African social scientists new opportunities for engagement and intellectual production. The chapter argues that the struggle for academic freedom is a broad one, encompassing struggles against the interventions of the authoritarian state, as well as the authoritarian power relations in academic institutions, societal practices and cultural norms that inhibit research, and the Euro-centric academic structures and traditions that tend to marginalize African scholarship.

The Pen and the Sword

The state has played a pivotal role in the social production of intellectuals in Africa. In most African countries, universities and research institutes, which train and employ intellectuals, are owned, financed, and controlled by the state. State control of education in Africa is just one manifestation of the statism of African social formations, which is an outgrowth of the authoritarian colonial state, the underdevelopment of capitalism, and the structure of internal class relations. Fiscal control over the universities, research institutes, and other centres of higher learning has provided a cloak of legitimation for the state's pervasive interventions in these institutions. The belief that the state has a right to control higher education is deeply ingrained among the intellectuals themselves. As Shivji (1990:6) notes, during the debate on the 'Dar es Salaam Declaration on Academic Freedom' held in April 1990, 'rather than assert the rights of civil society against the state, one heard the state being gratuitously defended. 'If the state pays for education, it has the right to set its modalities', some academics argued, forgetting that it is the people who pay, not the state'.

The relationship between the institutional African intellectuals and the state has been complex and contradictory. On the one hand, these intellectuals are state functionaries in so far as they are linked to the state through its educational or media apparatuses (Farah, 1990:7-10). On the other, they constitute a critical social category of civil society, whose power to contest the hegemonic pretensions of the state belies their relatively small numbers. The relationship, therefore, has been characterized by both collaboration and contestation. Collaboration was particularly pronounced in the years immediately after independence when the post-colonial state enjoyed immense popular legitimacy earned from the nationalist struggles. Also, during this period the ideologies of nation-building and development held sway, which the majority of the intellectuals themselves believed in. Thus, the state was able, with the intellectuals' acquiescence, to push an instrumentalist agenda for the universities.

Relations between the state and the intellectuals increasingly turned sour as the problems of nation-building and development proved far more intractable than originally anticipated. Research conducted by the social scientists, intentionally or not, began to raise awkward questions about the ideologies of nation-building and development. It was hard for the latter to ignore the ethnic and regional imbalances in the distribution of the fruits of *uhuru*, or the appearance of corruption, incompetence, and intolerance among members of the political class. The social scientists could not ignore these realities for, besides being state functionaries, they were also the representatives and interpreters of the various constituencies of civil society which were jostling for place and privilege in the emerging post-colonial order.

The widening rift between the social scientists and the state was also a product of growing radicalism in African studies. This was expressed in opposition to imperialism and neo-colonialism and commitment to the masses and 'preference for socialist economic and political strategies' (Waterman 1977:1). New approaches that sought to unravel the fundamental realities of Africa were adopted. Theories of modernization gave way to dependency and Marxist approaches. No discipline in the social sciences was left untouched, from history and sociology to economics and political science (see Gutkind and Waterman 1977; Hetlne 1983; Eke 1986). The radicalization of the social sciences coincided with the crisis of developmentalism, triggered by the global recessions of the early 1970s and the early 1980s and the structural adjustment programmes adopted by African governments at the behest of the World Bank and IMF (Onimode 1988, 1989a, 1989b; Mehretu 1989; Mkandawire and Olukoshi 1995; and Chapter 3 below). All this led to massive retrenchment and divestment from education and the other social sectors. By the 1980s 'academic infrastructures', notes Ake (1990:6), 'such as libraries, bookstores and research facilities (were) collapsing'. Thus, the economic crisis and the consequent fiscal crisis of the state ravaged many of Africa's once great universities and undermined the material security of the intellectuals (Morna 1991). As the crisis deepened so did the administrative regulation and political repression of the academic community which was increasingly 'targeted as merchants of subversive ideas' (Ake 1990:6).

While no group of intellectuals escaped the brunt of state repression, the social scientists were probably more vulnerable than other groups, such as lawyers, writers and journalists because they had little organized support (*Codesria Bulletin* 1990:1). They did not have the equivalent of Pen International, the Committee to Protect Journalists, Reporters Sans Frontiers, Index on Censorship, Physicians for Human Rights, and their African equivalents. State repression of social scientists and other academics took various forms, from arbitrary arrests, detention, and summary executions, and restrictions on the freedom of expression, assembly, movement and organization, to dismissal of faculty staff, expulsion of students, closure of universities, censorship of reading and teaching material, manipulation of curricula, and denial of promotions, research funds or scholarships to politically active faculty staff and students. In addition, the state exerted other subtler forms of repression and control, for example, by posting security agents in classrooms disguised as students, encouraging students and faculty to inform on each other, and coopting senior administrators into the ruling party. The catalogue of academic repression throughout the continent is a long and depressing one as amply chronicled in the reports by Africa Watch (1990), Amnesty International (1990), in the papers presented at CODESRIA's symposium on academic freedom[1], and in CODESRIA's (1990a) compendium of press cuttings on the abuses of academic freedom in some twenty countries between 1987 and 1990.

All these assaults on the academic community and the deteriorating material conditions severely undermined the capacities of African social scientists to produce knowledge. For many, repression and persecution bred self-censorship, which became necessary for survival. They tended to internalize repression, 'to police themselves thereby ensuring further erosion of academic freedom' (UDASA 1990:3). The result was that some social scientists virtually gave up on serious research and writing, others became adept at doublespeak or sank into the banality of recycling the developmentalist rhetoric of the state and the so-called international donor community. In short, the state socialization, subordination and incorporation of social scientists in Africa gravely undermined academic freedom and the production of critical social knowledge. These processes also hindered the development of effective links between the social scientists and civil society.

The Strings of Civil Society

Social scientists, and intellectuals in general, are linked to civil society through their socialization and identification with particular social, regional and ethnic groups whose interests they often translate and articulate. Civil society nourishes their research, giving them the linguistic and social keys to unlocking the inner workings of their groups which outsiders can only glimpse from the outside. Not surprisingly, many African social scientists opt to study their own societies, communities and social groups. These social units become the analytical treadmill through which the great national and continental questions of historical change, social transformation, economic development, and political power are examined.

Thus, African social scientists are 'organic' to various social groups. But just as the relationship between the state and intellectuals is a complex and contradictory one, so is that between intellectuals and civil society. To begin with, the relationship is mediated by the state, which, as noted earlier, controls intellectual reproduction. This weakens the strings of society on intellectual production and reproduction. The alienation between intellectuals and civil society is vividly manifested and reproduced through language. While the social scientist is often versed in the language of discourse of his social group, the latter is not versed in the language of discourse of social science. As Mamdani (1990:3) states: 'from the outset, there was a sharp rupture between the language of the school (reflecting that of the state) and the language of the family/community, between the language of instruction and the language of day-to-day social communication'. The result is that the

African intellectual, Mamdani continues, 'is doubly distanced from working people. Not only is there the social (class) distance that is a product of the break between mental and manual work in any class-based society, there is also the cultural distance because the language the intellectual writes and communicates in is not the language(s) in which working people interact'.

Notwithstanding the progressive domestication of European languages in Africa in the last few decades, communicating in these languages 'continues to be a privileged discourse, a linguistic curtain, on the other side of which are to be found the vast majority of working people, shut off from the affairs of the state, of higher education and of science. This is true of most African countries, 'conservative' or 'radical'' (Mamdani 1990:5).

Social scientists, unlike creative writers, have yet to take the question of language seriously (see Wa Thiong'o 1986). The 'linguistic curtain' not only undermines relations between African social scientists and civil society, but also devalues the relevance of their work. This reinforces the statist, elitist, and anti-democratic orientation of African social science research, whether conducted by supporters or opponents of the state, or whether using the paradigms of the 'right' or the 'left', for it is a discourse the social scientists share with the political class and not the 'masses'. This is not to suggest that African social scientists lack commitment to their societies, or that their level of social responsibility is low. Nothing could be further from the truth. 'Even a cursory reading', to quote Mafeje (1990:15), 'of what African intellectuals write would show that they are preoccupied with problems of development in their countries to the point of sounding hysterical in the case of the left'. Rather, it is merely to point out the poor links between intellectual reproduction and the broader realities of production, power, and consciousness in Africa.

Few would dispute that civil society in the continent is extremely complex and dynamic, although the Africanist discourse on civil society, Mamdani (1996:13) avers, 'is more programmatic than analytical, more ideological than historical', its contemporary prismatic lens borrowed, not from a reading of African political developments, but from the uprisings of Central and Eastern Europe of the late 1980s. The dominant conceptionalizations of civil society do not adequately capture the complexities of African political cultures and social formations. For the followers of Marx the realm of civil society is the market, while in the Gramscian paradigm it resides in public opinion and culture, and for Habermas it is embodied in the associational life of the public sphere. For one thing, the publics of subordinated classes, gender, ethnic and racial groups are often excluded from these definitions. A fuller conceptualization of African civil societies would have to see them as historically constructed constellations of competing and overlapping associational ties and social spaces, rural and urban, encompassing the cultural communities of language and ethnicity, the social solidarities of class and gender, and the prophetic traditions of religion.

It is easy to idealize civil society, to see it as in contradistinction to the heinous state, a repository of all that is benign and humane. It should be remembered that the state and civil society interpenetrate, thereby regulating and appropriating each other's functions, despite the demarcation between them (Mersha 1990). Moreover, civil society, no less than the state, is circumscribed by the material conditions from which it springs. This is to suggest that civil society has not provided a haven for African social scientists. Not only does it lack the capacity to reproduce the intellectuals, it has not always shielded them from the long arms of the state. Indeed, the various constituencies of civil society, to which African social scientists are tied, are themselves often inimical to the production of critical social knowledge. The cultural norms of these groups have often been used by the communities concerned, as well as the state, to circumscribe research inquiries.

The articulation of cultural norms inimical to critical social science research of course vary from country to country and have changed over time. In Egypt, for example, from the 1970s the Sadat regime sought to use the religious ideology of Islam, the dominant cultural norm in Egyptian society and long used by successive political systems as a tool of political legitimation and mobilization, 'to neutralize, weaken and eliminate the Nasserite social base', and to win support for its own policies of economic reform and capitalist accumulation (Farah 1990:15-6). The formation of radical Islamic groups to wage war on the Nasserite and other leftist groups was encouraged. These groups later turned against many of the regime's policies, including the signing of the peace treaty with Israel, and assassinated Sadat himself in 1981.

By then religion was not only the dominant cultural paradigm, but also the dominant ideology. The impact on social science research was profound. 'This highly ideologized context resulted in the emergence of a unique type of academic research; the so-called Islamic sciences of economics, medicine, sociology, engineering, etc. This type of research is highly rewarded by the dominant elites, not only in Egypt but in the whole Arab world, and is very attractive to the general public that condones it. Against this flood of ideologically influenced research, academic research run on more scientific bases, runs the hazards of censorship, harassment and loss of position' (Farah 1990:18). Under these circumstances undertaking secular and critical studies of religion itself is positively risky. And discussing gender inequalities as enshrined in the accepted interpretations of the religious texts is almost anathema.

Restrictions on gender research in Africa are, indeed, widespread because of the historical, cultural, social, and institutional marginalization of women, rooted in the conjunction of pre-colonial and colonial patriarchal ideologies and practices, which continue to be reproduced throughout society, from the family to the university and the state. These restrictions can take various forms, such as the belittling of women researchers and of gender research by the society and the academic community itself (Iweriebor 1990; Mama and Imam 1990). In some cases even physical intimidation is applied. In the Sudan, for instance, 'women academics have been particularly affected by Muslim extremists. Women are now forbidden to travel in the absence of a *muharram* - a close male blood relative to act as guardian. At least two senior female academics', Africa Watch (1990:40) reports, 'have been prevented from attending international conferences on account of this ruling. Women students are being encouraged to wear the veil and subjected to intimidation and harassment if they do not'.

Civil society imposes other types of restrictions against social science research. Researchers who are not local may be met with indifference, suspicion or hostility. This reinforces the tendency towards inbreeding in African social science research, whereby researchers concentrate on their own groups because they feel, or are, more readily accepted. Under repressive regimes social science research can be suspect; the researcher may be seen as a government informer and is given little cooperation. Sometimes resentment occurs because the social scientist is seen as a member of the elite class, shamelessly prying into the poverty of ordinary people's lives. The list could go on. But the point has been made that the state is not the only agency which obstructs social science research in Africa while civil society stands by watching from the sidelines. Nor are the social scientists themselves innocent bystanders.

Living in Glass Houses

For the most part, African social scientists are produced, and reproduce themselves, through the university. The institutional base for the social production of the African intellectual was established after independence, for while the colonial system had produced educated Africans, the latter neither controlled the educational institutions, nor could they develop 'an intellectual trajectory which was peculiarly their own' (Mafeje 1990:4).

With independence there was an explosion of education at all levels, including the universities. It is a remarkable testament to the social bankruptcy of colonialism that in 1960, the year of African independence, no more than 9% of the population was literate (ILO/JASPA 1989:5). University education was a rare privilege. Populous Nigeria only had one university, and so did the three east African countries of Kenya, Uganda and Tanzania combined, as was the case for the three southern African countries of Malawi, Zambia and Zimbabwe. Today, Nigeria has over three dozen universities, and the east African countries each have more than one university, with Kenya alone boasting four public and a handful of private universities. According to World Bank figures covering Sub-Saharan African countries, between 1965 and 1987 the proportion of young people attending primary school, secondary school, and tertiary institutions rose from 41% to 68%, 4% to 17%, and 0% to 2% (World Bank 1990:235, 79). This is a remarkable achievement by any standard.

But the rapidly expanding African university sector was weighed down by its own peculiar institutional constraints. In keeping with the monopolization of power by the post-colonial state, senior university administrators in most countries were state appointees. The president of the republic would appoint the vice-chancellor, usually for political reasons rather than administrative or academic abilities. The vice-chancellor, in turn, would appoint the deans of faculties or heads of departments. Thus the entire decision-making process is often discretionary and authoritarian. The authoritarian relations of power in the universities and research institutes are manifested through recruitment, screening, promotions, allocations of work loads, provision of leave and sabbaticals, scaling of staff, gate-keeping and policing campuses, surveillance, sexual harassment, and the administration of welfare facilities. Research, which is essential for the production of critical social knowledge, is enmeshed in the same contraption. Quite often research funding is employed as a weapon for punishing radicals, rewarding sycophants, and settling scores. Staff are also sometimes humiliated and harassed through the use of accounting procedures.

Thus authoritarianism, corruption, and discrimination on ideological, intellectual, national, ethnic, religious and gender bases are quite widespread in institutions dominated by the social scientists themselves. This breeds censorship. Collective self-censorship is, in fact, deeply embedded in the consciousness of the university community. It is generated by fear of the repressive political system and of upsetting the inherited university traditions and established intellectual orthodoxies.

Self-censorship by 'radicals', younger scholars, women, and foreign scholars is often their only protection against political harassment and intellectual persecution in the university. Collective self-censorship curtails the development of original and creative thought, which is a threat to authoritarian institutions. The university's bureaucrats and ossified intellectual elite are as threatened by probing thought and research as are the state functionaries.

In short, the structures which the intellectuals themselves control do not always operate in a manner that is conducive to academic freedom and the production and dissemination of critical knowledge. It is important for the African academic community to address the inadequacies and contradictions of their own organizations and practices, not only in order to make their critiques of the state and civil society morally justifiable and credible, but also to create new structures and foster a more propitious environment for intellectual research and academic freedom in general.

From its inception the African university jostled between its mission to promote modernization patented on westernization and to forge an African personality, to use Nkrumah's phrase, from the memories of an invented national culture. In its first role it acted, to use Mazrui's (1978) apt imagery, merely as a branch of the multi-national western university. It borrowed academic standards, curricula, and even faculty copyrighted in the west. 'Oxbridge in profile and cult', says Hagan (1990:9) of the university in former British colonies, 'it was alien and surreal'. But the borrowed clothes could not withstand forever the tropical winds of independence. The Africanization of the state apparatuses soon extended to other public spheres, including the universities. African intellectuals became more conscious of themselves and of their responsibilities. They were expected to study their societies and devise solutions for the problems of nation-building and development. In short, they found both an institutional base and a mission.

But it was a mission articulated in a distinctly western accent. Africanization did little to remove the veil of universality from western social science. Concepts and models were eagerly imported, tested, and accepted or rejected according to prevailing academic fashions in the North. Familiarity with northern intellectual fads, and publication in western academic media continued to be the rites of passage to the secret society of academia. It bestowed the distinction of erudition and the privileges of promotion, name recognition, and self-importance. At the same time, African social scientists were reluctant to publish in their own journals or to quote each other's work, preferring the better known Northern journals and 'authorities' for greater legitimacy. The homage to the North in terms of publications was more than a sign of wilful indulgence, however. It was partly a reflection of material constraints in the African universities themselves, where publishing facilities were either inadequate or lacking. But publishing in journals outside Africa simply reinforced the attrition rate of existing African journals and devalued their quality because they became the publication media of last resort. African social scientists also established themselves as hostages to editorial foundations they did not, and could not, control. Their chances of publication in western research media were poor, for it should not be forgotten that scientific research 'is, first and always, a social activity... infected with the same social stratification, class, racial, ethnic, and sex biases, lusting for power, and interpersonal frictions found elsewhere ... in society' (Zelinsky 1975:133), and 'what gets accepted for publication is a complex function of who wrote the paper, the attitude of the author toward the topic; and his weighing what referees say; the extent to which the paper is 'too hot', 'too controversial'... and ... the message' (Syzmanski and Agnew 1981:33). Papers by African social scientists rarely appear in western journals, including those that specialize on Africa. Indeed, except for a token African or two, the editorial committees of almost all the major Africanist journals in western Europe and North America hardly include African scholars, even those permanently working in these countries (Zeleza and Sindiga 1985; and see next chapter)[2].

By concentrating on securing publication space in such journals, African social scientists have been undermining their own intellectual freedom, their productivity, and the opportunity to establish vigorous social science traditions and communities. The rejection slips that many of them inevitably receive, thanks in part to, as Mkandawire (1989:10) puts it, 'our dated references, reflective of the parlous state of our libraries, (which) give the impression, often wrong, that what we say is out of date', only serve to discourage them and the younger scholars under their tutelage from publishing, and so inertia often sets in. Moreover, in the absence of local journals and awareness 'of what is going on in the continent, there is an enormous amount of useless inventiveness and one does not get the sense of being engaged in a cumulative process of ... intellectual interaction' (Mkandawire 1989:10).

Material constraints in themselves do not entirely account for the ritual genuflections of African social scientists to western scholarship. The fact that they and their universities maintain strong links with western universities, where many of them received their postgraduate training, cannot be ignored. These relations of dependence are also a product of the intellectual division of labour between African and Africanist social scientists, in which the former concentrate on narrow empirical studies of their societies and communities that the Africanists collect and process into 'macro' syntheses wrapped in currently fashionable theoretical packages. Research on regional and continental issues has largely remained a monopoly of the Africanists. This has meant that the methodological standards and theoretical paradigms have been effectively set by the latter. African social scientists have been reduced to followers, or at best critics, of each new intellectual fad emerging from the Africanist centres of the west. This division has not been of benefit to anyone, neither the Africans producing mountains of empirical data, nor the Africanists engaged in faddish theorizing and pursuing 'exclusively expatriate and ephemeral 'debates' that vanish as mysteriously as they emerge without due consideration of the historical specificities of the African condition' (Mkandawire 1989:8).

The imported paradigms pervade all social science disciplines from economics and political science to sociology, psychology and history. Western realities, practices and values are taken as the normative standard. In development studies Africa is measured according to European stages of economic growth, whether Rostovian as in modernization theory (Rostow 1971; Meier 1976; Meier and Seers 1985), or Marxian as in the theories of modes of production (Melotti 1981; Crummey and Stewart 1981). Theories of underdevelopment share the same predilection, for they often define underdevelopment as the absence of western capitalist development (Brenner 1977; Palma 1978; Roxborough 1984; Blomstrom and Hettne 1984; Harris 1986; Kitching 1989; Escobar 1995).

It is this universalism that leads some political scientists to assume the normalcy of the western state and see the African state as somehow abnormal and in need of a special, mostly derogatory, descriptive vocabulary. For a long time this universalism also allowed sociologists and anthropologists to place African societies at the lower rung of a unilinear evolutionary ladder on top of which was Europe. Some still portray African family and household forms as peculiar and characteristic of 'backward' social formations (Guyer 1981; Netting et.al. 1984; Mafeje 1991). Psychologists eagerly import methods and concepts and use 'standard profiles' constructed in the west and 'calibrated to favor white middle class males' in their psychometric tests of Africans (Mama and Imam 1990:13-14). And many historians still place higher value on written documents and write comfortably African history only using the written records of colonial adventurers and proconsuls buried in some stuffy European archives, as Illife (1987) has done in his monumental study of the history of African poverty (see Chapter 7 below).

Mama and Imam (1990:20-21) correctly point out that 'the problem with this kind of universalism is that it restricts the ways in which (African) social researchers are able to think and theorize. We are forced to take on board these norms and waste time tilting at windmills to find out why we deviate from these patterns instead of finding out what our own patterns and realities are'. Further, they indicate that the dominant social science paradigms are inadequate, not because 'they are Western per se. They are inadequate to Africa because, having been developed in a particular context (which happened to be European/North American, colonial-imperialist, racist, sexist), their theoretical frameworks and methodologies were influenced by this. Consequently, the forms of knowledge these frameworks and methodologies produce also are imbued with them' (Mama and Imam 1990:12).

Dancing to the Piper's Tunes

The domination of western social science in African intellectual discourse has also been facilitated by the growing reliance of African social scientists on western donor agencies and foundations for research funds. This reliance has grown as the fiscal crisis of the African state has deepened. Outside of the state, there are no major alternative sources of financial resources for research. The African bourgeoisie is too mired in 'primitive accumulation' to adequately support the cultural and intellectual enterprises, so when it is not being maintained by the state African social science research is funded by foreign donors and research foundations.

In many countries the 'Link', as Hirji (1990:9-16) calls relations with foreign donors, has become indispensable not only for research funding, but also for the provision of equipment, such as computers and photocopiers, and ordinary teaching materials, including paper, pencils and rulers. And through the 'Link' books are acquired for bookstores and libraries, and conferences are organized. The 'Link' is ideologically neutral, for it is pursued assiduously by 'both ultra conservatives and red hot radicals' (Hirji 1990:14). For the latter the 'Link' causes some pangs of conscience, for the donors represent the very 'imperialist forces' they love to denounce. But they pursue the courtship nonetheless, for its pecuniary benefits are so essential for their survival that they cannot be ignored. Moreover, their consciences can be assuaged by the social democratic outlook of the foundations that dominate research funding.

There can be little doubt that foreign donors have rescued the scholarly enterprise in some African countries from penury. The relative freedom that these donors enjoy from domestic political constraints enables them to fund research themes that no local authority would consider. Moreover, because of their high standing and influence they sometimes shield their research grantees from harassment and persecution by the local authorities. In fact, in certain circumstances they have been able 'to offer support, haven or flight to beleaguered scholars' (Court 1990:8). Apart from research, foreign donors have also played a major role in the training of African social scientists, organizing conferences, and funding scholarly exchange programmes.

But a high price has been paid in terms of academic freedom and fundamental social science research. To begin with, as David Court, the Rockefeller Representative in Nairobi, candidly admits, the relationship between the donor and the recipient is inherently unequal. 'One has resources, the other would like them. In order to gain access the applicant can hardly avoid adjusting the manner of his approach to accord with the known or perceived preferences of the donor in a process of self-restriction and hence reduction of freedom' (Court 1990:9). Needless to say, 'changes in donor interests are bound to provoke a corresponding response by scholars leading them to take on topics which are of lower personal or institutional priority than those on the external agendas' (Court 1990:10). Donors can also 'eliminate work in certain areas by merely indicating areas that they consider fundable... (and) research may be constrained by bureaucratization of evaluation procedures where 'doability' narrows the areas that can be safely funded to meet certain bureaucratic schedules and goals' (CODESRIA 1990b:13). Thus the donors often set the research agenda.

Despite periodic shifts in emphasis, the donors' research agenda has been towards applied social science, which is justified in the name of development. Since the 1970s when the crisis of development in Africa became more evident, utilitarian and instrumentalist views of universities have become predominant. Consequently, the importance accorded to applied social science has increased at the expense of fundamental research. The work of many African social scientists has been reduced to consultancy and short-term contract work, which 'usually appears in reports that do not become part of the public domain' or open to wider intellectual discourse (Court 1990:10). Undoubtedly this has contributed 'to the creation of fragmented and non-cumulative social science... the executive summaries and reports replace articles and books' (Mkandawire 1989:12). In the process the continent's ability to define itself and the quality of African scholarship may have suffered. On the other hand, 'intellectual dependency on foreign scholarship which has the benefit of a domestic resource base, a domestically valued profession and hence opportunity for fundamental work' may have increased (Court 1990:11).

Thus, foreign donors have helped reinforce the utilitarian and instrumentalist views of the state on universities, which have served to delegitimize or marginalize the production of critical social science knowledge in the continent. They have succeeded in turning many of Africa's brightest social scientists into what Petras calls, with reference to Latin American intellectuals, institutional intellectual entrepreneurs, who 'live in an externally dependent world, sheltered by payments in hard currency and income derived independently of local circumstances' (Petras 1990:7). Unlike the earlier generation of organic intellectuals who 'moved in the world of rank and file political activists and militants, with a global vision that challenged the boundaries of the bourgeois liberal market place', the institutional intellectual 'writes for and works within the confines of other institutional intellectuals, their overseas patrons, their international conferences, and as political ideologues establishing the boundaries for the liberal political class' (Petras 1990:8).

The intense interventions that foreign donors engage in Africa would simply not be tolerated in their home countries, where structures of respect and support for universities and intellectuals are far better developed. That they are able to do so in Africa, and other Third World regions, is a testimony to the weaknesses of such structures in these societies, and an expression of material relations of domination of Africa, and not simply a reflection of some malice. Thus it should not be surprising that despite the fact that African social scientists have become more involved in applied social science research, the field is still dominated by western Africanists. Mama and Imam (1990) quote Nkinyangi (1983) who gives the example of Kenya where more than two thirds of all researchers analyzing that country between 1979 and 1981 came from outside. The discrepancy in research funding between the Kenyans and foreigners was staggering. Evidence is cited of 'a Kenyan receiving $92 to carry out research for an MA in education', while an American was allocated $91,000 to study 'the biochronology of African prehistoric monkeys', and a British anthropology professor was allocated '$1.27 million to study the ecology of subsistence pastoralism among the Turkana' (Mama and Imam 1990:6). Ali (1990) relates the story of how an ILO team to the Sudan in 1987 not only produced a report on the country's economy without full consultation with Sudanese economists, but when the latter raised questions, their Development Studies Research Centre was subsequently blacklisted by other donors, including the EEC, who had projects in the Sudan. Far from liberating African social scientists, the 'consultancy syndrome' has reinforced their subordination, not only to western donors but also to western Africanists who receive the bulk of the funding for research on Africa and often get to conduct what little theoretical work there is in these projects.

The most extreme form of intellectual intervention by external forces has come from the IMF and World Bank. The latter's 'blueprints revealed at the 1986 Conference of African Vice Chancellors in Harare, stated that Africa did not need university education' (Mama and Imam 1990:4). The outrage which greeted these proposals forced the Bank to retreat. But it did not give up on its goals of downgrading university education in Africa. In a detailed study, Bako (1990) shows that in return for a $120 million loan for Nigerian universities, the Bank called on the Nigerian government, firstly, to cut the number and size of the country's universities, ostensibly to make them more efficient; retrench academic and administrative staff; and reduce student enrolment. Secondly, privatization was encouraged for the universities. Thirdly, the Bank asked for supervision over the books, journals and equipment purchased through the loan and the close monitoring of the 'adjusted' universities through the Ministry of Education. Interestingly, $12 million of the $120 million loan was set aside for 'topping' the salaries of expatriates. The World Bank and the IMF have had a crippling impact on education in Africa through structural adjustment programmes which they have forced on African states.

Conclusions: Struggling for Academic Freedom and Democracy

The challenges facing African intellectuals are immense. On the one hand, they have to contend with state tyranny and the restrictions imposed by civil society. On the other hand, their work is undermined by authoritarian power relations in their own institutions and dependence on external sources for research funding, publication, and legitimation. Thus, the struggle for academic freedom is a multi-faceted one. It entails a series of struggles. Intellectuals must engage in the democratic struggles of the wider society, in addition to fighting for the democratization of their own institutions and practices. At the intellectual level, there is need to fight against research structures that undermine

African scholarship, theoretical paradigms that inferiorize African experiences, academic traditions that marginalize African contributions, and development prescriptions that ignore African struggles and realities. In short, African intellectuals need to fight against the Eurocentricism that dominates academic cultures and discourses on Africa, and for the construction of domestic structures and traditions that promote, support, and respect African intellectual production.

It is encouraging to note that these struggles are presently being waged. African intellectuals have begun to take seriously the question of academic freedom, as can be seen from the spate of declarations on the subject[3]. It is also reassuring to see that a number of regional and continental research organizations and networks are promoting, coordinating, and disseminating social science research in a way that has never been done before[4]. It is not by coincidence that it is these regional organizations, rather than the national universities, that are in the forefront of social science research.

Equally impressive is the new mood of self-criticism and self-confidence which one detects among African intellectuals, as was evident at the conference organized by CODESRIA on academic freedom (*Codesria Bulletin* 1990:14:1991). African intellectuals are becoming more daring in discussing their countries' problems, and, in the words of Mkandawire (1989:16), CODESRIA's executive secretary and a keen observer of the African social science community, 'increasingly aware of (their) preeminent position in African studies', thanks in part to 'the paradigmatic crises of the social sciences in the metropolitan countries [which] have contributed both to the de-fetishization of African social reality and the de-mystification of metropolitan social science and opened new vistas to approaches that are more deeply rooted in African social reality' (also see Mkandawire 1996).

The struggles for academic freedom by African intellectuals are part of a much larger battle for democracy currently taking place in Africa. As has happened so many times before, imperialism is trying to coopt and neutralize popular anti-imperialist projects, in this case democracy, by appearing as their champion. In reality, 'as the ideology of pluralism triumphs, the diversity of systemic options narrows' (Mazrui 1990:6). In other words, while countries are being encouraged to become more pluralistic internally, globally uniformity reigns supreme. The god of capitalism, we are told, has won, history is over. The mass struggles in Africa since the 1980s constitute the second wave of revolutionary upsurge in the continent this century. The first began in the 1930s, in the throes of the world depression, which tore asunder the material base of colonial capitalism. These struggles culminated in the nationalist drive for decolonization. Decolonization did not entail the final defeat of imperialism, rather its retreat and reconstitution into neo-colonialism. Similarly out of the present struggles for democracy, or for the 'second independence', which have been spawned by increased state authoritarianism and deepening poverty due to the economic crisis and the structural adjustment programmes, there is no guarantee that new, progressive regimes will be instituted. Indeed, democratization seems to be giving a whole new lease on life to regressive forms of development ideology. The 'market' is seen as a panacea of all economic ills. Socialism is pronounced dead, and national liberation an embarrassing obsolescence. A democracy that is restricted to the political domain, as in western democratic regimes, while economic management is held captive to non-democratic principles of privatization, is an incomplete one. It does not take into account 'the social transformations demanded by the anti-capitalist revolt of the periphery' (Amin 1990:15). It is not enough to equate democracy with multi-party politics, or to see it as an absolute value, for that is ahistorical, nor to justify it simply in terms of development, for developmentalism has always been used to justify the authoritarian post-colonial state (Nyong'o 1988; Shivji 1986:1-13). 'The centrality of democracy in the present historical context lies precisely in the fact that it expresses or constitutes an ideology of resistance and struggle of large masses and popular classes of people' (Shivji 1989:13).

Intellectuals are important for the success of the current struggles for democracy in Africa. Their contribution should begin with the democratization of their own practices and the construction of academic structures and traditions that promote, support, and respect African intellectual production. African intellectuals have to challenge vigorously the Eurocentricism that dominates Africanist discourses. Africanists in North America and Europe also need to re-examine themselves critically. Their dealings with African colleagues often smack of intellectual paternalism, condescension, and indifference. They often do not critically engage the work of their African colleagues through reviews and citations. Moreover, mountains of Africanist intellectual production have not moved a molehill of popular racist perceptions of Africa and Africans in their countries. Indeed, many Africanists sometimes seem to find it easier to identify with the struggles of people in distant Africa than with those of their fellow citizens of African origin, the African Americans, African-Canadians, and the various groups of African-Europeans. If Curtin's (1995) vituperous polemic against the 'ghettoization of African history' reflects prevailing Africanist opinion, then the African-Africanist chasm may be as wide and dangerous as the Atlantic Middle Passage (for response see Black Historians' Response (1995). Since their formation in the 1960s, African studies associations in western countries have behaved like colonial expatriate clubs, with very few members drawn from, and linkages to, the local African populations of North America and Europe. The struggle for academic freedom is, indeed, a many sided affair.

Notes

1. These papers are too numerous to list here. Two stand out in my mind. One is by Sherman (1990) focusing on Liberia, and the other by Awiti and Ong'wen (1990) focusing on Kenya.

2. The paper referred to here was sent in 1985 to the *Review of African Political Economy* which had just published a special issue on the 'intellectual left in Africa'. We sought to open a debate on the 'Western Africanist left and Africa'. After receiving a one line note of acknowledgement nothing was ever heard from the journal again. I have heard similar complaints from many African academics. And I had quite an interesting experience with the original paper on which this chapter is based. The then editor of the *Canadian Journal of African Studies*, Rhoda Howard, approached me after an earlier version of the paper had been presented at the 'Canada /USSR/ Africa Symposium' organized by the Canadian Association of African Studies to submit it for possible publication in the journal. The paper was later accepted for publication. However, in the package returning the edited versions for revision the editor inadvertently enclosed a letter she had written to the associate editor, Barry Riddell, concerning the paper. The letter contemptuously dismissed the paper and suggested it be published only because of my links with CODESRIA. I decided to withdraw its publication in the journal despite embarrassed entreaties from the two editors and had it published elsewhere. Gladly, I told both of them that the letter 'proved' the very point I was trying to make about the unhealthy relationship between African and Africanist scholars.

3. See, for example, the declarations and statements adopted by the Arab intellectuals (1990), CODESRIA (1990c), the University of Dar es Salaam Staff Assembly (UDASA, 1990), and the Zimbabwe Association of University Teachers (1990).

4. Among them is CODESRIA which coordinates, funds, and publishes a growing volume of social science research in Africa; OSSREA (Organization for Social Science Research in Eastern Africa); AAPS (African Association of Political Science); and SAPES (Southern African Political Economy Series).

Chapter Four

TRENDS AND INEQUALITIES IN THE PRODUCTION OF AFRICANIST KNOWLEDGE

Introduction

Trying to map out the trends and shifts in the production of knowledge on Africa is not an easy task, for there are various knowledges produced in diverse languages, from many institutional sites and numerous social and ideological positions. This chapter examines the spatial and institutional locations of the leading Africanist academic productions in English and the national and gender identities of those who produce, categorise, disseminate, and safeguard this particular form of knowledge. Specifically, it looks at the publishing trends in five leading English-language Africanist journals between 1982 and 1992[1]. The journals chosen are published in Britain, Canada, and the United States. They specialise in history, political science, literature, and African Studies generally. For each journal I tabulated the total number of articles and book reviews and the distribution of the authors in terms of those based within and outside Africa, Africans and non-Africans, or Africans and Africanists[2], men and women.

The chapter begins with an analysis of the politics of academic publishing in North America and Western Europe, the centres of Africanist research and publishing, the assumption being that Africanist journals are immersed in the academic cultures of these countries. Second, it examines the implications of producing knowledge on Africa in the European languages, of which English is one of the most important. It is argued that the dominance of these languages in academic publishing on Africa, both within and outside the continent, facilitates the hegemony of Africanist scholarship in African Studies. Third, the chapter focuses on academic publishing in Africa, and suggests that the publishing crisis of the 1980s has undermined African scholarship and increased African scholars' dependency on Africanist publication outlets. Finally, detailed empirical data is presented on the publication practices of the five Africanist journals selected for this study[3]. Despite variations among them and some changes over time, the overall results show a marked under-representation of Africa-based, African, and female writers. These biases, the chapter concludes, have serious consequences on the forms and types of knowledge produced on Africa.

The Politics of Publishing and Perishing

In many parts of the world research and publishing constitute the lifeblood of the academic enterprise, often overshadowing teaching as the measure of intellectual excellence and productivity, so that those who do not publish must perish. Several factors are responsible for this apparent overvaluation of publishing in academia. Focusing on the historical evolution of the American university, Skiff (1980) traces it to the struggles in the nineteenth century between a clergy that dominated the universities and an ascendant secular faculty. 'Emphasizing research in general', he argues, 'the clergy's incompetence in science in particular, and by elaborating this ideology of competence, faculty were able to construct and legitimate a socially segregated subuniverse of meaning; a body of role specific knowledge which is altogether esoteric as against the common stock of knowledge. Thereby they established themselves as its objectivating subsociety' (Skiff 1980:179). In short, faculty gained control over universities by stressing research and publications for whose evaluations they were solely responsible.

In time, especially from the 1960s as university education exploded, the publish-or-perish doctrine increasingly became a screening mechanism, a means of maintaining class differences in the now congested world of academia, of distinguishing the intellectual elite from the rest of the crowd. It also fitted nicely with the corporatization of universities in management and affiliation, all justified in the name of maintaining American global competitiveness. The universities were transformed from the 'ivory towers' of teaching and learning into 'industrial parks' of research and publication (Jacoby 1991). The researching and publishing elite were rewarded with employment and tenure in the most prestigious universities, professional recognition in their disciplines, faster career mobility, and higher salaries. The intense pressure to publish resulted in perverse inflation of publications, in which dissertations were cannibalized and quantity mattered more than quality, and mountains of papers were churned out to be listed and indexed rather than read.

All this sound and fury of what Barzun (1987) has called excessive specialism, the mass-production of trivial articles and unsynthesized bits and pieces of research, hardly signified major advances in knowledge.

The fragmentation of research and scholarly writing, or 'involution into narrow and specialized solitudes', as Bourne (1988:1423-4) so poignantly puts it, was reinforced by 'the increasing commercialisation of academic journals, a trend which must be seen as part of the international restructuring of the entire publishing industry. That industry has been extensively reorganised, refinanced, and redirected, and now tends to emphasize quantity over quality, sales over content, market penetration over effective communication...and the packaging of personalities and publications, rather than the dissemination of new ideas'. All these pressures contributed to the corruption of intellectual ethics, including outright fraud.

As the number of journals proliferated, it was no longer simply enough to publish, but to publish in the 'respectable' journals, and to be cited frequently. Citations, in fact, became a seemingly objective and critical measure of productivity, so that 'the frequency of referencing a person's work, appears to be an important determinant of salary differences among academics' (Hammersmesh 1982:481). In reality, of course, citations are nothing but a crude measure of intellectual merit, for they indicate neither the 'quality' nor 'impact' of academic publications. Comparing the citations of men and women certainly shows that citations are far from objective, for as a number of studies have suggested, 'people tend to evaluate members of their own sex more favourably than they do those of the opposite sex', either because of strongly held prejudices, or due to the simple fact that 'most researchers belong to networks within which they exchange papers, thereby becoming most familiar with the work of these particular colleagues' (Ferber 1986:282, 388; also see Deaux and Taynor 1973:261-2). Needless to say, these networks are characterised by sex segregation to a considerable degree. Thus citations constitute what has been called 'an exchange act between fellow travellers in academia, 'you cite me and I cite you'' (Sussman 1993a:116).

In North America and Western Europe the same processes are at work in relation to the work of racial minorities, including those of African descent. As bell hooks (1989) has bluntly put it, there is a tendency to overvalue work by white scholars, a point that has been convincingly demonstrated by many others. For example, in a passionately argued paper, Leslie (1990:892) takes to task the editor and reviewers of *Social Science and Medicine* for publishing a paper by the notorious racist Canadian psychologist Philippe Rushton (Rushton and Bagaert 1989) who argues that Africans are the least intelligent, most aggressive and oversexed of all races, hence the assumed prevalence of AIDS in Africa and among Diaspora Africans[4]. Leslie asks what appealed to the editor and the scholars he asked to evaluate the manuscript 'so that garbled biology and sociology appeared to be 'sufficiently respectable scientifically to merit publication', and racism appeared to be 'reasoned argument?''. The answers lie in the existence of deep-rooted racism in western societies and among sections of the scholarly community, which is continuously reinforced by the popular media, and more specifically, it reflects the scientific pretensions of social science and the inherent weaknesses of an essentially incestuous peer-review system. To quote him again (Leslie 1990:903-4):

> Almost all of the contributors to *Social Science and Medicine* are positivists. We like what we call hard data, and Rushton's article, with its maps, tables and charts, looked like the work of a positivist. Perhaps, also, its scientific vocabulary sounded persuasive... Also, on the whole we trust each other. Our conflicted community is built on trust that peer review will be even-handed, that we will not violate confidentiality, that we will study and write in an honest manner. When someone is a member of our community, it is very hard for us to think of his work as disingenuous... If Rushton had submitted a paper on 'Astrological Susceptibility to Aids', [the editor] and the outside reviewers would have agreed that it was inappropriate for *Social Science and Medicine*... We no longer have the burden of refuting astrologers because we agree that their pretence to science is fraudulent. Clearly, the scientific tradition in which Rushton's article is written does not yet evoke this degree of consensus...

We can only speculate whether or not Rushton's article would have seen the light of day if the editor or the reviewers had been African or Diaspora African intellectuals.

It should be clear, therefore, that scientific research and academic production is fundamentally a social activity, which is deeply implicated and infused by the social hierarchies and inscriptions of class, race, ethnicity, gender, and other inequalities found elsewhere in society and the acceptance or rejection of a publication is filtered by the editors and referees through the prism of their intellectual traditions, ideologies, and networks. The fact that the rates of publication and citation tend be quite low for Diaspora African and other minority scholars, sometimes being even lower than their percentage in the field, can hardly be surprising. While often the result of editorial and criterial biases, this also reflects different rates of submission, which is itself spawned by the different expectations and choices of publication audience.

In North America Diaspora Africans and women not only tend to be cited less frequently than Diaspora European males, but also their citation rate is often lower than their production rate. For example, an analysis of the citation rates of four anthropology journals revealed 'that women produced about one-quarter of the literature of the late seventies but received about one-fifth of the citations' (Lutz 1990:620). Reflecting gender segregation in research interests, undervaluation of the approaches most taken by women, and the discounting of women's work by men, the production and citation gap also arises from the fact that citation practices occur in relatively private contexts where the official and public discourse that disallows evaluation on the basis of gender can be routinely disavowed without incurring any costs[5]. The 'old boy' network, in which a small group of researchers reinforce each other's work, is revealed not only in the citation counts, but it also influences reviewer recommendation for manuscript publication[6].

Networking seems indispensable to the pursuit of a publishing career, certainly for unestablished scholars. In a candid tip on how to become published, one editor suggests the following strategies: 'networking at professional meetings, giving presentations at conferences, volunteering for editorial assignments, getting colleagues or professional editors to criticize your work...'[7]. In many cases, networking involves, in the initial stages, mentoring from a seasoned and well-published researcher. As Berardo (1993:59), a leading figure in the sociology of the American family confesses, he 'was the beneficiary of some early mentoring from an experienced researcher who took a personal interest in [his] career and has remained a friend to this day'. The importance of the academic socialization of mentoring and sponsorship is emphasized by many other prominent scholars (Burr 1993; Gelles 1993; Glick 1993). Because of the historical domination of white males in university education and publishing, Diaspora African and women scholars have relatively fewer chances of rising from the crowded ranks of academic houseboys and maids through mentoring and sponsorship.

Evidently, then, academic publishing cannot be divorced from the structures of power and the inscriptions of gender, race, and class. An academic publication, like any text, 'plays an indispensable role in the administration of power, both in support and in opposition' (Lorimer 1993:204). Indisputably, in western countries men, whites, and the elite have enjoyed, and continue to enjoy, better access to textual communication than women, Diaspora Africans, and the lower classes. These gender, racial and class biases are of course not condoned in the official discourses of academia. Academic criterial systems pride themselves on their objectivity and fairness. They all rest on peer review, long regarded and widely accepted as the best method of evaluating scholarly work, maintaining standards, and regulating competition.

'Myths', Sussman (1993b:164) notes sarcastically, 'are essential to the maintenance of professional cultures... The myth that peers review one's submission of scholarly work in an objective manner has narcotic effects... [T]he traditional practice [is] that senior scholars judge junior ones, not the reverse'. Thus it is not peers who induct one into the sacred academy of the well-published, but the gatekeepers, at whose head table seats the editor and his appointed team of associate editors and reviewers. In most cases, it is the editors who make the final publication decision, so that the chances of one getting published depend to a considerable degree upon whose desk the manuscript lands[8]. Undoubtedly many editors try to be conscientious and fair-minded, but few of them can avoid the tendency to shape their journals according to their perceptions of quality research and scholarship, or indeed the temptation, if not to build academic empires, then at least fiefdoms[9].

Critics have increasingly pointed out that the reviewing process, whether it is double-blind reviewing in which the identities of both the writer and the reviewer are kept confidential, or single-blind reviewing in which only the identity of the reviewer is kept secret, does not always guarantee fairness or quality. Discrimination against a submission based on the author's personal characteristics or ideology is not eliminated, for there are numerous ways of identifying a writer even if his/her name is removed (Yankaner 1991; Schrank 1994; Hiatt 1994). The single-blind reviewing process lacks accountability and facilitates plagiarism and appropriation of other peoples work. Some have gone so far as to argue that peer review 'hardly does better than random choice' (Handlin 1987:217), that 'virtually no fraudulent methods or data have been detected by this process' (Ben-Yehuda 1985:177), and that it 'strongly discourages originality' (Singer 1989:132-3 and 1990).

While Singer attributes, rather simplistically, the criterial crisis to what he calls the rise of a formalised technology of assessment spawned by the proliferation of journals, itself due to the demands of the publish-or-perish syndrome, Hill (1990:301-2) convincingly blames it on 'the continuing and coercive control of the few well-funded journals by small, elite networks who demand conformity to narrowly-defined research agendas, approved methodologies, and centripetal bibliographic rituals'. It is not uncommon for editors and referees to 'use their positions to vent hostility and aggression, or suppress ideas, findings, or methodologies with which they disagree. The typical anonymity of referees and the control editors exercise over the review process make it very difficult for authors to substantiate such infractions' (Berardo 1989:252).

There is in fact considerable evidence that journal rejection rates have more to do with variation in consensus than space shortages. Hargens (1988) found that journals in the natural sciences in which there are high levels of consensus have lower rejection rates than those in the social sciences where the variation in consensus is wider[10]. In short, the chances of publication are enhanced if one's 'research falls within the paradigmatic guidelines currently accepted by the established members in a particular field' (Berardo 1993:62; also Berardo 1981). Some have argued that the expansion of small journals in the social sciences has reduced the power of the gatekeepers in the prestigious journals and facilitated the production of unrecognised scholars. But the corollary is that 'it takes longer for the latter's work to be appreciated and incorporated into the' mainstream (Hill 1990:299). Marginality from the mainstream entails fewer rewards from the limited reputational and pecuniary spoils of academia. This is to suggest that name recognition matters, for it often translates into less rigorous evaluations, and that first impressions, interpersonal skills, and social networking are important in determining the success or failure of an academic career (Paludi and Bauer 1983; Paludi and Strayer 1984; Sindermann 1982; Siow 1991).

The academic publishing establishment has developed various techniques of depreciation and dismissal of the work of women and minorities as intellectuals and scholars, one of the most prevalent being the denial of its 'originality'. 'The concept of 'originality'', Carroll (1990:136) writes, 'though essentially empty of meaning, is used today to justify and rationalize a class system based upon claims of property in ideas'. This class system of intellect, in which the word 'original' is a ritual stamp of approbation and reprobation, inclusion and exclusion, is also patriarchal and racist, for the work of women and minorities tends to be disproportionately portrayed as 'derivative', 'imitative', and 'unoriginal' as compared to the work of white males, among whom are to be found the 'fathers' of fields and traditions, the 'leading figures' of various 'schools of thought', the eponymous 'discoverers' of this and that, and the loyal sons who, as in 'paternity suits', compete vigorously 'over rights of inheritance to property in ideas and to whatever rewards, honours, or privileges may accrue to the successful heirs' (Carroll 1990:150).

What does all this mean for Africanist scholarship and the place of African production within it? In so far as it is anchored in the social, political, and institutional contexts in which research and publication are implicated in the hierarchies of gender, race, and class, Africanist scholarship cannot provide a neutral, let alone favourable, site for intellectual production and communication by Africans. Often separated by race, and sometimes gender, ideology, and social and physical distance, African scholars are largely outside the social and academic patronage networks of the Africanist establishments in North America and Western Europe. Africanists are no less slaves to the relentless productionist imperative of publish-or-perish that drives academic life in these countries than their compatriots in other fields. For them publications and citations also constitute social practices and resources, forms of cultural and symbolic capital, which often translate into the reputational capital of status and the actual capital of jobs, tenure and higher salaries. Nor are they immune from the racial, gender, and class biases that colour so much social science work in their countries. Moreover, despite their segmentation, the various cultural markets in which Africanist work is produced share common definitions of good work rooted in western epistemologies and resonant with current intellectual fads, and always adorned in universalistic theoretical packages[11].

In His Master's Voice

The question of language is central to understanding the nature and dynamics of this inequality. A lot of the writing on Africa, both academic and literary, is done in languages initially imported from Europe largely during the colonial period. The debate on language is an old and fascinating one, which we can only discuss briefly here. It has interested creative writers and literary critics far more than social scientists (Owomoyela 1993a). There are three contending positions: the rejectionists, neo-metropolitans, and evolutionist/experimenters, as Okara (1991) calls them. The rejectionists have opted to write in their indigenous languages arguing, like Wa Thiong'o (1986) in his 'farewell' to English, that true African sensibility cannot be expressed in the European languages with their baggage of imperialistic, colonialistic, and racist world views. For their part, the neo-metropolitans adopt the metropolitan language without consciously trying to adapt it to African circumstances and aesthetics, while the evolutionist/experimenters try to transform it in various ways so that it becomes a medium capable of giving full expression to African cultures[12].

The debate is often cast in terms of commitment or radicalism: who is more committed to African society — the writer in the local languages or the one in the European languages?. It can be argued that this way of problematising the debate is simplistic, for it ignores the fact that reactionary and repressive attitudes and values can also be articulated in local languages, and more importantly perhaps, that the European languages are part of Africa's historical experience and cannot be wished away (Mazrui and Mphande forthcoming). Indeed, some have argued that these languages have become increasingly domesticated and largely indiginized, for example through pidginization and creolization (Jones 1991).

While it is true that English is one of the languages in the former British colonies, the question of its assimilation into the social fabric and its role in intellectual discourse still remains. The increased production of academic and literary texts in the metropolitan languages in the ex-colonial world, including Africa, reflects as Ahmad (1992:76) has argued, the 'greater elaboration and deeper penetration of the state into all spheres of civil society' and the 'consolidation, expansion, increased self-confidence, increased leisure, increased sophistication of the bourgeois [and middle] classes in these countries'. The centralizing imperatives of the post-colonial state and developmentalist ideology have reinforced and extended the earlier colonial claims for English as a force for cultural unification, national integration, administrative efficiency, and modernization.

The privileging of English in the social processes as *the* language of *national* culture, bourgeois civility, and intellectual production not only reproduces, to borrow Mamdani's (1990:5) phrase, 'a privileged discourse' and maintains a 'linguistic curtain' of class power and separation between a small elite and the working masses, it also raises fundamental questions about the constructions, relevance, believability, and legitimacy of theories, concepts, and beliefs about Africa produced through the prism of 'inherited modes of understanding', as Appiah (1992:5) puts it. Isn't the expression of African modalities in non-African languages, asks Mudimbe (1988:186), not distorting in that the continent's peoples, societies, and histories are interpreted from the margins of African contexts, in languages and discourses limited by their exteriority, categories and conceptual systems rooted in a Western epistemological order?

There are no easy answers. But one consequence of the privileging of the European languages in academic productions of knowledge on Africa may be the capacity it gives western scholars for intellectual accumulation, appropriation, and domination. Owomoyela (1993b:354) notes, quite perceptively, that were Africans to write only 'in African languages their works would be inaccessible to all but a handful of the non-Africans to whom they are available at the moment'. It is rarely seen as problematic to study and write on an African society without knowing the language of that society, as Sklar (1993:100) so arrogantly confirms[13]. While the typical African scholar knows at least one European language, including his/her own, and sometimes other African languages, the typical Africanist scholar often has no competence in a single African language and relies on research assistants for field research. Needless to say, this is hardly tolerated in intellectual relations among the metropolitan countries themselves. An American scholar studying Germany without knowing German would not be taken too seriously by experts in the field.

But it seems different with Africa, with the imperialized formations of the ex-colonial world, where the indigenous languages, the languages of the vast majority of the people, of their cultural and often literary texts, wilt into inconsequence before the universalistic and totalizing gaze of western discursive hegemony. It is this 'power of definition', as Professor Muwonalero puts it in the opening story in Chapter 1 above, that allows many western Africanists 'to jump across the continent without learning a single African language', and to engage, as Mkandawire (1989:8) once bemoaned, in 'faddish theorising' and pursue 'ephemeral debates' without the voice of Africans[14].

Africa's Book Hunger

Publishing is critical not only for the academic enterprise, but also for the cultural identities of nations, peoples, classes, and groups. It provides the material basis for producing, codifying, circulating, and consuming ideas. The record of publishing in post-independence Africa has been a complicated and rocky one. At independence multi-nationals from the former colonial powers dominated the publishing industry in most countries. Soon, however, viable indigenous publishing established itself in some of them, either as a result of state or private initiatives, or both. In some cases the foreign companies were nationalized or bought out[15].

While it is dangerous to make generalizations for the whole of Africa, the 1960s and 1970s, to quote Zell (1993:386), one of the leading authorities on the African publishing industry, might be described as decades 'of boom and expansion', whereas, 'the 1980s can only be described as a decade of crisis for the African book industries'. The economic recession that hit many countries and the ill-conceived IMF- and World Bank-devised structural adjustment programmes took their toll on local publishing industries, as investments and sales plummeted, printing equipment and facilities deteriorated, production and retail costs escalated, and distribution networks and outlets atrophied, leading to what has widely been referred to as the 'book famine'[16]. Many once renowned periodicals and journals ceased publication or were reduced in size and frequency, while many new ones often did not survive beyond 'volume 1, number 1'.

'What is remarkable', Zell (1993:373) tells us, is that 'despite the overall gloomy picture... new indigenous imprints continue to mushroom all over Africa, and some privately owned firms have shown a great deal of imaginative entrepreneurial skill in the midst of adversity... There are particularly dynamic indigenous publishing companies in Zimbabwe, Kenya, Nigeria, and even Ghana, which was especially hard hit by the economic recession'[17]. And in the Francophone countries, Nouvelles Editions Africaines continues to expand, so that it 'is now a major force in all areas of publishing with a massive and impressive list, although it can be argued that their dominance and near monopoly has stifled the growth of small independent publishers' (Zell 1993:371).

Thus, the recurrent recessions have forced the restructuring of the publishing industry in many African countries. Liberalization has led to the emergence of a host of new publishers in countries as diverse as Zambia (Chirwa 1994) and Senegal (Faye 1994). Perhaps the most exciting development has been the creation of publishing companies by academics themselves aimed at the regional and continental markets. For example, the Zimbabwean political scientist Ibbo Mandaza and his colleagues set up Sapes Books[18], which already has an impressive list on Southern Africa. Most noticeably at the continental level was the decision by the Council for the Development of Social Science Research in Africa (CODESRIA) to undertake a systematic publishing programme of books and monographs, not just the occasional publication of select papers from its conferences as had been the case before. In addition to its flagship quarterly journal, *Africa Development*, Codesria also decided to assume publication of *Afrika Zamani*, a journal of African history, in collaboration with the Association of African Historians (Codesria 1993) and to launch two new journals in Sociology and International Relations[19]. Creative responses to the journal crisis have included the recently established African Journals Distribution Program, which seeks to facilitate the distribution of African journals to university and college libraries throughout the continent (Gidney 1994; Brickhill 1994; and see next chapter). On the marketing and distribution front, the most important development was the formation of the African Book Collective (ABC) in 1989 by African publishers to undertake joint promotion and distribution of African books outside the continent, especially in the critical markets of Western Europe and North America. This was followed three years later by the formation of the African Publishers

Network (APNET) to encourage intra-African publishing and trade in books. Within a few short years ABC has achieved considerable success in generating sales and slowly mainstreaming African books in some western markets, although it still has a long way to go to fully achieve its objectives (Jay 1994; Bgoya 1994).

The quality of the publications from these ventures has generally been high, thus confirming Graham's (1993:249) pithy observation that 'adversity, it seems, is good for the written word', and challenging Shaw's (1993:153) self-serving, arrogant, and fallacious contention that given the perilous state of African publishing Africanists not only have 'superior opportunities both to research and to publish', but that 'the most original research will continue to emanate from outside rather than inside the continent in the 1990s as in previous post-independence decades'.

Despite their heroic efforts, African publishers have not been able to satisfy the great demand for publication that exists among the continent's scholars. As university presses and journals collapsed, and commercial publishers turned their backs on publishing specialised and costly titles, many scholars were forced to either abandon publishing altogether, especially as this coincided with the deterioration in academic infrastructures and salaries and a rise in state authoritarianism and assaults on academic freedom, or they increasingly turned to the profitable pursuit of writing consultancy reports, which are often unpublishable for their lack of theoretical or intellectual stimulus. Others tried their luck publishing abroad. The latter became increasingly dependent on publishing outlets in Western Europe and North America controlled by Africanists. Among them were those who in the thriving sixties and seventies had deliberately chosen to publish in local journals and with local presses, but now found themselves hostage to editorial and scholarly networks they neither belonged to nor trusted[20]. As might be expected from our analysis above, they have been met with less than enthusiastic welcome as the data below demonstrates.

The Colour and Gender of Africanist Journals

An empirical analysis of the publications record of the five major English-language Africanist journals that were examined, demonstrates quite clearly the marginalization of African intellectual production in Africanist scholarship. The proportion of articles and book reviews published by scholars based in Africa, whether African or Africanist, and by African scholars both within and outside the continent is extremely low. There is also a glaring gender gap. The period examined was between 1982 and 1992, during which the African 'book famine' was particularly severe and the African need for 'publishing aid' could be expected to have been relatively higher than in the 1960s and 1970s. This cannot, of course, be entirely substantiated without a comprehensive survey of the number of *submissions* the journals received, relative to those they chose to *publish*. Indeed, it would be instructive to compare the submission and publication rates in the 1980s with data from the earlier two decades. Better still, an extensive survey of African scholars, to determine how they decide their choices of audience and outlets for their writing, would yield invaluable qualitative information that cannot be gleaned from the dry statistics of publication indexes. Unfortunately, this is beyond the scope of this analysis.

Altogether, the five journals published 1361 articles and 3010 book reviews between 1982 and 1992. African authors accounted for 24% of the articles and 15% of the reviews[21]. These figures amply demonstrate the limited involvement of Africans in Africanist journals, while the low percentage for reviews underscores the exclusion of African scholars from the intellectual networks through which the review process is organised. A more detailed analysis of the data (see appendices) indicates that Africa-based authors accounted for 20% of the articles and 13% of the reviews, while African authors based in Africa accounted for 15% of the articles and 10% of the reviews, which shows that a significant proportion of published submissions from Africa, 22% of the articles and 26% of the reviews, were by expatriate Africanists working in Africa. Conversely, large numbers of Africans who published in these journals, 38% of the article writers and 32% of the reviewers, were abroad at the time of publication. Thus many of those published from Africa are not Africans and many of the published Africans do not live in Africa. This can be seen in the Table 1.

An examination of the trends between 1982/3 and 1991/2 shows an overall increase in the number of articles published in the five journals from 268 to 281, a rise of 5%. Specifically, the number of articles by Africans nearly trebled from 38 in 1982/3 to 105 in 1991/2, and their share of published articles rose from 14% to 37%, respectively. As might be expected, the number of Africans who published articles while domiciled abroad rose sharply, from 24% of the total in 1982/3 to 48% in 1991/2. In the meantime, the numbers of expatriate Africanists publishing while in Africa hardly changed. Similar trends characterised the publication of reviews, although Africans continued to publish a far lower proportion of the reviews than articles. The total number of reviews increased by 19% between 1982/3 and 1991/2, while African contributions more than doubled, and their share rose from 10% to 20% of the total. Also, nearly 48% of the African reviewers lived abroad.

It can be seen that the publication space allotted to African scholars in Africanist journals grew quite considerably between 1982 and 1992, although from a low base and these scholars still remained under-represented. The increase reflected, in part, the growing migration of African scholars to the western countries. Submissions from Africans still based in the continent may also have increased due to the crisis in local publishing, as suggested earlier[22]. What we don't know are the rejection rates, and the gap, if any, between the submission rates and publication rates of African and Africanist scholars. It would not be farfetched to assume that the rejection rates would be higher for African as compared to Africanist scholars. The anecdotal evidence appears overwhelming.

Similarly, despite some modest improvements, women scholars fare poorly in the Africanist journals. They accounted for only 19% of the articles published between 1982 and 1992, and 18% of the book reviews. Between 1982/3 and 1991/2 the number of articles and reviews by women rose from 42 to 53, and 94 to 119, respectively, that is by about 26%, far below the rate of increase for African authors as a whole. As a percentage of all the materials published by the five journals the increase was small, indeed, from 16% to 19% for articles, and 19% to 20% for reviews.

Table 1: Comparative Publication Data, 1982-1992

Journ	Articl	AB	AA	ABA	Revie	AB	AA	ABA	MAt	FAt	MRev	FRev
JMAS	272	55	52	25	262	60	80	44	230	42	229	33
ASR	210	35	50	32	426	22	70	19	167	43	318	108
RAL	321	102	132	85	762	168	185	130	254	67	580	182
CJAS	304	40	67	37	764	104	140	86	233	71	639	125
JAH	254	35	32	28	796	48	28	19	220	34	703	93
Total	1361	157	233	197	3010	402	503	298	1204	157	2469	541

Key To Tables:
Journ: Journals
Articl: Number of Articles
AB: Africa Based Authors
AA: African Authors
ABA: Africa-Based African Authors
Revie: Number of Reviews
MAt: Male Authors
FAt: Female Authors
MRev: Male Authored Reviews
FRev: Female Authored Reviews

Journal Abbreviations:
JMAS - Journal of Modern African Studies; ASR - African Studies Review; RAL - Research in African Literatures; CJAS - Canadian Journal of African Studies; JAH - Journal of African History.

The only time women feature prominently is when gender issues are being discussed, as is the case, for example, in the 1988 *Canadian Journal of African Studies* Special issue on African Women[23]. Interestingly, African women scholars are conspicuous by their absence even in this issue. In fact, articles and reviews by African women are a rarity. So not only are Africans and women generally under-represented, African women are virtually invisible in the Africanist intellectual universe[24].

It stands to reason that there are some differences among the five journals, partly influenced by disciplinary, institutional, ideological, and locational affiliations. Table 2 gives some indication of the standing of these journals in terms of the publication space allotted to African and women scholars. It can be seen that the *Journal of African History* has the worst performance on all counts, while *Research in African Literatures* scores consistently high, claiming the first or second position in all categories.

Tables 3 and 4 illustrate the trends in the publication of articles and reviews by Africans and women in the five journals. Once again, it is quite clear that the *RAL* has made the most significant advances in incorporating the work of Africans and women, while the record of the *JAH* is mixed. As it increased its publication of African contributions, those by women fell. The *CJAS* was the only other journal to register a slight fall in the percentage of articles and reviews it published by women. The sharpest fall in women's reviews was in the *ASR*. In so far as Africans increased their publication share of both articles and reviews, it could be argued that the gains were made by African men at the expense of women.

The poor performance of the *JAH* as far as Africans and women are concerned could partly be explained by locational factors. The fact that the journal is based in Britain, which offers far less employment opportunities for migrant African academics than the United States where the relatively more accommodating *RAL* is located, may be important. It is, in fact, instructive to compare the location of the Africans who contributed articles to the two journals. For the *JAH* 88% of their African contributors were based in African countries, while the equivalent figure was 64% for *RAL*. Also, the other British-based journal, *JMAS*, ranks last in most of the categories in terms of the percentage of articles it published by Africans and women, between 1982/3 and 1991/2.

No less important, perhaps, are the varied disciplinary orientations of these journals. Despite the much-vaunted revolution that has occurred in African historiography, in which the *JAH* has played a considerable role, Africanist history has not engaged the questions of African cultural identities, consciousness, and values as literature has tended to do. While the Africanist historian uses archival and oral narratives that do not contest his reconstructions or constructions of various pasts, the literary critic works with cultural texts whose producers can, and often do, interpret, challenge, and interrogate interpretations of their work.

Universalistic and methodological pretensions are more difficult to sustain in a field where debates over the languages, infrastructures, and audiences of the intellectual artifacts are so central as is the case in literary productions.

Table 2: Publication Space Allotted to Africans and Women, 1982-1992
Percentage

Journal	Africans: Articles	African: Reviews	Women: Articles	Women: Reviews
Journal of Modern African Studies	19	31	15	13
African Studies Review	24	16	20	25
Research in African Literatures	41	24	21	24
Canadian Journal of African Studies	22	18	23	16
Journal of African History	13	4	13	12

Table 3: Trends in Publications of Articles by Africans and Women, 1982/3 - 1991/2
Percentage

Year	JMAS		ASR		RAL		CJAS		JAH	
	African	Women	African	Women	African	Women	African	Women	African	Women
1982/83	6	4	20	14	23	15	16	24	7	17
1991/92	22	19	24	24	52	20	34	22	26	9

There can be little doubt, therefore, that securing publication space in Africanist journals is conditioned by location, nationality, and gender, among many other variables. This is, of course, not peculiar to these journals.

The unequal intellectual relations between Africans and Africanists reflected in the publication practices of these journals impacts unevenly among the English-speaking African countries. Scholars from, or based in, Nigeria and South Africa, Africa's largest economies, dominate in accessing publication space. Needless to say, there are important differences in the composition of the scholars from the two regions: the South Africans are predominantly white, not black as in the case of Nigeria.

Table 4: Trends in the Publication of Reviews by Africans and Women, 1982/3 - 1991/2
Percentage

Year	JMAS		ASR		RAL		CJAS		JAH	
	African	Women	African	Women	African	Women	African	Women	African	Women
1982/83	11	4	9	38	16	26	11	22	2	10
1991/92	34	19	15	25	32	28	26	21	5	8

This further demonstrates the point that regions and groups with concentrations of economic and political power tend to dominate the production and dissemination of knowledge.

Conclusions: Beyond Provincialism

In analyzing the data for trends, one is struck by how modest the changes have been. To be sure, as Tables 3 and 4 and the appendices demonstrate, there have been fluctuations among the five journals in their allocations of space to the various groups under discussion. One notable change is the shift to North America in the location of the African scholars publishing in these journals. This undoubtedly is connected to the 'brain drain' analyzed in Chapter 2. But on the whole, if these journals are any indication, Africanist publishing is still largely a preserve of white male scholars. While the economic and educational crises in some African countries may have forced many scholars who did not migrate to seek publication outlets abroad, their accessibility to Africanist journals may have actually declined because the 'book famine' deprived them of the opportunity to read and cite the works of the gatekeepers, thus presenting the latter with impeccable excuses to reject their 'poorly researched' manuscripts.

The publication data from the five journals seems to show, therefore, that African scholarly production finds limited outlets in western academic media. This marginalization reflects the historical and contemporary relations of domination and dependency between Africa and the West. The production of knowledge is related to the structures of power which are articulated with spatial, social, gender, ethnic, and racial hierarchies. Thus, for African scholars to 'make it', sometimes even within Africa itself, they have to demonstrate familiarity with, indeed they must be immersed in, each intellectual fad that emerges from the Africanist capitals of the West. Once so baptised, they may be allowed to enter the secret society of academia, to partake in the crumbs of the scholarly enterprise: conferences, citations, and perhaps a visiting professorship here and there. It can be argued that these structures of reference, attitude, and legitimation have reinforced the marginality of African scholarship, for African scholars have ended up holding themselves hostage to editorial foundations they do not, and cannot, control. How many African scholars have complained of getting their manuscripts rejected only to see their work used and published by the gatekeepers who jealously protect their African research fiefdom. It is mistaken to think that Africanist journals were established as forums for open debates on Africa, let alone to provide African scholars with unfettered access. As with most journals, Africanist journals are provincial, all universalistic pretensions aside, set up to advance the intellectual traditions of the societies in which they are founded, and the academic careers of their patrons[25]. There is, therefore, no substitute for a vigorous publishing industry in Africa for African scholars.

In conclusion, it is more than evident that Africans and women are grossly under-represented in most Africanist journal publications. The chapter has argued that this is related to the structures of power, the prevalence of racism and sexism in western academic cultures, compounded by the publish-or-perish syndrome, and the way in which the field of African studies has historically been imprisoned by languages, epistemologies, and discourses that are externalist, so that Africa is often reduced to no more than an empirical lab to test pretentiously universalistic models, theories, and paradigms concocted in the academic factories of North America and Western Europe.

In the final analysis, the solution lies in Africans developing and sustaining their own publishing outlets, out of which can emerge truly African intellectual communities capable of directing and controlling African Studies. When that is done, and current developments seem encouraging[26], the exclusionary practices of the Africanist academic media will not matter much and the doors will be open for more fruitful and equitable intellectual relations between Africans, both at home and abroad, and the Africanists. Academic provincialism, imbued with racism and sexism, would be the only loser.

APPENDICES

Journal of Modern African Studies, 1982-1992

Year	Articl	AB	AA	ABA	Revie	AB	AA	ABA	MAt	FAt	MRev	FRev
1982	28	7	1	1	13	1	1	0	27	1	12	1
1983	23	4	2	1	32	6	4	2	22	1	31	1
1984	25	3	2	1	29	11	7	6	12	4	25	4
1985	15	2	6	2	7	0	0	0	13	2	6	1
1986	24	4	5	4	5	0	0	0	15	9	3	2
1987	24	2	2	0	22	4	4	3	22	2	17	5
1988	27	4	4	1	21	5	7	5	12	4	20	1
1989	26	6	7	4	30	11	13	7	22	4	28	2
1990	26	5	9	4	36	8	21	8	21	5	33	3
1991	26	6	8	4	35	11	14	11	22	4	31	4
1992	28	6	6	3	32	3	9	2	24	6	23	9
Total	272	55	52	25	292	60	80	44	230	42	229	33

African Studies Review, 1982-1992

Year	Articl	AB	AA	ABA	Revie	AB	AA	ABA	MAt	FAt	MRev	FRev
1982	26	4	6	5	19	0	1	0	23	3	11	8
1983	23	4	4	3	13	0	2	0	19	4	9	4
1984	22	8	8	7	17	6	7	6	21	1	15	2
1985	15	3	3	3	15	3	3	2	13	2	14	1
1986	24	2	5	2	24	1	2	1	14	10	14	10
1987	21	2	3	2	19	1	4	1	17	4	18	1
1988	25	5	8	5	28	0	3	0	23	2	20	8
1989	10	3	4	3	30	1	5	1	6	4	23	7
1990	19	3	3	1	105	4	10	2	12	7	77	28
1991	10	1	1	1	89	5	17	5	6	4	64	25
1992	15	0	5	0	67	1	16	1	13	2	53	14
Total	210	35	50	32	426	22	70	19	167	43	318	108

Research in African Literatures, 1982-1992

Year	Articl	AB	AA	ABA	Revie	AB	AA	ABA	MAt	FAt	MRev	FRev
1982	23	4	4	4	70	20	9	7	20	3	50	20
1983	17	5	5	5	53	12	11	9	14	3	41	12
1984	20	5	2	1	69	12	16	10	9	11	53	16
1985	22	10	10	6	76	13	13	8	17	5	58	18
1986	17	7	8	6	88	21	25	17	15	2	69	19
1987	27	6	8	6	93	19	27	19	22	5	73	21
1988	25	12	13	12	85	25	21	19	23	2	64	20
1989	23	11	9	9	89	17	20	14	18	5	71	18
1990	38	10	16	8	47	11	14	10	29	9	35	12
1991	51	14	23	13	46	8	13	7	37	14	35	11
1992	58	18	34	15	46	10	16	10	50	8	31	15
Total	321	102	132	85	762	168	185	130	254	67	580	182

Canadian Journal of African Studies, 1982-1992

Year	Articl	AB	AA	ABA	Revie	AB	AA	ABA	MAt	FAt	MRev	FRev
1982	35	4	2	1	72	8	6	27	8	55	17	
1983	39	5	10	5	68	7	6	29	10	54	14	
1984	34	2	5	2	51	7	6	29	10	40	11	
1985	23	1	1	1	74	8	10	20	3	65	19	
1986	16	1	1	0	77	5	5	14	2	63	14	
1987	20	5	9	5	97	13	10	18	2	88	9	
1988	40	5	9	5	55	6	3	19	21	46	9	
1989	25	5	6	6	65	14	10	22	3	55	10	
1990	17	4	6	4	68	13	11	15	2	65	3	
1991	26	5	8	5	65	7	5	22	4	50	15	
1992	24	5	9	5	72	16	14	17	7	58	14	
Total	304	40	67	37	764	104	86	233	71	639	125	

Journal of African History, 1982-1992

Year	Articl	AB	AA	ABA	Revie	AB	AA	ABA	MAt	FAt	MRev	FRev
1982	26	3	2	2	84	4	1	1	24	2	79	5
1983	28	2	2	2	81	4	3	2	21	7	69	12
1984	23	1	2	0	77	8	5	2	20	3	71	6
1985	17	5	2	2	84	3	2	2	15	2	73	11
1986	25	3	2	2	52	3	1	1	17	8	50	2
1987	20	3	3	3	78	3	2	0	18	2	64	14
1988	25	3	3	3	72	4	2	0	21	4	54	18
1989	24	3	3	3	56	5	1	2	22	2	51	5
1990	23	2	2	2	62	4	3	3	22	1	54	8
1991	21	2	4	2	65	4	2	2	20	1	60	5
1992	22	8	7	7	85	6	6	5	19	3	78	7
Total	254	35	32	28	796	48	28	19	220	34	703	93

Notes

1. The journals are *Journal of Modern African Studies (JMAS)*, *African Studies Review (ASR)*, *Research in African Literatures (RAL)*, *Canadian Journal of African Studies (CJAS)*, and *Journal of African History (JAH)*

2. These categories are problematic. The term African in this paper refers to those who by birth or nationality, or both, are from an African country. This obviously includes white South Africans and North African Arabs. The term Africanist is used here to denote non-Africans who study Africa. Included in the latter category are people of African descent, referred to in this paper as Diaspora Africans, who were born and are nationals of countries in the Americas and Europe.

3. This selection was based on personal judgements on what I consider the leading Africanist journals covering Africa as a whole in fields I work on: history, literature, and African studies generally. Reflecting the racist foundations of African studies, the Africa these journals normally cover is the 'sub-Saharan' concoction, excluding North Africa, which incidentally nobody calls 'supra-Saharan Africa'!

4. What Rushton says in this piece about the African origins of AIDS and the role of promiscuity for its apparent rapid spread within the continent has been stated in 'respectable' western magazines, from the *Economist* and *Newsweek*, to *Time* and *The Atlantic Monthly*, where Africa-bashing is common, and in some alarmist Africanist publications. The medical pathologization of African bodies, spaces and sexualities, and especially of African women, has a long history in Africanist scholarship, as Summers (1991) and Jackson (1996) so convincingly demonstrate. For an early critique of the racist reporting on AIDS see, Richard Chirimuuta (1989). And for a sample of the African media's response, see *New African,* December, 1993; October 1994.

5. For succinct feminist critiques of academic publishing, Spender (1981); and Ward and Grant (1985).

6. In a review of five behavioral journals it was found that gender was not a problem in the final editorial decision, but in the earlier stages of the review process, especially where women are working in male-dominated fields and are thereby seen as violating sex-role stereotypes. See M. E. Lloyd (1990).

7. This summation of one editor's advice, out of 93 who responded to a questionnaire, is in Hanks (1993).

8. In a survey of 400 editors of journals, 93 of whom responded, only one editor stated that the reviewer makes the decision, and many stated 'they read each article before it was sent to reviewers.. 'to screen out and deflect articles that are inappropriate before review'', see L. Matocha (1993:33).

9. I had a revealing experience with the *Canadian Journal of African Studies* in 1982 when I submitted a paper based on a chapter from my Ph.D. dissertation (what else!). My name was removed when the article was sent to the reviewer, who nonetheless knew I was the author, for he had been the external referee on my thesis defence, which he obliquely alluded to in his report to the editor. In the dissertation and the paper I attacked his interpretation of certain aspects of Kenyan labour history. In the relatively public context of a thesis defence he passed my thesis, but in the private process of reviewing he rejected the paper!

10. Cole, et.al. (1988) dispute Hargen's thesis and contend that rejection rates can be attributed to field-specific norms concerning publication, the diffuseness of a field's journal system, and differences in training practices in various fields. Hargen (1988) for his response.

11. Michele Lamont (1987) uses this concept of segmented cultural markets to provide an illuminating analysis of the different cultural and institutional contexts which facilitated the legitimation of Jacques Derrida's theories in France and the United States.

12. Owomoyela (1993b: 350) contends that 'the controversy about the choice between African and European languages is virtually confined to the Anglophone parts of the continent, because with regard to the Francophone areas, Arabization has made it irrelevant in the Maghreb while sub-Saharan Africa has shown little discomfort with the primacy of French.'

13. The exception being North Africa where scholarly writing in Arabic is long-established and Arabization, as noted above, makes the linguistic debate mute. That is perhaps one reason Africanists prefer to work on their beloved 'sub-Saharan' Africa, and leave North Africa to the 'Orientalists', who often have competence in Arabic.

14. This is quite evident in the papers in Bates, Mudimbe, and O'Barr (1993). Reading most of the papers in the book you would hardly know that there are African anthropologists, economists, political scientists, art historians, historians, or literary critics. The authors are preoccupied with western writings on Africa, and advertising Africa's potential in testing 'universal' theories.

15. In Malawi, for example, Longman was nationalised and incorporated into Dzuka Publishing, while in Kenya Heinemann was purchased and renamed East African Educational Publishers, see Chakava (1994). In many countries there have been tensions between state, multinational, and indigenous private publishing, especially over the captive and lucrative educational textbook market, see Czerniewicz (1993).

16. Not all African countries of course have been in a state of economic crisis. Of the 35 countries for which the World Bank had data for the period 1980-1991 about half grew at or above the average world growth rate of 3%, 5 of whom registered a growth rate of 4-5%, two 5-6%, one 6-7%, and one, Botswana, 9.8%, making the latter the fastest growing economy in the world, a distinction it enjoyed in the 1970s when its annual growth rate averaged an incredible 14.5%, see, World Bank (1993: 240-1). Botswana, however, does not have a large local publishing industry, while its sluggish neighbour, South Africa (growth rate 1.3% between 1980-1991), does, although its ownership and publications have been highly discriminatory in favour of the minority white population against the African majority in keeping with the then prevailing apartheid system, see Steve Kromberg, (1993). This underscores the need to avoid making a simplistic association between economic growth and publishing. Other factors, including literacy rates, reading and educational traditions, and book trade structures also play an important role.

17. The development of the publishing industry in Zimbabwe in fourteen short years of independence has been particularly impressive. In fact, the Zimbabwe International Book Fair has quickly become Africa's largest and one of the world's renowned book fairs. For a report on the 1994 Fair see, Mbanga, (1994: 1-2).

18. SAPES stands for Southern Africa Political Economy Series, a trust, established to facilitate scholarly exchanges in the region, disburse grants to graduate students working on Southern Africa topics, and that publishes research monographs, occasional papers, and a lively monthly magazine, *Southern Africa Political Economy Monthly (SAPEM)*.

19. See Codesria (1993: 21-22). It is a mark of the quality of CODESRIA's publication that one of its 1993 titles, my *A Modern Economic History of Africa. Vol. 1: The Nineteenth Century*, won the prestigious Noma Award for Publishing in Africa for 1994, out of over 140 titles that had been submitted, from 55 African publishers, in 17 countries.

20. In the 1960s and 1970s, for example, the leading Kenyan historians largely chose to publish their books and monographs with the newly established and vibrant East African Publishing House and research papers in *Kenya Historical Review* and *Transafrican Journal of History*. In fact, so prestigious did the *Review* become that it was almost mandatory for anyone, including the Africanists, who sought recognition in Kenyan historiography, to publish there. By the turn of the 1980s the *Review* had folded and the East African Publishing House ceased operations at the end of the decade, thanks in part to the collapse of the East African Community. Its operations were taken over by the Kenya Literature Bureau. See Zell (1993); information also based on personal communications with Professors W.R. Ochieng' and B.A. Ogot.

21. Locational identifications were made by examining the notes on contributors, institutional affiliation, and acknowledgements, depending on the format used by each journal. Continental, regional, and national identifications were made along the same lines, as well as through names. Identification by name is, of course, tricky for there are some Africans who use 'European' names; this is particularly problematic in the case of South Africa. In cases where it could not be determined otherwise, people using European names writing on South Africa were assumed to be South Africans and therefore African. In cases where one person wrote more than one book review, each review was counted separately. Where there was more than one author of an article or a review, but the authors were of the same gender/location/continental identification they were counted as one person, but in cases where this did not apply the figures were rounded up in favour Africa-based, African, and women authors. But the incidence of this was so low as to be statistically insignificant.

22. Throughout the 1980s there were many African scholars who continued to avoid publishing in the Africanist journals, preferring to publish in Africa-based and African-controlled journals as part of their efforts to build vibrant, autonomous, and self-sustaining intellectual traditions in Africa. I was one of those who deliberately chose to publish within the continent, a practice I have largely continued since 'migrating' to North America in 1990. This point was also emphasised by Mahmood Mamdani during deliberations on this chapter at the conference where the original paper was presented.

23. Out of the 18 articles and 8 book reviews, 17.5 and 5 are written by women, respectively.

24. When I started working on the paper on which this chapter is based, I intended to separate publications by African women and Africanist women, but I had to give it up because the numbers of African women published in these journals were so few as to be statistically insignificant. African women also fare poorly in journals

from Africa, except those that specifically focus on gender issues. This is a problem evident in most theoretical and empirical studies on Africa. Gender biases are of course deeply rooted in the social sciences in general. See Chapter 9 below, and Mhone (forthcoming).

25. Catherine Coquery-Vidrovitch made illuminating comments on the provincialism of French journals at the conference where the original version of this chapter was presented.

26. Ironically, the foreign exchange constraint in many African countries, reducing the imports of books and journals from the western countries, has meant that African scholars are not reading and referencing Africanist works as much as before, while the growth of regional academic organizations and publishing endeavours, such as Codesria and Sapes, has led to greater cross-referencing among African scholars. The result is that the African and Africanist scholarly communities have increasingly been moving in different directions in their research agendas and programs. Impressionistic evidence seems to suggest that many African scholars based in the continent do not generally hold Africanist scholars who work on their countries in high regard, often seeing them as 'academic tourists'. For a discussion on some of these trends, see Mkandawire (1993); and Diouf and Mamdani, eds., (1994).

Chapter Five

SCHOLARLY UNDERCONSUMPTION: THE STRUGGLE FOR LIBRARIES

Introduction

We live in the information age, so we are always told, in which information is apparently as vital as agriculture and industry once were, an age of infinite possibilities in education and scholarship, teaching and research, economic growth and political freedom; a brave new world blessed with the open intimacies of the village, where the boundaries of national isolation and intellectual provincialism are withering away as knowledge explodes in its relentless march towards human enlightenment. Extravagant claims, no doubt. Knowledge, as creed and commodity, as a proprietary privilege, reflects and reproduces the spatial and social divisions of power, old and new, material and ideological, between and within societies. The information highway is a dangerous place for those on foot or riding rickety bicycles. It is designed for, and dominated by those driving on the backs of powerful and prestigious publishing systems and academic enterprises of the industrialized North, who churn out the bulk of the world's books, journals, databases, computers and software and other information technologies, and dictate international copyright and intellectual property laws on the hapless, information-poor world majority. A harmonious global village, it is not. A feudal estate, hierarchical and unequal, it may be.

What is Africa's position on this information feudal estate? Where does it fit in the international political economy of knowledge production, dissemination, and consumption? To answer these questions we need to assess the development and state of the continent's basic infrastructures of knowledge creation and distribution, namely, the availability of publishing houses, technical expertise, printing facilities, electronic technologies, libraries, and bodies of capable writers. It is not enough, however, to bemoan the regional and social differentiations of access to information, or to chronicle the unequal patterns of information acquisition, outreach, and infrastructure. We need to unravel the content, the value, of the information. What social good has it generated? To what extent has the explosion of information led to more enlightened human relations within and among nations? Is the information highway all speed, noise, and fury leading nowhere, and leaving behind data glut and confusion, 'not wisdom, character, common sense' (Roszak 1993:4). In short, we must interrogate the ethics of information, the social and political morality of knowledge creation, consumption, and content, and assess its record in bettering the human condition, not just materially, but in ennobling social relations, in uplifting the human spirit.

These, then, are some of the issues discussed in this chapter. The first part offers an overview of the severe challenges specifically facing African research and academic libraries, crucial centres for the consumption and production of knowledge, and critiques the band-aid solutions that have been tried, whose effect has been to reinforce the continent's external dependency[1]. The second part argues that the plight of African research libraries as a crisis of scholarly communication can only be adequately tackled by developing and improving local academic publishing and information production capacities, so as to ensure the dissemination of knowledge that better reflects African realities. But we must avoid the pitfalls of either romanticizing indigenous knowledge, or fetishizing library holdings, for neither guarantees accessibility nor enlightenment. Africanist scholarship bears this out. Africanists in the major western universities have access to relatively well-endowed libraries. Yet their work rarely reflects the intellectual productions of the countries and societies they research on, either because African publications are not acquired regularly and aggressively[2], or they are not indexed in most of the existing western data bases, or they are simply not valued. 'It is interesting to note', writes Mkandawire (1996:14), 'that most of the reviews of books by Africans in North American journals are done by Africans often resident in North America (not by Africanists). Africans are responding by considering any work in which Africans' texts are not considered as suspect'. In short, whatever the cause, Africanist scholarship is largely self-referential and appears as pretentious and irrelevant as African scholarship is alleged to be unsophisticated and politicized. For an African academic in an underfunded university, there may not be enough scholarly bread on the shelves, for the Africanist in a wealthy university there may be nothing but layers of cakes. Both end up sporting the bloated bellies of scholarly kwashiorkor.

The Struggle for the Bookshelves

African libraries carry a heavy colonial imprint, even in those regions with long traditions of literacy and libraries, such as Northern Africa, Ethiopia and parts of Western and Eastern Africa, partly because virtually the whole continent, including Ethiopia between 1935 and 1941, was under colonial rule. After independence, a period that witnessed the fastest growth and expansion of libraries in the continent's history, the colonial traditions were reinforced by a scramble for modernization. African libraries heedlessly borrowed their architecture, collections, bibliographic and classification systems, training and staffing structures from the North without adequately tethering them to the stubborn local realities of poverty and illiteracy, on the one hand, and the rich media of oral culture and the voracious appetite for education, on the other (Amadi 1981; Ndiaye 1988; Sturges and Neil 1990)[3].

Research and academic libraries were the least domesticated, much like the universities themselves, whose institutional lineages and intellectual loyalties lay overseas (Mazrui 1978; Diouf and Mamdani 1994). All was well in the heady years immediately following independence, when healthy commodity prices and booming economies kept modernization hopes alive. The tentacles of information dependency grew tighter and thicker, despite the inchoate nationalist yearnings for cultural decolonization. Then from the mid-1970s many African countries fell into a spiral of recurrent recessions, which wreaked havoc on development ambitions and the bookshelves grew empty. The 'book hunger' joined the litany of Africa's other famines of development, democracy, and self-determination.

The prevailing library and information system was in a crisis of self-reproduction and relevance. This is amply borne out by the 1993 survey of 31 university and research libraries in 13 African countries conducted by the American Association for the Advanced of Science (AAAS). All but three of the libraries reported a sharp drop in their subscriptions to journals from the mid-1980s. Among the worst hit were the Addis Ababa University Library, the University of Yaoundé Medical Library, and the University of Nigeria Library, which in the late 1980s and early 1990s cancelled all their subscriptions to some 1,200, 107, and 824 journals, respectively, due to shortage of foreign exchange (Levey 1993:2-3). Currency devaluation, one of the linchpins of structural adjustment programmes, took its toll, too, in reducing the buying power of libraries. As the Librarian of Abubakar Tafawa Balewa University put it in 1993: at the current rate of 25 *naira* to the dollar, I should have about $229,000 for books. Ten years ago, I would have been swimming in dollars for at $1.50 to N1, the same *naira* would have equalled over $8 million' (Levey 1993:9). Compounding matters were unpredictable currency fluctuations which imposed additional and unanticipated expenditures.

It was a fatal concoction, this combination of currency devaluations and fluctuations, together with the escalating cost in the price of journals and books. Today journals that cost $1,000 are quite common, especially in the sciences. One study estimates that serial costs in North America, from where African research libraries import a lot of their materials, increased 115% between 1986 and 1994 and monograph costs rose 55%. As a result of these trends, serial acquisitions among members of the US-based Association of Research Libraries went down 4% and monographs 22% (Birenbaum 1995). If research libraries in the rich North were feeling the chill, those in Africa caught pneumonia. The case of the University of Ibadan Library is all too typical. Its number of subscriptions plummeted from over 6,000 serials in 1983 to less than a tenth of that a decade later (Levey 1993:3).

The three fortunate libraries that reported increases in the number of subscriptions, the University of Nairobi Medical Library, the National Mathematical Centre of Nigeria, and Abubakar Tafawa Balewa University, subscribed to no more than 200 journals each. Indeed, only seven libraries in the AAAS survey subscribed to more than 200 journals with internal funding. Out of the seven only three, led by the University of Zimbabwe Library with 1,578 journals paid for through the library's budget, could boast of more than 500 subscriptions. But even the University of Zimbabwe Library saw its foreign currency allocation decline from 65% of the funds requested in 1989 to less than 40% in 1991 (Levey 1993:4-5).

Aggravating the dire financial straits facing the libraries themselves, were the ill-advised government taxes on imports of books and journals[4]. Bureaucratic red tape often adds insult to injury as I discovered in November 1994 when I went to Tanzania to receive the Noma Award for Publishing in Africa. My publisher, CODESRIA, had brought a crate of my winning book for display and sale at the Tanzania Book Fair, around which the award was organized. But we were unable to get the books out of customs despite remonstrations from the organizers of the book fair in time for the award ceremony at State House. The universities themselves are also to blame. Their expenditure patterns are usually skewed in favour of salaries and privileges for the administrative elite, with their fleets of official cars, heavily subsidized housing, and numerous allowances, self-indulgent practices reminiscent of the political class. And so the universities seek to reproduce themselves, not as intellectual ivory towers, nor as locomotives of progress, but as sleepy state apparatuses, a mission that leaves little room for serious commitment to scholarly communication and critical pedagogy.

One response to the library crisis has been growing reliance on gifts and donations of books and journals from charitable organizations and foreign governments and their agencies. The AAAS survey found that only five of the libraries subscribing to journals in 1993 did so exclusively with internal funding. The rest were dependent to varying degrees on donor support. Five libraries were dependent for as much as 100%, and another five for 80% and more. Four had neither donor support nor their own funding. 'Thus without external funding', the AAAS report states, 'many libraries would have few current journals on their shelves. But donor support', it notes quite correctly, 'raises its own set of dilemmas, which revolve around the dreaded term 'sustainability' (Levey 1993:19). The donors do not underwrite projects indefinitely, which makes it difficult to pursue a rational programme of journal acquisitions. For example, the University of Makerere Library reduced its number of subscriptions from 700 serials to 200 titles when grants from the Overseas Development Agency and the European Community expired in 1991.

Another problem is that library aid, like all aid, has strings attached (Maack 1986). 'Book presentations', Clow (1986:87) writes, 'are usually restricted to items published in the donor country... training usually involves donor-country citizens as teachers; if a scholarship is awarded, the scholar usually travels to and spends most of the money in the donor country'. African libraries rarely choose the journals and books that they receive from the donors[5]. Predictable, also, is the fact that most of the journals donated are North American and European, not African[6]. In short, book aid tends to reinforce Africa's dependency on western values, languages, discourses, and institutions. Reluctant to bite the hand that feeds them, many librarians keep quite even when the donations are irrelevant and inappropriate for their needs. In the process the culture of silence and submission to imperialism, which is partly responsible for the African crisis in the first place, deepens. And so meekly they receive, and fill their shelves with, or quietly dispose of, propaganda materials from embassies, the discarded miscellanea of western libraries, grimy, out-of-date texts, and unwanted publishers' remainders. By filling the bare shelves of African libraries well-meaning, but sometimes misguided, philanthropists can display their altruism, and hard-nosed publishers can dispose of their unsold tomes, and thus save themselves warehouse charges and earn welcome tax relief.

From the 1970s donors and international agencies, especially UNESCO, also came up with a series of training and information development programmes. But most of these programmes, Sturges and Neil (1990:97) contend, 'failed to produce results commensurate with the attention that the information professions have paid to them'. They attribute the failure of UNESCO's national library and information development programmes to erroneous assumptions, inadequate planning, and poor design, problems often exacerbated by the lack of state support, sparse infrastructures, and excessive duplication and rivalry among the donor agencies themselves. Similar challenges have bedeviled efforts by Africa-based organizations to develop regional information systems. The most well-known is the Pan African Documentation and Information System (PADIS) begun in 1980 and administered by the Economic Commission for Africa (ECA). Its broad aims are, first, to help African countries strengthen their own internal information systems, and second, to set up a decentralized information network for the continent. While PADIS has made considerable progress, and publishes useful bibliographic indexes, especially concerning development, it has been criticized for its inability to develop effective sub-regional coordinating centres and to cooperate with other international agencies working on information in the continent (Aiyepeku 1983). In its first ten years it certainly achieved far less than the $160 million investment warranted, partly because of misguided emphasis on expensive information technologies for countries with poor telecommunications infrastructures.

This is not to suggest that the latest information technologies should not be acquired, for not to do so would be to reinforce Africa's marginalization. It is simply to point out that basic infrastructural development is essential, and that in themselves the advanced technologies offer no magic solution to the challenges of information dissemination and scholarly communication facing Africa. Many African research libraries, usually with donor support, are investing heavily in computer and CD-ROM capability, and electronic networking (AAS & AAAS, 1992). To its champions the CD-ROM is a wonder technology that is universally appropriate: not only can it hold huge amounts of data, it is durable, cheap to mail, requires no special handling, storage space, or telecommunication facilities, and can withstand climatic extremes and power cuts and the ravages of insects and fungi (Compton 1993; Kagan 1992a, 1994). The liberatory and repressive potentialities of advanced technologies is in serious dispute (Kagan 1992b; Buschman 1992). Lancaster (1978) has urged developing countries to seize on the new technologies and leapfrog to electronic libraries bypassing the book. His African critics have argued that electronic information service in Africa is only of benefit to a small, already privileged elite. African librarians, they state, ought to be concentrating on assisting the illiterate majority of their people learn to read and write (Mchombu, 1982; Olden, 1987; IFLA, 1995). Others argue for an integrated approach that combines the need to improve information delivery both to the poor and to the elites (Tiamiyu 1989; Sturges and Neil 1990: Chapter 5).

The 1993 AAAS report found that all but five of the 31 libraries surveyed had computers, about half of them purchased locally, and most of them acquired through donor support. Nineteen libraries had CD-ROM capability and two were expecting to acquire it by the end of 1993. African librarians have been keen to acquire CD-ROM technology 'for fear of being left behind' in the words of Newa (1993:82), the Director of Library Services, University of Dar es Salaam. At a workshop on new technologies for librarians from 17 libraries in 11 countries in eastern and southern Africa, including South Africa, held in Harare in 1993, sixteen of whom had CD-ROMs in their libraries, there was universal agreement on the importance of this technology, despite some of its perceived shortcomings. With a few exceptions, many of the libraries reported extensive use of the CD-ROM facilities. The University of Zambia Medical Library was even forced to ration time to 30 minutes per person. Most of the libraries in the AAAS report subscribed to databases in agriculture and medicine mainly because of the interest of donors, who largely pay for the subscriptions in these fields. The notable exception was the library of Cheikh Anta Diop University which had a significant number of CD-ROM databases in the social sciences (Levey 1993:13-16).

Computers and CD-ROM technologies have breathed new life into Africa's ailing research library systems, although they pose their own problems, and reinforce some old ones. Lack of the relevant technical expertise locally and among librarians often leads to poor product choices, and installation and maintenance difficulties. One study reports, for example, that 'the librarian of the University of Ghana Medical School had no one in Ghana to whom to turn when he had trouble installing his CD-ROM drive, for his is the first library with CD-ROM in the country. Ultimately he called New York to receive instructions over the phone' (Levey 1991:12). But long-distance advice can be costly and inappropriate as the librarian of the University of Zimbabwe Medical

School discovered after buying a non-compatible CD drive 'on the basis of advice from our New York software vendors' (Levey 1991:12). These technologies of course do not come cheap, so the question of funding remains. Besides the one-time equipment costs, which rise each time local currencies are devalued, there is the high recurrent cost of subscription to databases. Training costs can also be high and recurrent, especially given the fact that the technology is growing and changing rapidly. It is essential to budget for CD-ROM subscriptions for the long run because subscribers are usually only allowed to use the databases for the duration of the subscription and may be requested to return the discs should their subscriptions run out, unlike journals which a library keeps when its subscription lapses (Levey 1992; Keylard 1992). Not surprisingly, there are reportedly a handful of libraries with CD-ROMs who do not use them because they have no funds to purchase subscriptions. Of the 16 libraries that had CD-ROMs surveyed by the AAAS in 1991, only four indicated they had funding for subscriptions in the future. Also, literature searches do not guarantee the users access to the documents identified. Given the inadequacy of many African research libraries' serials collections, bibliographic databases that do not contain abstracts are virtually useless (Patrikios 1992:30-7). Few donors include document delivery as an integral part of their grants for database subscriptions, and supplying photocopies from Europe and North America, as is sometimes done, is costly and cumbersome. The document delivery barriers may ease as full-text literature is routinely published on disc as well as in print copy.

The Struggle for Knowledge

African librarians are fully aware of these problems and many realize the importance of national and regional cooperation, although declared intentions tend to be louder than concrete action[7]. But even if the problems of access to citations and documents were resolved, Africa's knowledge base would not necessarily improve, for these databases, like the bulk of the journals and books imported into most of the continent's libraries, primarily contain Northern scholarship. Production costs for CD-ROM databases are still quite prohibitive for any aspiring African database publisher, although efforts are being made to create local databases[8]. Besides, the publisher would have to develop extensive scholarly, marketing, and support networks. Northern database publishers are still largely unwilling or unable to incorporate bibliographic records from the South. By the mid-1980s there were an estimated 700 databases of direct concern to Africa located outside the continent, a figure that has most probably shot up with the explosion in electronic communications since then (Seeley 1986). These databases are not only difficult to access in Africa itself, their input of African research and publications is abysmal. For example, less than 1% of over 36,000 items on Africa contained in the FRANCIS data file (with 1 million items altogether), produced by the French Centre National de la Recherche Scientifique as of March 1986, were published in Africa (Sturges and Neil 1990:64-65). Even the best of these databases, FAO's Agricultural Information System (AGRIS), only has 25% of its content from the developing countries as a whole. The various major Social Science and Humanities online indexes do not include journals published in Africa[9].

The marginality of African knowledge is evident even in the Africanist intellectual system, which is firmly rooted in a western epistemological order and an academic culture driven by a ruthless ethos of 'publish or perish', and consisting of multi-national publishing houses, university presses, journals, peer review networks, citation, and bibliographic conventions, and has little room to accommodate the alien views, voices, and visions emanating from Africa itself. In this scholarly treadmill, Africa appears nothing more than a research object to verify faddish theories that emerge with predictable regularity in the channel-surfing intellectualism of Northern academies. And so we get the strange spectacle of books and articles being churned out containing no reference to the scholarship produced in the countries and regions concerned, thus re-enacting and re-enforcing the David Livingstone syndrome, whereby each Africanist researcher acts as if they are the first to discover the 'Victoria Falls' of this problem, and that theory's application to African data. It is work that often contains, as Mkandawire (1996:14) observes, the latest bibliographic references to Africanist research and rather dated facts, while the work of African scholars 'will contain dated bibliographic references and the latest facts. Obviously such a divide is unsatisfactory and does nobody any good'.

Detailed analysis of the contents of Africanist publications would be revealing. To what extent do the themes and topics in Africanist publications engage the realities and priorities of the communities studied and the 'genuine' research interest of the scholars from those communities, as opposed to research orientations dictated by the consultancy syndrome or careerist calculations in situations where publishing in western scholarly media carries more weight than publishing within Africa. There is some evidence to suggest that the research agendas of the African and Africanist research communities have grown more divergent over the years, which is attributable to the changing conditions for African studies in the western countries and the scholarly enterprise in Africa. On the one hand, Africanist scholars spend less time in Africa whether doing research or teaching than they used to, partly because of funding difficulties, reduced salaries in African universities, and fewer teaching opportunities because of the successful Africanization of faculties, while the proportion of African scholars studying for higher degrees in the North, especially in the social sciences and humanities has also fallen because of declining need and financial resources, diminishing returns and attractiveness of academic careers, and growing immigration restrictions. Contacts are especially poor for what Mkandawire (1995) calls the 'third generation' of African scholars, a point echoed by Guyer (1995) with reference to the younger crop of aspiring North American Africanists.

Mkandawire (1992:4), CODESRIA's former executive secretary and a keen observer of the two scholarly communities, has noted, for example, that in the 1980s while many Africanists were fashionably bemoaning or applauding the 'exit' of peasants and other exploited social classes from arenas dominated by the authoritarian post-colonial state, 'African social scientists moved in a different direction, casting attention more towards the study of social movements and democracy'. On the loudly trumpeted African crisis, too, there have been analytical divergences. Africanist political scientists and economists, especially, have largely accepted the diagnosis and prescriptions of the international financial institutions, which have been vigorously opposed in African intellectual circles (Mkandawire 1996:11-12). The descent in Africanist scholarship from multi-factor to single factor explanations of the African crisis, from analyses that tracked the crisis to a convergence of multiple external and internal factors to one in which primacy was given to internal factors, as embodied in a supposedly malevolent and malfeasant post-colonial state, gave rise to the intellectual fatalism of Afropessimism which, in turn, encouraged support for the structural adjustment programmes being imposed by the international financial institutions (Eyoh 1996:5-10).

In recent years, in yet another sign of the deepening divergences, post-modernism has been casting its spell on many in the Africanist fraternity, and some are anxiously covering their mouldy African data with its ephemeral fragrance, forgetting proclamations they made earlier in the 1960s that Africa was modernizing, in the 1970s that it was underdeveloping, and later that modes of production were being articulated. Sleeping its way through the lost 1980s, Africa somehow woke up in the 1990s to find itself in a post-modernist universe, or it should have, we are told (Vaughan 1994; Parpart 1995). To many

African scholars on the continent, such arcane preoccupations seem the nadir of intellectual solipsism and decadence. According to Aina (1995:2), the crisis of African Studies in North America and Europe is creating

> a process of intellectual reproduction about Africa that is characterized by sterility, outdated facts and information, casual and ad hoc observation, name-calling and sometimes wild speculation. It is our argument here that for an up to date realistic, correct and appropriate ... understanding of Africa, the most appropriate and relevant source is that scholarship and production emanating from or still directly linked to the continent in terms of research experience and reflection; from this living and challenging source and expression, no amount of post-modernist, post-industrialist, post-Marxist or 'post-Nativist' conceptualization or discourse can take away the relevance, immediacy and centrality.

In short, African disenchantment with Africanist scholarship is widespread and growing. The old African 'reverence for non-African writing', to quote Mkandawire (1996:15) again, 'is changing for a host of reasons. One trivial one is the growing inaccessibility of Africanists work to book-starved Africans. The second is the growing African literature. Africans are simply publishing more today than ever before. The third is the sense of decreasing credibility... of Africanist scholarship among Africans'.

The inescapable conclusion, therefore, is that importing knowledge from abroad is no panacea. For Africa to depend on external sources for knowledge about itself is a cultural and an economic travesty of monumental proportions. To use a phrase from the dependency paradigm, African libraries may grow from buying or receiving donations of tons of journals and books, and they may acquire the latest information technologies and the largest databases but without actually developing, without expanding and strengthening the continent's capacities for authentic and sustainable knowledge creation, information generation, and data collection. More often than not, knowledge produced on Africa elsewhere is distorted or irrelevant, and importing databases or receiving donations serves to deepen the ties of intellectual dependency. Sturges and Neil (1990:79) cheekily suggest that 'many of the donations that do arrive would be far better if they were pulped. This might at least provide some new paper, a basic resource which Africa needs more urgently than other countries' cast-off books'.

The real challenge, then, is not simply to fill empty library shelves and acquire gadgets for faster information retrieval, but to produce the knowledge in the first place, for Africa to study, read and know itself, to define itself to itself and to the rest of the world, and to see that world through its own eyes and not the warped lenses of others. There is no substitute for a vigorous intellectual system, of which publishing is an integral part. I note elsewhere (Zeleza 1994:238):

> Only by developing and sustaining our own publishing outlets, can there emerge truly African intellectual traditions and communities capable of directing and controlling the study of Africa, of defining African problems and solutions, realities and aspirations, of assessing our achievements and failures, our pasts and futures, and of seeing ourselves in our own image, not through the distortions and fantasies of others. Publishing is critical not only for the cultural identities of nations, peoples, classes, and groups. It provides the material basis for producing, codifying, circulating and consuming ideas, which, in turn, shape the organization of productive activities and relations in society.

The challenges of publishing in Africa and other Third World regions are well-known, and were described at greater length in the previous chapter. They include poor infrastructures, in particular shortages of skilled editors, designers, distribution experts, readily available and cheap supplies of printing equipment and paper, as well as low literacy rates, language problems, and meagre incomes and purchasing power, problems which have been exacerbated by the recurrent recessions. Promotion and marketing, at home and abroad, remains a critical hurdle for many African publishers (Zell 1995:16-18). In their study, Nyariki and Makotsi (1995:11) found that the promotional and marketing activities undertaken by many Kenyan publishers are ineffective and unprofessional because of lack of trained staff. Moreover, widespread government intolerance and censorship in many countries only makes matters worse. Nor does the existence of relatively small and fragile academic communities help, especially for scholarly publishing. Also, poorly-capitalized indigenous publishers must often compete with large multi-national publishing companies, and heavily subsidized state-owned publishing houses[10].

These constraints are real and serious, but they are not insurmountable. Literacy rates have risen quite remarkably in many countries, and 'the much publicized myth that the African mind is orally-oriented and therefore Africans do not read' is becoming more threadbare as evidence mounts that a lot of people actually read for pleasure, as Nyariki and Makotsi (1995:11) found out in their research in Kenya: 'a majority 39% of consumers buy books because of a love of reading'. They also show that the number of indigenous publishers in Kenya doubled to 72 between 1974 and 1994. Altogether, local publishers produced 60% of the books on the local market. These trends are confirmed by Hans Zell (1993:373), a seasoned watcher of the African publishing scene, who states that 'despite the overall gloomy picture ... new indigenous imprints continue to mushroom all over Africa, and some privately owned firms have shown a great deal of imaginative entrepreneurial skill in the midst of adversity'. And the formation in 1989 of the African Books Collective (ABC), referred in the previous chapter to undertake the joint promotion and distribution of African books outside the continent and of the African Publishers Network (APNET) in 1992 to encourage intra-African publishing and trade in books underscores the determination of African publishers to forge ahead[11].

Libraries must do their part. They constitute the backbone of scholarly publishing. In many parts of the world, including the industrialized countries, libraries provide the major market for scholarly products. In fact, in the United States, despite relatively high academic salaries and a large professorate, it is library purchases, not subscriptions by individuals, that sustain journals, so that 'journals would not survive without the library market' (Altbach 1987:63). Often libraries generate up to 90% or more of the income of journals, especially the expensive journals in the medical and scientific areas. Having fed for so long on western imports and donations, of information materials and technologies, African libraries have not always ventured with enough appetite to acquire local publications. For their part, publishers bred on the captive school textbook market are not always aggressive enough in promoting their wares. At the 1993 Harare workshop publishers and librarians took each other to task (Patrikios and Levey, 1993:3):

> Several publishers stated that few African imprints can be found in African libraries because librarians are reluctant to order materials, preferring instead to purchase books from England or the United States. Nana Tau [librarian of Fort Hare University] countered by telling of her experience in attempting to obtain information on African imprints in order to place an order for her library. The lack of response from the African publishers whom she wrote requesting catalogues forced her to place orders overseas.

On another occasion the librarian of the University of Makerere Library pointed out that 'most of the African journals are possibly not known by teaching staff who recommend titles to be subscribed by the library' (Levey 1993:11). Unfortunately, he may have been correct. It is a sad fact that in many African universities the processes of hiring, promotion, and research grant allocations are firmly tied to the legitimation structures of western scholarship. Familiarity with western intellectual fads and publication in the restricted western scholarly media bestow on the lucky few precious reputational capital that can be traded for lucrative consultancies and overseas visiting professorships and conferences. Local journals become publication outlets of last resort, repositories of second-rate scholarship. This must change.

African intellectuals need to shed their inferiority complexes about their own work by publishing, without apologies, in journals they control, by reading and citing each other, by demonstrating a greater faith in their own understanding of their complex and fast changing societies, for no one else will do that for them. They cannot continue being unwelcome guests at other peoples intellectual tables. Through their reward structures, facilities, and ethos universities should provide the major sources for intellectual production and markets for scholarly products. Where the scholarly communities are small, cooperative regional journal publication ventures should be encouraged. The mission, always, must be to promote the highest standards of research and scholarly exchange, to repossess the study of Africa, to define African realities, to understand and appreciate the African world with all the intensity, intelligence, and integrity it deserves.

Conclusions: Beyond Information Fetish

The manufacturing and distribution of scholarly knowledge and information is a major commercial and technological enterprise involving publishers, libraries, educational institutions, and communications companies, linked in elaborate networks requiring vast resources. While the news that we have entered a post-material age in which words matter more than goods is exaggerated, the importance of information technologies in the development process cannot be denied. But what kind of information, by whom, and for whom?

One of the factors behind the scholarly information explosion in the western countries, especially in North America, is the pressure to publish, the centrality of publications and citations in the academic enterprise. Publications have become a screening mechanisms for hiring, promotion, tenure, and research grant-awarding procedures. The system rewards those who generate large amounts of literature, however insignificant its intellectual contributions. Indeed, piles of paper are churned out to be listed and indexed rather than read. And so scholarly information doubles in volume every seven years. A decade and half ago it was doubling every 15 years (Birenbaum 1995). Information becomes an absolute good, an end itself, an intolerant, insatiable god that constantly spews data, 'hyperfacts', that require more powerful databases to keep track of the existing databases (Roszak 1993:4). In the process knowledge becomes incidental, a forgotten atavism. As the information glut grows, there is more pressure for excessive specialization, and 'involution into narrow and specialized solitudes' (Bourne 1988:1424). In the meantime, as the high priests of the information age pray at the altar of citations and chant 'jargons of an almost unimaginable rebarbativeness... society as a whole drifts without direction or coherence. Racism, poverty, ecological ravages, disease, and an appallingly widespread ignorance: these are left to the media and the odd political candidate during an election campaign' (Said 1993:303).

Thus beneath the apparent munificence of the western academy, behind the spiralling mountains of information, lies a profound flight from human connectedness, from meaningful social conversation; there is a yawning alienation from the gravity of human existence, from history. An almost infantile fascination with the innate and quantifiable, not the poetry of life, of words, seems to have taken over. Writing almost a decade and half ago, Amadi argued that 'information experts have mistakenly tended to approach the subject of information in purely quantitative terms: means, materials, manpower... This orientation of librarianship and formal education require both reassessment and redefinition' (Amadi 1981:51). The availability of more information is not an answer by itself. As Olden (1987:301) reminds us:

> the availability of information does not mean that use can be or will be made of it; that those who do use it are capable or willing to learn from it; or that what they learn will be used for the benefit of others. Taken together, United States libraries house what is probably the most comprehensive collection of recorded information and knowledge about other countries held by any nation in the world. Has the increase in the size of this collection since World War II been paralleled by an increase in the number of better foreign-policy decisions made by various administrations over the same period?

And one could add: are Americans much better informed about the rest of the world? Indeed, has more information helped them significantly transcend their own racial, ethnic, class, and gender divisions? Will access to Internet in every home and to a 500 TV-channel universe do it? Or will that simply lead to more fragmentation, to further descent into the abyss of cultural banality so evident on North American popular television today?

What, in short, do the terms information-rich and information-poor, which are so carelessly bandied about, actually mean in terms of the content of human relationships, the quality of social life, as embodied in the information being manufactured and consumed? To be sure, Africa needs to produce more information, its academic institutions need to reorganize themselves to encourage and reward scholarly production and productivity, and its libraries need to collect and make this information more accessible within and outside the continent. But the processes of production, acquisition, retrieval, and outreach cannot be ends in themselves, if the dangers of information overproduction and overload, currently engulfing the western world, are to be avoided. Africa must indeed repossess the word. But whose word, and to what ultimate purpose? It must be to elevate, not debase, our humanity.

Notes

1. The chapter does not address the wider questions of the creation of knowledge and the provision of information for the popular classes in the urban or rural areas. For a detailed study on the provision of information to rural African communities see IFLA (1995).

2. One often hears the complaint that it is very difficult to get African publications. With the creation of the African Books Collective (ABC) this is increasingly a self-serving justification. ABC regularly mails out catalogues to Africanist institutions all over the world and takes part in publishers' displays at African studies associations' conferences. I get African books on the areas that I work on without difficulty from ABC, which is headquartered in Oxford, England, a scholarly metropole, indeed. North Americans seem to prefer the fast fare from local publishing pizza houses, and if they have to go out it is not for African food; African 'exoticism' lies somewhere else. The supposed difficulties of getting publications from Africa also becomes a justification, even among African scholars resident in North America, not to publish in Africa.

3. Many of these analyses make the historically erroneous assumption that orality and writing in sub-Saharan Africa were sequential stages. For more convincing analyses of the antiquity and complex relations between the two in several sub-Saharan societies see Gerald (1981), Julien (1992); and for Africa as a whole see Scheub (1985) and Mudimbe (1994).

4. An interesting example is that of Côte d'Ivoire, where the Telecommunication and Postal Ministry was privatized. The AAAS stopped sending free journals to the university library because the latter could not afford to pay the ministry the levies charged on the journals! (Levey, 1993:9).

5. Many of those concerned about book dumping in the Third World have suggested that donations schemes should be request-led, see Abid (1992).

6. A remarkable exception is the program initiated by the International African Institute which in the early 1990s launched a project to distribute 12 African serials, which were selected after consultations with both African publishers and research libraries.

7. Only in South Africa do the efforts to integrate library systems and resources seem serious. For example, there is the Western Cape Cooperative Project and the Committee on Library Cooperation in Natal. See Darch 1993; Madly, 1993.

8. The Zimbabwe and Zambia Medical libraries, for example, in collaboration with other countries in the continent are producing an African Index Medicus, while the Bunda College of Agriculture in Malawi has created a bibliographic data base of Malawi's maize research. See Hill, 1993; Ngwira, 1993; Patrikios, 1993; Stunt, 1992.

9. Out of curiosity, I checked references to my journal publications in the Wilson Index: there were four book reviews and two articles published in North America. It was unflattering!

10. The multinational publishing companies can be quite opportunistic. For example, they all closed their businesses in Tanzania during the 1980s financial crisis and 'returned in the 1990s when they heard that there would be an allocation of US$60 million from the World Bank for educational supplies'! Mcharazo (1995:245).

11. For a discussion of these organizations and their activities in the last few years see the 1993-95 issues of the *Bellagio Publishing Network Newsletter*, published on behalf of the donors that support African publishing, APNET's organ, *African Publishing Review*, and *The African Book Publishing Record*.

PART TWO

THE PITFALLS OF
AFRICANIST
HISTORIOGRAPHIES

Chapter Six

THE RISE AND MUTATION OF AFRICAN HISTORIOGRAPHIES

Introduction

In 1976, Terence Ranger (1976) noted that there was 'a crisis for African history arising out of the collapse of the consensus of the golden age', a complaint that was echoed by others soon after (Ogot 1978; Flint 1982; Webster 1982). A year later, Ranger (1977) warned about the emerging 'romanticization about the "people"', which was prompted by the criticism of his work particularly made by the Isaacmans (1977) and Depelchin (1976, 1977). Such sad and nostalgic reflections on a supposedly receding 'golden age' in the face of new historiographical trends betrayed the crisis of liberal/nationalist historiography, of which Ranger was one of its major and ablest proponents. The challenge posed to nationalist historiography by theories of dependency and modes of production did not constitute a crisis in African historical scholarship, rather these critiques represented healthy and welcome attempts to ask more penetrating questions and to provide more satisfactory answers to problems that are central to a deeper understanding of African history and society. Less certain is whether the currently fashionable indiscriminate attacks against 'grand' narratives is entirely fruitful, although, as before, it may be mistaken and misleading to argue that African history as a whole is in 'crisis' (Adeoye 1992).

Nationalist and Liberal Narratives

It is beyond the scope of this chapter to review in any great detail the complex subject of nationalist historiography, or other 'schools' in the study of African history which sprang from imperialist or anti-colonial traditions (see Fyfe 1976; *Issue* 1976; Copans 1977; Kapteijns 1977; Roberts 1978; Eriksen 1979; Temu and Swai 1981; Wamba-dia-Wamba 1986, 1987; Jewsiewicki and Newbury 1986, Jewsiewicki 1989). Suffice it to say that the development of nationalist historiography in the 1960s was peculiarly fitted to an era marked by euphoria about the achievements of the nationalist movements and full of great expectations about the future. Cultural heroes were reclaimed from the Hegelian world of 'natural man in his completely wild and untamed state', and glorious empires exhumed from the Africa of the 'Unhistorical, Undeveloped

Spirit' [(Hegel 1944:93, 99) Hodgkin 1976:8]. It was discovered that Trevor Roper, the eminent Oxford don, had mistaken cultural resilience against the colonial onslaught for the 'gyrations of barbarous tribes' (Wilks 1976:7). The 'native agitators' of colonial rulers became the 'founding fathers' of the new nations, the 'modernizing elites' in the sanitized vocabulary of the development economists and political scientists who were scurrying across Africa with briefcases full of advice.

Chiefs, spirit mediums, and valiant warriors who had resisted the imposition of colonial rule were finally absolved of slanderous charges that they were 'backward looking', inspired by the atavistic instincts of their primitive past; they became the precursors, in fact, mentors, of the latter-day nationalists. Terms like 'native' and 'tribe' were finally hurled into the dustbin of imperialist history. The legitimacy of the nationalists was shored up, continuity in African history re-established, and colonialism became just one other episode in the long history of Africa, separating the idyllic and egalitarian past, and the post-colonial future of nation-building, development and equality, pride and dignity.

Thus, at last anti-colonial writers and critics, from Morel (1969a, 1969b) and Leys (1926), to Hodgkin (1956) and Davidson (1961a, 1961b, 1967, 1971) lost their marginality; nationalist historiography incorporated their critiques of colonial oppression and exploitation. The wandering prophets of panAfricanism finally reached the promised land (Blyden 1857; 1869; 1994; Dubois 1947; Hansberry 1977; Padmore 1936, 1949, 1956; James 1938, 1962; Diop 1974, 1987a, 1987b). The age-old nationalist cry 'Africa for the Africans' no longer echoed in the wilderness but became a clarion call to students of African history and society to resurrect 'African activity, African adaptations, African choice, African initiative' (Ranger, 1968:xxi) from the onerous weight of colonial oppression, overlaid by Eurocentric and sometimes racist imperialist historiography and ideology. It was a big challenge, but few historians seemed unduly daunted by it. Their enthusiasm carried them through. National histories appeared. Colonial policies were demythologized as the inherently exploitative and oppressive nature of the 'colonial situation' came to be emphasized. Other subjects, such as the study of messianic movements and independent churches, which in a bygone era would have raised the eye-brows of the imperialist historian as a confirmation of the barbarism of the 'Dark Continent', were carefully analyzed. Egypt was reclaimed and the Sphinx's Negroid nose finally reconstructed. The 'Hamitic factor' was questioned; the Zimbabwe ruins were after all built by the Shona and not Phoenicians. It was also shown that African traders had engaged in long-distance trade long before the introduction of so-called 'legitimate commerce' after the abolition of the European slave trade. And the scrolls of Timbuktu were resurrected from the expanding Sahara: the 'natives' had not, after all, been blissfully cursed with ignorance before Europe magnanimously undertook its 'civilizing mission'. The African ancestry of the Pushkins was revealed. And it was proclaimed that Christianity had traversed Africa long before reaching Rome. Vansina (1965) brought oral tradition out of the village. African history finally achieved institutional respectability.

Ironically, anthropology, which had produced so many detailed studies of African societies long before the study of African history was even recognized, found itself on the defensive. It was charged that its functionalist-positivist paradigms exonerated, if not actually extolled, colonialism (Gough 1969; Goddard 1969; Magubane 1971; Onoge 1973; Mafeje 1976). Anthropology went into a period of deep epistemological crisis, from which it only began to emerge in the late 1980s when many of the new born-again anthropologists rode on the flimsy wings of post-structuralism (Moore 1993). But, while the study of African history continued to thrive, from the early 1970s nationalist historiography fell victim to its own enormous success: nobody could any longer seriously contest that Africa had its own history. Cabral's (1969) impassioned call for the 'inalienable right' of Africans to have their own history, like other people, had been heeded. Students now began to ask new questions for which nationalist historiography, grounded as it was within the terrain of bourgeois social thought, with its idealism, empiricism, and liberalism, did not have the methodology or theoretical inclination to provide satisfactory answers.

Criticism began to flow (Denoon and Kuper 1970; Ranger 1971; Ochieng' 1974; Manning 1974; Saul 1977; Bernstein and Depelchin 1978-79; Swai 1979, 1980; Temu and Swai 1981). It was charged that the 'African voices' which nationalist historiography had reclaimed, were voices of the leaders, whether the kings of the pre-colonial era, or the 'new men' of the colonial period, or the nationalists who later became the rulers of the newly independent states. In short, nationalist historiography narrowly focused on, and universalized, the activities and interests of the 'traditional' and 'modern' ruling classes, and not the 'people' themselves, those beloved 'masses' of the nationalist demagogues. Nationalist historiography had proved all too susceptible to pressures to provide 'cultural heroes' and validation for myths of African classlessness propagated by African ruling classes in order to mask and legitimate their vested privileged interests. Students began to ask: What had happened to all those notorious African slave traders? And how did kings come to be kings, anyway? And if everybody was equal, who built the pyramids, and why?

Nationalist historiography had been too preoccupied with showing that Africa had produced organized polities, monarchies, and cities, just like Europe, to probe deeper into the historical realities of African material and social life before the advent of colonialism. As for the colonial period, nationalism was made so 'overdetermining' that only faint efforts were made to provide systematic, comprehensive, and penetrating analyses of imperialism, its changing forms, and their impact, not to mention the processes of local class formation and class struggle. By ignoring these themes, nationalist historiography over-stated its case: the overall framework in which the 'heroic' African 'initiatives' were taken was lost, and, in addition, African societies were homogenized into classless utopias.

Thus, nationalist historiography had failed to provide its own 'problematic', or at any rate, it took over questions as they were posed by imperialist historiography: to the latter's postulation of African backwardness and passivity, nationalist historiography counterposed with notions of African genius and initiative. In all this, 'politics' was emphasized at the expense of economic struggles for survival through the centuries. As the euphoria of independence disappeared into thin air with the failure of the much-vaunted 'political kingdom' to sustain its delivery of development, apart from the flags and national anthems, students began to ask why Africa remained so desperately poor despite its enormous natural wealth. Decolonization was re-examined. It was pronounced 'false'. The 'radical pessimism' of Fanon (1963), which Ranger (1971:53) had correctly predicted would become the main adversary of the 'Africanist historian', and not the discarded 'colonial school', was vindicated. Nationalism began losing some of its glitter. Conspiracy theories gained currency: the departing colonial powers had made 'deals' with the nationalist leaders to perpetuate the oppression and exploitation of the 'masses'. Neo-colonialism became the new catchword. Nkrumah (1963) was praised for his foresight.

This 'wind of change' in African historiographical circles soon crossed the Zambezi and shook liberal historians in the settler laagers of Southern Africa from their complacency. From the 1920s when liberal historiography became increasingly dominant in English-speaking universities in South Africa, the country was seen through the prism of race and culture, and its history was interpreted as a series of racial and cultural interactions between the Afrikaaners, Africans, and the British in the context of a changing and modernizing economy (Johnstone 1970, 1982; Trapido 1972; Marks 1972; Atmore and Westlake 1972; Kantor and Kenny 1976; Lipton 1976, 1985; Wright 1977; Wolpe 1978, 1990; Magubane 1979; Legassick 1979; Minkley 1986).

Liberal historiography elevated racial stereotyping and moralizing into a doctrine. Afrikaners became the eternal villains, the collective 'evil genius' behind the development of the vicious system of apartheid. In contrast, the British peeped from the pages of South African history as an enlightened people blessed with racial tolerance. That the British settlers developed similar racist attitudes towards Africans, that it was, indeed, largely British capital that built and sustained the pillars of racial separation with regard to the critical resources of land, labour, and political power, was conveniently forgotten. Africans, on other hand, appeared generally innocent, in fact, they were reduced to passive and pitiful lumps of human clay. Their herculean struggles against the settlers and their expanding colonial frontiers were left unacknowledged. In fact, until the late 1960s the history of Africans in South Africa was largely an adjunct of anthropology (Harries 1985:30).

The idealistic approach of liberal historiography to social relations and the racial system was partly based on the assumption that capitalism is inherently rational, efficient, and non-ascriptive (i.e. 'colour blind') so that over time its development in South Africa would marginalize the prevailing archaic and irrational racial attitudes. In short, liberal historiography preached that the relations between economic development and the system of white supremacy were essentially antagonistic and contradictory because of the liberalizing and integrative propensities of capitalism, industrial capitalism in particular. The liberal historians simply failed to see that South Africa's immense and rapid economic growth was not accompanied by any relaxation in the racial system. On the contrary, rapid economic growth, for instance, after the Second World War, went hand in hand with a more rigid application of apartheid and a widening social gulf between the races. The thesis that apartheid and capitalism were incompatible led to the illusion that apartheid would die a natural, if slow, death by the magical operation of economic forces. In short, it bred reformism.

By the turn of the seventies, liberal historiography had come under severe attack. The false political lull of the post-Rivonia years was over: Black Consciousness, Soweto, and mass strikes erupted shaking the material and ideological foundations of apartheid. Radical historians were energized. They began arguing that not only was apartheid compatible with, but 'has actually been an integral, functional component of South African capitalism and economic growth' (Johnstone 1982:8). Hence, revolution, not reform, would destroy apartheid. These historians started looking more systematically at the unsung heroes and hidden processes behind the history of South Africa. Africans lost their invisibility and their resistance to colonial conquest and adaptation to an expanding capitalism was acknowledged; the formation, expansion, impoverishment, and exploitation of the peasantry and working classes began to be analyzed; the pivotal role of the mining revolution in the late nineteenth century in the transformation of the political economies of Southern Africa was underlined; and political struggles among the various white groups and between them and Africans began to be seen in their bewildering complexity. Thus, at last, South African history moved away from magical political dates, like the so-called 'great watershed' of 1948. Settler colonialist perspectives, values, and myths were exploded: 'lusotropicalism' in the Portuguese colony of Angola, for example, was exposed as the cruel hoax that it was (Bender 1978). Class analysis was no longer shunned like a virus[1]. Themes of 'interaction' went out of the window, and in came analyses of the concrete realities of racial, national, and class struggles over land, labour, and political power. Apartheid and capitalism were finally consummated, and liberal historiography gave way to the history of South Africa's 'racially structured capitalism'.

Liberal historiography in Southern Africa, therefore, was very much like nationalist historiography to the 'black north' in that both tended to be highly empiricist, idealistic in their preoccupations, and betrayed deep bourgeois biases. It is significant that both began feeling the winds of discontent at the turn of the 1970s. By then, theoretical questions, political concerns, and the vagaries of time were coalescing into a profound critique of nationalist/liberal historiography. Its practitioners mournfully declared that there was a general crisis in the study of African history as a whole. There was never any crisis. Nationalist/liberal historiography had simply lost its 'hegemony' over African historical scholarship. Africa had now been subsumed into the developmentalist ghetto of the 'Third World'. Gunder Frank (1967, 1969) was being discovered and imported into Africa. The continent's material poverty finally became underdevelopment.

The Eternal Curse of Dependency

Notions of development and dependence developed out of dissatisfaction with prevailing bourgeois descriptions, analyses, and prescriptions for Latin America, as well as Marxist ideas about 'backward' countries[2]. Orthodox development theory saw underdevelopment as an *original* or *traditional* state. Consequently, the so-called underdeveloped countries could only wrest themselves out of this state by passing through a number of Rostovian stages (Rostow 1971), acquiring Parsonian value systems (Parsons 1951), and keeping their doors open to 'free' trade, and the diffusion of Western investment and technology[3]. Meanwhile, Marxists still clung tenaciously to Marx's optimistic prognosis that the expansion of capitalism through trade and investment would eventually break down all pre-capitalist modes of production and bring about capitalist economic development in the image of Western Europe (Marx 1978:91; Marx 1980; Marx and Engels 1976:13-14; Avineri 1969). Contrary to the positivist projections of both theories, however, the 'Third World' failed to break out of underdevelopment.

Raul Prebisch and the Economic Commission for Latin America (ECLA), which was formed in 1948, led the challenge against conventional theories of international trade and economic development. The ECLA showed how the international division of labour was not a *natural* outcome of world trade, and that it brought greater benefits to the centre than the periphery. The commission advocated the use of a structuralist and historical perspective in order to understand underdevelopment and devise solutions for its eradication (Prebisch 1971). But the apparent failure of the import-substitution industrialization model of the ECLA encouraged writers on Latin American underdevelopment, like Gunder Frank (1967, 1969), to seek more radical analyses and solutions. The reformulation of ECLA analyses and strategies almost occurred simultaneously with attempts by the Latin American left to reconceptualize obstacles facing capitalist development, particularly industrialization, in the periphery as a result of pervasive 'feudal-imperialist' alliances. It was left to Baran (1956) to provide the first systematic analysis of underdevelopment from a Marxist perspective[4]. He insisted that western development had historically taken place at the expense of the underdeveloped countries, and that the dominant interests in the advanced capitalist countries were profoundly inimical to economic development in the periphery. All these critiques were united by a common pessimism regarding the possibility of capitalist development in the periphery. Socialism, broadly and variously defined, was seen as the only real alternative to perpetual underdevelopment and dependency.

Fanon's radical pessimism no longer seemed so radical or strange any more; it assumed axiomatic familiarity. The dependency school found ready and eager students in poor old Africa, impoverished by centuries of imperialist exploitation. Frank's grand reconstruction and periodization of Latin American history was repeated for Africa by Rodney ([1972] 1982), Amin (1974, 1976), and Wallerstein (1976). It was demonstrated that from the time of the Atlantic slave trade to the era of formal colonization and, finally, the post-independence period, the history of Africa, like that of Latin America, was characterized by a constant siphoning off of 'social surplus' from the continent to the West through numerous mechanisms, principally the operation of unequal exchange, which was a product of an asymmetrical international division of labour. In short, the underdevelopment of the periphery and the development of the centre were constantly being reproduced through an interminable satellite-metropolis chain, in which the surplus generated at each stage was successively drawn to the centre. African or Third World underdevelopment was, therefore, simply one side of the same coin of western development. The dualist models of modernization theory, with their lethargic 'traditional' sectors and dynamic 'modern' sectors, were buried; the world had become a single integrated unit. Capitalism attained universal omnipresence, and the 'development of underdevelopment' assumed a Sisyphean inevitability.

For Samir Amin, accumulation on a world scale involves a continuous process of 'primitive accumulation' in the periphery for the benefit of the centre. He argues that, unlike expanded normal reproduction, the mechanism of 'primitive accumulation' is unequal exchange, that is, the exchange of products of unequal value, or rather whose costs of production are unequal. The dynamic of unequal exchange is rooted in the very structure of linkages between the socio-economic formations of peripheral capitalism and of capitalism at the centre. Unlike the latter, capitalist formations on the periphery are characterized by unevenness of productivity between sectors, disarticulation and extraversion of the economic system, and domination from outside. The combined and cumulative effects of these factors create the conditions for the drainage of surplus to the centre, thereby reinforcing and reproducing the commercial, financial, and technological dependence of the underdeveloped countries on the centre. Aghiri Emmanuel (1972, 1974), on the other hand, narrows unequal exchange to 'an unequal rewarding of factors', notably the labour factor, between 'poor' and 'rich' countries. In other words, wage disparities, even for the same productivity, between poor and rich countries are at the root of unequal exchange. Thus, the periphery is drained of much of the social value of its labour. International working class solidarity is thereby undermined. Henceforth, the proletariat of the periphery takes over from their privileged brethren the role of a vanguard in the global socialist revolution.

Underdevelopment finds its historic mission: it is the grave-digger of capitalism.

Wallerstein tried to systematize the dependency notions of 'incorporation', 'transfer of surplus', 'specialization', and others, into a 'metatheoretical' construct with which to explain the origins of capitalist development and underdevelopment and to locate the mainspring of their subsequent evolutions. He saw capitalism as a trade-based world division of labour in which a unique pattern of labour usage characterized the centre (free, skilled labour), and the periphery (coerced, unskilled labour). 'When labor is free everywhere,' he proclaimed, 'we shall have socialism' (Wallerstein 1974:127). According to Wallerstein, therefore, the development of capitalist production, which facilitated the growing division of labour, was itself made possible by the regional specialization of labour control. Capitalism is depicted as a system of labour rationalization and of unequal exchange. In short, Wallerstein's world system is a global Parsonian monster in which the peripheries are assigned specific economic roles, and all they can do is jockey either for semi-peripheral or core-status, until the system self-destructs sometime in the 'twenty-first or twenty-second century'. This is a world in which social struggles are spectacularly trivial, and historical processes are reduced to a series of ahistorical functionalist games of system maintenance. Pessimism finally matures into fatalism. Fanon is turned on his head.

Marxist critics charged that Wallerstein, Amin, Emmanuel, Frank, and others who constructed grand teleologies of development and underdevelopment, mislocated the dynamic of capitalist accumulation by concentrating on exchange relations rather than production relations (class structure, class struggle) (Laclau 1971; Pilling 1973; Bettelheim 1972; Brenner 1977; Smith 1980; Polychroniou 1991). The 'external' determination of dependency is so overemphasized that the role of 'internal' structures in reproducing dependence is obscured. Thus, set against the 'unequal exchange' of the underdevelopmentalists, is the 'comparative advantage' of the development economists, so ferociously attacked by the former (Hopkins 1975, 1976); both dwell on trade at the expense of production itself, disregard classes which emerge from the productive process, the ensuing class struggles, and the complex and contradictory effects of those struggles on social formations of the so-called peripheral capitalist societies. This is partly because, despite appearances to the contrary, underdevelopment analysis was focused almost exclusively on the economic terrain. In short, dependence writers miserably failed to delineate the specificity of the political in the reproduction of the economic conditions of underdevelopment (Goulbourne 1979).

By 'blaming' the metropoles and international capital for poverty, backwardness, and stagnation in the periphery, the local ruling classes were absolved, thereby misdirecting political struggle. Indeed, the tendency to portray the local bourgeoisie as 'lumpen', 'comprador', or 'auxiliary', incapable of rational accumulation and rational political activity, forces political activists to choose between immediate socialist revolution or surrender to a permanent state of capitalist underdevelopment. One leads to adventurism, the other to complacent pessimism. By territorializing poverty, dependency theory shared with development theory a tendency to define out of existence the rich of the South and the poor of the North. Thus, despite its apparent radicalism, dependency theory had conservative political inclinations. Moreover, dependency notions of 'unequal exchange' and international specialization undermined international working class solidarity and encouraged reactionary 'third worldist' nationalist ideology.

Yet the kaleidoscopic reality of 'Third World' countries strains any attempt to homogenize that world into a 'periphery', to see their history unfolding according to the lockstep of a predetermined Rostovian-like pattern. For Warren (1973, 1980; also see Hansen and Schulz 1981; Michael, Petras, and Rhodes 1975), the chances of successful capitalist development, that is industrialization, were quite good for a number of major underdeveloped countries. In fact, 'substantial progress in capitalist industrialization has already been achieved' in these countries (Warren 1973:3). Imperialism is actually declining as capitalism in the periphery grows. Reversing his earlier position, Leys (1974) was moved to say that the core-periphery framework was nothing but a 'polemical inversion' of well-known 'simplistic pairing' (Leys 1977, 1980, 1982). Swainson (1977:40; 1980) asserted that the much-abused national bourgeoisie was not merely an 'impotent class of intermediaries for international capital'. Independence did matter. Lall (1975) wondered whether the characteristics of dependent economies did not apply to capitalist economies in general since they were not exclusive to the former. Was it not, Palma (1978:908) asked, confusing a socialist critique of capitalism with analysis of the obstacles of capitalism in the Third World to talk of 'growth without development?' Kay (1975:ix-x) remarked provocatively: 'capitalism created underdevelopment not because it exploited the underdeveloped world, but because it did not exploit it enough'.[5] And Cooper ([1981] 1993:98) admonished: 'dissecting complex problems with concepts like underdevelopment, incorporation, unequal exchange, and core-periphery relations is rather like performing brain surgery with an ax: the concepts cut, but messily'. With characteristic certitude, Nabudere (1977) concluded that dependency theorists were propagating 'petty-bourgeois' ideology.

By the late 1970s, therefore, dependency theory was beginning to lose its intellectual seductions. Like nationalist historiography before it, it had 'proved' its case: development and underdevelopment were imbricated with each other. Africa or the Third World had been integrated into the capitalist world system and in the process its poverty had lost the veneer of exoticism smeared by development economics. Notions of dependent capitalist development began to be heard of. Writers like Cardoso (1972, 1977; Cardoso and Faletto 1979) tried to marry some of the dependency perspectives on unequal exchange, the changing international division of labour, and uneven development, with Marxist concerns with accumulation within the sphere of production, the processes of class formation, and class struggle[6]. The construction of a mechanical-formal theory of Third World underdevelopment, in which the dependent character of these economies is the hub on which the whole analysis of underdevelopment turns, was replaced by dialectical analyses of historical processes; the latter being conceived of as the result of struggles between classes and groups that define their interests and values in the process of the expansion of a mode of production with all its contradictions and disjunctions.

An increasing number of writers, therefore, tried to advance beyond the ubiquitous and homogeneous capitalism of dependency theory by positing the 'articulation of modes of production' (AMOP), whereby the supposedly pre-capitalist modes in the colony were articulated in their diverse relations with the capitalist mode. Thus, the introduction of capitalism does not eliminate pre-capitalist modes, but reshapes them. In other words, these modes continue to exist, but they are progressively subordinated to capital through a contradictory process of destruction, preservation, and transformation. The treacherous marshland of dualist theories and dependency's universal capitalism is thereby carefully skirted (Foster-Carter, 1978).

But to talk of articulation of modes of production presupposed a general conception of a mode of production and theories of particular modes of production, which proved no easy task. Hindess and Hirst (1975:9) saw a mode of production as 'an articulated combination of relations and forces of production', with 'relations of production' here referring to the mode of appropriation of the surplus product[7]. Bernstein and Depelchin, (1978/1979:4) insisted, however, that 'the categories of social relations (economic, political, ideological) and the relations between these categories cannot be theorized generally in the concept of mode of production itself, but vary according to each mode of production'. A mode of production, they continued, was only concretized through the social formation in which the mode was manifested. There was also no agreement on the term 'social formation'. The definitional difficulties were evident in Balibar's (1970:207) much-quoted formulation: the term could be used, he advised, either as 'an empirical concept designating the object of a concrete analysis, i.e., an *existence*: England in 1860, France in 1870, Russia in 1917, etc., or an abstract concept replacing the ideological notion of 'society' and designating the object of the science of history insofar as it is a totality of instances articulated on the basis of a determinate mode of production'.

While the capitalist mode of production could be specified, notwithstanding the tortuous debates, the same could not be done with the oxymoron, African pre-capitalist modes of production. By the time modes of production theory entered Africanist discourse in the 1970s, it was too late to talk of African states and societies in terms of 'primitive communism', and except for those working on Ethiopia and a couple of other highly centralized 'kingdoms', it did not make much sense either to describe them using the labels of 'feudalism' or the 'Asiatic mode of production'(Goody 1971:Chapter 2; Law 1978). Instead, Africanists tried their hand at conceptualizing, hoping to add to the crowded corpus of Marxist terminology. Coquery-Vidrovitch (1976, 1977) concocted the 'African mode of production', in which a patriarchal ruling class controlled long-distance trade. But it failed to capture anyone's imagination: why should relations of distribution be dominant over the more basic relations of production? Attempts to construct the 'lineage', 'tributary', or even 'slave' modes of production also tended to suffer from their own problems. For instance, in the 'lineage' mode kinship units played a key role so that the term 'kinship' was taken as a given, while it actually needed to be explained, or rather problematized. The 'tributary' mode, on the other hand, had a pronounced bias towards exchange relations and not the dynamics of the productive system. And the evidence was not compelling that there were many societies where other modes of production were subordinated to the requirements of reproducing the slave mode (Cooper 1979)[8]. Part of the problem lay in trying to construct a single mode or a few distinctive modes of production for the diverse and complex historical reality that is Africa. But chasing modes of production behind every tree in Africa's savannah hinterlands proved no solution either.

Despite the difficulties of specifying Africa's indigenous modes of production, many jumped on the bandwagon of articulation of modes of production. In a widely quoted paper, Berman and Lonsdale (1980:60) defined the process of articulation as involving:

Extracting surplus product from and/or forcing labor into capitalist or quasi-capitalist formations.... The form of articulation varied according to the particular character of capitalist penetration, the nature of indigenous modes of production, and the local ecology and resource endowment. The resulting variations in the subjugation and transformation of local societies and the degree to which capitalist forms of production were introduced also determined the differing patterns of class formation within and between colonies[9].

The process of articulation, Aidan Foster-Carter (1976) stressed, was accompanied by violence, certainly during the phase of 'primitive colonial accumulation', when the capitalist mode of production was being introduced.

To many, articulation came to be seen solely as a continuous process of interaction through which the so-called pre-capitalist modes paid the costs for the reproduction of the labour force (Cooper 1980). 'Yet in recent years', Mafeje (1981) wrote, 'we have witnessed in South Africa the dumping of unwanted labor in the reserves, not to reproduce their labor-power, but to perish'. This was a sobering reminder that at one stage in the process of articulation, indigenous modes of production could be used to 'subsidize' capital accumulation, while at another to provide dumping grounds for the 'rejects' of capital, especially the unemployed. It was back to stages. Three stages of articulation were distinguished. In the initial phase a link is established in the sphere of exchange, 'where interaction with capitalism *reinforces* the pre-capitalist mode'. In the second phase 'capitalism "takes root", subordinating the pre-capitalist mode but still making use of it'. In the third stage, which he believed had not yet been reached in the Third World, there would be 'the total disappearance of the pre-capitalist mode, even in agriculture' (Foster-Carter 1976:56). Thus, capitalism's capacity to look after itself increased with each stage. Articulation assumed unilinear progression towards the certainties of capitalist modernity.

In reaction to this, others insisted that it is crucial to emphasize that different capitals, at various times, require different things from pre-capitalist societies, so that there should be no 'bland talk of 'capitalism' doing or being this and that, in relation to other modes of production' (Foster-Carter 1976:69; Bradby 1975). The process of articulation was too complex to be interpreted mechanically as referring to sharply defined and sequential stages. Moreover, lest modes of production become actors in themselves endowed with their own inexorable logic, it was pointed out, the articulation process essentially involved a struggle between *classes* these modes defined.

But the sceptics saw functionalism lurking behind all this articulating. The degree and forms of 'dissolution/conservation' of the indigenous modes of production seemed to be determined by what was functional to capital (Bernstein 1977; Bozzoli 1983). Neo-Marxists smelling of dependency theory, such as Kitching (1977), insisted that it was futile to see a country like Kenya, for example, as a social formation of articulated modes of production, rather than 'a satellite of the world capitalist mode of production'. The search was on for new concepts. Some came up with the notion of the colonial capitalist mode of production (CCMP), which was advanced, its proponents claimed, to capture the specificity of imperialism during the colonial era (Magubane 1976; Alavi 1975). But the trajectory of the CCMP was never quite clear, whether or not it was a transitional mode from pre-capitalist modes to capitalism, or to something entirely novel, like, perhaps the post-colonial mode of production? The confusion was palpable.

In short, the theory of articulation of modes of production soon developed the sclerosis of imponderable jargon, and it went the way of its sibling, dependency theory, also conceived out of Africa. In the mid-1980s, the *Canadian Journal of African Studies* (1985) held the appropriate discursive rituals to bury the theory that had met its demise so early. Prominent Africanist historians gave their heartfelt eulogies. None was an African scholar, and there were only a few women. Never mind, such grave theoretical matters were not for them. Klein (1985), Clarence-Smith (1985), Freund (1985), Newbury (1985), Cordell (1985) and Geschiere (1985) applauded AMOP for providing useful questions and hypotheses for historical enquiry; Alpers praised it for focusing scholars' attention on production; Kimble (1985) vowed to continue using its insights; and Harries (1985) and Beinart (1985), a little grudgingly perhaps, commended its concepts and metaphors, respectively, for assisting in the radicalization of South African historiography. Kitching (1985:100) was less sanguine, insisting that modes of production literature was 'a piece of massive conceptual overkill', a message that was lost in the din of his own conceptual overkill.

It was the last of the great 'metanarratives' in Africanist historiography, at least for a while. Some went back to the traditional delights of empiricist history, others sat waiting for the next theoretical inspiration, and before long they fell for the dazzling parade of post-structuralisms, and yet others plodded along with a little of theory, some speculation, and a lot of empiricism. More often than not, the theories were borrowed from the dismal Social Science disciplines - economics, sociology, and political science.

Discovering Associational Ties

All along, as dependency and modes of production theories made their tour of African studies, the 'natives' were being reconfigured. One of the great discoveries was that they could enjoy the modernized associational ties of class. This was an intellectual breakthrough, for in the heady days immediately after independence, African politicians, bureaucrats, and academics declared that Africa was classless. Consequently, for them class analysis represented slavish importation of 'foreign ideology', if not something actually worse. They gloried in Africa's uniqueness, its nationalist achievements, and natural genius for 'socialist harmony.' Western intellectual tourists, on the other hand, brought up on a fulsome diet of bourgeois social science and tales of African primitivity, had refused to see Africa in class terms either. Some simply asserted that, unlike in Europe, African societies had not 'developed' enough to generate distinctive and antagonistic social classes, or that the 'elites' and the workers were numerically so small that neither could constitute a coherent stratum. Others contended that, in naturally 'tribal' Africa, 'tribalism' undermined the growth of class consciousness and class-based action, and since the great majority of Africans resided as an undifferentiated mass in the rural areas, it was futile to take class stratification seriously. Under such circumstances, it was argued, it was better to use the theory of cultural pluralism (Kuper and Smith 1969).

Denying the existence of social classes in Africa because they did not seem to exhibit the same subjective characteristics as in Europe, the radical scholars railed, was tantamount to placing .'concepts in a historical deep freeze, embalmed around a particular historical conjuncture conditioned by an image of an ideal or pure form of the social object' (Cohen 1976:155-6). The obscurantist cultural determinism of the reigning pluralist model was challenged for painting conflicts in Africa with the broad strokes of 'tribalism'. Judging by the reputed primordial characteristics of a 'tribal' society, the critics pointed out, then such societies had been extinct in Africa for a very long time indeed. Calls began to be heard for the substitution of 'tribe' with the less 'offensive' and more 'objective' concept of ethnicity (Southall 1970; Enloe 1973). Historians added their discoveries that many of Africa's contemporary 'tribes' were, in fact, invented by colonialism (Ranger 1983; Chanock 1985). Colonial capitalism, Leys (1974:199) argued, politicized ethnicity not only because of the deliberate colonial tactics of divide and rule, but also because people from different ethnic groups were forced to compete against each other for the limited opportunities and resources of work, land, and education. After independence, ethnic consciousness, or 'tribalism', was further promoted, Mafeje (1971) argued in his seminal intervention, by the new leaders as a mask of class privilege and exploitation. Thus, 'tribalism' was 'unreal', it merely served an ideological function. It did not represent a primordial political force in Africa, rather it developed in response to imperialism in the context of articulated modes of production, in which certain aspects of the pre-capitalist modes were reinforced and perpetuated. 'Tribalism', in fact, was an integral part of the class struggle in the ideological sphere. Far from being Africa's 'natural condition', therefore, the ethnic 'interpellation' (together with the national 'interpellation'), as Saul (1979) put it, was spawned by internal class contradictions, as well as the centre-periphery contradiction.

So it was that ethnicity was banished from African studies for almost two decades (Osaghae 1994). It was a conspiracy of silence that satisfied the interests of diverse, ordinarily opposed, elements: nationalist politicians, conservative modernizers, and revolutionary intellectuals. Class analysis became all the rage. New lines of battle were drawn. One group borrowed from bourgeois social categories and talked of 'social stratification', in which class is determined by behavioral patterns: indeed, it becomes trivialized as one aspect, and a descriptive category, of social differentiation subject to empirical observation. After using this framework, some came up with such banal conclusions as: 'most contemporary African societies are 'one-class-societies' (Jackson 1973:393; Chodak 1973), or they divided African societies into the amorphous layers of 'elite', 'middle', and 'lower' classes on the basis of status and the acquisition of 'civilized', i.e., 'European' values (Mitchell 1956). The second group consisted of those who saw class in materialist terms in the context of the social relations of production. Unfortunately, like missionary zealots, some of them took Marxian categories of the western capitalist countries and mechanically imposed them on African societies, thereby turning these societies into caricatures of themselves (Cohen 1972; Wallerstein 1973; *Review of African Political Economy*, 1975; Katz 1980).

In particular, debates raged about 'peasantization' and 'proletarianization'. What is a peasantry, a working class? Can one even realistically talk of 'peasants' or a 'proletariat' in Africa? Who are more reactionary or progressive, peasants or workers? Is there a peasant mode of production? Are African workers 'labor aristocrats'? For a long time, many writers had resisted calling Africa's rural masses 'peasants', preferring to see them as 'husbandmen', 'rural capitalists', or even 'protopeasants' (Post 1972; Bernstein 1979). As for those who dared to call them 'peasants', they tended to define peasant in cultural and sociological terms, so that they saw peasants essentially as 'primitive' cultivators living in self-sufficient, kin-based communities, which had progressively been made dependent on external structures and sanctions (Fallers 1961). Their critics insisted on drawing a distinction between 'peasant' and 'subsistence' economies on the basis that peasants produced primarily for the market, rather than for subsistence (Dalton 1964). But this distinction appeared overdrawn; peasants, some argued, produce partly for an external market and partly for their own consumption, so that they were subject to both external incentives and controls and local requirements and regulations (Middleton 1966).

Saul and Woods (1979) tried to provide a more systematic conception of African peasants and carry the debate further. They argued that a peasantry, which consists of people who enjoy access to a portion of land and use the family household or homestead as a production-consumption unit, could only be understood within the wider political economy and in the context of historical change. They also stressed that in the process of their development, peasants become internally differentiated. For this reason, and the fact that the circumstances and degree of incorporation of African societies into the world capitalist system differed, one could talk of African peasantries, which included pastoralists. Subsequent writers elaborated on the distinction between 'poor', 'middle', and 'rich' peasants, culminating at the turn of the 1980s in Bundy's (1979) important study of the development and demise of the South African peasantry, which others tried to transplant to other colonies, rather unsuccessfully (Palmer 1977; Palmer and Parsons 1977; Leys 1971; Atieno-Odhiambo 1974). Peasants had come a long way; they were now not only internally differentiated, they were also structurally distinguished from pre-colonial agriculturalists, capitalist farmers, and the rural proletariat. And in the case of South Africa they had apparently risen and fallen[10].

Debate now centred on how to analyze relations between peasants and capital. On the one hand, there were those who, like Bernstein, argued that the production and reproduction of peasants was determined by the predominance of the world market, capital, and the state. Consequently, 'peasants have to be located ... within capitalist relations of production mediated through forms of household production...[I]n this way peasants are posed as 'wage labor equivalents' (Bernstein 1977:73). In other words, peasants were 'semi-proletarians' producing surplus value for capital, but located outside the direct capitalist labour process. It was therefore fallacious, according to this view, to talk of a 'peasant economy', or a 'peasant mode of production' articulating with a dominant capitalist mode. But these controls that Bernstein discusses, Cooper (1980:309) noticed, 'circle around the point of production rather than enter it directly'. Moreover, to 'proletarianize' a peasant who was not divorced from his or her means of production, and did not sell labour power as a commodity for a wage, was to fly in the face of Marxist theory and assume automatic worker-peasant class unity instead of the negotiated possibilities of popular class alliances.

Other writers, therefore, contended that 'the contradiction between peasants and capital was a contradiction between different modes of production, and between classes within different modes of production, not between antagonistic classes within one mode of production' (Boesen 1979:159). That seemed like constipated articulation smelling of dualism. Hyden (1980, 1983), indeed, went so far as to claim that African peasants, unlike those of Latin America and Asia, had not been sufficiently made dependent on the market and the dominant social classes. Sustained by their 'economy of affection', they had the power to abandon commodity production and return to subsistence and self-reproduction. While this might have served as a useful correction against those who emphasized the 'rule of the market' in shaping peasant choices, Hyden's notion of an 'uncaptured peasantry' evoked modernization theory and articulated the developmentalist ambitions of both the post-colonial state and international capital for the effective control and management of the recalcitrant subaltern classes. It ignored the fact that peasants were not exploited a little, but a lot, hence their persistent struggles, including partial withdrawals from the extractive markets controlled by foreign and local capitalist interests.

Exhausted by these endless debates, which reflected conceptual confusion among scholars borrowing promiscuously from other discourses and disciplines, carelessly importing theories from different historical contexts and generalizing from isolated case studies, some researchers descended from their theoretical high horses and began to look at the changing organization of peasant work, especially its complex articulations with gender relations and divisions, environmental conditions and changes, and the complex patterns of rural cultural construction. As they paid more attention to what peasants actually did and listened closely to their voices, they began to appreciate peasant knowledge, admire their resilience, and perceive the intricacies of peasant politics and struggles at various levels, from the household and the local community, to the national and global system, along the structured inscriptions of gender, ethnicity, and class, and the national and international hierarchies of economic and political power. As much focus was now given to everyday struggles and other covert forms of resistance, including withholding production and consent, as was previously accorded to the heroic, but episodic, peasant rebellions. Also, peasant ideology was disentangled from the monopolistic claims of nationalism, and the complicated connections and contestations between peasant and nationalist ideologies and visions began to be extricated. The antiquity of the peasants also came to be recognized as pre-colonial agrarian histories were reconceptualized and rewritten (Isaacman 1990).

In the meantime, major analytical shifts were occurring in the study of African labour history as well. Before the 1960s, when serious scholarly interest in African labour began, the field was dominated by managerial studies conducted by colonial government officials, or reformist studies produced by visiting ILO (International Labor Organization) and ICFTU (International Confederation of Free Trade Unions) missions, and other movements, such as the Fabian Colonial Bureau, that were sympathetic to colonial peoples. These works, mostly in the form of reports, were concerned with either providing the colonial governments with data on the supply, control, cost, and productivity of labour, or, as in the case of the latter, they sought to effect policy and practical changes in colonial labour markets and working conditions. After the Second World War, attention focused on labour stabilization (Orr 1966; Friedland 1974; Allen 1969, 1972; Sandbrook and Cohen 1975; Gutkind Cohen, and Copans 1978; Zeleza 1982; Freund 1984, 1988).

It would be misleading, of course, to assume that the subject of African labour was confined to institutional studies. One has only to recall those beleaguered humanitarian critics of forced labour who tried to analyze the fate of people moving from, in their imperious language, the 'traditional' African world into the 'modern' western one. Despite their obvious sensitivity, albeit paternalistic, to the gross abuses of colonialism, their critiques were against the excesses, not the essence, of the coercive colonial labour control system. There were, of course, hardly any historians during this period who focused on labour issues. The subject was left to anthropologists, then happily trying to dress their 'natives' in functionalist garments. And the central problematic was labour migration. They debated the factors which generated it and its impact on both rural and urban life. Some attributed the growth of labour migration to the coercion of state policies, others believed Africans were voluntarily responding to the bright lights of economic opportunity. Opinion was similarly divided on its impact, with many arguing that it led to the decay of rural life, to the destabilizations of 'detribalization', a position that was challenged by those who detected 'retribalization' in the urban centres and approved of the benefits of migrants' remittances and their modernizing influence on the supposedly stagnant rural hinterlands (Wilson 1941; Wilson and Wilson 1945; Read 1942; Schapera 1947; Gulliver 1955; Watson 1958; Elkan 1960; Skinner 1960; Mitchell 1956, 1961). The debates and disputes aside, Africans entering wage employment were primarily seen as 'labour migrants', rather than as *workers*, who went to work to achieve a particular purpose, largely to acquire cash to pay taxes, perhaps to improve farming or set up a simple business, and usually to finance the simple pleasures of African rural life: polygamy and the endless ritual ceremonies. And so they were described as 'target workers', and later as 'lumpen-proletarians', or 'worker-peasants'. It would take a long time before the system of labour migration was understood as a complex process rooted in the 'logics' of both labour and capital, and the perennial struggles between them over the control of space, work, and culture.

At independence anthropologists came down together with the colonial flags. In the new era of nation-building and development, it was the stars of political scientists and economists that rose. Institution-building was everywhere. Naturally, interest shifted from colonial labour managerialism to the role of trade unions. Lingering concerns from the anthropological past, and the preoccupations of the nationalist movement, appeared in the heated debates over the relative importance of 'tribalism' in African trade unionism (Scott 1966; Friedland 1969; Smock 1969; Heisler 1970). Cold warriors added their opinions for what they were worth, depicting African trade unions largely as 'alien' institutions founded by political agitators, and manipulated by trade union internationals (Meynaud and Bey 1967; Lynd 1967). Many turned their research sights to the institutional links between trade unions and political parties. Some emphasized the role of trade unions in mediating between urban 'elite nationalism' and rural 'mass nationalism' (Hodgkin 1956). Others saw trade unions as mere appendages of nationalist parties (Belling 1968). Berg and Butler (1966:40) noted the failure of the labour movements 'to become politically involved during the colonial period, their limited political impact when they did become involved and their restricted role after independence'. Workers' struggles were declared 'economistic', devoid of political content.

The ground was clear for the 'labour aristocracy thesis' to strut its theoretical wares, and rob African workers of their newly acquired progressive image. Fanon's (1963) categorical dismissal of the African working class as a privileged segment of the colonial population and, therefore, abysmally lacking in revolutionary potential, was taken up and popularized by Saul and Arrighi (1968, 1969), who argued that this 'labour aristocracy' of skilled workers, largely groomed by relatively high-paying multi-national corporations, was divorced from the rest of the working class and developing economic and political interests essentially congruent with those of the national bourgeoisie[11]. The 'radicals' were outraged. The obscene disparities in the living standards of the workers and the national bourgeoisie, they pointed out, were too obvious for there to be much congruence of interests between them. But even if the upper echelons of the working class had 'one foot in the steeply rising embourgeoisement ladder', in Jeffries (1975:276) memorable phrase, political attitudes could not simply be reduced mechanically to standards of living. The workers and the national bourgeoisie, it was emphasized, occupied essentially antagonistic positions in the production process. Moreover, it had to be recognized that neither the colonial nor neo-colonial economic system had the capacity, let alone the will, to buy off the working class (Waterman 1975). The notion that workers constituted a privileged group in relation to the urban unemployed and the rural peasantry was 'a classic example of the 'displacement' of the 'primary contradiction' between the interests of the exploiting and exploited categories on to a 'derived' contradiction between exploited classes' (Peace 1975:300). It was simplistic to focus on the spatial dimensions of exploitation; the fact that the urban-based industrial sector expropriated surplus from the rural peasant-based sector did not mean that workers benefited. They were themselves exploited. Moreover, there were numerous mechanisms through which income was transferred from the workers to the so-called 'informal sector' and the rural peasantry. The theory, the critics continued, diverted attention from imperialism and the fact that skilled workers in Africa, Asia, and Latin America were paid lower than their counterparts in the western industrialized countries for the same work. Despite its populist Fanonist garb, the theory justified state assaults against labour, to hold down wages in the interests of developmentalism and imperialism. Saul (1975:308) was forced to recant: 'the African working class should not be *prematurely labeled* ...

[because] the role of this class is far from being frozen by history or by any internal logic of the current African socio-economic structure'. The search was on for a new hegemonic narrative. Sandbrook (1975) came up with the concept of clientelism as an explanatory model of the relations both between the labour movement and the post-colonial state and within the labour movement itself. There were few pickers. 'Contemporary patron-client relationships', Saul (1979:135) waived dismissively, 'are themselves contingent upon the established hierarchies of a neo-colonial economy [so that] to propose a description based on clientelism as an *alternative* to class analysis ... is patent nonsense as soon as one moves from the most limited micro-analysis to ask questions about the system as a whole and in whose interest it works'.

In short, African labour studies were long on speculation and prescription, often masquerading as theory, and short on historical analysis of the complex and changing dynamics of the labour process. Those that eschewed the convoluted debates and concentrated on investigating the old managerial issues, tended to reduce industrial relations to a set of formal contractual arrangements between unions, employers, and the state, with little reference to the social and economic context in which they operated (Roberts 1964; Roberts and Bellecombe 1967; Smith 1968).

Until the mid-1970s, historians had been conspicuous by their absence in these debates, partly because many were too preoccupied with reconstructing the pre-colonial past, or demystifying colonialism and celebrating African nationalism, to worry too much about labour history. As the lustre of nationalist historiography began to fade, and the 'masses' acquired distinctive class names, as peasants and workers, through the discursive largesse of dependency and modes of production theories, historians began paying attention. The editors of the landmark collection, *African Labor History* (Gutkind, Cohen, and Copans 1978), urged labour historians to unravel the unfolding consciousness of the African working class, to plot its grand Marxist march from a theoretical category of class-in-itself, to a living and self-conscious class-for-itself.

The call was heeded and historians began to examine systematically the impact of colonial policies on labour, even if they might reach different conclusions, some sympathetic to the colonial regimes (Clayton and Savage, 1974), others critical (Arrighi 1973; van Onselen 1976). Few saw the need any more to attribute strikes and other forms of labour militancy to cultural and psychological dysfunctions resulting from 'detribalization', or to 'subversive' outside influences. Now labour history was analyzed in terms of underdevelopment, and working class formation was periodized in the context of the changing conditions of peripheral capitalism (van Zwanenberg 1975; Jeffries 1978; Parpart 1983; Stichter 1982; Shivji 1986). As the broad outlines were filled, the picture became more nuanced, and detailed portraits began to emerge about the changing contexts of workplaces, working class communities, cultures, and discourses, most notably in the works by van Onselen (1982a, 1982b) and in Cooper's (1977, 1981, 1987) trilogy on the transformation of work in Mombasa (also see Burawoy 1972, 1982, 1985; Cooper 1983; Zeleza 1995; Freund 1995). As befitting historians, the analysis was projected progressively backwards, and the historical contours of African workers' experiences in the centuries long before the colonial conquest began to appear from the lifting mists of ignorance and assumptions that workers were moulded from the clay of colonial capitalism (Coquery-Vidrovitch and Lovejoy, 1985). Labour histories of Ghanaian and Mozambican workers were revealingly told from the second half of the nineteenth century by Crisp (1984) and Harries (1994). Atkins' (1993) subtle, elegant, and perceptive study of the cultural construction of a Zulu work ethic in the last half of the nineteenth century marked a brilliant apotheosis to this phase of African labour historiography.

But not all exhibited Atkins's panAfrican sensitivities and superb analytical skills. Some offerings were dreary textual excuses for convoluted Marxist theorizing, and many more presented teleological narratives that saw working class consciousness lying at the end of the tunnel of proletarianization. Particularly weak were the attempts at synthesis: they were marked by poor periodization, imprecise inter-regional comparisons, over-generalizations, unsubstantiated assertions, and erratic theorization (Stichter 1985; Freund 1988). Above all, with a few notable exceptions (White 1990; Mandala 1990), many of the studies ignored the question of gender, a void that is slowly being filled as more work is done on women's labour history (see Chapters 9 and 10 below) and the construction of violent masculinities among male workers (Maloka 1994).

Conclusion: End of Syntheses?

Thus, in a little over three decades Africa had moved from a continent without history into one with long, multiple histories of ethnicities and nations, state formation and decline, migrations and invasions, kings and peasants, merchants and workers, production and exchange, religious and social movements, colonialism and nationalism, European power and

African agency, accumulation and exploitation, urbanization and rural transformation. By the 1980s, new themes were beginning to be explored systematically: environmental history, medical history, women's and gender history, just to mention the prominent ones. The African and Africanist historiographical achievement was truly impressive. But there was little celebration. Gaping holes still remained to be filled, the energy and enthusiasm of the early days was gone, and there was little sense of direction or common purpose. The era of syntheses, of Dubois, Dike, Diop, and Davidson, of asking the big civilizational questions, appeared over, replaced by that of excessive specialization, petty enquiries, and endless self-indulgent critiques of 'theories' whose lifespan was the proverbial fifteen minutes of western pop culture[12].

The proliferation of histories, or the disintegration of the old 'grand' narratives, both the nationalist and Marxist-inspired ones, is of course not confined to African history (see Chapter 22 below), nor is it an entirely negative thing. Despite their universalist pretensions, these narratives were narrowly political and economic, leaving out large segments of the human experience, especially issues concerning culture and the environment. They were also androcentric, so that the history of half of humanity was invisible. However, abandoning all attempts at synthesis, at trying to tell 'large and long stories', has its own dangers. Preoccupation with post-structuralist concerns especially threatens to subsume the crucial narratives of oppression and exploitation and resistance at whatever level except, perhaps, at the most atomistic. Also, it forces greater focus on the contemporary, which diverts research attention from the more distant historical periods. Already, it would appear that outside the ranks of North American Afrocentrists, who are more inclined to rhetoric than research, the current generation of African and Africanist historians is doing comparatively less work on the ancient past than the generation of the Ajayis and Ogots, Vansinas and Curtins. There are several reasons historians committed to Africa's liberation have to continue pursuing, or trying to construct, the 'big picture'. First, imperialist historiography, which never disappeared even in the heyday of nationalist historiography, has been experiencing a resurgence, carrying the same messages of African inferiority and pathology in new thematic and theoretical packages. Related to this, second, is the fact that the historical knowledge that has been produced over the last thirty to forty years has done little to contain the suffocating grip of developmentalist discourse, which is closely tied to imperialist historiography. This suggests, third, that historians' knowledge has either not been adequately and effectively disseminated to the various publics outside academia, or it has failed to engage their historical imaginations. Thus, historians have their work cut out, without dissipating their overwrought energies in either apolitical theorizing or the pedantic chase for the insignificant. These issues are pursued in greater detail in the next few chapters.

NOTES

1. But class also became a veil for some 'radical' white South African scholars to hide from serious analysis of the questions of race, ethnicity, and culture both within the country, and in their own production as a racially privileged, almost exclusive, intelligentsia. Now that apartheid has ended, the veils have come down and the true colors of cultural conservatism among some of them are becoming apparent, especially over 'affirmative action' in 'their' own universities. 'Radicalism' was, of course, more valuable in the anti-apartheid international academic market in the 1970s and 1980s than conservatism. This partly explains its popularity among exiled and internationally conscious South African academics at the time. Some are now apparently finding refuge in the currently fashionable depoliticized posturings of post-modernism.

2. This is the conventional genealogy of dependency theory. Kitching (1989) proposes a radically different one, locating it in the tradition of Russian populism, although his argument is unconvincing (Zeleza 1991). The literature on dependency theory is vast. For succinct summaries, see Rhodes, 1970; Oxaal, et. al., 1975; Bernstein, 1976; Palma, 1978; Roxborough, 1979; Blomström and Hettne, 1984; Harris, 1988; Escobar, 1995.

3. The literature on development economics is equally vast. Early influential essays can be found in Agarwala and Singh (1958); and later expositions in Meier (1976), Meier and Seers (1984), Seligson and Passé-Smith, 1993).

4. That Marx himself changed his optimistic prognosis of opportunities for capitalist development in colonized countries is shown by his later views on Ireland, see Davis (1979; also see Mohri 1979). Subsequent Marxist writers like Bukharin, Luxemburg, and, of course, Lenin, were keenly aware of the destructive and distorting effects of imperialism or capitalism on the colonies (Kiernan 1974).

5. For a perceptive critique of Kay, see Bernstein (1976). A version of Kay's argument is Kitching's (1989: 155) contention that the argument that imperialism and colonialism underdeveloped the present-day Third World can be dismissed on the grounds that 'in order to make such an assessment *one must know what would have happened if circumstances had been other than they were'*. Well, they were not. History cannot be wished away.

6. It is not entirely correct, however, to assert, as is so often done, that dependency writers such as Frank, Amin, and Rodney entirely ignored class analysis. Certainly their later writings have few equals in their analysis of the class structures of dependent capitalist formations, see Frank (1978, 1981, 1984); Amin (1974, 1980), and Rodney (1981).

7. A couple of years later, Hindess and Hirst (977) wrote an auto-critique of their book; also see Asad and Wolpe's (1976) long and incisive critique of the Hindess and Hirst's (1975) earlier book.

8. Since the original paper on which this chapter is based was published in 1983, there has been an explosion of studies on African slavery, some of it quite good, but most is pure drivel and seems to be inspired by that age-old European attempt to whitewash the Atlantic Slave Trade by saying that Africans, after all, enslaved each other. The loudest noises have been made by Lovejoy (1981, 1983, 1986, 1993). Much of this work centers on the nineteenth century, not earlier, which as Rodney (1980), Inikori (1982), Mandala (1990) and Manning (1990) have

convincingly argued, suggests that internal slavery in Africa was a byproduct, and did not develop independent, of the Atlantic slave trade.

9. Later Lonsdale (1981) elaborated and tried to distinguish between four different forms of articulation, namely, transitive, intransitive, tributary, and syncretic.

10. Bundy's thesis was subsequently challenged and his analysis overtaken by others. For a more detailed historiographical discussion of this subject, see Zeleza, 1993: Chapter 1; Klein, 1980; and Issacman, 1990; for an early attempt to analyze the East African peasantry, see Cliffe 1977.

11. Its Fanonist inspiration aside, the use of the concept of labor aristocracy was one more instance of writing African social processes by imitation and anecdote; the imitation in this case came from Engels' and Lenin's designation of English workers, in the context of global imperialism, as a 'labor aristocracy' (Nicolaus 1970: Hobsbawm 1970); and by anecdote because none of the proponents of the labor aristocracy thesis actually conducted empirical research, except perhaps to observe the workers from their hotel rooms, or talk about them in university Senior Common Rooms.

12. Some seek solace in postructuralism, others like Cooper (1994: 1518), are looking to the 'Orient' for 'subalternean' inspiration, after discovering that 'African historians' use of the concept of 'resistance' is less subtle, less dialectic, less self-questioning than Indian historians' deployment of the idea of subaltern agency. 'But all he can come up with after sampling the 'sophisticated' analyses of the Indian Subaltern Studies Group is a call for revisiting, and tease out the complexities in the murky, contested spaces between the old binaries of colonial history: colonized and colonizer, collaborators and resistors, engagement and autonomy. Hardly a new or, indeed, a very stimulating agenda.

Chapter Seven

THE NINE LIVES OF IMPERIALIST HISTORIOGRAPHY

Introduction

Historiographical traditions have a way of going into hibernation, shedding their aged and hideous scales, and beginning life anew, ready to spit the same old poison. That is what seems to have happened to imperialist historiography, dealt crushing blows in the 1960s and 1970s by nationalist and Marxist historians. The partial, but powerful, images that dominated the world's media in the 1980s and early 1990s, of a continent wracked by coups, chaos, and carnage, of people dying from drought and disease, and countries collapsing under the weight of state corruption and coercion, provided a fertile ground for the resurgence of crude, rabidly racist perceptions of Africa. And like vultures smelling death, supply-side bankers, with IMF or World Bank attaché cases, sanctimonious western politicians, award-seeking journalists, self-appointed 'aid' missionaries, and even publicity-starved 'pop stars', descended on the continent for their pound of flesh. In this atmosphere, imperialist historiography and neo-classical developmentalism flourished. This chapter explores the startling connections between the two in current Africanist discourse on colonial and post-colonial Africa.

A hundred years before, during the 'Scramble for Africa', the world heard of Kipling's (Wright 1976:113) 'white man's burden' to come and 'civilize' those 'half-devil, half-child' peoples of Africa, a call that was answered. Imperialist discourse was widely shared by both the unlearned masses and the enlightened savants, the street mobs and the members of parliament. Similarly, the current discourse of neo-imperialism, that Africa needs to be saved from itself by the international financial institutions and other external agencies, is a staple of the western mass media and policy establishment, and is articulated in Africanist scholarship, among the triumphant 'men of the right' and disillusioned 'men of the left', as Illife (1987) calls them. There is almost universal consensus that 'things have fallen apart again' (Riddell 1992) since independence, that the considerable 'endowment of social and economic assets' left behind by the 'colonial overlords' have been gratuitously squandered (Kilby 1983:246). The analytical and prescriptive conclusions are obvious: first, independence was a mistake; and second, the externally-imposed structural adjustment programmes are correct in their diagnosis, prognosis, and cure of the African 'crisis.' It is a discourse that marks a return to the bigotries of imperialist historiography and biases of colonial developmentalism. Both share a pathologization of African societies and histories, past, present, and future.

Pink Skins, White Masks

In his magnum opus, *A History of Africa* (1978), J. D. Fage, one of the most renowned Africanist historians, forcefully resurrected the 'white factor' in African history. Using scanty, unscientific, and contradictory evidence he saw Caucasoids behind every nook, creek, and tree in Northern and North-Eastern Africa during ancient times. To the nationalist historians, this represented an insidious attempt to divorce Egypt from Africa, the Sphinx from the Negro, the continent from civilization. Fage's racialization of African history also threatened to foster endless debates about pigmentation, instead of concentrating historians' attention on more fruitful studies of material, social, religious, and intellectual developments among the diverse peoples of Africa. As Ogot (1984) concluded, African historical scholarship was being forced back to its ignoble beginnings.

Fage's ancient Caucasoids and their exploits were progenitors of the latter-day imperial decolonizers[1]. In other words, the 'white factor' which was resurrected for ancient Africa was also invading that sacrosanct area of the nationalist historians' turf: decolonization. The thesis: decolonization was planned after all. The verdict: nationalism was merely of nuisance value. Now, to be sure, imperialist historians were not the first to dismiss nationalism. For many 'radical' writers decolonization was false, a moment full of sound and fury as the guards were changed, but signifying nothing for the masses. But nationalist historians, who worshipped at the altar of tedious archival research, often backed by assiduously collected oral traditions, could ignore the dependency writers for their 'empty theorizing'. Not so with the imperialist writers who rested their case on similar esoteric sources.

In the years immediately after independence, it was commonly held wisdom in many circles that decolonization largely came about as a result of nationalist pressures, whatever one thought of the content of independence. To be sure, nationalist narratives of African insurgency and European surrender were too heroic and oversimplified, for decolonization was a complex process in which European power and African agency confronted each other and made compromises, whose content, of course, varied from country to country, depending on the strength and composition of the nationalist forces, and the policies and interests of the imperial power, not to mention the constellation of regional and global forces. The imperialist narrative veers to the other extreme; it tells the story from the citadels of empire, rather than the trenches of colonial opposition. It is a revisionist celebration of metropolitan policies and plans, of imperial calculations, not concessions, foresight rather than fright[2].

So it is said that the remarkable speed of decolonization, or the 'transfer of power', to use their preferred term, can be attributed either to events in India, the USA, or the prescience of imperial officials sitting in the stuffy chambers of the Colonial Office. It was, according to Low (1982:3, 28), 'the epic struggle of the Indian National Congress' which loosened 'imperial grips in tropical Africa'.[3] Louis and Robinson find their explanation in America's 'historic tradition of anti-colonialism'. It was 'dependence on the US since 1941', they assert, which 'profoundly influenced the official mind of British imperialism' to begin making plans for decolonization (1982:47; Louis 1977). By 1947 the plans were in place and Andrew Cohen of the Colonial Office had emerged as 'a master planner in the style of a Platonic Philosopher King', whose 'constitution-mongering (finally) awoke a slumbering genius of nationalism in West Africa' (Robinson 1980:60, 66). In short, 'whatever persuaded the empire in 1947 to plan its demise in tropical Africa, it was not the fear of black freedom fighters. It was not the black, but the white freedom fighters in Kenya, Rhodesia, in England and the United States that were jolting their assurance in the years 1941-1947' (Robinson 1980:52-3). In the case of France, it is claimed in the standard textbooks, 'especially those intended for Africa ... that decolonization was a plan long pursued by General de Gaulle following the Brazzaville conference in 1944, and which triumphed when he took power for the second time in 1958, despite setbacks caused by the ability of public opinion to understand what was going on' (Suret-Canale 1988:193)[4].

The Second World War is the critical watershed in these narratives. According to Hargreaves (1982:132), it was during the war that the transfer of power to 'African hands, formerly a vague aspiration for an indefinite future was specifically envisaged as the culmination of comprehensive programmes of social engineering designed to reconstruct African societies to accord with the ideas and interests of a changing British Commonwealth'[5]. Perhaps they should not have been so hasty and hopeful for, in Pratt's (1982:259) words, they should have anticipated 'the political fragility that would be so prominent a feature of independent states of Africa'. Fieldhouse (1982:512) put it more bluntly: 'the eventual transfer of power in colonial Africa by marked contrast with that in South and South-east Asia came before the indigenous people had the experience or training necessary if they were to meet the needs of autonomous nation-states'. In short, the decolonization plans were implemented 'prematurely'.

The message was too loud. Dennis Austin (1980, 1982) and Tony Smith (1982) tried to tone it down, but without changing its substance. To assume that the 'transfer of power' in India opened the way to a total abandonment of empire is, Austin (1980:10) proclaimed, 'to interpret history in reverse'. Smith (1982:89) added: 'What is lacking in these accounts is a sense of the conflicts, hesitations, and uncertainties of the past, and of the attempts to reinterpret or renege on the promise of eventual independence for India'. Yet, Smith (1982:115) could, in the same breath, argue that 'it was largely because Kenya was so unimportant that the British could arrange for the sale of the European farms to the Africans at full value and so created virtually overnight, an export elite on whom they could base their post-independence relations'. With 50,000 British troops sent to suppress Mau Mau, one wonders how many more would have been sent if Kenya had been more 'important'. And being a good Englishman, Austin (1980:30) could not resist the temptation to praise British 'pragmatism', in contrast to French 'illusions', during decolonization, arising out of the fact that 'the British have always been an exclusive race', for whom 'the empire was always kept therefore at arm's length'.[6] In other words, the British could not have cared less about decolonization, just as three quarters of a century earlier they apparently acquired their African empire in 'a fit of absence of mind'.

The congruence between the two imperial narratives on the partition and decolonization is quite remarkable. Both emphasize 'politics' at the expense of 'economics', and the workings of the inscrutable 'official mind', rather than the operation of broad historical forces. When the 'economic factor' is broached, it is to ridicule and repudiate it. The story is that after the Second World War, Britain and France were apparently more intent on domestic economic reconstruction than in looking after their colonial 'slums' and 'cinderellas', so that they were only too glad to relinquish them. Indeed, in the case of Britain, it was a mark of its concern and magnanimity that before throwing these territories into the uncharted stormy waters of independence, it sought to make them more viable regional federations, in addition to embarking on a comprehensive programme of colonial welfare and development[7]. While France split each of its West and Central African federations into separate territories, the promise was that any possible balkanization would be offset by the interdependence of *la Francophonie*. This ignored the fact that after the war both Britain and France actually embarked on the 'second colonial occupation', that they became more, not less, dependent on colonial resources for their economic reconstruction, notwithstanding the discourse of 'colonial development and welfare'[8]. In its more dependency-oriented version, the thesis states that the imperial states chose to end formal colonial rule in the fifties and sixties because 'for the first time [they] felt confident that the European economic stake in Africa would be safe without a continued political presence' (Fieldhouse 1982:514)[9]. It was an effortless transition from 'formal' to 'informal' empire, just as a century earlier there had been a transition from 'informal' to 'formal' empire (Wright 1976).

For students of history this was a familiar thesis, echoing Gallagher and Robinson's (1953; Robinson and Gallagher 1961) argument that the colonization of Africa at the end of the nineteenth century signalled a transition from 'informal' to 'formal' empire because collaborative arrangements in Africa had broken down, which threatened the security of Britain's sea route to India. Thus, Africa was not conquered for sordid economic gain, but on strategic grounds. In short, the continent was a gigantic footnote to India. Robinson and Gallagher rested their case on that most reified of historical actors, the 'official mind'[10]. This was the imperialist historians' response to the naughty nationalist and especially Marxist interpretations of imperialism. It was a logical culmination in the long tradition of liberal thought on imperialism started by Hobson, whose 1902 book, *Imperialism,* set off the debate about the 'new imperialism'. He underscored the importance of economic forces behind the 'new imperialism' but he argued that nothing in the logic of capitalism demanded imperialism. On the contrary, imperialism benefited only certain trades and classes, especially the financiers, but colonies were virtually worthless for Britain as a nation. In fact, imperialism, which was generated by surplus capital accumulated from the underconsumption of the workers, positively threatened to subvert popular democracy at home, encourage militarism, and further the pauperization of the working class masses, and through colonial wars lead to the 'degradation of Western States and a possible debacle of Western Civilization' (Hobson 1972:138).

It was a short trip from Hobson to Schumpeter's doorstep. Once underconsumption, the linchpin of Hobson's thesis could be disproved empirically, imperialism could be absolved of scandalous economic charges. Schumpeter (1951) revealed that there was nothing new about the 'new imperialism'. Imperialism was as old as human society, a product of those irrepressible instincts of fear and national pride, the desire for conquest and domination. The so-called 'new imperialism' was simply a resurgence of these atavistic instincts, not the product of modern capitalism which, if anything, was anti-imperialist, for it thrived best with peace and free trade. The revival of these atavisms was made possible by a peculiar and an 'unnatural alliance' between a declining but still powerful 'war-oriented' nobility and a rising, but not yet dominant, bourgeoisie. Predictably, this 'unnatural' alliance was located in the newly industrializing countries of Central Europe, not the already industrialized countries of Western Europe. Thus at a stroke, Britain, France, Belgium, Holland, and the USA, among the leading imperial powers of the late nineteenth century, were exonerated of imperialism. In fact, Schumpeter confidently predicted that the USA would exhibit the weakest imperialist tendencies because it had the weakest pre-capitalist structures. The path was cleared for his warm embrace by some so-called American 'liberal-leftists' (Bell 1979:Chapter 6), and the seeds were sown for the thesis that the United States would be in the forefront of the decolonization drama (Louis 1977).

The argument, then, came to be that while there were no economic motives and certainly no economic gains to be made, the colonies were acquired to divert attention from the social crisis in Europe, mostly in Central Europe, and especially in Germany, arising out of rapid industrialization and the consequent social dislocations (Wehler 1969). Others argued that the masses in Europe were gripped by nationalistic fervour and were hungry for national prestige and glory. It was the masses, therefore, who forced their 'reluctant' governments to embark on the economically costly road to colonial conquest and overlordship (Hayes 1941). These 'reluctant' governments also happened to stumble into Africa because it provided a diversionary diplomatic chessboard for a Europe suddenly run out of political space following the unification of Italy and Germany, and France's defeat at the hands of the latter in 1871. Italy and Germany wanted their places in the sun as their rites of passage to nationhood and greatness. France, smarting under the humiliation of defeat, wanted to redeem her prestige in Africa as well. Bismarck apparently encouraged France in order to divert its attention from Alsace-Loraine and embroil it in conflicts with Italy, and reorient the whole European alliance system in Germany's favour (Taylor 1938; Mansergh 1949). Thus it was Bismarck, an operator in the grand Machiavellian tradition, who was behind the partition of Africa.

Imperialism is reduced to the machinations of great individuals, into a narrative of the 'official mind', just as Andrew Cohen in Britain and de Gaulle in France dominated analyses of decolonization. The African freedom fighters of the late nineteenth and mid-twentieth centuries only appeared occasionally in the background.

The British 'official mind' was, of course, quite 'reserved' and uninterested in an African empire, Robinson and Gallagher (1961) proclaimed in their dense text. As revealed in their diaries, British leaders were preoccupied with more lofty matters than commerce and the imaginary economic assets of Africa[11]. If only those misguided Egyptian and Boer nationalists had not risen in revolt, threatening Her Majesty's sea route to India, then Britain would not have occupied Egypt and the Transvaal, thus raising the ire of France and the pro-Boer sentimentality of Germany, both of whom sought redress in Western and Eastern Africa respectively, thereby unwittingly plunging Africa into colonialism. Thus, it was crisis in the 'periphery' that brought about the partition. Perhaps against its better judgement, and certainly against its best interests, Europe was lured and forced into a continent that was in a Hobbesian state of nature, where life was truly nasty, brutish, and short. It follows that since industrializing Europe was reluctant to get into Africa in the first place, devastated post-war Europe, preoccupied with its own reconstruction, was only too happy to get out. After all, it had tutored the 'natives' in the arts of 'good government', laid the material and institutional infrastructures of 'civilization', now called 'development' as a concession to the sensitivities of the colonial and ex-colonial 'natives'.

The European powers, therefore, did not plan to colonize Africa, but they planned to get out . They were pulled in by Arab and Boer nationalists, but they were not pushed out by African nationalists. In good faith they arrived, and with pride they left. The thesis of 'uneconomic imperialism' denies that Africa's economic resources enticed Europe to come, while that of 'planned decolonization' dismisses the role of African nationalism in forcing Europe to depart. Both seek to mask the true face of imperialism. Postulating that African nationalism had little to do with decolonization is a repudiation of African agency and a reaffirmation of European authority. It represents an attempt to delegitimize the post-colonial state, demobilize popular politics, and derail the search for alternative social systems, for the tradition of social and political struggles in the inter-war and post-war years, which is what 'nationalism' during this period was really all about, is disavowed in an effort to clear discursive space for neo-classical developmentalism. Hargreaves states categorically that the thesis of 'planned decolonization' was inspired by the 'failure' of post-colonial Africa:

It may be thought that this interpretation devalues the importance of those African initiatives which at one time occupied the centre stage of contemporary historiography. Yet the most effective critics of racial injustice were always Africans who drew from their reading of European history a Mazzinian faith in the capacity of the independent nation-state to promote material progress and cultural renewal; in the euphoric 1950s the political parties they founded seemed natural heirs to colonial authority... Those who followed Nkrumah's advice to seek first the political kingdom did not find all else added to them. That 'Third World' which idealists once hailed as a new source of creative energy has become a heavy burden on western consciences... (Hargreaves 1988:3).

That is all that independence, aroused by European ideas, not the realities of colonial rule and African struggles against them, has come to: *'a heavy burden on western consciences'*. The West, of course, has nothing to do with it, neither through the legacies of colonial rule, nor the contemporary international division of labour. It is an *internal* African crisis. The functionaries of the World Bank and the IMF could not put it better.

It is quite evident, therefore, that historians' debates, once stripped of their disciplinary fetish for 'facts', are really about the present and they reflect and reproduce contemporary discourses. It is not surprising, then, that latching on to the discourse of African post-independence 'crisis' and 'failure', a revitalized, imperialist historiography seeks to disrobe nationalism of its glories. By saying that Europe deliberately engineered the demise of its African empires and rebuking it for having done so, Europe, and the West more generally, is given the historical initiative for managing the two most crucial moments in African history this century, first, the 1940s and 1950s when the colonial 'civilizing mission' collapsed, and second, the 1980s and 1990s when the post-colonial developmentalist project crumbled, both due to deepening internal contradictions and mounting resistance by African social movements.

A Dying Post-colony

The language and conclusions of the imperialist historians are echoed by political scientists. The writings of Crawford Young, one of the doyens, amply bear this out. Contemporary Africa's 'profound and dispiriting crisis', he writes, 'is, in part a crisis of the state itself... reflected in its problematic relationship with civil society, its propensity to over-consumption, and its inability to effectively organize the quest for development'. He suggests that 'the roots of the crisis may lie in the nature of the colonial state legacy' (Young 1985:1)[12]. The unusual disjuncture between the state and civil society and the rampant authoritarianism in Africa arises out of the manner in which the colonial state was created. The scramble in Africa was far more concentrated, intense, and competitive than in the other regions. Moreover, the colonial state-building venture in Africa included a far more comprehensive cultural project than was the rule in Asia or the so-called Middle East. Finally, colonial expansion in Africa occurred when the European states were fully developed and consolidated and, therefore, less likely to experiment with indigenous political structures. In other words, the problem of hegemony, security, autonomy, legitimation, and revenue, the five reasons of state, were more pressing and required the constant application of coercion, certainly during the first phase of colonial state construction, when the doctrine of state was also primarily directed toward the metropolitan and external audiences. Although this was no longer exclusively so during the second phase of consolidation, during which the ideologies of colonial development, good government, and trusteeship were articulated, the stirrings of civil society hardly influenced the 'official mind'.

It would not be until the final phase of decolonization after the war, that attempts were made to foster 'a constitutionalized state-civil society relationship, mediated by open political competition, [which] served as legitimating myth for the power transfer process itself' (Young 1985:58-9). The official classes in the metropole had planned decolonization because they were confident that the 'native elites' they had trained in the mysteries of 'good government' would proceed to maintain stability and sustain democracy. Before long, however, 'the ephemeral nature of the graft cuttings of parliamentary democracy upon the robust trunk of colonial autocracy' became all too apparent (Young 1985:59). The nationalist parties degenerated into intransigent political monopolies of exclusive oligarchies, coups became an institutional mechanism for succession and populist ideologies could not camouflage tendencies towards the personalization of power and the patrimonialization of the state. Why? What had gone wrong?

Hegemony and legitimacy eluded the new rulers essentially because the post-colonial state emerged from the fact that the colonial state had succeeded 'in organizing its own metamorphosis' (Young 1985:58). Indeed, it was unfortunate that the decolonization period was too short to consolidate and institutionalize the 'constitutionalized state-civil society relationship'. Or as Richard Joseph put it (1985:23, 24): 'the third of the three stages was a mere parenthesis, a pause, in the process of state formation and articulation. Following its own logic, once the metamorphosis was successfully engineered, this state-in-formation was returned to the process of construction and institutionalization, with subsequent parentheses no longer requiring a time-specific term of 'decolonization', but the more generic one of democratization'. Imperialist historians would simply say that decolonization came too soon, before the 'natives' were ready. And where will the new democratic political kingdom come from? 'Part of the answer', Young submits, 'probably will lie in a reconceptualization of the state'. How? 'Possibly out of the mood of anxiety and foreboding will emerge a formula for the decolonization of the state' (Young 1985:64). *Mood*, not *social movements*. This was in 1985. Africanist scholarship had yet to discover African civil society.

Many resorted to prayer: if only the 'managerial capacity of the leaders' could be improved (Kaba 1985:22), or more African leaders could adopt the 'statesmanship exemplified by Senghor's voluntary retirement and Nyerere's announcement to the same end'; yes, if African leaders could be convinced of the need to 'end terror as an instrument of government, and the democratization of government and political structures', then Africa might just be saved from its crippling crisis of governance and subsistence (Selassie 1985:37). It was a crisis of leadership, a failure of will and reason, a moral challenge to the world. Africa and its endless crises was becoming a burden on western consciences, but the West could not shrug its shoulders and watch a whole continent sink into oblivion because of its ruthless and rapacious rulers and the unfortunate legacies of decolonization plans implemented in good conscience. For a start, the World Bank and the IMF could help them put their economies in order. After all, we all inhabit the same nice little globe.

These discourses represented, and still do, discursive attempts to arrest and appropriate the profound struggles and transformations taking place in Africa. They articulated the ethos of a resurgent free-market developmentalism, which required subservient states and placid populations. In Africanist discourse, this has been expressed in growing contempt for the administrative capacity of the state, and concern for the associational civility of society. Gone are the days when the post-colonial state was seen favourably as the 'factor' of cohesion and the 'motor' of development. This was in the 1960s when nation-building and modernization were all the rage in political science. The support for state intervention in the development process was derived from many sources. First was the powerful heritage of the interventionist colonial state and its belated, but self-conscious, developmentalist mission, which resonated with the statist and developmentalist ideologies of the nationalist movements themselves, a reflection, in large measure, of the underdeveloped nature of the indigenous capitalist classes. Second, the Keynesians were still influential in economic policy circles; development economic theory called for 'strong' interventionist states; and development economists had not yet abandoned reading history, which showed the crucial role played by the state among all 'late industrializers'. Third, the Soviet model of planned industrialization looked attractive. Finally, the mood of the times, reflecting no doubt the long post-war boom, was optimistic, and the new states tended to be viewed in a more charitable, if often uncritical and paternalistic, manner (Mkandawire 1995:16-19).

As these conditions changed in the 1970s, following the end of the boom and the onset of the global recession, the overthrow of the Keynesians by Friedman's 'Chicago Boys', and as evidence mounted that all was not well in the Soviet socialist paradise, the post-colonial state fell from grace and came to be seen no longer as the locomotive of modernization, but as a millstone around the neck of development, which, it was now proclaimed, could only come through the unfettered operation of the market. Claiming to be the high priests of this infallible god, development economists stopped reading about the societies they sought to catapult into modernity, and they zealously preached and imposed the commandments of neo-classical developmentalism. In the meantime, the post-colonial state had lost friends on the left, whose denouncements of it as 'overdeveloped', 'compradorial', 'clientelist', 'unsteady' sealed its fate (Alavi 1972; Saul 1974, 1976; Lamb 1975; Leys 1976; Langdon 1976, 1977; von Freyhold 1977; Ziemann and Lanzendorfer 1977). The African state came to be viewed in Mkandawire's (1995:18) memorable metaphor, as 'a veritable bull in a Chinese shop, distorting markets, creating monopolies, blunting incentives and generally being a bane on society'.

Epithets, anecdotes, and caricature replaced sober analysis. The race was on to coin the most denigrating labels. Some viewed the African state as impotent: Callaghy (1987) called it a 'lame leviathan'; Rothchild (1987) preferred the term 'soft'; Young and Turner (1985) saw it as 'decadent'; Bates (1983) believed it was economically 'irrational'; Graf (1988) said it was 'incomplete'; Hyden (1980, 1983) thought it was 'weak' and 'suspended above society'; Chazan (1988) found it 'omnipresent' but hardly 'omnipotent'; and Shaw (1982) described it as 'anarchic'. Others, and sometimes the same people, condemned the state for being too strong and authoritarian, for its 'patrimonialism' (Sandbrook 1985), 'prebendalism' (Joseph 1984, 1987), and 'corporatism' (Nyang'oro and Shaw 1989). Then there were those who tried to sketch out a universal

Africanist theory of 'state collapse' (Zartman 1995), or reduce African politics to grotesque metaphors. And so Bayart (1993) talked of 'politics of the belly', and Mbembe (1992a, 1992b) called attention to 'the banality and vulgarity of power', a depraved world in which politics centres around an obscene 'obsession with orifices, odors and genital organs' (1992a:6). This discourse, we are told, is used by the state to erect and ratify power, and by the abused population to distance and domesticate the power of the 'hollow', 'unreal', 'lecherous', 'ceremonial' state. These vivid descriptions of the vulgarity of power in the post-colony, of Africa as grotesque and Africanity as pathology, themselves reek of intellectual vulgarity. With friends like these, Owomoyela (1994) rightly wonders, who needs enemies.

By the early 1990s, therefore, Africanist scholarship, as can be seen from analyses of the state, had returned to some of the worst excesses of the imperialist and Eurocentric mindset. Mamdani (1995) has correctly argued that the 'new political sociology', as Eyoh (1995) calls it, was characterized by a unilinear evolutionist perspective, so that the African state was often interpreted as representing an earlier moment of the European state, hence descriptions of it as 'absolutist', 'Bonapartist', and the pervasive tendency to see it as a deviation from some fictitious European normalcy. Also, its methodological thrust was anti-democratic in that the fixation with state capacity and efficiency betrays a conceptualization of the state

> as an exclusively institutional category, with its coherence, logic and capacity; it is not seen as simultaneously a condensation of social relations and a relatively autonomous arena which in turn shapes these same relations. The state interest is seen as purely the interest of its managers; it is not put in the context of a wider galaxy, struggles within 'civil society'. (Mamdani 1995:609).

By focusing on state formation to the exclusion of other political processes, especially the history and complexities of popular struggles and the internal and external realities of African political economies, the state is abstracted either as embodying the will of its managers, or failing to realize some universal Hegelian idea of state behaviour. Those who have recently discovered African 'civil society', often do no better than the 'state-centrists' in capturing the complex interpenetrations of state-civil society relations, and they betray their instrumentalist and modernist preoccupations by emphasizing the associations of 'urban' civil society and ignoring, for example, the complex articulations of ethnicity, which is often seen as a reservoir of explosive and anti-developmentalist cultural particularisms and fundamentalisms.

This 'new' Africanist political sociology plays a discursive space clearing function for the resurgent 'free-market' capitalism. It is not coincidental that it assumed hegemony in the 1980s, Africa's 'lost decade', when one country after another succumbed to the dictates of the international financial institutions to undertake 'free-market' economic reforms. The narrative of African state failure absolved the international economic system and the international financial institutions of responsibility for generating the structural conditions and contexts behind Africa's economic crisis and for devising many of the policies that the African states were now solely being blamed for. The invention of a fake genealogy of African developmentalism was launched by the World Bank (1981) itself in its famous Berg Report, which put the African state in the dock for causing the crisis almost single-handedly and with apparent premeditation. Africanists generally endorsed the sentence and proceeded to eagerly chronicle the crimes of failure and stagnation in their beloved states[13]. The neo-classical interpretation of the 'African Crisis' evidence won, not because it 'was intellectually superior, but because the international financial institutions and the monetarist governments that came to power in the main industrial countries saw a radical reassertion of market principles as an essential means of protecting the value of their accumulated assets' in the face of the global economic crisis. 'Accordingly, they used their power over research funding, over publications and especially over credit to propel their interpretation of the facts to a dominant position' (Bienefeld 1988:70).

Intellectual fatalism, in the form of Afropessimism, reigned supreme. Shaw (1993:140) predicted that come the twenty-first century, Africa will assume a new importance, no longer as an exporter of precious minerals and cash crops, but rather of 'AIDS, drugs, migration or pollution'. Fundamentally, Afropessimism represented a discursive closure against conceptions of different futures for Africa outside of the prescriptions of the hegemonic 'neo-liberal' developmentalism. It encouraged, as Eyoh (1995:9) has put it, broad support for the structural adjustment programmes of the international financial institutions 'as a means of disciplining Africa's malevolent ruling classes'. The agenda was more comprehensive: it was to tame Africa's re-energized social movements, revitalized in part by the effects of those very structural adjustment programmes.

The Wretched of the Earth

As with imperialist historiography, by the 1980s development theory was returning to its roots. As a discursive formation, developmentalism emerged after the Second War following the discovery and problematization of poverty and backwardness in Africa, Asia and Latin America. These countries were crystallized into the 'Third World', a concept which quickly lost its original and more honourable reference to geopolitical non-alignment. The conjunctures of mass nationalism in Africa and Asia, and revolutionary pressures in Latin America, as well as the expansion of the socialist bloc, gave developmentalism a combative ideological flavour from the very beginning. Representations of the Third World, and the proposed development prescriptions, privileged and universalized Western knowledge and idealized models. Development was conceived of as an economic problem amenable to value-free technical and technological solutions. This enabled the professionalization and institutionalization of development practice, and the discursive repudiation of the broader social, cultural, and political contexts and consequences of economic transformation. The development experts were born. They would descend upon the impoverished and homogenized 'Third World' with the unshakable convictions of western superiority, the knowing smiles of modernity, and the reformist zeal of an earlier generation of missionaries. Their briefcases would be full of stylized facts about the operations of labour and financial markets, the appropriate role of the state, and the benefits of comparative advantage (Mkandawire 1996). And, of course, they would bring along pre-cooked packages of solutions for whatever 'crisis' of underdevelopment was the flavour of the month.

Development economics, a child of neo-classical economics, inherited its parent's veneration for the 'invisible hand' of the market, belief in the two commandments of perfect competition and perfect rationality, indifference to the classical concerns of growth and distribution, and fondness for dualities, for which the 'Third World' proved quite fertile: 'modern-traditional' societies, 'market-subsistence' economies, 'formal-informal' sectors. Needless to say, this discourse underwent structural changes in subsequent decades, 'but the architecture of the discursive formation laid down in the period 1945-1955 has remained unchanged, allowing the discourse to adapt to new conditions' (Escobar 1995:42).

The global ambitions of bourgeois development economics did not, of course, go uncontested, both in theory and practice. Dependency and neo-Marxist theories and the African socialisms were among those challenges, and so were the various 'actually existing socialisms' of Central and Eastern Europe, Asia, and the Caribbean. It was also vulnerable to occasional fits of self-doubt. At the turn of the 1970s all these forces seemed to coalesce and conspire against neo-classical developmentalism. There were concerns about the availability and interaction of global physical resources. Could global 'modernization' be sustained in the face of finite resources? Was global development to be equalized at American levels of consumption, or stifled at a standard between a Portugal and an India? These fears were partly fuelled by growing environmental consciousness and insistent demands from the restive 'Third World' for a New International Economic Order (NIEO), which received tremendous material and ideological impetus when OPEC raised oil prices in a rare instance of concerted political defiance by the 'Third World'. These developments coincided with the end of the long post-war boom in the West. A period of prolonged and deepening crisis had set in for the global capitalist system. Suddenly, the future dimmed. Liberalism was in crisis.

Doomsday scenarios were painted for the world, and the entourage of epistemological 'posts' began to appear or gained ascendancy: post-structuralism, post-modernism, post-industrialism. Malthusian fears of overpopulation were whipped up. World futures theories proliferated. Despite varying models, techniques of extrapolation and simulation, and doses of pessimism, it was shown that under the weight of 'uncontrolled' population growth, accompanied as it was by limits to agricultural growth, depletion of mineral resources, dropping world water table levels, and spreading ecological and psychic pollution, the 'world system' would buckle and collapse. Unless, of course, every regional component displayed 'responsibility' to maintain 'equilibrium' and ensure the survival of the whole. Thus, it might be more prudent for the Third World, apart from considerations of its comparative advantages, to forsake industrialization with its dangers of resource depletion and global pollution. Redistribution, not more growth, at least not at the global level, was the answer to the North-South dichotomy (Hoogvelt 1982)[14].

The notion of redistribution offered western developmentalism a possible outlet from the ideological and conjunctural crisis in which it found itself by the mid-1970s. It not only appropriated some of the 'Third World' discourse about a New International Economic Order, it also sought to sugar-coat the structures of the international division of labour. A stream of articulate and morally appealing reports appeared towards the close of the 1970s spelling out the international reforms and institutional changes necessary to create a prosperous, harmonious, and peaceful 'world community'. Multi-nationals were called upon to change heart and adopt codes of conduct which would synchronize their profit motives with the development interests of the 'Third World' countries. The Superpowers were exhorted to abandon their murderous and obscenely expensive arms race and divert their resources to peaceful ends. 'Third World' states did not escape the admonitions either. They were asked to abandon autarchist tendencies, control their population growth, and take the question of domestic redistribution seriously (Brandt 1980).

The World Bank, in one of its many rhetorical reincarnations, embraced the new concept and overnight became a champion of the poor, insisting on targeting international 'aid' to meet their 'basic needs' (Streeten 1977; Green 1978). This became an ideological conduit to justify its supervision and management of African and other 'Third World' economies. Thus, while the right of African countries to own and control their natural resources and determine their own economic policies was still formally recognized, the basic needs approach queried the right of these states to act as the ultimate arbiters of their peoples affairs. In short, it allowed for the further internationalization of poverty and its management by western developmentalist institutions. By the turn of the 1980s the 'Third World' revolt had been contained as symbolized by the tumbling oil prices and many of the countries' growing indebtedness to western banks. Then Margaret Thatcher was elected in Britain, followed by Ronald Reagan in the United States, Helmut Kohl in Germany, and Brian Mulroney in Canada. The era of the Right had arrived. The World Bank abandoned 'basic needs' and adopted 'structural adjustment'. Its infatuation with the poor was over.

It was back to the neo-classics of developmentalism with a vengeance, perhaps in retribution for all that insolence from the 'Third World' in the 1970s, those dangerous illusions that an alternative development path could be found. The case had to be restated, boldly, that 'Third World' economies and peoples were wards of the world capitalist system. The collapse of the Soviet Union finally freed the western countries from making concessions to the political sensitivities and economic aspirations of the 'Third World', and stripped the term of any lingering meaning of opposition to western hegemony. As if to signal the end of this history, the 'Third World' was rechristened 'post-colonialism', 'post-coloniality', 'post-colony' in rebarbative 'new' cultural and literary theories, a discursive resumption of its colonial subjectivity. Historians joined in drawing the new cartographies of knowledge and power. In line with their calling, they sought to trace back the African crisis to its roots, to show that poverty in Africa was an aboriginal condition, a structural, primordial presence, having very little to do with external forces, certainly not those emanating from Europe. So said Illife (1987) in his monumental volume on the African poor.

Illife set himself to challenge both the nationalist and dependency approaches contending that the former ignored poverty in their search for great states in pre-colonial Africa and celebration of the activities of the nationalist heroes during the colonial period, while the latter subsumed it under the frozen dialectic of the development of underdevelopment. His central thesis was that structural poverty in Africa, as he called it, had not changed from the pre-colonial to the colonial and post-colonial periods. In other words, Africans have always been poor throughout their history. If that is so, then poverty in contemporary Africa could not, and should not, be blamed on colonialism and capitalist exploitation as the 'radicals' maintained, nor, indeed, could it be attributed to the weather or population growth, or to the incompetence of African governments, as the 'conservatives' believed. No, poverty in Africa was a primordial condition, oblivious to the movement of time and the organization of space. It is this omnipresence of poverty that makes Africa unique and fascinating. He enthuses:

Africa's splendor lies in its suffering. The heroism of African history is to be found not in the deeds of kings but in the struggles of ordinary people against the forces of nature and the cruelty of men. Likewise, the most noble European activities in Africa have been those — often now forgotten — who have cared for the sick and starving and homeless (Illife 1987:1).

Illife's history of the African poor is a story of indigent Africans and charitable Europeans.

This portrayal bears uncanny resemblances to imperialist historiography, in which Africa was seen as a 'dark' continent stuck in a moribund state of nature, populated by a primitive race wallowed in depravity, cruelty, and poverty. In contrast, European adventurers in Africa were portrayed as harbingers of progress, and colonialism was celebrated as a civilizing mission. Illife tells us that there was a scarcity of formal institutions dealing with poverty in pre-colonial Africa. Anxious not to be overwhelmed by the African poor, the first European visitors introduced 'poor relief institutions from their own countries, so that sub-Saharan Africa — itself so lacking in formal institutions — now experienced early modern Europe's diverse approaches to poverty' (Illife 1987:95). To eulogize Portuguese and Dutch relief to the poor, as Illife does in Chapter 7, when the former were busy plundering Africa and carting away shiploads of slaves and the latter were decimating the Khoisan and other Africans through commando raids and disease epidemics, is the height of historical perfidy. The missionaries come from Illife's deodorized story as paragons of virtue, humanism, and boundless charity. Forgotten are their seedy activities as employers of forced labour, kidnappers of children, grabbers of land, racist ideologues, and accomplices of the colonial conquerors.

There are other affinities between Illife's book and imperialist historiography. Like the imperial historians who were anxious to divorce North Africa, and especially Egypt, from the rest of Africa lest the continent exhibited some light of civilization, Illife's Africa excludes North Africa. Also, Europe is used as the yardstick against which to judge Africa's alleged inadequacies and abnormalities. The alleged scarcity of formal charitable institutions is a case in point. Such institutions are defined from the context of idealized European history. Their scarcity in Africa leads him to make the untenable conclusion that care for the poor in pre-colonial Africa was poorly developed in comparison to Europe. This history by analogy has a long tradition in Eurocentric scholarship. A more careful historian would have sought to analyze systematically institutions and networks of poverty alleviation in the pre-colonial era, as has indeed been done by some historians with reference to the problem of food shortage and hunger (Zeleza, 1993:Chapter 1).

Then there is the question of sources. One of the methodological strengths of nationalist historiography lay in its use of other sources apart from, and in addition to, the written sources, the most well-known being oral tradition. Illife makes no use of oral sources, whose impressions of poverty, he avers, can be misleading. His study has also not used, he says, 'unpublished sources surviving in Africa. It rests largely on published sources and certain documents available in Europe' (Illife 1987:3). The problem of sources cannot be underestimated. But that does not call for an excessive reliance on European adventurers for descriptions of poverty in Africa. Many of these descriptions, moreover, are taken from the travelogues of nineteenth century wanderers, most of whom were unrepentant imperialists and racists, for whom 'savage' and 'backward' Africa was only good for European colonization[15]. In relying on such sources, Illife failed to decipher the differentiated, changing, and socially constructed meanings of poverty. But that was, of curse, not his intention. He was content to throw in one or two African words from a single African society and make generalizations about the nature of poverty across time and space for the entire continent.

Convinced that the case for African poverty was self-evident, he did not bother to define it. Poverty was in fact conflated with disease, especially leprosy, notwithstanding his own concession that 'modern' understanding of this disease only dates to the late nineteenth century, and its nature is still not fully understood. Illife's thesis rests on a distinction which is made between *structural* and *conjunctural* poverty. Structural poverty refers to 'the long-term poverty of individuals due to their personal or social circumstances, and conjunctural poverty ... is the temporary poverty into which the ordinary self-sufficient may be thrown by crisis' (Illife 1987:4). The sharp dichotomy between the two is contrived: conjunctural poverty, to use his terminology, often has structural underpinnings, while structural poverty is likely to be the accumulation of conjunctural poverty. In short, poverty as an integral part of the social process cannot be broken into neat binary categories. Thus, to argue that structural poverty in Africa has remained the same throughout the centuries is tantamount to saying that the social and economic process has remained unchanged, which is, of course, patent nonsense.

Illife also contends that there is a distinction 'between the structural poverty of societies with relatively ample resources, especially land, and that characteristic of societies where such resources are scarce. In land-rich societies the very poor are characterized by those who lack access to labor needed to exploit the land both their own ... and the labor of others' (Illife, 1987:4). Unfortunately he does not see it fit to discuss the structures and processes of labour organization, or the land tenure systems which were extraordinarily complex and diverse, in pre-colonial Africa. We may ask: if structural poverty in Africa in the 1980s was still what it was centuries back, does that mean Africa is still as land-rich now as it was then? The absurdity of such a proposition is only too obvious. Illife's claim that the nature and causes of poverty in pre-colonial Africa, and, indeed, in most of Africa to this day, 'had little to do with technology, landownership, intensive agriculture, or even in a direct sense the pattern of social stratification, although these did affect the behaviour of the poor' (Illife 1987:4) sounds like a miserable joke. If none of these things had nothing to do with poverty, then what did? This is not serious history.

Predictably, the story begins in Ethiopia, the scene of so much suffering during the famines of the early 1970s and 1980s. Ethiopia's poverty appears timeless, unaffected by changes in the productive system, social relations, and political institutions, none of which are discussed in any coherent manner. Christian Ethiopia, we are told, placed poverty at the centre of its culture. Yet, there was such little institutionalized care for the poor. The explanation is sought in the bilateral family structure of the Amhara. We are reminded that 'bilateral societies are characteristically individualistic and mobile, both socially and geographically' (Illife 1987:15). This is the language of a sophomore anthropology student, not a seasoned historian. Not surprisingly, the author jumps from the sixteenth to the twentieth centuries freely, unencumbered by the inconveniences of historical time and process. The discussion of colonial poverty is prefaced by a chapter on 'Early European Initiatives' concerning poverty relief. Having sufficiently comforted us with these initiatives the author then introduces us to 'poverty in South Africa, 1886-1948' in the next chapter. Much of the evidence presented actually challenges the argument that structural poverty has not changed in Africa over the centuries. Certainly Africans in colonial and apartheid South Africa lost the privilege of being 'land rich' and their underpaid labour power was abused by racial capitalism. But Illife's cynicism is never far from the surface. 'Ironically, South Africa's racialism', he says, 'probably helped African townsmen to endure their poverty: at least they need not see themselves as failures' (Illife 1987:139). And he concludes the chapter in a similar vein: 'Ironically, during the next forty years the National Party was to elaborate the most extensive welfare system in Africa, a system which like the Apartheid programmes, was borne of urbanization, inequality, state power, and rampant technocracy' (Illife 1987:142). No comparative evidence is presented. One wonders what the 'natives' were fighting and dying for. With only a few minor revisions, apartheid South Africa's propaganda chiefs would take no objection.

His discussion of poverty in 'colonial Africa' in subsequent chapters, as if South Africa was not a colonial formation, continues the analytical charade. Rural conjunctural poverty, he says, changed quite considerably, while structural poverty did not. By conjunctural poverty changing he means that 'the great famines which in the past had periodically decimated populations ceased in the mid-colonial period and were replaced by more subtle problems of nutrition and demography' (Illife 1987:10). Surprisingly, nowhere in the previous chapters has Illife discussed any of the pre-colonial 'great famines'. One would have expected Illife to substantiate this point since there have been many historians who have argued that the scale and intensity of famine in colonial Africa increased over what it had been in the pre-colonial period because colonial capitalism disrupted the commodity composition and the productive and ecological organization of African agriculture, as well as famine relief systems. Feebly, he explains away the widespread hunger and famine of the early colonial period by stating that 'tropical African rainfall as a whole was low and exceptionally erratic during the first half of the colonial period...' (Illife 1987:156). False claims pile on top of each other like rotten fish. It is certainly not true that 'wages were usually quite high during the early colonial period' (Illife 1987:149). This period was dominated by poorly paid forced labour.

Urban poverty does not escape his revisionist imperialist gaze. Colonialism did not create urban poverty, despite all those appearances of unemployment, prostitution, and delinquency. It only made the poor more visible.

The poor of precolonial Africa were bred in the countryside but seen in the town. That is why they were so often overlooked: precolonial Africa had few towns. During the colonial period towns grew quickly. Observers, white or black, noticed more and more poor people and assumed that their numbers were increasing and that towns created them (Illife 1987:164).

This was obviously wrong: the colonial towns, as spaces of European modernity, could not possibly have generated Africa's eternal structural poverty. Towards the end of the book he returns to the question of institutional care for the poor in colonial Africa. Pride of place is given to the role of the missionaries, and the heroes of colonial folklore like Albert Schweitzer whose 'self-sacrifice caught European imaginations and validated European civilization' (Illife 1987:195), despite acknowledging, quite grudgingly, that institutionalized care for the poor was limited, so that they mostly relied on families and neighbours. But 'how far they (the families and neighbors) met these obligations is unknown' (Illife 1987:212).

Independent Africa provokes his indignant wrath. Forgetting his admonitions earlier that Africa's poverty was beyond social or political causes, he attributes the post-independence growth of structural poverty to 'the demographic expansion which had begun between the wars' (Illife 1987:237), and 'policy failures [which] were probably even more important' (Illife 1987:233). External factors are exonerated. Not only did structural poverty grow after independence, conjunctural poverty returned in the form of mass famines. This was caused by two main factors. 'One resulted from warfare and political conflict' (Illife 1987:250). 'The second reason for the return of mass famine was drought' (Illife 1987:252). The World Bank got the historical version of the Berg Report. But Illife did not, of course, think that he was a right-wing ideologue, nor did the liberal friends of Africa who honoured the book with the Herskovits Award in 1988, calling it 'a fascinating and moving study that breaks new ground' (ASA News 1988:22). Indeed, he criticized both the left and the right for misunderstanding the nature of African poverty.

Men of the left commonly misconceived it as a recent phenomenon due to colonial and capitalist exploitation. Men of the right misconceived it as a recent phenomenon due to the weather or population growth or the incompetence of African governments. Few realized how much structural poverty had not changed at all (Illife 1987:259).

Conclusion

Like the proverbial sinner who could not see the logs in his own eyes, Illife could not see that his analysis put him squarely among the 'men of the right'. His entire thesis was, in fact, the final vindication of their contention that Africa's crisis of underdevelopment and poverty was a product of unfavourable internal factors. Illife had carried the argument to its logical conclusion: the internal factors are deeply rooted in Africa's history. Independence brought to an end the brief interlude when the 'conjunctural' poverty of famine was contained, and Africa returned to her isolated and splendid eternal poverty.

The 'men of the right' are, indeed, back with a vengeance. From Fage's ancient Caucasoids, to the fantasies of planned decolonization, the narratives of African state failure and stagnation, and Illife's primordial African poverty, the discursive architecture of neo-classical developmentalism has been bolstered, and the exploitative and oppressive charters of colonial modernity are reconstructed. It is a late twentieth century scramble for the appropriation of African histories past, present, and future. The struggle against imperialist historiography, and other disempowering Africanist discourses indeed continues. One of the crucial sites of the struggle lies in the schools. To that we now turn.

Notes

1. See Fage's (1989, 1993) fascinating and revealing personal accounts of the intellectual evolution of his generation of British Africanists. That Fage was not alone in this endeavor is shown by the fact in 1978, the year his book was published, the theme of the African Studies Association of the United Kingdom's conference was ʿWhites in Africa Past, Present and Future'. The 'whites' considered were mostly the European settlers of colonial Africa, and while there were nostalgic and laudatory presentations, it was 'an occasion for humility rather than a renewed pink pride in expertise', Ranger (1979: 469) assures us. For some of the papers presented at the conference, see the special issue of the *Journal of African History* (1979).

2. The literature on decolonization is quite large. In addition to the works mentioned below, see Kirk-Greene (1979) and Morgan's (1980) massive 5 volume official history of colonial development and welfare, especially volume 5; and Wilson (1994) which tells the story of decolonization from the imperial metropole, see my critical review (Zeleza 1996). Also see the extensive bibliographic reviews by Gardinier (1982, 1988) and Kirk-Greene (1982, 1988). For earlier writings praising the planning and foresight of the Colonial Office and the activities of the colonial governments during decolonization, see Carrington (1961), Robinson (1965), Kirkman (1966), Lee (1967), Perham (1963), and Low (1973).

3. In a subsequent piece, Low (1988) moderated his position. While he noted the crucial lessons that the three major imperial powers, Britain, France, and Belgium had each been learning from Asia, he gave more attention to the role of African nationalism, and argued that, with the sole exception of Britain, of course, the postwar policies of these powers were more deliberate with regards to programs of colonial economic development than decolonization. But even in the case of Britain imperial plans 'were invariably broken by the force of nationalist developments within Africa itself' (Low 1988:71).

4. Suret-Canale (1988:193) himself is in no doubt that this 'is not in accordance with historical fact and therefore has to be rejected'. Panter-Brick (1988: 76) sees French decolonization as 'the combined handiwork of overseas political forces determined to secure independence and those French politicians prepared to brave domestic ideology and vested interests'. Where these politicians derived their convictions and the strength to carry them through is, however, not clear.

5. Also see Hargreaves (1980). In his book-length study of African decolonization, Hargreaves (1988) tries to have his cake and eat it: he concedes that at the end of the Second World War neither Britain nor France envisaged immediate independence for their colonies, ʿor a general lowering of flags', as he ridicules it, but they 'did set out to change political relationships to substitute collaboration for force ... counsel for control'. In fact, he continues, as early as the 1930s when 'the assumed stability of colonial rule became more questionable ... a certain number of persons in both countries conceived programmes of reform and renewal which would eventually lead towards the independence of their African colonies' (Hargreaves 1988:2). He cheerfully notes that 'this interpretation devalues the importance of those African initiatives which at one time occupied the centre stage of contemporary historiography' (Hargreaves 1988: 3).

6. A more illuminating comparative analysis of the decolonization of British and French empires can be found in von Albertini (1971), although he ignores the role of nationalism almost entirely. Also see Morris-Jones and Fisher (1980); Kahler (1984); and the essays in Mazrui (1993:Section II) which pay more attention to the role of African nationalism.

7. See, for example, official justifications for incorporating Nyasaland into the Central African Federation as presented by Clegg (1960), Creighton (1960), and Keatley (1963).

8. See Chapter 11 below for a detailed discussion of the evolution of Britain's postwar Colonial Development and Welfare programs, and the pivotal role they played in the formulation of development discourse.

9. Aware that this interpretation is too close to dependency theory for comfort, Fieldhouse (1988: 142) insists that by the 1950s 'Britain was free to decolonize because, for the moment, she did not depend on colonial economic support'.

10. The dominance of this thesis in African historiography on the partition diverted attention from developing more comprehensive and theoretically sound explanations for European imperialism. The debate is summarized in Louis (1976), and for a cogent critique see Hopkins (1986).

11. Uzoigwe (1974) read the same entries quite differently.

12. For another, perhaps historically richer, conceptualization of the colonial state, see Lonsdale and Berman (1979), and Lonsdale (1981). Berman (1990) presents an extended, if rather structuralist, analysis of the Kenyan colonial state. For an early and influential analysis of the evolution of the South African state, see Davies, et al. (1976).

13. The literature on the 'internal' dimensions of the African crisis has been a veritable academic growth industry, proving that as death is to the mortician, crisis is to morbid academicians. For a recent sample see Sahn (1994) and my review of the book (Zeleza 1995). The issue is discussed at greater length in chapters in the next section.

14. Among the doomsday scenarios, see Meadows (1972), Ward and Dubois (1972), and Heilbroner (1976). For a more detailed discussion of these issues, see Zeleza (forthcoming).

15. The need to deconstruct these texts is necessary because as Elbl (1992: 189) has argued in the case of the narratives of Portuguese trading relations in West Africa in the fifteenth and sixteenth centuries, most of them 'present a triple danger of distortion: the primary sources on which the account is based may have been unreliable, they may have been interpreted out of context, or they may have been used idiosyncratically to suit the author's literary purposes. Renaissance historiography was, after all, a form of literature in which telling a story well, in an erudite fashion, and with the literary embellishments, was at least as important as relating the facts or adopting a critical approach to sources'.

Chapter Eight

THE PRODUCTION OF HISTORICAL KNOWLEDGE FOR SCHOOLS

Introduction

It is more than evident from the previous two chapters that the production of historical knowledge is conditioned by a wide range of social, political, and economic forces in society, as well as by internal methodological and epistemological considerations. This chapter seeks to analyze the production of historical knowledge for schools. It is argued that there is a yawning gap between the knowledge produced by academic historians and that consumed in the schools. School textbooks have yet to adequately incorporate and reflect the methods, approaches, and findings of modern African historiography. The factors behind this disjuncture are traced to the contemporary political and economic conjuncture in Africa, and the contradictions in the process and practices of producing historical knowledge. The chapter begins with brief discussions of the growth of historical writing and the historical profession in general, before focusing specifically on the African trends. The third part examines the historiographical limitations of existing history textbooks used in schools throughout the continent. Finally, the politics involved in the production of historical knowledge for schools is analyzed.

The Production of Historical Knowledge

History means different things to different people in the same society and in different societies, at as specific moment or at different moments in time. It is therefore not easy to provide a definitive answer to the question that has exercised the minds of historians for age: 'What is history?' Of course, there have been no shortage of definitions[1]. But each definition implies a philosophical commitment, an ideological orientation, whether or not it is consciously articulated, for it rests on existing concepts, beliefs, and values derived from complex social practices and relations, shared memories and cultural imaginations. The definitions can be idealist, empiricist, or materialist, among various discursive possibilities. To idealists, the 'idea' is what is real, it is the motive force of history, while experience is merely its representation. Empiricists focus on empirical instances as the essence of history. Materialists believe historical processes are rooted in material reality, from which they abstract theoretical propositions. Post-modernists claim to distrust all metanarratives and celebrate the contingencies, if not the nihilism, of lived human realities and experiences (Cohen 1994; Jenkins 1995).

Many historians would probably agree that history is not simply a representation of the past but a process of reconstruction in which certain aspects of the past are abstracted and are acted upon and lived by people in the present. The making or production and reproduction of historical knowledge is conditioned by the structures, forces, and struggles in society. The material conditions of social life produce, and are simultaneously reproduced by, the political, ideological, and philosophical values which are reflected in the processes and practices of constructing and reconstructing history.

Thus the production and reproduction of historical knowledge is an integral part of the social production of knowledge. It matters a great deal who produces the history, where, when, and why. The questions that historians ask and try to answer, the themes they examine and the methods they use, the sources they rely on and the way they evaluate them, and the manner their findings are organized and disseminated, cannot be divorced from dominant social processes and discourses, whose transformation brings about changes in the practices of history producing and reproducing and in the prevailing perspectives.

Historians, whose job it is to organize historical knowledge, are, therefore, trapped in the station of their class, gender and culture, and the demanding boundaries of location and time, all of which condition their ideological and theoretical inclinations. The dominant historical methods and practices reflect the dominant material relations and hegemonic discourses; they represent the certainties of the social imaginary. The theories that historians use, sometimes unconsciously, in setting up the problem of enquiry, organizing and evaluating the data found, and in their analysis and interpretations are defined by the prevailing political, ideological, and philosophical orientations, as well as by organizational structures of social knowledge production.

There are, of course, many theories or approaches to the study of history which, as might be expected, have changed with time. For example, when history was mainly written by churchmen and religious ideology was hegemonic, providential history predominated. The historical process was seen as the unfolding of the divine will. History, according to St. Augustine, the great African Church father, is inevitably universal and metahistorical in that it envisages an eternal goal beyond the temporal order. The first serious challenge to providential history came in the fourteenth century in the works of Ibn Khaldun, another African and one of the greatest historians of all time. His history was interpretative, critical and profound, for it examined and advanced the fundamentals of historical method and meaning.

The rise of capitalism in Europe from the fifteenth century led to the spread of critical and systematic historical study. New approaches and techniques were developed. Archives and national libraries were established. Many of the historians became rationalists who were anticlerical and antimedieval. They themselves were later opposed by the romanticists, who stressed the supernatural, tradition, and individual differences, against conceptions of reason, science, and the common nature of humanity. It was not until the nineteenth century, in the midst of the European industrial revolution, that the pursuit of 'scientific history', as presently understood, gained currency. Many historians began making systematic criticism of earlier historians, and showed a marked preference for analysis rather than narrative. The technical progress brought by industrialization and the scientific ideas of evolution led to the development of positivist theories of history, according to which progress came to be seen as inevitable as the seasons; an infallible law that applied both to the natural world and human society. Human society was seen as moving unilinearly from a backward and primitive condition to the advanced state of modern civilization. This provided a new method of selecting, organizing, and interpreting 'facts'. European history, and its North American offshoots, assumed the universalistic and objectivist pretensions of modernity and the histories of all other regions and peoples were reduced to transitional narratives, in which the absences of 'European modernity' constituted the central historical problematic (Chakrabarty 1996).

Modernist historians came to believe that it was possible, indeed their duty, to discover 'new facts', determine the causes and sequences of events, and decipher the meaning and unity of history through the use of the scientific method, with its stress on exhaustive research, criticism, and 'objectivity'. They were a cheerful, optimistic lot, some of whom even thought it was only a matter of time before 'ultimate' or 'definitive' history could be written. Modernist history rested on the pillars of institutional professionalization and ideological imperialism. The emphasis on scientific research gradually turned history into a profession confined to the universities. The institutionalization and professionalization of history was reflected in new practices of producing historical knowledge. Perhaps as a reflection of the emerging dominant proprietary relationships the use of bibliographies and footnotes became the canons of historical scholarship, while plagiarism was turned into an intellectual crime. Historical associations were formed, journals established, and the provision of graduate and specialized training for future historians started. Collective production of historical works on large subjects also began as source materials expanded and the scope of history was widened from political and military affairs to cover many other aspects of society. Ideologically, modernist history was unapologetically Eurocentric. It celebrated the singular achievements of the West, invented as an encompassing civilizational space linking Western Europe and its cultural offspring across the Atlantic in North America, and denigrated, indeed denied, the histories of all others where it did not see fit to appropriate and incorporate their contributions into an eternally rising Western Civilization. African history was dismissed, and the histories of diaspora Africans in the Americas were excluded from the inspiring narratives of western modernity. In short, modernist history, like western modernity itself, arose in the womb of western global imperialism.

At the turn of the twentieth century, the unqualified faith in modernist history of the late nineteenth century began to wither as a result of growing doubts within the historical profession itself and wider changes in the world at large. Critics charged that history was becoming increasingly arid, with its celebrated search for facts turning into a pedantic chase for the insignificant. The outbreak of the First World War began stripping western historians of their universalistic and objectivist proclivities as historians from the different countries dressed up historical 'facts' in glittering national colours. The collapse of this historical model, which would continue unfolding throughout the twentieth century, was facilitated and followed by the rise of new approaches in the West itself and in the colonial and post-colonial peripheries. Among the epistemological critiques and shifts in the imperial heartlands, were historical relativism, cyclicism, and most importantly, Marxism. Those who believed in relativism emphasized that the historian was himself a historical being, so that objectivity was a myth. Indeed, as Croce, the Italian philosopher and historian put it, all history is contemporary history, by which he meant that history is not merely a record of what happened, but what present people thought happened. Bury, the English historian, went so far as to argue that the judgements of the historian are as relative as his facts. For their part, the cyclicists saw the historical process in terms of cyclical movements, in which there is a continuous recurrence of historical experience, from growth to civilization to decline. Arnold Toynbee, the British historian, made the most ambitious attempt to analyze world history in terms of cyclic theories. He portrayed civilizations going from genesis and growth to decline and disintegration. It has been suggested that the upsurge of cyclic theories reflected the crisis of western civilization following the pulverizations of the two world wars and the ascendancy of anti-western and anti-capitalist ideologies and movements at the global level.

Marxism, although itself doggedly modernist and western in its vision, produced the most potent challenge to the established western theories and approaches of history. Marxism affected the production of historical knowledge in a number of ways. It insisted upon the analysis of complex and long-term social and economic processes instead of the conventional descriptions of political and isolated events. Its emphasis on the need to examine the material conditions of people's lives brought the masses into history alongside the 'great men' of traditional history, and also turned the processes of class formation and class struggle into fruitful subjects of historical study. Finally, Marxism stimulated enquiry into the theoretical premises of historical methods and the character of theories of history. Marxist philosophy advanced the concepts of dialectical and historical materialism as the most efficacious analytical tools of historical processes. Marxism has exerted an enormous influence on twentieth century historiography.

The expansion in the production of historical knowledge in this century has been nothing short of phenomenal. The *Annales* school in France popularized the extension of the dimensions of history to many previously ignored aspects of human activity, and stressed the need to use a broad range of human artefacts as sources, from language to all forms of material culture. Historians readily borrowed findings and methods of other disciplines, from geography and economics, to sociology and psychology. As a result new branches of study emerged, such as the history of science and technology, historical demography, industrial archaeology, environmental history, and psycho-history. More recently, quantitative history has assumed prominence, putting emphasis on the measurement of quantitative information, rather than simply qualitative evaluations.

Since the end of the Second World War far-reaching changes have occurred in the social, economic and political fabric of virtually every society in the world and in the nature and patterns of global relationships. This has helped open new dimensions in history, as manifested, for example, in the great interest shown in the western countries themselves in the history of the non-western world, including

Africa, the rapid advances made in archaeology, which have extended historical time and geography quite considerably, and in the study of women, which has brought humanity's invisible half to the historical stage. In the process many historiographical certainties have crumbled, and a basis has been established for producing more relevant and meaningful historical knowledge at the local, regional and global levels long-scarred by elitist, sexist, and Eurocentric biases. The explosion and disintegration of modernist western historiographies, including Marxism, owes a lot to the intellectual and anti-imperialist revolts of the so-called South: Africa, Asia and Latin America. Many nationalist intellectuals from these regions, including historians, vigorously attacked western discourses, and began producing counter-discourses and counter-histories of their own societies and their deadly encounters with western modernity. These revolts, combined with the multiple crises of contemporary western societies and the growing marginalization of the intellectuals, produced the apparent uprisings against modernity embodied in post-structuralist and post-modernist paradigms. But these new 'posts' have their own totalizing, imperialist illusions: denial of the narratives of oppression and the possibilities of liberation for the affected collectivities. History becomes pastiche.

The post-war historiographical changes partly reflected transformations in the organization of historical work. Just as the size of the historical profession had never been larger, thanks to the rapid expansion of university education, so was the explosion in the quality and quantity of historical sources, as a result of the bureaucratization of social life, greater awareness and use of non-documentary sources, such as oral tradition, and the widening horizons of the historians' interests. It became increasingly impossible for an individual historian, however talented, to sift through all available sources on say a country's history or any major topic all alone. This led to specialization, indeed, excessive specialization. And perhaps, that is what lies behind some of the fulminations against 'grand narratives'. Historians, like most contemporary academics, especially in the historically-tired West, not only became technical tinkerers, they also seemed to have lost faith in the future and, therefore, the meaningfulness of the past, for history is ultimately an attempt to peer into the foggy future through the prism of the past. The answer to the explosion of historical sources does not lie in obsessive specialization and defeatist post-something whining, but in undertaking critical syntheses and creative collaborative work.

African Historiography

As demonstrated in the last two chapters, the study of African history since the late nineteenth century has undergone radical change. With the exception of Ethiopia and Liberia, Africa entered the twentieth century under colonial bondage. Not surprisingly, therefore, imperial or colonial historiography held sway until the Second World War when the forces of decolonization gathered momentum. On the rare occasions when leading western thinkers referred to African history at all, it was to deny its existence, to assert that Africans were 'a people without history', to borrow Eric Wolf's (1982) unfortunate phrase, except of course the northern part, which came to be seen as the southern Mediterranean shores of Europe. Any historical movement in it, any civilization, was an external imposition, representing the footprints of Caucasoid invaders. In imperial historiography, then, history in Africa began with the arrival of the Europeans, whose colonial exploits against the forces of nature, which included the African peoples themselves, dazzled the reading public in the West (Curtin 1964). Predictably, the geographical focus of the imperial historians was largely confined to the history of the 'white settler dominions'.

The production of historical knowledge was of course not an imperial monopoly even in the darkest days of colonialism. This was because colonialism and its capitalist project were contested from the very beginning. The colonial state was both authoritarian and fragile because it lacked legitimacy. Its ideological apparatuses were truncated for they were all extraverted and embedded in metropolitan practices and traditions. And capitalism was not an omnipotent presence that could inscribe its will on a *tabula rasa*, but a set of social relations and practices which had to be produced and reproduced in societies that had their own social practices and traditions. In short, colonialism and capitalism were processes involving struggles between opposing and contradictory social forces, classes, and histories.

The perennial struggles over the organizations of the economy, politics, and culture created space for the production of anti-imperialist historical knowledge by both the 'traditional' historians and the western-educated historians. In other words, the *griots* did not die, and their children who went to the colonial schools later turned into anti-colonial historians. There were also colonial critics in the imperial metropoles themselves. The relationship between these groups were complex and contradictory and varied from place to place and over time. Their methods, audiences, and objectives also differed in some cases and overlapped in others. Their very existence was in itself a challenge to imperial historiographical hegemony. More importantly, their work, in method and content, affirmed the historicity of Africa, the humanity of Africans, and the criminality and culpability of colonialism. This critique crystallized into nationalist historiography after the Second World War, whose development was a by-product of, and a significant factor in, the intensification of the national liberation movement. Nationalist historiography represented an ideological and methodological revolt and advance over imperial historiography. Independence created favourable conditions for the production of nationalist history. National universities were established. The ranks of professional historians swelled. Research funds were provided by governments, private foundations and other agencies keen to promote and exploit Africa's intellectual bloom. Historical associations were formed, journals launched, and publishers scrambled for the latest research findings. Students of African history flocked from abroad where African history was incorporated into university history syllabuses, and specialized African studies centres mushroomed. Nationalist historians celebrated the rise of the new states by eulogizing the activities of the African nationalists. They gave the fragile new states historical identity by writing national histories stretching into the remote pre-colonial past. In this way continuity in national history was conjured up, national memories invented. Thus nationalist historiography provided the African nationalists and the new states with a legitimizing ideology.

The methodological forte of nationalist historiography lay in its discovery of new methods of data collection. Oral tradition, historical linguistics, and historical anthropology joined written and archaeological sources as valid methods of historical research. Oral sources eventually gained acceptance and were employed to enrich western histories. Oral methodology served as a link between 'academic' historiography and 'traditional' historiography, although the relationship between 'traditional' and 'academic' historians, both African and non-African, was an instrumental one, with the latter using the former solely as informants and suppliers of source materials, rather than as active collaborators in the production of collective historical knowledge. This relationship reflected the alienation of expatriate historians and the African educated elite from mass culture, and the centralization of historical research in universities patterned on, and often linked to, western universities.

Thus, the institutionalization of the production of historical knowledge in Africa, paradoxically, undermined the prospects of decolonizing

African history. Departments of history were not only modelled on those in the western universities, but they also maintained close contacts with them through the exchange of students, teaching staff, and materials. In fact, African academic historians were more responsive to the methodological shifts and research agendas developed by their Africanist colleagues than to those of their compatriot 'traditional' historians and the fermenting historical consciousness within their own societies, from which a methodologically and epistemologically more tasteful home brew could be distilled. The strong transnational links maintained by African academic historians and other scholars betrayed the fact that after independence African universities continued to act as conduits for the dissemination of western modernity. Independence did not lead to a profound reorganization of the process of social production of indigenously rooted and relevant social knowledge since the structures of internal underdevelopment and external dependence persisted and even deepened. In short, the production of social knowledge in Africa was as disarticulated, extroverted, and dominated from outside as the continent's economies themselves.

This might explain the rapid unravelling of nationalist historiography from the early 1970s, the same period that many African economies entered a prolonged period of crisis, thanks to the global recession. The premature demise of the nationalist historiographical revolution was caused by the growing irrelevance of its problematic and the rise of new paradigms. It had neither anticipated neo-colonialism, nor had the analytical tools to deal with it. Aside from its ideological demonstration, nationalist historiography like its nemesis, imperialist historiography, was rooted in the liberal tradition with its empiricism and idealism. For its part, the dependency approach had no concepts with which to analyze African history before the emergence of the capitalist world system, or internal processes since then without subordinating them to external agency. In this sense, dependence historiography shared the interpretative logic of imperial historiography.

The other approach that rose to challenge the hegemony of nationalist historiography was Marxism. Marxist influence grew with the triumph of radical national liberation movements in the early 1970s, and the adoption of Marxism as a developmentalist tool by some political parties and states, and by western intellectuals who were dissatisfied with their bourgeois privileges and sought the excitement of foreign revolutions. The Marxist historians examined the processes of production, social formation, and class struggle in the societies that they studied, as well as the complex mediations and contradictory effects of imperialism in modern Africa. The focus of Marxist history on 'internal' processes mirrored the preoccupations of nationalist historiography.

The apotheosis of the African historiographical revolution was the publication of the eight-volume UNESCO *General History of Africa*, undoubtedly one of the greatest achievements of historical scholarship this century, an opinion shared by Vansina (1993), who was involved with the project almost from its inception. He calls it not only 'a unique venture in twentieth century historiography', but also 'the most impressive venture of this century, not only because of its size or complexity, but because it involved authors from the most diverse origins and belonging to all the schools of thought then active in international academic circles' (Vansina, 1993:350). It confounded the critics, including many in UNESCO itself, who, brought up on the cherished virtue of academic individualism, neither relished nor believed that history could be produced by 'committee'. It succeeded where UNESCO's earlier venture, *The History of Mankind*[2], failed. The historiography of the *General History* is a truly remarkable story[3]. It brought together the largest group of historians ever assembled to work on a research project, and besides the volumes themselves, it generated numerous crucial symposia and the publication of invaluable archival guides, which will have a lasting impact on African historiography. Needless to say, the *General History* has some serious shortfalls, including the poor coverage of women's and gender history, a subject discussed at length in the next chapter.

The project was born out of panAfrican nationalism. According to Vansina (1993:307), 'at its founding meeting in 1963 the Organization of African Unity asked UNESCO to create a general history of Africa to replace the books then used in African schools, which were castigated for teaching nonsense'. The project was launched in 1965 and the first volumes started coming out in 1981. The *General History* (Ogot, 1981:xxiii-xxv) seeks to present 'the true history of Africa' long-obscured by Eurocentric preoccupations, methods and referents. It treats Africa as one historical unit, thus discarding spurious divisions based on geography and colour. Its 'method is interdisciplinary and is based on a multifaceted approach and a wide variety of sources'. It aims 'at the highest scientific level' and seeks to be 'a work of synthesis avoiding dogmatism', although it does not intend to be exhaustive, but rather to demonstrate 'the present state of knowledge and the main trends in research', thus preparing 'the ground for future work'. It is meant to be 'a faithful reflection of the way in which African authors view their own civilization'. Consequently, all the editors and the majority of the writers are African.

No sooner had the project taken off the ground than some Africanist historians who had not been included, or who felt threatened by it, embarked on their own mammoth eight-volume study of Africa known as the *Cambridge History of Africa*[4]. These volumes were quickly churned out to pre-empt the *General History*. In fact, many of those involved with the *Cambridge History* confidently predicted that the *General History* would never see the light of day, but when it finally did they turned to writing vicious reviews. This is as clear a demonstration as any of the complex struggles involved in the production of historical knowledge in, and on, Africa today[5]. The *General History* provoked furore in some Africanist quarters because it threatened to overturn the international division of intellectual labour under which African historians narrowly concentrated on their ethnic groups, and at most on their nations, while western scholars provided regional and continental syntheses. By Africa, most Africanists often meant 'sub-Saharan' or 'Black' Africa, definitions which were intended for various ideological and political reasons to divorce North Africa from the rest of the continent. Since the emergence of modern western historiography in the mid-nineteenth century, as Bernal (1987, 1991) has demonstrated, strenuous efforts have been made to deny the Africanness of North Africa and its peoples, especially the Africanness of the great civilization of ancient Egypt. It is revealing that Africanists have not seriously engaged Bernal, who has been attacked by the unholy trinity of Egyptologists, Orientalists, and Occidentalists, on the one hand, and appropriated with cultural desperation by American Afrocentrists on the other. As for the designation 'black', prefixed as a badge of African racial identity and contrasted to the Europeans' spurious 'whiteness', it evokes the binary oppositions of evil and good that are so deeply embedded in European religious iconography and mythology.

The emergence of the intellectual division of labour between African and Africanist scholars was of course not simply an imperialist scheme hatched by the latter. It arose out of the ideological imperatives of nation-building in Africa itself. The post-colonial nation-state set the boundaries of research and intellectual discourse for history and other disciplines. As noted in Chapter 3 above, scholars were implored to show commitment to the problems of their nation, to study its institutions and values, in short, to provide solutions to the national problems of economic development and political integration. Many African social scientists, including historians, responded to the call, and in so doing not only helped in advancing the ideology of nation-building, they also increasingly narrowed the scope of their concerns, analyses, and expertise. As research on regional and continental issues largely became a monopoly of the Africanists, the latter effectively began setting the terms of debate, and African scholars were reduced to followers, or at best critics, of each new intellectual fad that emerged from the Africanist capitals of the West, and less frequently, the East. The *General History* bucked the trend towards intellectual balkanization and dependence in the production of social and scholarly knowledge on Africa. This was not a fortuitous development. The *General History* was rooted in the panAfricanist tradition, which peaked in the late 1950s and early 1960s, before it was engulfed by territorial nationalism, its own offspring. PanAfricanism promoted a continental view of African history, indeed, a transatlantic vision of African pasts and futures.

The goal of the *General History* is declared to be the raising of historical consciousness in Africa and about Africa in the world at large. Youths and school children are singled out as a target group whose historical consciousness needs to be properly developed. As evidenced by the *General History* the level of scholarly research in African history at the tertiary level is relatively high. To what extent has this percolated to the schools? In other words, how far have school history textbooks incorporated the methods, approaches, and findings of modern African historiography?

The Limitations of Existing History Textbooks

In the 1980s a number of conferences were organized by UNESCO in different parts of the continent to discuss the content and writing of history textbooks used in African schools. They included the conferences held in Nairobi in 1984, 1985 and 1989, in which I participated (Zeleza 1984, 1985a)[6], and in Dakar in 1986. From the deliberations at these conferences, general agreement emerged that the existing school history textbooks did not adequately reflect the state of African historical research, as represented most concretely in the *General History*.

The shortcomings identified in these textbooks are many and varied. To begin with, the importance of history teaching itself seems to have been diluted in many countries with the adoption of the social studies model, in which history is incorporated, at the primary and junior secondary school level. The result has been that students are not taught the long-term historical perspective, or the basic information about the evolution of African societies over time. Thus, the social studies curriculum, whatever its other merits, is widely seen by many leading

African historians as a threat to the cultivation of historical consciousness among students.

Even when history is taught separately, the Dakar meeting noted, in most cases Africa accounts for no more than 10-15% of the contents of history syllabuses (UNESCO, 1986). Sometimes only national history is taught, and occasionally regional history. The rest of Africa is ignored or subsumed in reference to the history of the other parts of the world, which often receive greater attention. As a result of this, students are not sufficiently exposed to the ebbs and flows, patterns and links, that differentiate and bind African histories.

When it comes to comparative history, Africa's contributions in world history tend to be understated or ignored. This arises in part from a tendency to use idealized developments in European history as the points of reference. In particular, incorrect comparisons are made of societies at different levels of development, for example, comparing pre-industrial Africa with industrial Europe rather than pre-industrial Europe. Indeed, students are usually taught the history of modern Europe, of Europe at the pinnacle of its historical development and global hegemony, while they encounter other civilizations in their moments of decline and subjugation to Europe. Intended or not, students are left with the notion that Europe is dynamic and the fountainhead of innovations, while the rest of the world, including Africa, are static and passive recipients. Comparative history needs to be freed from the trap of Eurocentricism, and a new methodology developed that truly depicts the contributions of various civilizations to the knotted tapestry of world history[7].

At the various Nairobi meetings it was noted that in existing textbooks Africa is often inadequately defined and conceptualized. In textbooks used in East and Southern Africa, for example, there is an assumption that North Africa is more Arab than African, so that it receives casual treatment. Similarly, in North African textbooks only a negligible amount of space is assigned to the rest of the continent. In the Algerian textbook for final year classes, for instance, only three lessons out of thirty-seven are wholly or partly devoted to Africa. North Africa sees itself, first and foremost, as Muslim-Arab, and seems to prize its Mediterranean links more than its trans-Saharan ones (Cherif 1989). In short, in textbooks used virtually throughout the continent, Africa and African history tend to be fragmented along questionable geographical, racial or religious lines. The implication is that the various regions are historically distinct and unrelated, except peripherally.

Then there is the problem of classifying Africans. In many of the textbooks used in East and Southern Africa, Africa's peoples tend to be classified on a rather dubious basis. The most commonly used are linguistic-cum-racial classifications, whose scientific basis and historical validity is questionable. For example, in East African textbooks there is the distinction between the Bantu and Nilotes. 'Bantu' and 'Nilotic' are linguistic classifications but in these textbooks they are presented as distinct 'racial' or 'cultural' types, and it is also assumed that they are historically unrelated peoples. Within each of these groups there are further distinctions based on irrelevant features of their environment or a few peculiar and ephemeral customs. For example, distinctions are made between the 'plain Nilotes', the 'lake Nilotes' and the 'river Nilotes'.

Existing textbooks also suffer from confusing definitions of sub-regions as historical units. The problem arises because designations based on colonial divisions are not always compatible with historical and current geopolitical realities. For example, there is the problem of defining Central Africa. In Anglophone circles the reference is to the former British Central Africa, comprising Malawi, Zambia, and Zimbabwe. In the geopolitical context of modern Africa, Central Africa consists of the region from Cameroon and the Central African Republic to Zaire, while Malawi, Zambia and Zimbabwe belong to Southern Africa. The result is that Central Africa 'proper' hardly features in East and Southern African history textbooks, and Southern Africa is virtually reduced to South Africa and occasionally includes the bordering states. Similarly, East Africa is confined to the former British-ruled territories of Kenya, Tanzania and Uganda, excluding Somalia, Ethiopia, Burundi and Rwanda, which tend to be ignored. Sudan is left in a limbo, unsure whether it is a part of eastern or northern Africa. In short, there is confusion as to whether the definition of sub-regional historical units should be based on geographical expression, or modern political boundaries, or the similarity and inter-connection of historical developments depending on the period. For example, in the nineteenth century Mozambique had more trading links with East Africa, while in the twentieth century these have gravitated to South Africa, so that it makes sense to treat Mozambique as part of East African economic history in the nineteenth century and of Southern Africa in the twentieth.

The same problem crops up in covering 'national' history. As is well-known, colonial boundaries, upon which modern African nations are based, cut across pre-colonial nations, ethnic groups, and linguistic and cultural collectivities, so that, with a few notable exceptions, 'national' history only makes sense with reference to the colonial and independence periods. Attempts to stretch the historical personality of the modern nation-state into the pre-colonial period often end up in presenting a disparate collection of stories of 'migration' of the different ethnic groups into the 'nation' space, and a catalogue of their institutions, relations, and conflicts. For the pre-colonial period the discussion of historical processes in a regional and continental context makes more sense than conjuring up 'national' history.

This brings us to the question of periodization, which is often poorly tackled. There is a tendency to distinguish between the pre-colonial, colonial, and post-colonial periods, and to give each of these periods, especially the first two, equal treatment in terms of coverage. It does not make much sense to compare the pre-colonial period, a vast stretch of historical time, from the beginnings of human society to the nineteenth century, with the colonial period which in much of Africa lasted no more than 70 years, less than the lifespan of my great grandfather who preceded and outlived British colonial rule in Malawi, and the post-colonial period which is only three decades old. The periodization of African history into the pre-colonial, colonial, and post-colonial in itself makes colonialism the pivot around which African history spins. This subtly reinforces the view that dynamic movement in African history started with colonialism, that the multitude of generations that lived before the Berlin Conference were preparing for this great moment[8]. Alternatively, some try to periodize African history in terms of the divisions of European history, and glibly talk of the Dark Ages or the Middle Ages in African history. This serves to deny African history its autonomy and reduces Africa into a historical appendage of Europe against which it is judged.

Existing history textbooks, in fact, tend to portray the history of the pre-colonial era as static. The term 'traditional' is bandied about carelessly, implying a changeless order of things. The only movement seems to be that of migration seen literary as the movement of people from place to place, rather than as complex expansion of production, political, and cultural frontiers. Changes in the social, economic, and political spheres are treated in an idealistic fashion merely as products of external stimuli and not as endogenous developments. In East African history textbooks, there is a continued use of the Hamitic myth in the form of the 'cushitic' factor. Long-discredited notions continue to be reproduced, for example, that agriculture and metallurgy diffused from the 'Middle East' (middle from whom, one may ask? Imperial Britain, of course!) to Africa, when research has shown conclusively that the domestication of plants and animals and metal production had independent origins in different regions of the world, including Africa. Subsequent development involved complex processes of local innovation and interchanges within and between continents. Similarly, the universal religions, such as Christianity and Islam, continue to be portrayed as alien, when both were in fact introduced in some part of Africa long before they were imported into many parts of Europe and Asia where they are considered indigenous today.

When discussing the colonial era many existing textbooks have two main shortcomings. First, there is a tendency to compartmentalize colonialism on the basis of the nationality of the colonizing power and to depict the other colonial systems with the prejudiced paint-brush of the ex-colonial power. For example, in the textbooks used in the countries formerly under British rule in East and Southern Africa, the British are portrayed as 'liberal and pragmatic', the Germans as 'authoritarian and brutal', the Portuguese as 'backward and ruthless', and the French 'romantic and racist'. Colonial ideologues are taken at their word, and British and French systems of administration, for instance, are contrasted as 'indirect rule' and 'assimilationist', respectively, when, in practice, they had more similarities than differences. Second, in pursuit of some fraudulent 'objectivity', the balance sheet approach is used, whereby the 'bad' and 'good' aspects of colonialism are counterpoised. Oppression and exploitation are counterbalanced with the benefits of modernization. Colonialism may not have been the 'one-armed bandit' of Rodney's (1982:205) provocative imagination, but neither was it an amiable creature amenable to the facile dichotomies of balance-sheet analysis. Colonial policies, institutions, and transformations are rarely presented as processes, rooted in complex material realities and arising out of contradictory social struggles and discourses.

The treatment of the period since independence also leaves a lot to be desired. Historians always seem uncomfortable with writing about contemporary issues, which are usually left to political scientists and others whose perspectives are often lacking in historical dimension. When writing about the post-colonial period, the textbooks tend to echo the official ideological position of the regime in power, and since in some countries regimes have changed quite frequently through coups, writing meaningful contemporary history becomes virtually impossible. History is often reduced to empty developmentalist platitudes. This is as dangerous as the pathological Afropessimism found in Africanist texts in the western countries.

One of the major shortcomings in many existing African history textbooks is that they are written as if theory does not matter. They are either catalogues of dry facts, or they present theoretical interpretations as fact. Rarely is theory woven throughout the text critically and imaginatively[9]. Students are also not made aware of rival historical interpretations on particularly contentious issues, or conflicting interpretations are juxtaposed haphazardly. Such textbooks are usually lacking a guiding panAfricanist perspective, a critical understanding of the complex ties that bind Africa and its diasporas. As a result of all this, the 'meaning' of history is lost on the students and a critical historical consciousness among them is hardly developed.

The poor articulation of theory has led to a situation where there is careless use of terminologies borrowed from other historical contexts or disciplines, which are inadequately defined and conceptualized. For example, terms like 'feudal', 'empire', 'kingdom', 'aristocracy', 'class', just to mention some of the most common ones, are used with abandon when referring to all types of social and political formations in African history. This creates considerable confusion and promotes the writing and teaching of African history by analogy, as transition narratives.

Moreover, the quantitative method is hardly used even for periods and areas where quantitative data is available. The result is that historical interpretations and conclusions become no better than enlightened opinions at best. The creative use of quantification, wherever possible, would greatly animate discussions of trade, urbanization, migration, warfare and many other areas of economic and social history.

Many of these themes are, in fact, ignored in school history textbooks which tend to concentrate on political issues. Students are given little or no exposure to economic, social and cultural history, let alone the history of science and technology, environmental history, gender history, and other branches of contemporary historiography. When issues of economic, technological and cultural change are mentioned they are presented simplistically as derivatives of political processes. In the case of Africa, emphasis is put on the notion of diffusion and borrowing from the other continents.

Many history textbooks also still suffer from the use of derogatory nomenclature. Terms like 'Bushmen' or 'Hottentot' for the Khoisan people are still widely used. Concepts like 'Negro', 'native', 'animism', 'pagan', and 'tribe', with or without inverted commas can be found in many textbooks. Students end up internalizing contemptuous images of Africa, its peoples and cultures, in short, of themselves. Negative descriptions and references, often Eurocentric in their origins, to Asians and the indigenous peoples of the Americas, also creep up in many African history textbooks.

Finally, many existing textbooks are poorly illustrated. The illustrations are sometimes unimaginatively done and reinforce communal, racial, religious, and gender prejudices. For example, farmers are usually portrayed in poses that make them appear superior to pastoralists, Europeans superior to Africans, followers of Christianity or Islam to followers of local religions, and men superior to women. The power of visual images in the formation of perceptions and prejudices which can last for life cannot be overemphasized.

These shortcomings demonstrate that African historical scholarship as contained in the *General History* has yet to be fully incorporated and reflected in the textbooks used in schools. Thus, there is a disjuncture between the knowledge produced by academic historians in the universities and that consumed by students in primary and secondary schools. Indeed, the general level of historical consciousness in many African countries appears low relative to the enormous advances that have been made in African historical scholarship. What accounts for this?

The Political Economy of Producing Historical Knowledge for Schools

The factors behind the failure of scholarly research in African history to raise the levels of historical consciousness among the general public are quite complex. They are determined by general economic and political developments in each country and the specific processes and practices involved in the production of social knowledge.

During the struggles for independence, and immediately after, African leaders were in love with history. They used history to legitimize nationalism, to authenticate the nation-state they had inherited. They gave priority to the expansion and decolonization of the educational system. The impatient masses expected no less for their struggles; they wanted their children educated to enjoy the fruits of *uhuru*. For their part, African scholars were an enthusiastic, energetic lot, anxious to contribute to the dreams of development and nation-building with the inspired rhetoric of their pens (Mazrui, 1978). They, too, believed in the new nations, if only as a prelude to the panAfrican vision of continental unity. This intellectual ferment and faith was underpinned by buoyant economic growth.

Before long, however, the promise of economic development began to wither in the glare of the global economic crisis. Also, the more consolidated the nation-state seemed to become the more distant the vision of panAfricanism receded. Politics deteriorated into the art of manipulating ethnic constituencies and class coalitions. The new political culture undermined the development of historical consciousness in several ways. It encouraged the production of parochial histories of ethnic groups, which 'traditional' historiography, with its myths and traditions of origin, was better qualified to handle than academic historiography. But the 'traditional' historians were not a part of the theatre of power as were their academic counterparts who shared class interests with the new ruling elites. This is to suggest that the ethnic histories produced by the former did not have the same political resonance as those produced by the latter, whose texts could be, and were, selectively disseminated throughout the rapidly expanding educational system. Ethnic histories facilitated the production of power by the political class by fostering the divisive power of communal consciousness and foreclosing the disruptive power of class and other social solidarities, while at the same time the consumption of these histories threatened the integrative imperatives of nation-building. History became highly politicized, a subject treated with great suspicion and even hostility. Indeed, Africa's political leaders soon lost patience with scholars and historians asking awkward questions about neo-colonialism.

Thus, African academic historiography became marginalized because the erosion of its panAfricanist preoccupations deprived it of a critical epistemological anchor, so that it became vulnerable to the assaults from both the 'traditional' historians, from whom it sought to steal its ethnic narratives, and from Africanist historians, to whom it increasingly looked for theory. The failure to find meaningful collaboration with, as well as its growing subordination to, Africanist historiography is what narrowed the intellectual space of African academic historiography. And hovering above were the rumbling clouds of state authoritarianism. The economic recession that gripped many countries further compounded the problems of intellectual production and productivity as the academics and their universities began to face unprecedented hardships. In the meantime, the economic and social progress of the 1960s and early 1970s, and the search for self-reliant development and African autonomy upon which the post-independence contract was built, faced serious threats, as Africa was incorporated deeper into the world capitalist system and draconian structural adjustment programmes were imposed by the almighty World Bank and IMF

The impact of all this on education in general, and history in particular, is not hard to imagine. The drive to expand and decolonize the educational system lost steam in the most affected countries. Teaching facilities became overstretched and shortages of foreign exchange led to a book famine as it became increasingly difficult to import books and the materials necessary for local publishing. As research grants dried up, research dissemination media, such as journals, either folded or appeared irregularly. Some historical associations either died or continued to exist only on paper. Intra- and inter-regional scholar communication and exchange lessened. In short, the institutional and infrastructural base for the production and dissemination of historical knowledge for schools deteriorated. Lack of adequate foreign exchange and falling real incomes have meant that such indispensable books as the *General History* are not readily available or affordable to the producers of school history textbooks. Moreover, economic difficulties, as noted earlier, discouraged original research and forced many academics to concentrate on applied consultancy research projects. For their part, cash-strapped publishers became even more reluctant to publish costly academic titles. In short, textbooks were increasingly written based on previous textbooks, resulting in the recycling of discredited and discarded historiographies, stereotypes, and interpretations.

The demise of historical associations has adversely affected the development of history writing and teaching. Historical associations, rallying academic and traditional historians and school teachers, public officials and individuals interested in history, constitute an essential part of the basic infrastructure for the production and consumption of historical knowledge. An active association can disseminate the latest historical research findings through its journal and other outlets for publication, including the mass media, by organizing public lectures, conferences and symposia, as well as refresher courses for history teachers. It can also raise historical consciousness by identifying and drawing public attention to significant historical monuments and the anniversaries of major historical figures. Moreover, an active association can act as a watchdog to monitor the curriculum and content of history teaching at all levels, and historical accounts and images portrayed in various publications, including literary works, official handbooks, and the mass media in general. Given that historical associations can do so much, the absence or disintegration of these associations in many African countries is a cause for alarm. It is one of the factors behind, and a manifestation of, the growing disjuncture between scholarly research at the tertiary level and the kind of historical knowledge found among the school population and the general public.

It can be seen that the nature and development of the political economy in each country, and within Africa as a whole, conditions the production of historical knowledge in general which, in turn, affects the production of history textbooks. The process of producing school textbooks also involves the design structures of syllabuses and examinations, the availability of writers and publishers. The nature of the relationship between these agencies, of course, varies from one country to another. With the notable exception of some West African countries which still follow common syllabuses and examinations, there is little coordination between countries in the designing of syllabuses. The result is that the quality of history syllabuses varies greatly among countries. This also means that the scope and content of historical knowledge imparted in history textbooks is extremely uneven.

The responsibility for designing syllabuses and setting examinations generally rests with the ministry of education, which often creates special boards, commissions, institutes, committees or councils to deal with the various tasks. The composition of these bodies also varies from country to country. In some countries membership is restricted to ministry of education officials, while in others professional historians from the universities and school history teachers are included. The method for selecting the historians and teachers is often *ad hoc*, based on personal connections[10]. Zimbabwe has, however, made attempts to institutionalize and democratize the selection process. In each district school teachers elect representatives to fill the quota of seats allocated to teachers on the syllabus and examination panels. From the deliberations at the Nairobi meetings it clearly emerged that in many African countries decision-making on syllabuses is highly discretionary. Government bureaucrats, who in many cases are poorly informed about the latest findings and trends in historical research, tend to make the final decision concerning the structure and content of the syllabuses.

Syllabuses form the basis for writing school textbooks. So if they are poorly conceptualized or outdated this will inevitably be reflected in the kind of textbooks that are written and used in schools. In some countries history syllabuses have remained almost unchanged since the 1960s. For example, by the mid-1980s a number of Francophone countries were still using syllabuses designed in 1967. Such syllabuses were obsolete and irrelevant given the great strides that had been made in world and African history research in the previous two decades.

But even in those countries where syllabuses have been recently reformulated many problems still remain. Let us take the example of Kenya. In 1985 new syllabuses for primary and secondary schools were adopted. The new history syllabus is divided into two parts. One part deals with the history of Kenya, and the other with themes in world history. The syllabus moved away from emphasis on political and sub-regional history. Themes dealing with economic, social, cultural, and scientific history were given their due weight. In many ways the syllabus represented a considerable advance over the previous syllabuses, but it had some shortcomings. Most significantly, perhaps, is the fact that Africa as a continent lost its historical identity, and was dissolved into the amorphous context of world history, largely seen from the vantage point of Europe. The period emphasized coincides with the period of modern European-based world system. Also, the thematic approach is not adequately counterbalanced by a chronological approach.

The syllabus, moreover, contains some outdated historiography. One example will suffice. In the section dealing with African responses to colonial conquest, students are required to know about societies that 'resisted' and those that 'collaborated'. The examples of the 'resisters' and 'collaborators' are listed. The latter include the Buganda and the Lozi states and the Krios of Sierra Leone. As some chapters in Volume 7 of the *General History* amply make clear, the contrast between 'resistance' and 'collaboration' is too neat and contrived. Not only is 'collaboration' Eurocentric by implication, it misconstrues the objectives of the so-called 'collaborators' and the trajectory of their relations with the European invaders. The same Kabaka Mwanga of Buganda who is deemed to have 'collaborated' with the British imperialists later waged war against them[11].

This example shows that the thematic content of history syllabuses needs to be carefully looked at by historians. Outdated and backward historiography can also be perpetuated through the examinations. This can happen in two ways. First, examination questions are based on the themes and topics outlined in the syllabus. Second, the marking schemes prepared by the examination council are subsequently used by teachers as guides of correct historical interpretation. So if the themes and topics on which the questions are based are historiographically 'wrong' the questions and the marking schemes will reflect this. A vicious circle is created that is difficult to break.

The actual practice of writing history textbooks in Africa also varies considerably among countries. Three main forms can be identified. First, in countries such as Ethiopia, Ghana, and Uganda, for the first cycle schools, textbooks are specifically prepared and published by the government itself. The ministry of education commissions local authors to write them. The criteria for selecting the writers are not always based on professional considerations. Second, there are countries where textbooks are not specifically commissioned by the government, but are written by local or European historians on their own initiative, and then recommended for use by the government itself, as in Malawi, or by the authorities of each school, as in the Gambia. Third, there are countries like Kenya where government-commissioned and published books compete with non-commissioned and privately published books, although lists of recommended books are issued. Despite these differences, in all these countries the ministry of education has decisive powers on the textbooks that are actually used in the schools. It is an open secret that sometimes ministry officials are bribed or manipulated by publishers to recommend history textbooks of worthless value.

It was noted at the Nairobi meetings that school teachers, who should be actively involved in the writing of history textbooks, are often unable to undertake the task because they have little time or inclination for historical research and writing. Indeed, school teachers are often unaware of new trends in African historiography. It has to be remembered that in some countries a large proportion of the teachers, especially at the primary school level, are untrained and even when they are trained they may not necessarily have received training in history teaching methods, let alone modern historiography.

For their part, university lecturers and professors are reluctant to involve themselves in the writing of school textbooks because they are worried about their careers since such texts do not enhance their CVs. This is only one illustration of the elitism of academic scholarship. The vacuum is then filled by unscrupulous writers whose primary concern is making quick money. For example, when the new history syllabus was released in Kenya in 1985, Malkiat Singh, who writes on every subject imaginable, quickly churned out history textbooks without concern for recent research findings and the presentation of accurate information in a lively manner. His books unfortunately set the standard in the schools. Biases, distortions and inaccuracies could be reduced or eliminated if a more collective approach to the writing of textbooks was encouraged. Collective not simply in terms of the number of people involved, but in their proficiency as historians and capacity to undertake progressive and collective intellectual production.

The practices of publishers also make life difficult for serious writers. Publishers want to publish a book as quickly as possible once a new syllabus has been introduced in order to corner the market. This considerably affects the quality of the publication, for instance in terms of illustrations, lay out and editing. More importantly, perhaps, is the practice of many foreign based publishers to send manuscripts overseas or to people who are not professional historians for evaluation. Let me give a personal example. In the mid-1980s, I was contacted by a local subsidiary of a British publishing firm to write two, later turned to four, secondary school history books on the world history sections of the new Kenyan syllabus. Once completed, the first manuscript was sent to England for evaluation. When I got the galley proofs I was shocked to see the reader had added some points to the text. On the section dealing with the impact of the slave trade he inserted Curtin's disputed estimates, and added the following sentences: 'The Europeans were not responsible for capturing the slaves. This was done by the Africans themselves, and although some slaves were mistreated by their white masters, many were well looked after'. Clearly the intention was to whitewash European culpability. We fought over this for some time until they agreed to remove it[12]. On this side of the Atlantic, I have also encountered self-righteous, but wrong-headed, editorial decisions, whose effect is to reproduce stereotypical and negative African representations. This time I was approached by a New York publisher working on a series of African peoples for high school students to write three books on some of Kenya's major ethnic groups. Despite my misgivings about the ethnic model, I agreed to do so, hoping to present narratives that were historiographically critical and complex enough to challenge that very model. The dispute centred around the illustrations used in one of the books which I saw at the galley proof stage and found demeaning, inaccurate, and in some cases unconnected, or in contradiction, to the narrative in the text. I fired off an angry letter and after several unhappy exchanges the publisher promised that the offending pictures would be removed.

It is quite clear, therefore, that working with foreign publishers has its risks and frustrations. In cases where the local publishing industry hardly exists or is poorly developed they cannot be avoided. But even in situations where the opposite is the case, there is often excessive dependence on imports of raw materials, such as paper and machinery, so that import constraints due to foreign exchange shortages and the ill-conceived structural adjustment programmes, undermine local publishing activities. Also, the requisite editorial skills may be unavailable locally or quite low[13]. The bottom line, of course, is that textbooks must be produced cheaply, for in the final analysis it does not matter how up-to-date a book is in its historiographical content and methods, or how well it is written and illustrated if school teachers and students cannot afford to buy it. It will remain on the shelves as a monument to irrelevance.

When discussing textbooks the conventional wisdom is that colleges and universities do not need them. This is unrealistic. In any case, students in African universities already use textbooks written for North American and European students and audiences. History texts written for the tertiary level have to conform to high standards of scholarship. The practice of writing college and university textbooks can help set the standards for school syllabuses and textbook writing. Tertiary institutions also enjoy greater autonomy in setting their syllabuses and examination systems. Opportunities for coordination and the collective production of historical knowledge at this level are, therefore, much brighter than at the lower levels of the educational system. These opportunities need to be exploited. But to do so would require the regeneration of the vision and spirit of panAfricanism among African scholars.

Conclusion

The basis for overcoming the disjuncture between the relatively high standards of historical research at the academic level and the low historical consciousness at the school and popular levels is there. UNESCO's plan to produce, publish, and distribute abridged versions of the *General History* and model textbooks that reflect its content and spirit is an important point of departure. UNESCO's project needs the support of African historians. UNESCO could also help through its conferences of ministers of education to push for the adoption of history syllabuses that are patterned on the *General History*.

The ultimate responsibility for the promotion of accurate and relevant historical knowledge rests on the shoulders of the historians themselves. Impressionistic evidence seems to suggest that more university historians are now writing school textbooks. Perhaps the persistent recessions have something to do with the erosion of the intellectual conceits of university academics: the remunerative rewards of writing textbooks for captive school markets are quite attractive. The need to revive or strengthen historians' associations cannot be overemphasized[14]. There is also need to reinvigorate historical research at the tertiary level itself. Perhaps it is time to start thinking seriously about establishing a new institutional base for historical research. The universities, which are overburdened with teaching and poor research facilities, can no longer realistically be expected to provide the sole basis for scholarly research. The creation of independent research centres, such as CODESRIA, SAPES, and CBR, among many others, shows that this is already, in fact, being done for the social sciences. The creation of a few strategically located specialized institutes for historical research, which could consolidate what has already been achieved and move in new directions, is long overdue. Such institutes would need to explore ways of integrating 'traditional' historiography and overcoming the domination of Africanist historiographies. Already the gulf between African and Africanist historians appears to have grown as serious scholarly communication has diminished due to institutional and intellectual developments within both the African and the western countries. It is encouraging to see that many African historians on the continent remain unimpressed by the posturings of the post-politics paradigms parading in the western academies.

As the saying goes, a journey of a thousand miles begins with one step. In our case a lot of mileage has already been covered in historical research and the reconstruction of African history. But it is no time for rest. As we continue on this arduous journey towards our liberation as a people, we cannot afford to leave anyone behind, certainly not half the population. On the question of women, African and Africanist historiographies, as with history writing everywhere, the journey may only just be beginning. This is the subject of the next two chapters.

Notes

1. For succinct overviews of the development and philosophy of history, including Africanist history, see Barraclough (1978), Temu and Swai (1981), Feierman (1993), and Wamba-dia-Wamba (1987).

2. This project launched in 1950, and entrusted to a handful of 'eminent' European historians, was a disastrous failure and this may have induced some of the scepticism against the *General History* within many circles, including UNESCO. *The History of Mankind* was also intellectually bankrupt, insisting on distinguishing the 'elite civilizations' from 'plebeian cultures', a reflection of the unabashed Eurocentricism and racism of the time, see Vansina (1993: 351).

3. As is evident from Vansina's (1993) personal account, and the accounts I heard from Professor B. A. Ogot, a longtime president of the scientific committee, a colleague and mentor for the six years I taught with him at Kenyatta University. I also had the opportunity to meet and talk to several other editors of the volumes in Nairobi. One hopes they will produce their memoirs.

4. Vansina (1993:338) suggests that the British historians, Roland Oliver and John Fage, had been approached by Cambridge in the late 1950s for a multi-volume project on African history, but they declined, saying it was 'premature'. A few short years later, in 1966, a year after the launching of the Unesco project, the time suddenly became 'ripe' for the undertaking. Just a coincidence, perhaps? Many people involved with the Unesco project read an incriminating subtext in the latter endeavor, whose editors, and the majority of its writers, happened to be Europeans or European-North Americans.

5. Ogot correctly predicted that of the two, the UNESCO series would be used more widely by students in African universities. Impressionistic perusal through various African university libraries showed that the Cambridge series were squeaky clean from underuse as compared to the heavily marked UNESCO volumes.

6. The original version of this chapter was presented at the 1989 Nairobi Conference (Zeleza 1989), which brought together the editors of the UNESCO *General History*. This section is based on a paper I gave at the 1989 Nairobi conference and the two reports I produced on this and the 1984 and 1985 Nairobi conferences (Zeleza, 1984, 1985a, 1985b); and the papers presented at the 1989 Nairobi Conferences, especially those by Ajayi (1989), Boahen (1989), Cherif (1989), and Ki-Zerbo (1989).

7. This became abundantly clear to me when some colleagues and I in the History Department at Trent University embarked on developing a world history course. We perused dozens of syllabuses and textbooks used in many North American universities. We found three main trends. First, world history as a glorified western civilization course, in which there are a few 'non-western' embellishments on an unchanged Eurocentric narrative. Second, world history as the history of international relations, focusing mostly on the twentieth century, the role of international conflicts, organizations, and events, in which there is a little history and a little political science that does not add up to much. Third, world history as a hotchpotch of regional histories bound together by the book covers, and the occasional Eurocentric threads. In all these histories, Africa fared poorly. We tried to design a course that gave the students a complex understanding of the different continental histories, their interconnections, and the emergence of a global order.

Finding appropriate texts was not easy, since our course constantly 'battled' against most of them. The 'battles' made the tutorials intellectually stimulating.

8. It is such considerations which made me decide against using the term 'precolonial' in the title or descriptive analyses of volume one of my book on African economic history (Zeleza 1993).

9. This is of course apparent in many scholarly books, especially those converted from dissertations, which often wear their theoretical credentials loudly in the introduction, and sometimes in the conclusion, where they usually denounce previous texts for all manner of theoretical sins, only to present descriptive analyses in the rest of the text that resemble the maligned texts. Theory becomes, in this sense, a means of clearing space for oneself, not a mark of original thinking, and so it is swallowed and spit out undigested with the data.

10. That is how I came to serve on the History Panel of the Kenya National Examinations Council in the late 1980s through introduction by a friend who was serving on the panel.

11. For some interesting discussions of the different social and class dynamics of African resistance, see Isaacman (1976) and Crummey (1986).

12. This project has had a checkered history. After the galleys of the first textbook were ready, the Kenyan shilling began falling sharply against the major foreign currencies, including the British pound. This threw a spanner in the works: the practice was for the parent company in London to publish the books and then export them to its Kenyan subsidiary. In this way, Kenya's then stringent foreign exchange regulations were circumvented and the parent company could reap most of the proceeds. The sharp devaluation of the Kenyan shilling meant that the books would now be too expensive in local currency. They decided to suspend publication 'until the situation improved'. In the meantime, the local subsidiary was only allowed to publish an examination guide based on my work and that of my colleague who had written on Kenya (Zeleza, Sharman and Williams, 1989). I was left hanging with four galley proofs and unprinted manuscripts. It was a costly lesson in the political economy of multinational publishing, which reinforced my inclination to stick to African publishers for my scholarly work.

13. One of the frustrations I have found publishing in some African journals is the poor level of editing. Some do not even have the courtesy of sending galleys for proof-reading and it is not a pleasant experience to see one's sentences mutilated and grammar mangled.

14. It is encouraging that the continental Association of African Historians is being revived and that CODESRIA has agreed to underwrite and regularize the publication of the history journal, *Afrika Zamani*.

Chapter Nine

GENDER BIASES IN AFRICAN HISTORIOGRAPHY

Introduction

In the last two decades the literature on African women has grown rapidly. This can be attributed to several factors, including the political impetus of the women's movement and the crisis of conventional development theory and practice, and the consequent rise of the women-in-development project. For the discipline of history, more specifically, interest in women's history has been spawned by the widening horizons of historical epistemology and research, especially the growing interest in, and the development of, new approaches to social history. Until recent times historians preoccupied themselves with political history. They tirelessly described political developments, wars and battles, and celebrated the lives of great men (Barraclough 1978; Conkin 1989; Himmelfarb 1987).

Despite the proliferation of the literature on women, including women's history, women remain largely invisible or misrepresented in mainstream, or rather 'malestream', African history. They are either not present at all, or they are depicted as naturally inferior and subordinate, as eternal victims of male oppression. Alternatively, the romantic myth is advanced that the roles of women and men were equal and complimentary in good old, harmonious, pre-colonial Africa, or the lives of notable, exceptional, heroic women are celebrated (Imam 1988). In short, in most institutions of higher learning in Africa women's history is still marginal and lacks recognition and academic respectability (Awe 1991:211)1. This situation is, of course, not peculiar to African history. It applies worldwide, and to the social sciences in general[2].

This chapter seeks to do four things. First, it will demonstrate the inadequate representation of women in African history by looking at some of the most frequently used texts. Second, an attempt will be made to identify some of the reasons for this by examining the dominant paradigms in African historiography. Thirdly, the chapter outlines the reconstructions of women's history made by feminist historians. These historians face two interrelated challenges. The first is to recover, empirically, the lives of women and restore their story to history. The second challenge is theoretical, to deconstruct the conventional historical paradigms and devise new ones which will rid history of its inherent androcentrism, in order to redefine and enlarge the scope of the discipline as whole, to make historical reconstructions more inclusive, more comprehensive, and more complex. The final part, then, suggests some ways of gendering African history.

The Invisible Women

The authors of African history textbooks differ in their approaches and research methods, in the subjects they examine, the interpretations they advance, and in their ideological outlooks. But they have two things in common: they are predominantly male and sexist in so far as their texts underestimate the important role that women have played in all aspects of African history. In more extreme cases women are not even mentioned at all, or if they are, they are discussed in their stereotypical reproductive roles as wives and mothers. The language used often inferiorises the women's activities, or experiences being described. Also, women's lives are usually cloaked in a veil of timelessness: the institutions in which their lives are discussed, such as marriage, are seen as static. In viewing them as unchanging, as guardians of some ageless tradition, women are reduced to trans-historical creatures outside the dynamics of historical development.

A survey of some of the most widely used history textbooks clearly demonstrates these biases. The chapter will examine three categories of texts: general histories that are continental in their coverage, regional histories, and histories of particular themes, such as political, economic, and social history. With each text, the chapter tabulates the space devoted to women in the text and in the illustrations, if any, and the general thrust of those references in terms of content.

The General Histories

Eight sets of general histories were examined. They are all written by prominent historians of Africa, both African and Africanist. None is a woman. Some of them do not even mention women in their indexes. This is true of Tidy and Leeming's (1981) two volume text, *A History of Africa* and Afigbo, et al. (1986) *The Making of Modern Africa*, also in two volumes. Volume One of the latter book looks at Africa in the nineteenth century and Volume Two at Africa in the twentieth century. I looked at the revised edition published in 1986. 'This very popular text', the blurb at the back proclaims, 'has been thoroughly revised to include the most up-to-date developments in research and historiography'. The two texts have 372 pages each, making for a total of 744 pages, none of which is specifically devoted to women. The illustrations are hardly any better. Out of the sixty illustrations in Volume One women appear perfunctorily in two. Volume Two is a little better. Out of 80 illustrations women appear in 13, mostly in the background. Only in three are they the central focus of attention.

The most comprehensive studies which seek to summarise current significant knowledge in African history are the UNESCO *General History of Africa* (1981-1993) and the Cambridge *History of Africa* (1975-86) both published in eight thick volumes[3]. Both studies have very little to say about women. An examination of Volumes 6 and 7 of the UNESCO *General History* and volumes 5-8 of the Cambridge History dealing with the nineteenth and twentieth centuries, periods upon which reconstructions of African women's history have concentrated, amply bears this out. Volumes 6 and 7 of the UNESCO *General History* have 861 and 865 pages, respectively. Women are mentioned only on 4 and 14 pages, respectively. In Volume 6, the women are mentioned with reference to Chokwe women who followed their husband traders (p. 302), provision of education for Egyptian girls by the Coptic Church (p. 347), women as gold washers in Asante and Lobi (p. 690), and sexual relations between diaspora African men in Europe and European women (p. 759), while in Volume 7 they are mentioned with reference to their fertility patterns (five pages) and polygyny (four pages). It needs to be noted that the references to women on these pages are mostly restricted to a sentence or two. As for visual representation, out of 125 illustrations in Volume 6 women appear in 20. Only in ten of them are women represented alone. The women depicted are mostly either slaves or queens. In Volume 7 women appear in 11 out of the 96 illustrations. Only in two of these illustrations are the women the central characters.

The same pattern can be seen in the Cambridge History. Women appear on three pages out of 517 pages in Volume 5; ten out of 956 pages in Volume 6; 30 out of 1063 pages in Volume 7; and nine out of 1011 pages in Volume 8. Of the three references in Volume 5, one is to Creole women traders, the other to Chokwe acquisition of slave women, and the third is to the growing numbers of European women in the colonial enclaves towards the end of the nineteenth century. Interestingly, in this volume marriage is mentioned on 8 pages without even referring to women at all! In Volume 6 the references are to women's agricultural work (on four pages), women as 'assets' or 'pawns' for chiefs, local lords, and elders (on four pages). The last two are on young women migrating to towns in southern Africa and the importation of British female domestics to South Africa. The bulk of the references to women in Volume 7 are to women's resistance against colonial rule, specifically pass laws in South Africa and taxation in Nigeria. Next comes references to the increased agricultural burden on women as a result of expanded cash crop production, the imposition of forced labour and male labour migration. Interestingly, most of the references are to women in Southern Africa. Women in Central Africa are referred to only on one page, East African women on two pages, North African women on three pages, and West African women on six pages. It is quite remarkable that in Volume 8 which deals with the period 1940-1975, for which there is abundant literature on African women in development, there is only one reference to women as producers! Indeed, in this volume women are largely mentioned in passing, with reference to urban migration, employment, seclusion, and apartheid pass laws.

The single volume general histories are no different. Basil Davidson's (1991) revised and expanded edition of his celebrated *Africa in History*, has only one reference to women, in which the author states rather blandly that 'generally, all women in Africa suffered, as most of them have continued to suffer, from more or less gross forms of discrimination imposed by men' (p. 191). In Curtin, et al.s (1978) African History which, we are told, 'celebrates the coming of age of *African history*, representing a quarter of a century of research by scholars from Africa, Europe and America', and in which 'less emphasis is given to political history and more to social, economic and intellectual trends', women are mentioned only on nine out of the 612 pages, and appear in one out of 25 illustrations. On five of the nine pages, women are mentioned or alluded to in relation to polygyny, in which they are depicted merely as commodities that were circulated. In the remaining references, a paragraph is devoted on page 161 to discussing, in static terms, gender inequality in early East Africa. This is followed, on pages 559 and 566 by sketchy discussions of two paragraphs each, first of the impact of male migrant labour on women during the colonial period, and second, of gender imbalances in settler and non-settler colonial cities. The longest section dealing with women, tries to examine, in three paragraphs, the 1929-30 women's 'riots' in Nigeria known as the Aba 'women's war'. The lone illustration with a woman's representation is a piece of sculpture, whose caption reads: 'Kneeling woman holding a bowl, from Luba, Zaire, Buli workshop. Such statues were used by Luba kings. White porcelain clay with supernatural powers was kept in the bowl. This is a utensil of sacred kingship'. This is all the authors have to say about gender relations in this society![4]

There are more references to women in Robert July's (1992) latest edition of *A History of the African People*. They appear on 20 out of the book's 593 pages and in seven out of the 78 illustrations. But the descriptions and depictions are very sexist. Women are portrayed either as high status queen-mothers or merely as pawns and commodities that were distributed by male elders. According to the author, they were valued in pre-colonial societies primarily for their fertility (p. 548), and by the Europeans as concubines (p. 146), for they were otherwise part of the rural 'unproductive population' (p. 405). Indeed, in July's account women are discussed in the same breath as children, debtors and slaves in the pre-colonial era (p. 125), and as children, the aged, and the infirm in the colonial era (p. 406). Women's lives are seen as static, as shown by the fact that the longest section on women, which revealingly comes towards the end of the book (pp. 546-7), discusses women 'in traditional African society', thereby glossing over the impact of colonialism, and then jumps to contemporary discrimination against women, which the author attributes largely to 'widespread ignorance among African women concerning the specific details of their own rights'. In the illustrations we mainly see the women walking. When they are doing something, like pounding grain, it is before a background of a drought-stricken landscape, the effect of which is to reinforce the futility of their efforts. The ravaged landscape becomes a metaphor of their utter helplessness and victimisation by, and in a perverse way affinity to, nature.

The victimisation, indeed infantilization, of women is no less explicit in Freund's (1984) self-proclaimed radical book, *The Making of Contemporary Africa*, which is written, it is claimed, 'from a materialist perspective [that] provides a refreshing reinterpretation of the complex events in sub-Saharan Africa since the eighteenth century. It also serves', the blurb continues, 'as a succinct introduction to the history of modern Africa, incorporating in the text a critical appraisal of the best scholarship in recent years'. However, women, who are mentioned on 22 out of the 357 pages, but hardly shown in any of the 12 illustrations, are treated no better than in the other books examined above. Almost invariably, they are mentioned as 'dependents', together with youths, clients and slaves whether in the pre-colonial period (p. 63), or the colonial period (pp. 129, 131, 134). Women and youths are mentioned interchangeably when examining their entry into wage labour (p. 147) and colonial cities (p. 183). For a study claiming to be informed by historical materialism, it is rather strange that before the nineteenth century men and women are shown to have lived in an oversimplified, static, and homogeneous world, in which the men hunted and the women grew and prepared food (pp. 19-20), until, behold, the Europeans brought cassava which 'may have freed women from agricultural labour', never mind that 'the evidence for this is very limited' (p. 45). The marginalisation of women extends to the bibliography. Publications on African women are given only one paragraph in a fifty-page select bibliography.

The Regional Histories

The regional histories display the same tendencies. There are those that totally ignore women, and others that mention them in passing. The few that discuss women in slightly more detail still betray androcentric biases. I have examined five regional histories, covering each region of the continent. Needless to say, regional history is unevenly developed, reflecting no doubt different historiographical traditions, patterns of colonization and decolonization, and the varied constructions of regional identities5. By comparing different editions, some of the regional histories under survey clearly demonstrate that women's history has yet to penetrate the thick walls of androcentrism that encircle African historiography.

An example of a regional history that does not mention women is Abun-Nasr's (1975, 1987) *A History of the Maghrib*. In the second edition published in 1975 women are not even indexed. In 1987 the author published a revised volume that 'supersedes' the previous two editions. He was compelled to do so, he states, because 'our knowledge of Maghribi history has advanced rapidly and new perspectives for interpreting it were opened by research in which Maghribi historians have participated in an outstanding way' (p. xi). The new book is certainly more detailed: it has 455 pages compared to 422 pages for the 1975 edition. But it resembles the earlier editions in one fundamental way: women are still totally ignored. So much for the 'new perspectives'.

Women are also largely absent from the regional histories of southern and eastern Africa that I looked at. They are not mentioned in the first edition of Denoon and Nyeko's (1972, 1984) *Southern Africa Since 1800*. Neither are they mentioned in the second edition of 1984, which was undertaken, the authors tell us, because of the 'very great changes in the quality and quantity of information available. In order to accommodate the new evidence, and the new ideas which have been circulated', they conclude, 'we could not simply make the small changes which are often introduced into the second edition of a book. Instead, we found we had to re-write the book, developing a new framework for this evidence and for these ideas'. This new evidence and the new ideas apparently have yet to discover women or gender. As for the illustrations, out of 23 in the first edition, only three show women, one of a woman barely discernible in a group of men, another of semi-naked women, and the third of women and girls smiling to the camera before a background of a shanty location. In the second edition, the offending picture of naked women has been removed, but the other two retained. In a third picture a handful of school girls are shown as part of the Soweto uprising; they are walking behind a large group of school boys. Thus there are still three pictures depicting women, but now out of 26 illustrations.

Omer-Cooper's (1987) textbook is not much better in terms of the illustrations. Women appear in 18 out of the 115 illustrations. They are prominently featured in only six out of the 18, and only in one do they appear alone. This is a picture of women leaving jail with their fists raised in defiant gesture. In the actual text, women are mentioned on seven out the 297 pages, with reference to marriage (on three pages), Zulu military settlements, royal women, and pass laws (on one page each).

The same skewed coverage of women is evident in the standard history texts on East Africa. Women are not mentioned in Ingham's (1965) study, or Ogot's (1973) widely used text, *Zamani*. Women are also notable for their absence in Volumes I and II of the three volume *Oxford History of East Africa* (Harlow and Chilver 1965). In Volume III women are mentioned on ten out of 691 pages, mostly in connection to their marriage patterns, fertility, and morals as perceived by missionaries and other colonial ideologues (p. 405-8). Women's political activities are mentioned very briefly on two pages, noting the formation in Tanzania in the 1950s of a Council of Women by a certain Lady Twining and a women's section in the Tanganyika African National Union (TANU), respectively (p. 185, 187). As for women's productive roles, the book is largely silent, except to note, in a sentence put in brackets, that '(women, except for those who had found freedom, at a price, in the towns, did what they had always done)' (p. 512).

The situation is not much better with Ajayi and Crowder's *History of West Africa* (1985), the standard textbook on West African history. According to the index of Volume 1 of the 1976 edition[6] women are mentioned on four pages out of the book's 649 pages. The textual material is confined to fleeting statements on the institution of women chiefs in the Ondo area of the Yoruba, and the active role played in political life by women relations of the king in the Wolof and Serer kingdom. There are 26 additional references to women which can be culled from the text. They include the three references to Queen Amina, and the 11 and 12 references to matrilineal and patrilineal systems, respectively. On 'the legendary exploits' of Queen Amina the author murmurs that 'her conquests and achievements may have been exaggerated' (p. 561). As for the statements on the matrilineal and patrilineal systems, they are often presented in the anthropological present, and no attempt is made to analyse how they developed, or the content of gender relations they embodied. For example, we are told (p. 464), without explanation, that in Djoloff the predominant matrilineal system gradually gave way to the patrilineal system. Volume 2 of the 1976 edition, which covers the nineteenth and twentieth centuries, has, surprisingly, even fewer references to women. There is only one reference to the category 'patrilineal', none to 'matrilineal'. No remarkable woman is mentioned. Half of the references to women, made on six pages in a book of 764 pages, are on the impact of the nineteenth century jihads. The famous 1929 Women's Aba riot is given short shrift in two sentences.

A comparison between the 1976 and 1985 editions shows little improvement in terms of gender coverage and analysis. The example of Volume 1 will suffice. In the 1985 edition, according to the index, there are two additional pages that refer to women. The additions are on women as slaves (p. 640-1). In the meantime, references to matrilineages and patrilineages have been reduced to two pages each, and if one adds references to marriage and family, there are 12 other references to women. In addition to those directly referring to women and Queen Amina in the index, women are mentioned on 22 pages, less than the number in the 1976 edition. And yet the 1985 edition is 93 pages longer than the former edition!

The most extensive coverage of women among the regional histories I examined was found in Birmingham and Martin's (1983) *History of Central Africa*. The fact that it was first published in 1983 may have something to do with it. Also, unlike the texts examined above, one of its editors is a woman. Volume One deals with the pre-colonial period, while Volume Two focuses on the colonial and post-colonial periods. In the first volume women are mentioned on 59 out of the book's 315 pages, and in the second volume on 53 out of the 432 pages. In both volumes, however, women are mostly referred to in relation to marriage. References to women and marriage can be found on 35 out of the 59 pages where women are mentioned in Volume One and on 30 out of the 53 pages in Volume Two. The bulk of the remaining references deal with women as timeless victims of a ferocious patriarchal order. In Volume

One women are mentioned as subordinate agricultural labourers and as slaves on nine pages each. In Volume Two women's labour, whether in the agricultural or the urban economy, is mostly discussed as an appendage of male migrant labour. Predictably, the remaining contexts in which women are mentioned centre on women's infertility and prostitution.

The Thematic Histories

It would appear that women's invisibility is no less marked in the historical studies dealing with specific themes. It is most apparent in studies dealing with political history, and slightly less so in texts on economic and social history. Out of the seven studies on nationalism and decolonization that I examined, four do not mention women at all (Davidson 1978; Mazrui and Tidy 1984; Hargreaves 1988; and Gifford and Louis 1988). In Rotberg and Mazrui's (1970) massive collection on *Protest and Power in Black Africa*, women are not indexed, but one of the contributions is on a woman religious and nationalist leader, Alice Lenshina of Zambia (Roberts 1970). That is one out of 35 contributions. In Gifford and Louis's (1982) *The Transfer of Power in Africa*, which is 654 pages long, women are mentioned only once, not in the actual text, but in the bibliographic essay, where a study on women's involvement in the Algerian revolution is noted and the point made that this involvement 'did not lead to an improvement in their condition in a Muslim society. Once independence was achieved, a traditional reaction scuttled the advances they had started to make' (p. 534). De Braganca and Wallerstein's (1982) three volume reader on African liberation movements only contains two documents by women: one is by Zanele Dhlamini on women's liberation in South Africa prepared on the occasion of the South African Women's Day in 1972 (Dhlamini 1982), and the other by Sinclair (1982), President of the South African women's organization, Black Sash, replying to a newspaper article disputing claims that conditions in South Africa in 1970 were improving. The cover of Volume 2, in which there is no document by a woman, shows a male soldier with a gun receiving a pumpkin from a woman, who is balancing another pumpkin on her head while holding a third by her other arm. The message is clear: men are the fighters, women the food providers. So much for the transformative power of liberation struggles!

Three of the six books on economic history that I looked at also do not mention women or deal with the question of gender (Munro 1976; Wickins 1981; and Issawi 1982). The other three make very feeble efforts to do so. In Rodney's (1982) renowned *How Europe*

Underdeveloped Africa women are mentioned on six out of 312 pages. Brief references are made to the exploitation and oppression of women in the Maghreb (p. 55), the women Amazon warriors in Dahomey (p. 121), and women's limited access to education during the colonial period (p. 251, 266). The most detailed treatment of women comes in the last chapter on the impact of colonialism on Africa. Ironically, it outlines the role of women in 'independent pre-colonial Africa'. The author discusses the 'two contrasting and contradictory tendencies'. On the one hand, women, especially 'in Moslem African societies', were exploited and oppressed by men through polygamous arrangements. But they were also accorded respect and enjoyed a 'variety of privileges based on the fact that they were keys to inheritance', on the other. Indeed, 'women had real power in the political sense, exercised through religion or directly within the politico-constitutional apparatus' (p. 226). It is quite strange that in an economic treatise women's economic roles are hardly addressed.

In Hopkins' (1973) *An Economic History of West Africa* and Austen's (1987) *African Economic History* only the barest allusions are made to women's economic roles. Hopkins refers to women on six out of 337 pages in two contexts: in connection with household labour and local trade. He notes that in the (timeless) pre-colonial era, West African 'societies distinguished between the labour of men and women, though the line was not always drawn at the same point' (p. 21). As for trade, women's involvement is portrayed as having been restricted to local trade on the grounds that 'local trade was a convenient adjunct to household and, in some societies, farming activities' (p. 56). Recent studies have shown that women were also involved in long distance trade (Afonja 1981; White 1987; and Amadiume 1987). Despite its publication almost a decade and half after Hopkins' study, Austen's book is far less satisfactory both as an economic history text and in its coverage of women. Women are mentioned on ten out of 294 pages, either in passing (sometimes even in brackets as on p. 180), or invoked to support dubious contentions. For example, Austen denies that the Atlantic slave trade had a negative demographic impact on Angola because women 'who are the key determinant of reproduction in any human population' were left behind (p. 96). He also disputes that colonial cash production undermined domestic food supplies for women continued their 'traditional' food producing activities (p. 139, 145).

The most extensive coverage of women in the studies I examined was found in books on labour and social history published in the 1980s. Earlier labour history studies tended to ignore women. For example, women are notable by their absence in the two renowned labour history studies published in the 1970s: *The Development of An African Working Class* (Sandbrook and Cohen 1975) and *African Labour History* (Gutkind et al. 1978). Two relatively recent labour histories compare favourably to this. One is by Stichter (1985) and the other by Freund (1988). In Stichter's *Migrant Labourers*, women are discussed on 82 out of 225 pages. In fact, two of the seven chapters are specifically devoted to women. In Freund's *The African Worker*, women are featured on 28 out of 200 pages. Stichter's analysis on women centres on two main issues. First, the effects of male labour migration on women where it is argued that male labour migration led to changes in the traditional division of productive labour between men and women. Women's workload increased as they took on tasks previously done by men and became heads of households. They showed initiative by adopting new agricultural strategies and trading roles, or by migrating to the cities. Secondly, in Chapter 6 Stichter examines women as migrants and workers by looking at the factors behind female labour migration, the patterns of women's employment, and the forms of women's consciousness and struggle.

Stichter seeks to celebrate women's active involvement in the labour process, but in the end she idealises colonialism as a force that liberated African women from ruthless patriarchal control. In 'African pre-capitalist societies', she asserts, women's 'status was not dissimilar to that of slaves and serfs' (p. 148). This contention is based on an uncritical acceptance of anthropological theories on 'domestic', 'lineage' or 'patriarchal' modes of production according to which male elders controlled the labour of junior males and women of all ages[7]. Not only is the conceptualisation of modes of production problematic, as demonstrated in Chapter 6 above, but gender relations in pre-colonial Africa cannot be generalised[8]. As Freund states, 'the rights of male elders to appropriate surplus in African societies varied immensely' (p. 6), so that 'it is a tricky business to generalise for sub-Saharan Africa as a whole on the question of women and labour exploitation' (p. 83). However, Freund's own examination of women and the labour process (concentrated on pp. 81-90) is far less satisfactory than that provided by Stichter. It lacks any systematic historical analysis, for unconnected and undeveloped points are thrown around on women's labour in the household, informal sector, factory work, and domestic service. That says something about the author's valuation of women as historical subjects.

A similar problem can be seen in the books on social history that I examined. While efforts are made to incorporate women, they are still depicted either as marginal or weak. For example, although several authors in *Peasants in Africa: Historical and Contemporary Perspectives*

(Klein 1980) refer to rural women, women are never depicted as central to the peasant production systems, societies, struggles, and transformations being analysed. In Feierman and Janzen's (1992) collection, *The Social Basis of Health and Healing in Africa* in which women are considered on 58 out of the 487 pages, the women are largely discussed with reference to their fertility patterns, rather than their role as healers, unlike men. We are also told of male perceptions of disease rather than female perceptions. In a rare comparison of male and female medical practitioners, we are informed that among the Zulu women practice medicine in a 'clairvoyant' manner while men practice in a 'nonclairvoyant' manner (Ngubane 1992). In Illife's (1987) ambitious, but disappointing, tome *The African Poor: A History*, women are discussed on about 100 out of 387 pages. But Illife's poor women, like his poor in general, are timeless victims of Africa's seemingly primordial structural poverty. They are invariably 'unsupported' or 'unattached' women, that is, women without men, the unmarried, widowed, and sterile women. Nothing could save them from poverty, neither wit nor informal sector activities. And they could not turn to poverty relief institutions or their own social welfare and support networks for these institutions and networks were poorly developed or non-existent. Their only salvation lay in marriage. In short, married and dependent women are invisible from the ranks of Illife's poor.

African Historiographies and Women's History

The relative underdevelopment of African women's history can partly be attributed to the fact that, as Bolanle Awe (1991:211) has argued, 'compared with the history of many other parts of the world, the writing of the history of Africa itself is a fairly recent development'. Few would dispute that history as a discipline is intrinsically empirical. That does not mean, however, that historical reconstructions are not based on deeply held philosophical assumptions, or specific theoretical frameworks often borrowed from the other social sciences. In the last three decades, as demonstrated in earlier chapters, three paradigms have dominated mainstream African historiography: the nationalist school, which was dominant from the time of decolonization to the early 1970s; the underdevelopment or dependency perspective, which held sway from the late 1960s to the late 1970s; and the Marxist approach which gained ascendancy in the 1970s and early 1980s. This periodization is not meant to denote neat sequential stages, for elements of all three paradigms have coexisted at any one time in the last three decades and, indeed, continue to do so, as shown in Chapter 7 on imperialist historiography.

As noted earlier, in reconstructing African history, the nationalist historians were preoccupied with eradicating imperialist and racist myths that Africa had no history prior to the coming of the Europeans, and in devising new methods of research to recover African history (Ki-Zerbo 1981; Vansina 1985; Henige 1982). This fixation with celebrating and laying the empirical framework of African civilizations not only consumed the historians' energies, but also blinded them to gender analysis. These historians sought to reclaim and glorify Africa's great states, cities, and leaders. In short, nationalist historiography was primarily political and elitist. It had little to say about the 'masses', whether men or women, or social and economic history. Almost invariably, exploitation and oppression were discussed only in reference to colonialism. Thus in its epistemology, nationalist historiography had neither the conceptual tools nor the ideological inclination to deal with class or gender hierarchies, exploitation and struggles in African history.

For their part the historians using the dependence paradigm focused primarily on the economics of exploitation, but in spatial, not social or class terms. Development and underdevelopment were seen as integrated and dialectical processes, linking and reproducing the differentiated spatial configurations of Europe and Africa, 'metropoles' and 'peripheries', 'centres' and 'satellites', the 'North' and the 'South', 'developed' and 'developing' countries, the 'First' and 'Third' worlds. Consequently, the central problematic of dependence historiography was to unlock and explain the process by which surplus from Africa and the peripheries in general was drained, expatriated, or appropriated by Europe or the metropoles in this integrated world capitalist system. Unequal exchange, whether of products or labour costs, became the pivot around which the entire process of western development and 'Third World' underdevelopment spun. The dependence paradigm produced a static, frozen history of Africa, one in which external forces played the predominant role. It is a history of inter-national, not class, relations and struggles. Whenever class is alluded to, it is often used as a derivative and functionalist category, simply as one among the many factors that mediate dependence and underdevelopment. If dependence historiography ignores class, it has proved stubbornly blind to gender analysis.

On this score, Marxist scholars were hardly any better, despite their vigorous critiques of both nationalist and dependency historiographies. Marxist historians were too preoccupied with fitting African histories into the Marxian modes of production, or inventing tropicalized varieties, and articulating them with the capitalist mode during colonialism, to delve seriously into gender analysis. Besides, class, not gender, is the central problematic of traditional Marxism. Women's oppression is seen as a secondary phenomenon, a symptom of capitalist oppression. As argued in the classic Marxist study on women, *The Origin of the Family, Private Property, and the State* by Engels (1972), women's oppression originated with the introduction of private property. Contrary to popular perceptions, this study does not offer a concrete historical analysis but an abstract model based on dubious anthropological data (Lane 1976). The inadequacy of the traditional Marxist paradigm has given rise to other feminist frameworks, including radical feminism and socialist feminism, which seek to comprehend the role of class as well as gender, race, and nationality, among other social constructs, in the creation of women's oppression and liberation (Jaggar and Rothenberg 1984; Hirsch and Keller 1990; Hutchful 1996).

It can be seen, therefore, that none of the three dominant paradigms used in reconstructing African history takes women's history and women's oppression seriously[9]. Not surprisingly, women are either absent or marginal in the historical studies examined above, which were in one way or the other inspired by these frameworks. Thus the challenge that faces feminist historians is not only one of recovering women's history, of redressing balances, but also one of developing new theoretical frameworks that better explain the real world. In this endeavour, feminist historians have been busy deconstructing the hierarchical conceptual dualisms that seek to encase women's lives in the worlds of 'nature' and the 'family', and the 'private' and 'domestic' spheres, as distinct from the supposedly male worlds of 'culture' and 'work', and the 'public' and 'political' spheres. To begin with, the binary vision contained in these dualisms, such as the private/public divide, misrepresents the interdependence and interconnectedness of social reality and processes. Moreover, these distinctions and dichotomies are not universal, whether as empirical realities, or as conceptual categories. They arose in a specific European historical context[10] and are derived from Enlightenment thought (Foster 1992:3-6).

Historians concerned with gender analysis have to guard against both essentializing and universalising the experiences of particular, mostly white middle-class western, women. 'There are startling parallels', writes Spelman (1988:6), 'between what feminists find disappointing and insulting in Western philosophical thought and what many women have found troubling in much of Western feminism'. All too often race, ethnicity, and class are inserted as 'additive analyses'. The unfortunate result is a discourse that is patently racist, especially when spurious comparisons are drawn between racism and sexism and the latter is depicted as being a more "fundamental" form of oppression, for it distorts and ignores the reality of Black women who experience both forms of oppression.

In North America the ethnocentrism and 'white solipsism', as Rich (1979) calls it, of western feminist scholarship has come under sustained attack from African-American and African-Canadian feminists and other so-called 'women of colour'[11]. These criticisms have caused white middle-class feminists considerable discomfort, guilt, and sometimes reappraisals of their intellectual and political practices. The problems of feminist ethnocentrism or Eurocentrism are even more blatant when it comes to studies of women in the so-called 'Third World' (Sievers 1989; Afary and Lavrin 1989; Reinharz 1992). In African studies the Eurocentric virus afflicts not only women's studies but all the social science disciplines and the humanities, especially when it comes to the construction of 'theory' and the writing of regional or continental surveys and syntheses (Imam and Mama 1994). Western Africanists, who are by their very existence implicated in western dominance, have often not displayed the necessary reflexivity and 'epistemic humility', to borrow Pierson's (1991) term. African scholars, including feminists, have fought vigorously against this 'intellectual imperialism'. Despite their criticisms, ethnocentric practices are still alive and well in western feminist scholarship on Africa as can be seen in the recent special *Signs* issue on Africa which blithely justifies the absence of contributions from African women scholars[12].

Our review of the literature has so far been derived mostly from the criticism of *content*, the poor coverage of women, the tendency to view women's lives as peripheral and unchanging, all of which reflect the absence of concepts that tap women's historical experiences. Little has been said about *methodology*, that is, the actual techniques and practices used in the research process. How do the methodologies of the three historiographical frameworks compare with the trends in feminist research?

Feminist researchers use a variety of methods. But they all arise, according to Fonow and Cook (1991a:2), 'from a critique of each field's biases and distortions in the study of women'. Their work tends to display, they argue, reflexivity, action-orientation, and attention to the affective components of the research, among other things. Feminist historians, more specifically, have embraced oral history as a key method to recover women's experiences and voices from androcentric notions, assumptions, and biases which dominate 'malestream' history everywhere. As one author has put it, 'women's oral history is a feminist encounter because it creates new material about women, validates women's experience, enhances communication among women, discovers women's roots, and develops a previously denied sense of continuity' (Reinharz 1992:126)[13]. Women's history is also unusually interdisciplinary in its approach.

Of the three paradigms, it would seem that nationalist historiography, has more in common with feminist history in terms of *methodology* than with either the dependence or Marxist perspectives, both of which rely on traditional social science research methods. Nationalist historians prize oral tradition, which they believe enables them to recover African experiences and 'voices', that is, African perceptions of their lives, their consciousness, often silent in the arid and self-serving written records of colonial functionaries. Oral sources remove the cloak of invisibility enveloping many aspects of African history. Confronted with limited or non-existent written sources, nationalist historians were also unusually open to the use of a wide range of sources, from oral traditions and historical linguistics, to the findings of anthropology and the natural sciences. This made interdisciplinarity an important feature of nationalist historical scholarship. Thus feminist and nationalist historians tend to privilege oral methods in their efforts to dismantle deeply entrenched biases and recover the history of long suppressed, exploited, and humiliated groups of people.

The goal of nationalist historiography was to bring Africa and Africans back into history. In this sense it was an emancipatory project. But nationalist historiography did not deviate from the contours of western historiographies, from which it borrowed most of its questions and assumptions. It sought to demonstrate that Africa had built civilizations ·comparable to those of Europe. To what extent can women's history escape such a fate? Is restoring women to history enough? Is women's history to develop as an autonomous field of research, or is its aim to reformulate and transform history as a whole? Women's history is slowly gaining ground in many countries but there are already signs of its ghettoisation[14]. Those who would wish to avoid this trajectory suggest going beyond writing women's history by writing gender history. Women's history focuses specifically on women's experiences, activities and discourses, while gender history provides analyses concerning how gender operates through specific cultural forms (Newman 1991:59).

Restoring African Women to History

In African history feminist historians are still largely at the stage of restoring women to history, of writing what Lerner (1979:Chapters 10-12) has called 'compensatory' and 'contribution' history, rather than of writing gender history[15]. The last two decades have seen rapid growth in the literature on African women. Most of it is the work of anthropologists, sociologists, and development specialists. The number of historians writing about the historical experiences of African women is still relatively small but growing[16]. Already the days when African women were painted with the brush of exotica and seen as a monolithic group afflicted by eternal victimisation seem to be long gone. Explanatory models of women's oppression derived from European and American history and racist anthropology have come under challenge and been stripped of their universalistic pretensions. African women are no longer seen as being cloaked in veils of 'tradition' from which they were gradually liberated by 'modernity', for the concepts 'tradition' and 'modernity' have been exposed for their ahistoricity and ethnocentrism[17].

The themes that preoccupied anthropologists for ages, such as kinship, marriage, fertility, sexuality, and religion are being re-examined as historical processes. Moreover, feminist historians are beginning to examine more systematically the historical development and construction of women's culture, solidarity networks, and autonomous social spaces. The importance of women's economic activities is being demonstrated, whether it is in agriculture, trade, or crafts and manufacturing. Researchers have also shown that women actively participated in pre-colonial politics, both directly as rulers and within arenas viewed as the female province, and indirectly as the mothers, wives, sisters, daughters, and consorts of powerful men. Women's involvement extended to military participation, both as individuals accompanying male troops and as groups of actual combatants. It can no longer be doubted that during the colonial era women actively participated in nationalist struggles. They either organised their own groups and fought against colonial policies which they saw as inimical to their interests, or they joined male-led nationalist movements. Colonialism is seen to have had a contradictory and differentiated impact on men and women, as well as on the women themselves. The more nuanced accounts reveal that while the position of most women declined during the colonial era, women also took initiatives that reshaped their lives and challenged the colonial order.

In terms of periodisation, most of the literature concentrates on the nineteenth and twentieth centuries. Women's history before 1800 is still largely tentative. The rest of this section presents a brief bibliographic survey of women's history in different parts of the continent[18]. For the period before 1800 the few works on women in the Western Sudan focus mainly on three themes, first, the political role played by women leaders, such as Amina; second, the impact of Islam on the gender division of labour and women's position in society; and third, the growth of women's slavery with the expansion of the trans-Saharan slave trade (Sweetman 1984; Callaway 1987; Robertson and Klein 1987). For the West Coast and its hinterland the literature has dwelt on women's active participation in trade, production and state formation, and increased social stratification among women (Afonja 1981; Awe 1977; Brooks 1976). The historiography on eastern and southern Africa has featured the role of queen mothers, marriage and kinship systems, and the role of women in production (Young 1977; Leacock 1991; Kaplan 1982; Mbilinyi 1982; Sacks 1982; White 1984; van Sertima 1985; Kettel 1986).

The historiography on women becomes more voluminous for the nineteenth century. The analysis tends to be richer in empirical detail and displays more theoretical sophistication. For Western Africa Aidoo (1981) emphasises the central role that Asante queen mothers played in the nineteenth century. Wilks (1988) looks into the life of one remarkable woman in Asante. Hoffer (1972) and Boone (1986) discuss how female solidarity among the Mende enabled some women to become chiefs and exercise political power. White (1987) sensitively charts out the development of women traders in Sierra Leone. Carney and Watts (1991) show that the intensification of agricultural production in the Senegambian region from the mid-nineteenth century was both a social and gendered process. Mann (1985, 1991) explores women's urbanization in Lagos by looking at the changing forms of marriage and social status for elite women and their access to landed property, capital, and labour in the second half of the nineteenth century. Roberts (1984) suggests that the growth of local slavery freed elite Maraka women from agricultural labour and allowed them to expand textile manufacturing which they controlled. In her penetrating study, Amadiume (1987) delineates the changing constructions of gender and sex roles in Igbo society. Boyd (1986) writes of the Fulani women intellectuals produced by the jihads, while Imam (1991) brilliantly charts out the development of seclusion in Hausaland before and after the establishment of the Sokoto Caliphate as well as during and after the colonial period.

The nineteenth century was also a period of rapid change in eastern and southern Africa. The expansion of commodity production, which sometimes included the slave trade, appears to have facilitated the subordination of women in some societies. Such appears to have been the case among the Mang'anja in southern Malawi (Mandala 1984), the southern Tswana (Kinsman 1983), the Maasai (Talle 1988), and in southern Mozambique (Isaacman 1984). In other societies, women's productive roles, economic autonomy, property rights, and household relations were transformed by the adoption of new technologies, such as the plough, as has been demonstrated in the case of Basotho women (Eldridge 1991), or as a result of political change, such as the reorganization and expansion of the military system as has been demonstrated in the case of the Nandi of Kenya (Gold 1985), which led to the progressive removal of male labour from the homesteads, and the intensification of female labour time in household production. Women responded to these changes in various ways. Their solidarity, as well as opposition and accommodation to their growing subordination, was articulated through song and poetry (Gunner 1979), the formation of spirit possession cults, dance, improvement, and puberty rites associations (Strobel 1979), the manipulation of ritual and prophetic power and conversion to Christianity (Comaroff 1985). In addition, some resorted to casual labour and prostitution, selling and buying land, or tried to put their role as food producers to good effect (Clark 1980; Crummey 1981, 1982; Spaulding 1984; Alpers 1986; Kapteijns 1985).

Analyses of women in nineteenth century North Africa have also become more sophisticated as historians abandon the idealist biases, according to which the status and role of women in these societies is primarily attributed to the ideas and values contained in Islamic religious and juridical texts. It has become quite clear that this approach ignores the fact that the formal texts do not tell us much about the changing realities of women's lives in the extremely diverse societies and countries that make up the so-called 'Muslim world' (Beck and Keddie 1978; Keddie 1979; Keddie and Baron 1991; Tucker 1983; UNESCO 1984; Jansen 1989; Ahmed 1992). The literature on Egypt makes it clear that the exploitation of peasant women increased in the course of the century thanks to agricultural 'modernization', state centralization, labour and military conscription, and the progressive decline of the extended family as a semi-autonomous unit and the consequent consolidation of family property around men. At the same time, however, some elite women acquired land either through purchase, inheritance, usually in the absence of male children, or grants from male relatives, especially a father (Tucker 1985). Seclusion of middle class women appears to have increased as the old merchant classes became marginalised due to the imposition of state trading monopolies and as the wives of the 'new' urban-based petite-bourgeois professionals were increasingly cut off from their husbands' professional lives and relegated to the domestic sphere (Cole 1981). All these changes provoked debate about the position of women in society. The feminist discourse was conducted among the intellectuals, including men (Cole 1981; Kader 1987; Philipp 1978; Cannon 1985).

For women's history in the twentieth century, the impact of colonialism has, predictably, featured prominently. Many of the writers already referred to in the preceding paragraph examine how African women were affected by the imposition of colonial rule. They demonstrate that colonial patriarchal ideologies combined with indigenous patriarchal ideologies tended to reinforce women's subordination, exploitation and oppression. Many elite women were progressively marginalised as they lost their political power and control over trading and manufacturing activities. But there were other women who took advantage of the expanding petty commodity markets (Ekejiuba 1967; Johnson 1978), or who sought to retain their autonomy by migrating to the rapidly growing colonial towns and cities where they often engaged in trading activities, beer brewing, domestic service, and sometimes prostitution, thanks to the acute demographic imbalance between the sexes (Little 1973; Bujra 1975; van Onselen 1982; Gaitskell et al., 1983; Robertson 1984; White 1990). The expansion of cash crop production and male labour migration increased women's workloads, while at the same time their ability to appropriate the products of their labour declined (Boserup 1970). Migrant labour was particularly prevalent and its negative effects on women especially evident in Southern Africa (Muntemba 1982; Wright 1983; Walker 1990). There were, of course, some societies where women did succeed in retaining and even improving on their previous autonomy, if only temporarily (Hay 1976; Mandala 1984).

All these developments produced acute tensions in gender relations, to which the colonial state responded by tightening restrictive customary law, which led to important changes in family structure and created new forms of patriarchal power (Chauncey 1981; Hay and Wright 1982; Chanock 1985; Roberts 1987). By far the topic that has attracted the most attention is that of women's resistance to colonial rule (Denzer 1976; Rogers [Geiger] 1980, 1990). The studies range from those that examine specific activists (Denzer 1981, 1987; Okonkwo 1986a;

Rosenfeld 1986; Brantley 1986) and events, such as the Aba Women's War of 1929 (van Allen 1976; Ifeka-Moller 1975), the Anlu's Women's uprising in the Cameroons (Ritsenthaler 1960), the spontaneous uprisings of South African women in the late 1950s (Bernstein 1985) and their participation in the struggles against apartheid generally (Goodwin 1984; Mandela 1984; Kuzwayo 1985; Barret 1986), to general analyses of women's involvement in nationalist struggles in various countries (Steady 1975; Denzer 1976; Mba 1982; Walker 1982; Weiss, 1986; Geiger 1987). It is now abundantly clear that women were actively involved in the wars of national liberation, such as Mau Mau (Likimani 1985; Kanogo 1987; and Presley, 1991), and those in Algeria (Gorden 1972), the Portuguese colonies (Urdang 1979, 1984), Namibia (Cleaver and Wallace 1990), and Eritrea (Wilson 1991). Studies are also beginning to appear on women's active involvement in labour movements and struggles (Robertson and Berger 1986; Zeleza 1988a; Mashinini 1991).

For the post-colonial period much of the literature has focused on whether or not women's position and status has improved or deteriorated with independence. The scope of subjects covered is wide, ranging from women in the rural and urban economies and women's participation in state politics and development projects, to changes in the structure of marriage and kinship. The literature shows that in many countries women's rural production has become more commodified since independence. In addition to farming, women in regions afflicted by the growing crises of subsistence have increasingly resorted to petty trading and wage labour to make ends meet. Commodification has increased the differentiation of rural women and made it more complex (Afonja 1981, 1986; Guyer 1984; Okali 1983; Crevey 1986; Newbury and Schoepf 1989).

Research on African women has privileged rural over urban women, perhaps because the vast majority of African women are still rural dwellers (Simmons 1988; Davison 1988, 1989). But it is quite clear that the number of women migrating to and living in cities has risen considerably (Sudarkasa 1977; Adepoju 1983; Perold 1985; Stichter and Parpart 1988). Much of the literature on urban women has tended to focus on their activities as traders or informal sector operators. Those studies that deal with women in wage employment have demonstrated that while women's employment has grown rapidly in many countries since independence due to economic expansion, increased women's access to education, changes in family structure, and struggles by the women themselves for economic independence, women still tend to be crowded in low-paying service jobs and have to juggle with the burdens of the double day (Selassie 1986; JASPA/ILO 1986b, 1986c, 1986d, 1986e, 1986f, 1988; Zeleza 1988b; Stichter and Parpart 1990).

The studies done on women's participation in state politics demonstrate that women have been excluded and marginalised from the political process, despite their active involvement in the independence struggles. In some countries women, especially petty traders, have been targeted as scapegoats and attacked by states facing acute economic problems[19]. The literature has also amply demonstrated that until quite recently most government and international aid organizations primarily focused on men rather than women in their development projects. This was gradually changed thanks to the growth of the feminist movement and the food crisis in many African countries. The 'women in development' movement and ideology was born. But it has done little, to date, to empower the vast majority of Africa's economically exploited and politically marginalised women (Brain 1976; Nelson 1981; Lewis 1984; Mbilinyi 1984; Overholt et al., 1985; Swantz 1985; and Munachonga 1989). This is true even in the self-styled 'socialist' regimes (Haile 1980; Urdang 1983; Fortman 1982; Seidman 1984).

But African women in the post-independence era have not been passive victims. They continue to struggle both individually and collectively against their exploitation, oppression, and marginalisation, and to push open the doors to economic, political, social, and cultural empowerment (Obbo 1980, 1986; Stamp 1986; Dolphyne 1991).

Gendering African History

It is quite evident that a lot of work has been done to recover women's history, but much more needs to be done. Also, the history that has so far been recovered has yet to be fully incorporated into the mainstream of African historical studies. Feminist historians, therefore, have to pursue a two-pronged agenda: writing women's history and gender history. Women's history, or 'herstory', is often seen as a reconstruction, a retrieval, of women's experiences, expressions, ideas and actions. Gender has been defined as the changing social organization and symbolic representation of sexual difference, the primary field within which or by means of which power is articulated or signified. As a concept it offers an epistemological redefinition of historical knowledge as construction rather than reconstruction (Scott 1988:Chapters 1 and 2). To put it simply, it is said that in women's history the primary focus is on women, while in gender history it encompasses both men and women as gendered subjects.

Apart from its explanatory power, the growing importance of gender as an analytical category reflected growing frustration among feminist historians at the relatively limited impact that women's history was having on mainstream historical studies[20]. There were also those who may have adopted the term 'gender' merely as a synonym for women because it sounded more objective and neutral than 'women', and thus gave their work academic legitimacy. Moreover, its popularity was probably helped by the proliferation of studies on sex and sexuality. It can further be argued that the concept of 'gender' offered the reductionist paradigms of Marxism and psychoanalytic theory a much-needed face-lift. Unfortunately, women's history and gender history have increasingly come to be seen in oppositional and hierarchical terms. This reproduces the very binary thinking and dichotomous models feminist historians have been at pains to discard[21].

The elevation of gender history over women's history may appear more 'radical' and inclusive, but can in fact play into the hands of anti-feminists and legitimate exclusionary practices in academia. Courses in women's history can be opposed on the grounds that gender is integrated in the mainstream courses when that is in fact not the case. This is, for example, the situation in Canada where, Pierson (1992:138) points out, there is no 'positive evidence that the paucity of women's history courses results from mainstream adoption of gender as "a useful category of historical analysis"', leading to an integration of gender history and the history of women's past experiences into non-women's history courses, undergraduate and graduate'[22].

Women's history and gender history, are mutually reinforcing, and need to be pursued simultaneously by feminist historians. In concrete pedagogical terms this means devising curricula that contains specific courses in women's history and consciously incorporating feminist perspectives in mainstream courses. Creating and maintaining specific courses in women's history is based on a recognition that women's history represents 'a field of knowledge production which has its own history, formed by both the politics of women's liberation and intellectual developments within history and in associated disciplines' and that there are methodological frameworks that are specific to women's history and women's studies in general (Allport 1993). Women's history, in short, must not be seen as a temporary necessity, something that is not 'real history'. Women's history is, both on an empirical and theoretical level 'one of the most exciting historical specializations today' and by its very existence is instrumental in 'deconstructing mainstream historiography. By emphasizing the 'other side' of history, women instead of men, the implicit male perspective of historiography that has obliterated women becomes explicit. This process is 'pivoting the centre' of dominant historiography. It exposes normative and expressive rules of both historical writing and teaching' (Grever 1991:77).

The actual content of the courses in women's history, and the teaching methods, will of course vary, reflecting, no doubt, different national histories, women's experiences, and intellectual traditions. Underpinning courses in women's history, epistemologically and pedagogically, should be feminist theorizing that recognizes difference and the gendered nature of all social relations and works on the immediate environment to achieve political action (Foster 1992:10-25). These courses must not only be offered at the university level, but at the primary and secondary school levels. Needless to say, this is likely to be met by resistance from the educational authorities in many countries. The strategies to overcome such resistance will necessarily vary. But such endeavours and struggles are unlikely to go far without organization. Feminist historians need to make women's history visible by organising all kinds of activities, penetrating the councils that design syllabuses and set examinations, and by publishing new material. Without new course books the case for women's history is unlikely to be advanced. In other words, in addition to publishing sophisticated articles, monographs and books on women's history for use at the university level, feminist historians have to undertake the far less glamorous task of publishing new material for schools.

Advancing gender history and mainstreaming entails gender-balancing courses and making gender as fundamental as, say, class as a category of historical analysis. Taking gender seriously as a conceptual tool for understanding the human past challenges the conventional periodizations based on political events and cultural and religious shifts in which men were preponderantly involved[23], and transcends the traditional questions and problematics, constructs of significant events, and the theories and explanatory models of social change (Scott 1988; Kelly-Godol 1984).

Gendering history is a process that involves a series of curricular changes, whose ultimate objective is a balanced and inclusive curriculum, in which women's and men's past experience can be understood together. A number of stages have been suggested in developing a gendered history curriculum (Schuster and van Dyne 1984; Schade 1993). Confronted with a curriculum in which women are absent, the feminist historian could begin by searching for and incorporating the missing women within the conventional paradigms. This would essentially be a story of the heroines, of the great women leaders, warriors, traders, thinkers, and so on. This could be followed, or accompanied, by offering specific lectures within the course on women experiences during the period under discussion.

This gradualist or additive approach is problematic. Introducing women's history into the curriculum through a few 'exceptional' examples does little to change the existing paradigms. In fact, a subtle, and perhaps unintended, message may be imparted to students: that since some women did succeed the failure of others to do so may be ascribed to their lack of motivation, ability, and other individual attributes. This serves to deny the reality of oppressive structures. Adding a couple or so lectures may make women seem anomalous, the material about them marginal to the core knowledge covered in the curriculum. This is merely to suggest that the larger goals of curriculum transformation must not be lost in well-meaning, but token, gestures which do not challenge the conventional paradigms[24].

The questions of gender, class, and other social constructs that shape historical change, such as race and ethnicity, must be discussed explicitly. One way of confronting androcentric historiographical biases and promoting gender history is to use 'battling readings' throughout the course. This involves using readings from 'regular history' and 'women's history' for every topic discussed. This forces students to confront different constructions of history and the differentiated participation of men and women in historical processes. For example, in discussing trade in nineteenth century West Africa one can pit Hopkins's (1973) *An Economic History of West Africa* and White's (1987) *Sierra Leone's Settler Women Traders*. In studying the pre-colonial iron industry Haaland and Shinnie's (1985) *African Iron Working* can battle it out with Herbert's (1993) *Iron, Gender, and Power*. For a general survey of African history the 'battling textbooks' can be *African History* by Curtin, et al. (1978), and Johnson-Odim and Strobel's (1990) *Restoring Women to History*.

This enables the students and the teacher to systematically question the existing paradigms, the validity of the conventional definitions of historical periods, causality, and normative standards of what constitutes significant knowledge, and the incorporation of gender as a category of analysis[25]. A gendered curriculum would embody an inclusive vision that explores history as 'ourstory', a complex, ambiguous, and contested story of the human experience, a story based on difference, diversity and inequality, rather than sameness, uniformity, and generalization.

A gendered historiography would, for example, demonstrate that migration, one of the beloved themes in African historiography involved more than the heroic adventures of male warriors and leaders, that essentially it entailed the expansion of productive, distributive, and demographic frontiers in which both men and women played a fundamental, but differentiated role, and gender relations, divisions of labour and ideologies were often reconstructed in the process. Migrations would no longer be depicted as dramatic but simplified events, rather as complicated, if prosaic, social processes. Gender would also help decode the symbolisms, ideologies and structures of state formation and the changing nature of hegemony and social struggle. Analysis of imperialism and colonialism would certainly be deepened, for imperial conquest articulated the misogynist constructions of 'manliness' and 'otherness' and the reconfiguration of African gender relations and sexuality featured centrally in the justificatory baggage of the colonial project. For its part, economic history would lose its neat and dualistic analytical categories that strictly separate productive from distributive activities, 'traditional' from 'modern' societies, 'subsistence' from 'market' economies, 'informal' from 'formal' sectors, 'unproductive' from 'productive' labour, 'private' from 'public' spheres, for it would be shown that women either straddle both, or their involvement in one reproduces the other. The male labour power that is mobilised for the 'modern', 'market', 'formal', 'productive' and 'public' spheres would hardly exist without women working in the 'unremunerated' ('unproductive' in the lexicon of neo-classical and Marxian labour theories) 'traditional', or 'subsistence', or 'informal', or 'private' sectors. Thus it would be clear that the dualisms of conventional economic historiography do not represent distinct, separate spheres, but integrated activities structured by gender and class.

Conclusion

The examples could go on. But the case for gender history, I believe, has been made. Gendered history offers an opportunity both to bring women to the historical centre stage and to make history a truly comprehensive study of the human past in all its complexities. The pursuit of gender history should not, however, be at the expense of women's history as a separate and distinct branch of knowledge and history. Feminist historians can, and wherever possible should, work on both fronts simultaneously. Privileging one over the other is to fall into the very binary dichotomies and hierarchies of 'malestream' historiography and western philosophy that feminists and African historians have been struggling against all these years. Gender history cannot go far without the continuous retrieval of women's history, while women's history cannot transform the fundamentally flawed paradigmatic bases and biases of 'mainstream' history without gender history. Ultimately our goal is both to understand women for their own sake, much as we try to understand workers or peasants for their own sake, as separate windows into aspects of the human past, and also to probe and capture our shared, but varied, diverse and unequal, historical experiences and relations as human beings.

Mainstreaming African women's history and gendering African history are immense tasks. It needs the collaboration of both female and male historians who are informed by feminist perspectives and committed to a deeper and broader understanding of the human past than is possible by using the conventional androcentric paradigms. More concretely, there is need for comprehensive and up-to-date surveys of women's courses offered in the Social Sciences and Humanities and Arts departments in African universities, as well as of faculty hiring by gender. Also, the importance of developing and disseminating bibliographic guides and syllabi cannot be overemphasised. Bibliographies of works by African scholars and published in Africa would help significantly: African feminist researchers need to be more aware of each other's work and use that to build relevant paradigms instead of always borrowing theories manufactured in the West. Many of the existing bibliographic surveys mostly contain works published in the western countries by Africanists[26]. Moreover, systematic work needs to be undertaken to generate national, regional, and continental syntheses and other materials on various aspects of women's history. The compilation of source materials on women's history, both written and oral, for research and which could also be used as primary readers in history courses, would be particularly useful.

Notes

1. It would be interesting to find out how many departments of history in African universities offer specific courses in women's history

2. This is quite evident from the papers in the collection by Offen et. al. (1991), which cover about 25 countries in Europe, North and South America, Africa, and Asia. See Kleinberg, (1988); Carroll (1976); Angerman, et al.. (1989). In the social sciences the usual practice is for women to be taught largely in segregated women's studies departments. See Hess and Ferree (1987); Nielsen (1990); Reinharz (1992).

3. All the editors of both series and almost all the contributors are men. At the time research for this chapter was conducted Volume 8 of the UNESCO series had not yet come out, hence its omission in the analysis that follows. This volume shows a slight improvement over the previous ones, with women being mentioned on 38 out of the 934 pages of text. Nothing to brag about.

4. In the 1995 second edition there is an expanded coverage of women, but only to 17 pages out of 530 pages of text.

5. In the Orientalist constructions of North Africa, for example, the region is often seen as part of the 'Middle East', the 'Arab' or 'Muslim' world, rather than an integral part of Africa. See Said (1979). Attempts to divorce North Africa, especially Egypt, from the mainstream of African history, were spawned by nineteenth century European racist historiography. See Bernal, (198, 1991). For problems of defining regions in Africa as historical units during different periods, and in relation to colonial configurations, see Zeleza (1984, 1985).

6. Volume one has no woman contributor, while volume two has one out of 16 contributors.

7. See especially the work of Terray, 1972; Meillassoux, 1981; and Seddon, 1978.

8. This point is made, and demonstrated powerfully, in Mandala (1990); also see, Zeleza (1993). This will also be demonstrated below when we examine the reconstructions of women's history attempted to date.

9. To be sure, as Foster (1992: 3) has argued, 'the Liberal and Marxist discourses have been stretched to include women but the dominant assumptions still exclude a feminist perspective. They cannot accommodate feminist interests which threaten the very foundation on which these theories rest'.

10. Bock (1991) observes that the old dichotomies are simply being replaced by equally problematic new ones, notably, gender/sex (social construction of male and female roles/biological differentiation between men and women), equality/difference, and integration/autonomy. She argues that the dichotomy between 'social' gender and 'biological' sex 'does not resolve but only restates the old 'nature' versus 'culture' quarrel. Again, it relegates the dimension of women's body, sexuality, motherhood and physiological sexual difference to a supposedly pre-social sphere, and it resolves even less the question of precisely what part of women's experience and activity is 'biological' and what part 'social' or 'cultural' (p. 8). It is often also not realised that 'the dichotomous distinction between sex and gender is largely specific to the English language' (p. 9). Also see K. Offen, R. Pierson and J. Rendall, 'Introduction', in K. Offen, et. al.

11. See, for example, the influential work of Bell Hooks (1981; 1984; 1988). The anguished debates between white women and women of colour can be seen in some of the books on women's history and feminist methodology already referred to, such as Offen, et al. (1991); Jaggar and Rothenberg, (1990); Hirsch and Keller,

(1990). Also see, *Feminist Review*, Nos. 22 and 23, 1986; Joseph and Lewis (1986); Lerner (1990); Stasiulis (1990).

12.Discussed in greater detail below. It is this attitude that leads Parpart (1992: 171-79) to argue (after noting that African women have challenged the widespread habit of western Africanists at conferences to discuss African women's experiences without engaging African women scholars themselves) that the question of who does research on African women's history 'is a red herring'.

13.The author notes that there are, of course, many types of oral history and various reasons why feminists use them. Also see Gluck (1979). Some feminist historians note that oral historians sometimes do not adequately question the concepts they use. For example, they may want to demonstrate women's marginality, when the women concerned may not see themselves as marginal, see Geiger (1990). Others are not convinced that oral history helps in 'liberating' the voices of oppressed women, see Personal Narratives Group, eds., 1989.

14.For the ghettoisation of women's history and marginalisation of gender history in Britain see Jane Rendall, in Offen, et al., 1991.

15.This history seeks, she argues, to write about women missing from, and describing their contribution to, traditional history. This constitutes, in her view, 'transitional women's history', which she distinguishes from women's history that studies the actual experiences of women in the past on their own terms, and what she calls 'universal history', a holistic history synthesising traditional history and women's history. The latter is what increasingly came to be referred to as gender history.

16.For detailed bibliographic surveys, see Robertson (1987); *Canadian Journal of African Studies*, 22 (3), 1988, *Special Issue on Women*; and the well-written monographs on so-called sub-Saharan Africa and the so-called Middle East, a large part of which covers North Africa, in Johnson-Odim and Strobel (1988).

17.Historians have amply demonstrated that many practices and values which are considered 'traditional' today, including those in the sphere of gender relations, were invented during the colonial period, see Ranger (1989) and Chanock (1985). Increasingly anthropologists have come to the same view, but in typically convoluted post-structuralist deliberations, see Comaroff (1980) and Moore (1986).

18.This section relies heavily on Johnson-Odim and Strobel (1988), and Zeleza (1993).

19.For example, in the 1980s the Nigerian military government increased its attacks on market women as Nigeria entered a period of economic crisis partly brought about by declining oil revenues. The women traders were blamed for high inflation and shortages, see Dennis, 1987. On relations between the Nigerian military and women see Mba, 1989.

20.Scott (1988: 3) gives this as one of the main reasons she turned to gender as an analytical framework in feminist history.

21.For a compelling critique of Scott's post-structuralist feminist historiography, see Hall, 1991; also see Schwegman and Bosch, 1991; and Newman, 1991. Bock (1991) and Sangster (1995) have argued forcefully for the deconstruction of the dichotomy between women's history and gender history. For her argument that gender history is not more encompassing, does not offer more profound insights, and is not theoretically more sophisticated than women's history, Sangster, a distinguished Canadian feminist historian, has been widely condemned by her younger colleagues (personal communication), one more indication of how vicious sectarian academic battles can be.

22. Pierson's (1992) data shows that the number of women's history courses in Canadian universities remains abysmally low, accounting for less than 3% of the total number of courses offered.

23. For example, in European history, the glory that was the Renaissance, the period during which men (elite men) saw their intellectual horizons widen, loses its glow with revelations that women became more subordinate and restricted than in earlier centuries, see Kelly (1984).

24. This point is made particularly well in the American context by Higginbotham, 1990. The *Women's Studies Quarterly*, has done several special issues on incorporating feminist perspectives in various Social Science disciplines, including Economics and Psychology with their rigid, positivist, and pseudo-scientific paradigms and models. See Vol. 18, Nos.1&2 devoted to curricular and institutional transformation; Vol. 18 Nos. 3&4, to 'Women's Studies in Economics'; Vol. 20, Nos 1&2, to 'Feminist Psychology: Curriculum and Pedagogy'.

25. In my third year African history class that I taught at Trent University, I experimented with this method, and it was fascinating watching the students becoming more aware that 'doing' history is as gendered as the historical processes they were trying to understand. For example, in my tutorial on the 'Islamic Revolutions in West Africa in the Nineteenth' century I used Chapter 12 in Curtin, et. Al. (1978) ('The Commercial and Religious Revolutions in West Africa'); Murray Last (1985) ('Reform in West Africa: the Jihad Movements of the nineteenth century', in Ajayi and Crowder, eds.); Boyd (1986) ('The Fulani Women Poets', in A. H. M. Kirk-Greene and M. Adamu, eds.); J. Boyd and M. Last (1985) ('The Role of Women as 'Agents Religieux', in Sokoto'); Kapteijns (1985) ('Islamic Rationales for the Changing Roles of Women in the Western Sudan' in Daly, ed); and J. Carney and M. Watts (1991) ('Disciplining Women? Rice, Mechanization and the Evolution of Mandinka Gender relations in Senegambia').

26. See, for example, the recently published bibliographic guide by Fong (1993). The absence or under-representation of works on women by African scholars in standard Africanist historiographic surveys is staggering as can be seen, for example, in Robertson (1987).

Chapter Ten

REPRESENTING AFRICAN WOMEN

Introduction

Judging by the rapidly growing list of publications, there can be little doubt that African women's history has come a long way. It is no longer permissible to write African history as his-story, to be blissfully ignorant about the activities, experiences, and contributions of women in humanity's long historical drama. In the process of incorporating women's history, of turning glorified androcentric narratives into more generous gendered history, many conventional interpretations and theories have lost their paradigmatic prestige and have been forced to be refined, reformulated, or rejected altogether. But the day of gender-neutral history and social science, as demonstrated in the preceding chapter, will take time to be fully realised partly because of concerted ideological resistance by the disciplinary gatekeepers. This does not mean that feminist historians are all on the side of the angels, or that the production of feminist history is free from the prejudices and perversions associated with modernist patriarchal historiographies. Feminist histories are also filtered through the layered lenses of place, race, ethnicity, class, and ideology. They are infused with the hierarchies of power within and between societies and among intellectual communities. This is certainly the case in the charged and contested contexts of African studies. Africanist feminism bears many of the afflictions that plague Africanist scholarship: tendencies towards the overgeneralization and essentialization of Africa, and the inferiorization and pathologization of its peoples and cultures.

This chapter seeks to look more closely at the epistemology and production politics of African women's history by examining four recent texts by Parpart and Staudt (1989), Stichter and Parpart (1990), White (1990), and the *Signs* (1991) special issue on African women[1]. These studies seek, in their various ways, to advance the process of integrating gender analysis into African historical and political studies. They re-examine, among other things, the questions of state and family formations, the development of rural peasantries and urban working classes, and the constructions of, and struggles over, gender. Some display the highest standard of scholarship, others should have been left in their authors' drawers to gather dust. The chapter begins by examining how these studies deal with the issue of women and politics, followed by the subject of women and peasant agriculture, then women and urbanisation, and ends with a discussion of the politics of producing feminist knowledge in African studies as exemplified in these texts.

Women and the State

Parpart and Staudt argue, quite correctly, that conventional analyses of the state in Africa have tended to ignore women. The book seeks to change that, to demonstrate that the process of state formation is a gendered one. The various contributions in the book demonstrate that women have been excluded and marginalised from the political process. But despite that women have organised and fought back. However, the discussion tends to be heavier on the political subordination of women than on their struggles. Mbilinyi's (1989) paper on colonial state intervention over beer brewing in Dar es Salaam in the 1930s is the only one that deals with women's struggles at length. It is also the only one that focuses almost entirely on the colonial period. The rest of the papers mainly examine the post-colonial period, and deal with such issues as state policies towards market women (Hansen 1989), women and land resettlement (Jacobs 1989), the impact of development policies on women (Munachonga 1989), relations between women and the military (Mba 1989), and the effects of economic crises on peasant women (Newbury and Schopf 1989). While the various authors conceptualise and problematise the process of state formation and women's access to state structures and resources differently, there is general agreement that the 'female experience in African politics during the past century... [has been] one of exclusion, inequality, neglect, and subsequent female consolidation and reaction' (Chazan 1989).

It can be seen that the range of topics covered is relatively wide. But the same cannot be said about the regions and time periods. Only five countries are examined. Two of the papers are on Zambia. West Africa is short shrifted with one paper and North Africa is not even mentioned. The same skewed coverage can be observed in the *Signs* collection, whose papers deal with six countries. Two of the papers are on Uganda. North Africa is also left out, quite deliberately, the editors tell us, because in their view, 'North Africa represents quite different patterns from Africa south of the Sahara' (*Signs* 1991:646). Why and how? Certainly nothing in Lazreg's (1990) paper on Algerian women in the Stichter and Parpart (1988) collection, bears out the contention that women's situation in North Africa differs markedly from 'patterns from Africa South of the Sahara'. Incidentally, Lazreg's paper was probably included less because Algeria is seen as a part of Africa, than a part of the so-called Third World, which is the focus of the collection, despite its grand sounding title. It is a shame that the imperialist and racist construction of Africa as 'Black' has invaded studies of African women as well.

As already mentioned, only one of the papers in Parpart and Staudt (1989) deals exclusively with the colonial period. For a book claiming to represent 'the first systematic effort to introduce gender into the analysis of the state in Africa' (Chazan 1986:185), this is a great pity. But the book is less informative on the role of gender in the origins or formation of the modern state in Africa than it might otherwise appear. Given its limited geographic and temporal coverage, it is not surprising the book's theoretical contribution is rather desultory. The introduction is a rudimentary outline of the dependency and modes of production approaches, and various perspectives that inform discussions on the state. The two so-called theoretical chapters are particularly unsatisfactory. Most of the generalisations they make about Africa are untenable because they are constructed on thin empirical data. Fatton's (1989) paper gives political science a bad name. It is full of fatuous and unsubstantiated assertions masquerading as theory. African societies and states are portrayed as peculiar deviations from some universal norm. In Africa, he tells us, 'women's subordination is more pervasive, acute, and accepted' (Fatton 1989:51). Than where? Women's struggle for emancipation in Africa, he continues, 'is replete with contradictions, ambivalence, and silence' (p. 54). Silence, no. Contradictions and ambivalence, yes. But where is that not the case? And he variously describes the African state as weak, fragile, non-integral, non-hegemonic, and authoritarian without authority. It is because the African ruling classes are non-hegemonic, he argues, that popular resistance 'is seldom frontal and revolutionary. Resistance takes the form of withdrawal from the public realm...' (p. 55). Are we to conclude that the American ruling class is non-hegemonic because Americans are not 'revolutionary' and the majority of them have withdrawn from the electoral process? Fatton mistakes conjunctural appearances for structural realities. In the aftermath of the tumultuous struggles for democracy that have rocked Africa in the last few years his analysis appears dated. Lovett (1989) displays the same tendency towards overgeneralisation. She constructs her theory from the cases of Nairobi and the Copperbelt. One wished both Fatton and Lovett had heeded Hansen's (1989:143) admonition that the processes and effects of state formation 'on distinct social segments are highly variable in historical terms and often differ from one country to the next'.

Musisi's (1991) paper in the *Signs* collection amply demonstrates the complex historical processes at work in gender and state formation during the pre-colonial era, which defy the kind of simplistic generalisations made by Fatton and Lovett. It examines the relationship between state formation, women's status, and marriage forms in Buganda from the thirteenth to the nineteenth centuries. She examines what she calls 'elite polygyny' and argues that 'polygyny in precolonial Buganda must be distinguished from colonial and post-colonial polygynous practices and viewed, most critically, in the context of elite strategies to create and ultimately to control not only economic but political and social components of a state apparatus as well' (p. 758). It is a fascinating, richly textured analysis, reconstructed through a feminist rereading and reinterpretation of Buganda oral and written sources. It is demonstrated that 'elite polygyny' was an integral part of the three processes of state, class, and gender formation. The point is made that women became more differentiated. While the majority were excluded from 'direct involvement in Buganda's political process', the wives of the elite, the *bakembuga*, 'played an important role at the state level in balancing internal and regional politics. The *bakembuga* became not only the mothers of kings but king-makers as well' (p. 786).

Musisi leaves her story in 1900. One wishes she had carried it forward to show how the superimposition of the colonial state on Buganda transformed the patterns of gender, class, and state formation. There can be little doubt that colonialism transformed these patterns in complex and contradictory ways. It is often assumed that Indirect Rule shielded Buganda from intensive state interventions witnessed in settler colonies like Kenya. The paper by Summers (1991) challenges that assumption. It argues that the British colonisers devised and tried to implement a highly intrusive policy on reproduction itself, ostensibly in response to Uganda's population decline in the early twentieth century. At first the decline was blamed on the trypanosomiasis epidemic which killed 250,000 to 330,000 between 1900 and 1920. After the epidemic, official attention turned to the continuing low birth rates, which were attributed primarily to sexually transmitted diseases (STDs), especially syphilis.

Out of this, the colonial state and its functionaries, built a crisis, against which they resolved to intervene medically. Before long, medical intervention was accompanied by moral intervention as the crisis of disease was turned into a moral crisis, and the incidence of syphilis, which Summers shows was exaggerated, became a barometer of African immorality. From here to intervention in the reproductive practices and choices of African women was but a short step. African women came to be viewed as 'clumsier, stupider, and dirtier than African men'. The state felt justified 'to intervene in the private sphere of pregnancy, birth and infant care' (Summers 1991:800). This was colonial social engineering at its most intimate, an unashamed attempt to reshape African families and reproductive behaviour. Summers indicates that women avoided the STD programmes, but the overall impact of colonial state interventions over reproduction and motherhood is not adequately drawn out.

The colonial state did not always work single-handedly to impose its capitalist patriarchal will on African women. As Schmidt (1991) argues in the case of colonial Zimbabwe, African chiefs and elders colluded with colonial officials and functionaries to control the behaviour of African women. The agendas of the two groups were of course varied. The former were trying 'to reassert their waning authority over women, their services, and their offspring', while the latter 'were concerned with obtaining cheap African male labour. If it took the regulation of African women's sexual practices to achieve this objective, the state was prepared to pass laws to that effect' (Schmidt 1991:756). Schmidt believes that the European colonisers strove to control women not only because they suffered from deep-seated racial and gender prejudices against African women, but also because controlling the sexuality and mobility of African women offered a means of mitigating the disruptive impact of migrant labour on African family life, and preventing the collapse of indigenous authority structures, especially in the face of growing women's resistance. Apart from the various restrictions imposed on women's opportunities and mobility, the colonial state, relying on age-old European misconceptions about African society and the selective memories of African chiefs and elders, created 'customary' law, which turned flexible custom into inflexible law.

The collusion between the African rulers and the colonial state does not mean that the two groups share equal responsibility for the construction of women's subordination in colonial Africa, as some writers, including Schmidt, seem eager to suggest. It should be remembered that the African chiefs and elders were themselves subordinate to the colonial authorities. The tendency to talk indiscriminately of 'African men' must be resisted, for African men, no less than African women, were not homogeneous. They were differentiated according to class, status, and occupation, so that they did not share similar interests with regards to women's position in society.

Much of the new revisionist literature which rejects the functionalist argument that the colonial state instituted migrant labour because it was functional to capital, by showing that migrant labour was also a product of domestic struggles between African men and women (Bozzoli 1983; Parpart 1986), often slips into the same functionalism, except now the argument is that migrant labour was functional for men. Just as migrant labour was not functional for all fractions of capital at all times, women's subordination was not functional for all men at all times.

Women as Peasants

Parpart's and Staudt's (1989) collection contains two papers which seek to examine African women as peasants. Both of them are on the contemporary situation. One is on the agrarian crisis in Zaire and the failure of remedial policies devised by the state and international lenders to target women and incorporate their needs and concerns. This will result, the authors argue, in the crisis deepening because women are central to the agrarian economy (Newbury and Schoepf 1989). The other paper examines land resettlement in independent Zimbabwe and argues that while legislation has tried to remove some of the worst aspects of colonial legislation which discriminated against women, women's needs continue to be neglected 'in many spheres of state policy, including the resettlement program' (Jacobs 1989). In the individual family resettlement programmes women have been marginalised, while progress in the cooperative resettlement programmes has been hampered by poor government funding and men's appropriation of the use of advanced technology such as tractors and harvesters.

These studies may be interesting, but they only scratch the surface of women's experiences as peasants. This reinforces a point made earlier that this collection lacks a long-term historical perspective and is excessively narrow in its coverage. The 'theoretical' paper by Lovett (1989) also helps to underscore the importance of treating the pre-colonial period historically, rather than merely as a static backdrop against which changes brought by colonial capitalism are set. Her analysis of gender and class formation in modern Africa is predicated on a cursory and misleading review of the sexual division of labour in what she calls 'precapitalist' societies. It is simplistically assumed, for example, that patriarchy was universal, unambiguous, and uncontested. Also, social age is defined only in relation to men, the omnipotent male elders of anthropological folklore. This ignores the varied and complex situation in the matrilineal societies, which existed in the regions she discusses.

There is one paper in the *Signs* collection which attempts to begin filling the gender gap in the historiography of rural production in the pre-colonial period, and another that covers the last 150 years, spanning the pre-colonial, colonial, and post-colonial periods. Using Lesotho as a case study, Eldridge (1991:708) seeks to demonstrate that women 'were the primary agents of accumulation and growth in the nineteenth century economy of Lesotho'. Thus the economic growth and prosperity that Lesotho enjoyed in the nineteenth century cannot be understood without considering women's contribution. She shows that in the agricultural economy women were responsible for gathering, caring for pigs and poultry, cultivation, bird-scaring, harvesting and threshing. In addition to food production, women also looked after food processing, they fetched water and collected fuel for cooking, and made many of the household goods, such as baskets and pottery. Many of the goods traded outside the household were produced by women. Despite this, however, women enjoyed little political power.

What is admirable about this paper is the way it captures the changes that took place in the gender division of labour. It is noted, for example, that as building in cut stone spread, home building ceased to be a female task and was taken over by men. In particular, it is shown that the widespread adoption of ox-drawn ploughs changed the labour time of both men and women. Women's workload in farming increased, in exchange for which they 'gave up other activities such as weaving, pottery and home building'. Men's labour was also 'reallocated; that is, they began to help women. It became more common to hold work parties for weeding, harvesting, and threshing, at which married and unmarried men helped with the agricultural tasks that women usually performed' (Eldridge 1991:723). As men became more involved in agriculture, and the goods that they previously produced in the household, such as blankets, clothing, and wooden and iron tools, weapons and utensils became readily available in the markets, they gave up on, or spent less time, producing these goods. For their part, the young and the old were allocated new tasks. These changes altered the old forms of women's subordination, empowerment, struggle, and differentiation. From Eldridge's study Basotho women are not pawns of some ubiquitous patriarchy, but historical actors who consciously shaped the changing world in which they lived.

The term 'traditional' has been widely abused in African studies. What often appears 'traditional' were practices and ideologies invented at specific moments in the recent past. Carney's and Watts' (1991) exemplary paper on agrarian change in Senegambia amply bears this out.

They argue that the agricultural system that currently operates in the Senegambia is not 'traditional' in the sense that it is ancient. It emerged in the mid-nineteenth century, a product of the growing commoditization of peasant production. In the last 150 years repeated attempts have been made to intensify rice production. Intensification was both a social and gendered process and one that was, moreover, continually negotiated and struggled over. Before the mid-nineteenth century, the gender division of labour was based on tasks. But as the commodity production of groundnuts expanded, there was a change from 'task- to crop-specific gender roles'. Rice, grown mostly for household consumption, became women's work, while groundnuts, largely produced for export, became men's work. Also, 'as groundnut cultivation expanded on upland fields' away from the floodplain and swamps, 'male and female agricultural labour became increasingly spatially separated between upland and lowland zones, giving rise to a much more rigid sexual division of labour by crop' (Carney and Watts 1991:657).

The withdrawal of male labour power from rice production intensified female labour time in food production. But women, burdened as they were with other household activities, were unable to produce enough rice, and so the region became increasingly dependent on food imports. When the colonial state was established in 1889 it was alarmed by the growing food imports and thus began attempts to reestablish household food self-sufficiency by increasing rice production. The measures included the introduction of improved Asian rices, the clearance of mangrove lands, and the establishment of a series of large- and small-scale irrigation schemes. Rice acreage and production increased. But it could not be sustained unless the gender division of labour was transformed. Further intensification depended on bringing men into rice cultivation. But men 'successfully resisted efforts to intensify their labour on the grounds that rice was 'a woman's crop'' (Carney and Watts 1991:661). In the meantime, the colonial rice development projects, generated conflicts between men and women over the control of land and crops. Mandinka men and women could claim the individual ownership of land if they cleared that land themselves; if it was collectively cleared it fell under common household ownership. In the 1940s and 1950s women's attempts to assert control over the land that they had cleared from the newly opened mangrove swamps and lay claims on the output, were resisted by men, who impressed upon the colonial authorities that women's land ownership contravened 'tradition'. The latter obliged and 'determined that the new rice lands were household, not individual property' (Carney and Watts 1991:664). The struggles between men and women, and women's resistance against further intensification of their labour time and appropriation of their surplus, led to the failure of the post-war rice schemes based on ambitious irrigation projects. For example, women were reluctant to work as wage labourers for the mechanised and large-scale irrigated rice project initiated by the Colonial Development Corporation (CDC) in 1949. When sharecropping was introduced to save the project 'women systematically under-reported their harvests to appropriate larger shares of the crop' (Carney and Watts 1991:667). The CDC abandoned the project in 1958, and at independence the new government returned the rice lands to local cultivators.

The post-independence government was no more successful in resolving the productivity and labour crises in rice production. Initially it tried to encourage the development of small-scale, irrigated units controlled by households and based on the double-cropping of high-yielding green revolution rice varieties. Although men's participation increased, their attempts to claim women's labour for year-round cultivation was resisted, for historically, 'the farming system was attuned to a five-month cycle' (Carney and Watts 1991:670). Following the failure of small-scale irrigation projects to lessen the country's food dependency, a large-scale irrigation project was introduced in the mid-1980s, based on a coercive labour regime, which reinforced patriarchal family relations. As before, the new project provoked resistance and contestation along social and gender lines. It can be seen that the development of rice production in the Senegambian region involved a complex interplay and reconstructions of gender roles and property rights, state interventions, and household conjugal relations and struggles.

Women and the City

Research on African women has privileged rural over urban women. This partly reflects the fact that, to date, the vast majority of African women have been rural dwellers, primarily engaged in agricultural work. When urban women are discussed, the focus has been mostly on their activities as traders or informal sector operators. The literature on colonial West Africa concentrates on women as traders, and that on East and Southern Africa on women as prostitutes. This divide can be seen in the works under review.

Mann (1991) explores women's access to landed property, capital, and labour in the city of Lagos in the second half of the nineteenth century. During this period Lagos expanded rapidly, thanks to the commercial revolution brought about by the end of the slave trade and the growth of the palm trade. The result was that an urban real estate market developed and land prices escalated. The alienation, privatization, and commercialisation of urban land affected men and women differently. Although women tried to take advantage of the changes in land tenure, 'many fewer women than men purchased or were granted land' (Mann 1991:691), due to the discriminatory policies of both the monarchy and the colonial government. However, there was a minority of women household heads who were able to acquire land. Other Lagos women inherited privately owned land from relatives and husbands. Limited access to landed property, undermined women's access to capital, credit, and labour, which, in turn, circumscribed their trading opportunities. Consequently, women became increasingly dependent upon men for land, housing, and capital. At the same time gender conflict in Lagos households over labour and resources intensified. Thus, from the mid-nineteenth century, in the face of expanding trade and changes in the land market, 'women in Lagos faced economic disadvantages that limited their ability to take advantage of new commercial opportunities and weakened their economic position relative to men. The final decades of the nineteenth century were no golden age for Lagos women' (Mann 1991:705), as studies which celebrate the penetration of European commercial capital tend to imply. At least for Lagos women it was not a time of expanding opportunities and increasing autonomy.

If commercial opportunities for Lagos women were closing at the turn of the twentieth century, wage employment opportunities opened up for some of them, as Parpart (1990) shows in her paper on the growth of women's wage employment in southern Nigerian cities from the late nineteenth century to the 1980s. She demonstrates that the size of women wage workers remained minuscule until the Second World War. There was rapid expansion in women's wage employment after independence, thanks to the expansion of the economy, increased educational opportunities for women, changes in family structure, and struggles by women for economic independence. Despite their increased participation, women continued to be clustered in a few occupations, principally nursing, secretarial work and teaching, which were seen as extensions of women's work in the domestic sphere.

The main objective of Parpart's paper is to examine the manifestations and impact of the double day on Nigerian working women. She demonstrates, quite convincingly, that these women were not immune from the burdens of the double day, especially as domestic labour became scarcer and more expensive following the oil boom, and day care centres remained in short supply. The dearth of paid and unpaid household labour was aggravated by the introduction of free universal education for children in 1976; 'parents who had previously been happy to send unschooled children to work in affluent relatives homes, now felt they should send their children to school' (Parpart 1990:170). She concludes that although Nigerian working women have developed various coping mechanisms, from the careful regulation of time and fertility, to spurning marriage altogether, the reproductive burdens of the double day have clearly undermined their career mobility. All this is true. Unfortunately, Parpart tends to weaken her case by focusing excessively on 'middle class' women, whose 'domestic labour may be more managerial than manual' (Parpart 1990:177). The basis of the distinction between her 'middle class' and 'working class' women is not made clear.

The only other study that examines women's urban wage employment in the works under review is Lazreg's (1990) paper on Algeria. It focuses on the period after independence, and shows that until 1978, despite the attempts to construct a socialist economy, women hardly increased their share of wage employment, although official statistics have tended to underestimate women's labour force participation. Algerian wage earning women work predominantly in the urban centres, and they are concentrated in the professions and services, rather than industry. They also tend to be relatively more educated than working men, although 'their education does not beget positions of responsibility' (Lazreg 1990:185). The major value of the paper is that it explodes the idealist myth that in 'Muslim' societies like Algeria Islam is the explanation of gender inequality. She argues, and proceeds to demonstrate, that the relatively low participation of women in the labour force can be attributed to a number of structural factors, especially the patterns of economic growth, demographic growth, and family formation.

Stichter (1990) concurs with Lazreg that in order to understand the levels and patterns of women's wage employment, one has to take into account the organisation of production and market factors, as well as the family or household structures, which entails examining such issues as household incomes and resources, power and decision-making patterns. Her paper offers an extensive and impressive survey of the literature on both the growth of female employment in the 'Third World' and theoretical approaches to the household. It is pointed out that 'despite the gender gap, female employment in the developing world has shown surprisingly rapid growth in recent years'. Needless to say, there are important sectoral and cross-national differences. Some African nations are among the few in the world where general female rates of labour force participation approach those of men. She argues that the prevailing and competing neo-classical and Marxist theories on households are both based on the nuclear or conjugal units of industrialised societies, so that they do not adequately explain the extremely diverse and complex household and family systems found in Third World societies. Consequently, our understanding of the dynamics of female employment in these societies remains patchy.

None of the studies under review analyze female wage employment in Eastern, Central and Southern Africa for any period. The papers discussing urban women in Parpart and Staudt, and White's book, focus primarily on prostitution in the former settler colonies of Eastern and Southern Africa. The argument is made that because urban wage labour opportunities for women were negligible in these countries until after the Second World War, prostitution offered women a chance, in the words of Fatton (1989:32), to carve 'a niche as petty-bourgeois accumulators by providing reproductive labour services to migrant workers'. It was the primary means by which many African women 'established themselves as integral parts of the emerging urban petty bourgeoisie during the twenty-five to thirty-five years of the twentieth century'. The thesis of prostitution as a means of petty-bourgeois accumulation for African women is fully developed by White (1990).

White's book synthesises and expands on a series of papers she published in the 1980s. It argues that prostitution was reproductive work, a form of family labour, not an activity to be decried in the moralistic language of deviancy, degradation, and depravity. She distinguishes three main forms of prostitution, each of which had 'its own characteristics, behaviour, rate of accumulation, and organization of labour time' (p. 13). First, there was the *watembezi* form, akin to streetwalking. The second was the *malaya* form, in which the prostitute stayed inside her room and waited for men to come to her. Finally, there was the *wazi-wazi* form whereby women sat outside the doors of their rooms or on the porches and called out for men. The study meticulously delineates the emergence, growth, and transformation of each form, the social and ethnic origins of its practitioners, their relationships with their communities and neighbourhoods, and their earnings and investment strategies. These processes are linked to housing arrangements and policies, changes in the colonial economy, fluctuations in male wages, patterns of labour stabilization, and the eruptions of the two world wars and the Mau Mau war of national liberation. She demonstrates that the colonial state was not opposed to prostitution, although it tried to control it. But state control over the city and prostitution, or over urban space and working class reproduction, was far weaker than is often assumed. White also sketches the solidarities and antagonisms among women engaged in the different forms of prostitution, the development of new family formations around them, and the constructions of ethnicity and sexuality, religion and respectability, class and feminist consciousness.

She writes well and persuasively. But appearances can be deceiving. She extols oral sources, but provides no list of the interviews she conducted. She seeks to offer a radical reinterpretation of Kenyan economic and social history, but often lapses into the colonial language of 'tribes' and 'detribalization'. The distinctions between the various forms of prostitution, and the changes that they underwent, are sometimes given a sharpness and temporal exactness that appears contrived. For example, the evidence is not compelling that *wazi-wazi* did not exist 'before 1936' (White 1990:104), or that *malaya* women 'began to ask for payment in advance' by 1933, as compared to before when they asked for payment after rendering their services (White 1990:84). Indeed, the author is forced to abandon discussing each form distinctively for the post-war period, arguing that 'after 1946 it is really not useful to examine the forms of prostitution individually; instead, postwar Nairobi had a fluidity - of streetwalkers, customers, and absentee landlords - that gave the regional variations of prostitution within the African locations a meaning they had not had since the 1920s' (White 1990:195).

There are several key planks of the analysis that do not stand up to closer scrutiny. She argues that Kenya's first prostitutes were from the pastoral communities devastated by the ecological disasters of the late nineteenth century. That may be true. But no concrete evidence is presented to support the assertion that women from these societies used their earnings to replenish their fathers' households' livestock. The contention that women 'from households engaged in subsistence farming... practised the malaya form', while those who chose the wazi-wazi form 'were generally women from families engaged in cash crop production' (White 1990:125) also lacks substantiation, apart from the problem of defining what is meant by subsistence farming. And how sustainable was the prostitutes' accumulation? According to her own evidence few women who entered prostitution after the Second World War were able 'to build a house with [their] earnings' (White 1990:202).

White seems so anxious to celebrate prostitutes that she sees prostitution everywhere in colonial Nairobi. She summons us to 'recognize working prostitutes as Kenya's urban pioneers, the first urban residents' (White 1990:34). Whatever happened to the residents of Mombasa, that ancient city on the Kenyan coast? White's prostitutes were strong, enterprising women. They were victims neither of pimps, nor weak, dysfunctional families. Indeed, they were dutiful daughters, driven to prostitution out of loyalty to their families, to support rural production and accumulation. Their activities had little to do with 'sex ratios' in Nairobi (White 1990:58), and did not depend 'on men's needs but on the woman's labour form' (White 1990:225). They were not unduly worried about diseases and violence, for they were in control of their lives. There is no moral ambiguity here, little sense of the way the wider society perceived prostitution. Were the rural families from which the prostitutes came only concerned about their daughters' earnings? If prostitution was new, as we are told, it surely must have elicited strong cultural responses. This is a laudable defence of prostitutes that turns into an idealistic defence of prostitution.

There is a split in the feminist literature on the acceptability of prostitution. Some see prostitutes as agents and prostitution simply as work, while others view prostitutes as victims and sex work as the highest form of patriarchal oppression. White would seem to belong to the first group. It is true the usual condemnations of prostitution are often misguided. Certainly, danger, injury, and indignity are not confined to prostitution. Neither is the lack of choice and the presence of coercion, nor the surrender of personal power and control and loss of independence. Similarly, indignity and non-reciprocity are not unique to prostitution. Prostitution is objectionable because, as one author so aptly puts it, it 'is an inherently unequal practice defined by the intersection of capitalism and patriarchy. Prostitution epitomises men's dominance: it is a practice that is constructed by and reinforces male supremacy, which both creates and legitimizes the "needs" that prostitution appears to satisfy as well as it perpetuates the systems and practices that permit sex work to flourish under capitalism. What is bad about prostitution, then, does not just reside in the sexual exchanges themselves, or in the circumstances in which they take place, but in capitalist patriarchy itself' (Overall 1992:724; also see Vance 1984).

From White's book one would be forgiven to conclude that Kenyan women in colonial Nairobi were all prostitutes. No serious effort is made to trace the development of women's wage labour and other types of informal sector activities, such as beer brewing, food processing, and trade. There are fleeting allusions to women wage employment in 1937 and 1944 (White 1990:152). Surely by 1963, when Kenya got its independence and White ends her story, wage employment opportunities for women were not as limited as they were in 1937 and 1944 (Stichter 1975-6; Zeleza 1982). Admittedly, this is a study on prostitution, but prostitution emerged and grew in the context of the changing economic and labour market opportunities for women, just as it was influenced by many other changes in Kenya's political economy that White chooses to discuss. What is the difference between these analyses that privilege prostitution, and colonialist views which portrayed African women as nothing but a bunch of prostitutes? Jacobs even makes the inane assertion that in contemporary Zimbabwe 'for women, urban life is still associated with prostitution' (Jacobs 1989:164). Associated by whom? And for which women?

The papers by Mbilinyi (1989) and Hansen (1989) show that women in the colonial cities of Eastern and Southern Africa were indeed involved activities other than prostitution. Mbilinyi examines beer brewing in Dar es Salaam in the 1930s, and women's struggles to retain their control over this industry. She argues that beer brewing was part and parcel of the dynamic and resilient informal sector, or what she calls 'the off-the-books' sector, which provided income generating opportunities for both men and women who could not find permanent employment. The struggle over beer brewing between women and the state was not simply over revenues and accumulation, but also competing gender constructions, and of women's place in the city, which was deemed by colonial ideologues to be for non-Africans, males, and wage workers. Women in the city were disapproved of, except for prostitutes who were grudgingly tolerated because in the official view prostitution 'was a necessary amenity in the township which provided a diversion otherwise filled by more political forms of action' (Mbilinyi 1989:116). This underscores the fact that the state treated African women differently according to their particular class and specific occupational category. In the end, the women won their struggle against the state over beer brewing by effectively mobilising their informal networks and tapping community support. Their struggle reinforced the oppositional ideology of popular culture.

Hansen notes that married African women, together with single women, in colonial Lusaka monopolised the sale of home-brewed beer and alcohol, and traded in prepared foods as well. These women were attempting to accumulate capital and at the same time challenging the gender definitions and demands of colonial and indigenous cultures. Although the bulk of the paper deals with Lusaka women traders in the 1970s and 1980s, it shows that these women did not spring out of nowhere. They had a history, one which they made, despite restrictions imposed by a colonial state with settlerist pretensions. From the mid-1970s when the bottom fell out of copper prices, the country's main export, the Zambian economy entered a period of severe crisis and the state was forced by the mighty IMF and World Bank to undertake structural adjustment programmes. These developments facilitated the rapid growth of what Hansen calls the 'black market' and reduced state intrusiveness into the activities of women traders, and recast gender struggles within households.

Whose Voice?

Parpart begins her paper on southern Nigerian female workers with a brief comment on the clashes between western feminists and Third World women at a number of forums on the construction of the gender debate. The latter, she avers, 'rejected western feminists' preoccupation with patriarchy and insisted that global inequities, not men, were the main enemy facing Third World women... The force of these arguments', she continues, 'alarmed western feminists, many of whom drew back from the apparently dangerous business of cross-cultural feminist analysis' (Parpart 1990:161). The issue, however, goes far beyond the question of 'cross-cultural feminist analysis'. It is about power: who sets the agenda? Who speaks for Third World Women, who represents African women in scholarly discourses?

The studies under review point to a disturbing reality in studies on African women. The voices of African women themselves are largely absent. In the three collections analyzed above there are only three. papers by African women that I could identify out of the 26 contributions. The Editors of the *Signs* collection justify the absence of papers from African women scholars with a long self-serving litany bemoaning the fact that the 'prevailing socioeconomic conditions in African universities are not conducive to the production of knowledge'; 'scholars situated in impoverished or beleaguered institutions lack the time or resources... to produce scholarly work'; 'African men are more likely to go to the university and become researchers than are African women'; 'the few women scholars situated in African universities often lack a supportive environment to do critical feminist work' (*Signs* 1991:645). The list goes on. This is the language of exclusion, of privilege and power, of intellectual imperialism.

I simply cannot believe that they could not find a single African woman scholar who could have written a piece of 'critical feminist work' for this issue. I just know too many able African women scholars in various parts of the continent to believe such balderdash. Yes, there are

African women scholars who do research and write against great odds. But who said that 'knowledge' can only be produced in comfortable surroundings, from the ivory towers of American universities? This is the voice of an arrogant, institutionalised American academic feminism. It fits into an old pattern. In an insightful paper on intellectual practices and the production of knowledge in African studies, Imam and Mama (1990:9) note the curious fact that 'it is possible to have, as recently as the last five years, at least three books which are collections of articles on African women, which appear to have no contributions at all from an African researcher...or, a review article on studies of African women published in 1987 in which possibly 15 articles by African women were referenced (out of maybe 200?) and where AAWORD (the Association of African Women for Research and Development) was mentioned favourably, none of its published papers were'[2].

The rationalisations contained in the *Signs* editorial remind me of the British colonial practice, whereby Africans used to be represented in some Legislative Councils (Legco) by a European, usually a missionary. It was believed the 'natives' could not speak for themselves. They were simple, illiterate people, who did not understand the workings of government. The educated elite among them were too busy scheming or aping the European to understand the 'native' mind. But the missionaries did. After all, they dealt in souls. As true representatives of the 'natives', the Europeans would periodically ask their hapless wards to submit memoranda listing their concerns, which the Europeans would sometimes table in Legco.

Traditions die hard. The colonial tradition of Europeans representing Africans lives on in African studies and the *Signs* collection. After feasting on the 'critical feminist work' of North American feminists, we are given a light African dessert consisting of 'Reports From Four Women's Groups in Africa' (*Signs* 1991:846-69). We are told this project was four years in the making. And yet, all the Africans could produce were reports informing the North American feminist fraternity that they have been keeping busy. That they are trying. And so the Africanist feminists can feel good about themselves. They have magnanimously given the 'natives' a chance to speak. Like the colonial missionaries, Africanists often act like evangelists out to save some benighted souls. They see themselves as not simply writing *about* Africa and Africans as, say, an American scholar might write about China and the Chinese, but as seers writing *for* Africans. The feminist Africanists in these collections arc following a well trodden path.

The tensions and conflicts that Parpart talks about are not simply between 'western feminists' and 'Third World women', for neither group is homogeneous. There are many feminisms, diverse women's voices in both the so-called western world and the 'Third World'. The 'western feminism' Parpart is talking about is white-middle class academic feminism. These feminists are as disconnected from the realities of the African women as they are from the realities of racial minorities in North America, including women of African descent. In recent years the 'women of colour', as the racial minority women are sometimes called, (an oxymoron, since all people, including those of European descent, have colour), have been vigorously challenging the right of white-middle class women to speak for them and define their agenda. The hegemony of western white middle class academic feminists over Africa should also be challenged, together with all forms of western intellectual hegemony. The endeavours to gender African history, and African studies generally, must continue. But let African feminist scholars speak for themselves, represent themselves. Only then can genuine feminist solidarity, extensive and empowering conversations, take place between African and Africanist feminist scholars.

Notes

1. This chapter originally appeared as a review essay of the four books.

2. The three books they refer to are by Hay and Stichter (1984); Robertson and Berger, (1986); and Stichter and Parpart, (1988). The review article is by Robertson (1987). The same observation can be made of the *Canadian Journal of African Studies* 22, n° 3 (1988) *Special Issue on Current Research on African Women*. None of the 15 articles and 'research notes' were written by Africans. Reading the articles and perusing their bibliographies one would think there is hardly any gender research being conducted by African scholars, either within or outside the continent.

Part Three

Encountering Development

Chapter Eleven

COLONIAL DEVELOPMENTALISM

Introduction

As an ideology of colonial and neo-colonial modernity, developmentalism was born during the Great Depression and bred into a hegemonic discourse in the immediate aftermath of the Second World War. The seeds were sown with the 1929 British Colonial and Welfare Act. They turned into sturdy developmentalist weeds under the Colonial Development and Welfare Act of 1945. It was in colonial Africa that most of these seeds and weeds were nurtured. It was there that the term *development* lost its naturalistic innocence and acquired the conceited meaning of economic growth modeled on the West[1]. Discerning the historical contexts in which colonial developmentalism emerged is essential for unravelling its discursive foundations which have proved enduring and adaptable to different terrains. But it also enables us to go beyond the deconstructionist proclivity of undressing the outer ideological layers of discourses without fully exploring their underlying historical conditions and processes. This is particularly important since the current popularity of the discourse analysis of development proposed by several writers, such as Escobar (1995), threatens to throw out the babies of development conceived in the struggles of social groups in Africa, Asia, Latin America, and indeed, in the industrialized countries themselves, with the bathwater of western developmentalism. The failure to engage historical realities and social struggles in the construction of developmentalism, the chapter argues, has resulted in an insufficient critique, which may partly explain the recent surrender of progressive nationalists and socialists to the market fetishism of contemporary neo-conservatism, their acceptance of the view that there are no conceivable alternatives to capitalist development as decreed by the triumphalist 'Right'.

Planting the Seeds

All too often, 'Colonial Development and Welfare' is seen either as marking the dawn of enlightened colonial 'development', or dismissed merely as a Machiavellian, if belated, attempt to overshadow the naked realities of colonial exploitation. In the orthodox development literature, it is maintained that the formulation of the post-war Colonial Development and Welfare programme represented the beginning of a 'new deal' for colonial peoples, signalling the dawn of 'welfare colonialism'. In explaining the origins of this new policy we are generally left with misleading impressions of imperial generosity (Kimble 1960:Chapters 15 and 23; Amsden 1971:Chapter 1; Goldsworthy 1971:Chapter 1; Clayton and Savage 1974:Chapter 9; Low and Lonsdale 1976; Crowder 1978:610-21; Crowder 1981:502-4; Lee 1967). Fieldhouse argues that by 1945 classical development theory had been jettisoned in favour of one stressing 'growth' and planned economic development for both the developed and colonial countries. The new colonial policy, reflecting this shift in emphasis, 'aimed to make', particularly in British Africa, 'African territories autonomous economies controlling their own instruments of economic policy' (Fieldhouse 1971:600; Fieldhouse 1986, 1988). Hopkins goes further and states that the infusion of 'substantial' metropolitan public investment in the colonies after 1945 represented an abandonment of the doctrine of colonial self-sufficiency in an attempt to 'remove, for the first time, the constraint by which the level of demand in West Africa had been determined almost entirely by the size of the export proceeds' (Hopkins 1973:267-8).

The assumed benevolence of metropolitan investment is dismissed in the nationalist and dependency literature. As early as 1948 George Padmore, the great panAfricanist, argued that Britain was faced with a shattered economy at home and the loss of old markets and extensive investments abroad, so that it was forced by sheer necessity to 'search for new untapped sources of raw materials and foods. And where more naturally than the still vast British Colonial Empire in Africa, with its apparently unlimited potential resources of labour and raw materials' (Padmore 1949:155). He concluded that Britain was, therefore, simply displaying her 'genius for making virtue of necessity' in proclaiming 'Colonial Development and Welfare'. Writing about two decades later, Walter Rodney charged that 'Colonial Development and Welfare' was nothing but a hoax designed to whitewash colonialism, 'to mask and deny its viciousness' (Rodney 1972:233).

This debate over the origins and aims of British Colonial Development and Welfare reflects a broader historical debate between development and dependency writers concerning the nature and evolution of relationships between colonial and imperial economies. As briefly noted in Chapter 7, orthodox development theory portrays 'Third World' economies, including those of Africa, as backward, traditional, static, and undifferentiated, so that they can only be liberated and developed through the infusion of Western capital, technology, ideas, and cultural values. Rostow (1971) propagated his 'universal' stages of growth, while Parsons (1951) unlocked the psychological secrets necessary for development, and Lewis (1966) with his 'dualism' models offered 'industrialization by invitation'[2]. Their followers dutifully built empirical tomes out of these revelations and injunctions. They maintained that the international division of labour in the world capitalist system was a natural outcome of world trade, as inevitable as night and day. They argued that changes in colonial agrarian and labour systems could be explained in terms of vent-for-surplus theories (Myint 1971; Szereszewski 1965; Helleiner 1966; Hopkins 1973:Chapter 6).

Underlying all this was faith in the unguided, invisible, but unfailingly rational, hand of the market. When it came to international trade these writers worshipped at the altar of 'comparative advantage'. For development economists, therefore, underdevelopment was a *state*, not a *process*, which could be overcome by an unhindered utilization and deployment of resources in an 'open economy' under the aegis of 'western investment'. The term 'development' inspired confidence in the inevitability of 'progress' even for the 'noble savages' of the 'dark continent', in the same way as the notion of the 'civilizing mission' had done in earlier decades of Africa's colonial history. In short, concepts of 'exploitation' and 'oppression' were as strange to the vocabulary of development economics as Orwellian Newspeak.

Not so for dependency writers who were inspired by moral indignation against the West and radical pessimism about the prospects of 'capitalist development' in the 'Third World'. As noted earlier, the dependency school had its roots in the Economic Commission for Latin America's critique of the orthodox development perspective and subsequent reformulations of that critique by Latin American social scientists as result of the failure of the ECLA's import-substitution industrialization model (Prebisch 1971; Cardoso and Faletto 1979). The dependency perspectives 'quickly spilled outside Latin America and the Caribbean to embrace Africa and social science theory in general' (Stern 1993:27). Dependency theory was cooked and consumed in a variety of ways, from Frank's (1969) simplistic offering of 'metropolis-satellite' chains and Wallerstein's (1974) more sophisticated world systems menu, to the hot Marxist platter of Baran (1956), Laclau (1971), and the heated Dobb-Sweezy debate on capitalist transition, continued by Brenner (1976, 1977) and others. Once in Africa, the dependency stew was seasoned and served in different nationalist and Africanist dishes by

Amin (1972, 1974, 1976) and Rodney (1972), Leys (1974) and Wallerstein (1976), among others (Harris 1976; Gutkind and Wallerstein, 1976; Cooper 1993a).

These writers insisted that the existing international economic system was structured in such a way that most of the 'social surplus' produced in the Third World is siphoned off through numerous mechanisms, principally the operation of 'unequal exchange', whether understood specifically as a product of wage disparities between poor and rich countries (Emmanuel 1972), or in a broader theoretical context of exchange relations between different social formations, and not within the narrow confines of exchange inside the capitalist mode of production (Amin 1974, 1976), or more generally in terms of western trade, financial, and investment controls. For dependency writers, therefore, the world was basically a single, integrated unit; development and underdevelopment were dialectical sides of the same coin. Reproduced historically, the phenomenon of external surplus extraction, appropriation, or drainage entailed that economic development for the centre and structural underdevelopment for the periphery were simultaneously and continuously generated by the same historical process of the expansion and development of the world capitalist system. Thus, contrary to the ahistorical perspectives of development writers, for the dependency writers underdevelopment was a structural condition, a product of historical processes. While the form of incorporation changed from period to period, the *dynamic* or *structure* of underdevelopment, they insisted, did not, indeed, it was constantly deepening.

Debates about development swirled around these two perspectives. The desirability of modernity for the 'Third World' was never in question by either side. This is because all the dominant theories of development, liberal or Marxist, shared in the underlying premises of what Berman (1994) calls the 'western paradigm of modernity'[3]. The dispute was whether it would be in the image of capitalism or socialism, through the idealized paths of the West or the East. Some Africans debated whether Africa could carve out its own trajectory from the imagined memories of past communalism or from the imagined futures unleashed by prevailing anti-imperialist confrontations, although their voices were not heard in the boardrooms of the international development agencies, or in the Africanist seminar rooms. The lines of battle among the western-based protagonists were often crossed, increasingly so by the 'Chicago Marxists' of Bill Warren's (1980) ilk, as Bienefeld (1988:79) labels them, who discovered that rapid capitalist development was actually taking place in many parts of the 'Third

World', and socialism would have to wait until capitalism had matured and fulfilled its progressive historical mission sometime in the indeterminate future. In Africa, the 'Chicago Marxist' safari found Kenya a most attractive destination. Swainson (1977, 1980) tried to show that an independent Kenyan national bourgeoisie, using the post-colonial state to further its own accumulation and often caught in competition with metropolitan capital, was developing and gradually moving from the sphere of circulation to that of production. Inspired by her and others, especially Cowen (1976, 1979, 1982), Leys (1977, 1980, 1982) abandoned his earlier dependency position and found that sustained capitalist development was, indeed, taking place in Kenya. The loudest Warrenite in the largely expatriate 'Kenyan debate' proved to be Kitching whose embrace of capitalism was coated with convoluted Marxist verbiage: 'I believe', he wrote, 'that the formation of a sophisticated working class in Africa will take a long time and that a prolonged period of struggle against a developing capitalism there is one of the important prerequisites of its creation. To that extent I am 'happy' to see continued capitalist development in black Africa' (Kitching 1985:148)[4].

By the late 1980s, therefore, it would appear that the disillusioned 'left' had been defanged by the resurgent 'right'. The former increasingly shared and echoed the latter's developmentalist assumptions, evidence, and conclusions. The ease with which this happened can partly be explained by the spatial and institutional locations of the Africanist producers of dependency and neo-Marxist perspectives. Located in the western universities, and only going to Africa on the occasional safari trip, the Africanist left drew its theoretical cues from the periodic paradigm shifts in their home institutions and the domestic ideological mood swings, rather than from African realities, let alone African intellectuals, whom they largely did not read, for they approached Africa, in Mamdani's (1995a:609) poignant phrase, 'as one would approach a zoo. For the collectivity of animals, no matter how interesting and intricate its internal organization, has no capacity for self-reflection and self-transformation!'[5]. The Africanist left's concession to neo-classical developmentalism, also reflected the fact that from the beginning it endorsed the universal mandate of western modernity. They never paid much attention, therefore, to the archaeology of developmentalist discourse, that it was rooted in the confrontation between imperial power and African resistance, which was latched on to intra-imperialist rivalries and intra-colonial struggles.

A fuller understanding of colonial developmentalism, and its ensuing mutations, requires paying attention to and integrating, analyses of the structures and processes of imperial-colonial relations, the dynamics of social movements and struggles, and the development of discursive regimes. Such an approach would demonstrate that while the discourse of 'Colonial Development and Welfare' sought to appropriate for the empire the virtues, representation, and definition of 'development', its genesis signified neither wilful 'benevolence' nor 'malice', colonial 'welfare' nor 'manipulation', but was an outcome of complex and often contradictory forces emanating from both Britain and the colonies themselves, underlined by deepening national and class struggles within and between the colonies and the imperial power in a post-war imperialist system characterized by a new international division of labour. Needless to say, this new international division of labour was both the reflection and mirror of the changing imperial-colonial relationships.

The material and ideological seeds of the development idea lay in the hardships and struggles of the Great Depression, whose causes are rooted in a complex web of transformations that took place in the world economy from the late nineteenth century and were accelerated by the First World War, principally the growing instabilities in international trade, finance, and production as manifested in growing protectionism, rising inflation, and price fluctuations. The depression came on the heels of a temporary boom among the major industrialized countries in the late 1920s, including the United States, the leading lender, whose economic recession, symbolized by the October crash of 1929, some economic historians believe, sent an already sick world economy into a tailspin (Kindleberger 1973; Fearon 1979; Bruner 1981; Kenwood and Lougheed 1983). Commodity prices crumbled, unemployment swelled, the gold standard collapsed, and extensive exchange controls and trade regulations were imposed. 'Free' trade gave way to regulated trade as tariffs were raised and import quotas imposed, so that by the end of the 1930s close to half of the world's trade was restricted by tariffs alone. The decadal average in per capita trade declined from a growth rate of 34% in the period 1881-1913 to 3% between 1913 and 1937 (Kenwood and Lougheed 1983:223).

The political and economic effects of the depression were quite contradictory among and within countries. On the political front, almost everywhere, nationalism was strengthened and given a powerful economic basis. On the one hand, growing nationalism in Europe sowed the seeds of the horrendous tragedies of the holocaust and the Second

World War, and on the other, in Africa and Asia it led to the emancipation of hundreds of millions of people from colonial subjugation. Because of their dependency on primary commodities, whose prices fell more sharply than for manufactured goods, and the lack of social welfare programmes, the colonies were hit hardest. The colonial 'civilizing mission' looked more threadbare than ever to workers losing their poorly paid jobs and peasants fetching paltry prices for their cash crops. As they sank deeper into colonial squalor, opposition to colonialism spread, although the intensifying African nationalisms were articulated in the specific idioms of location, class, and gender. In the meantime, growing militancy among workers and trade unions in the industrialized countries and the real threat the depression posed to the continued legitimacy and survival of the capitalist order led to the economic and social reforms that eventually became embodied in the post-war welfare state.

The African colonies, unlike the imperial powers, could not, of course, find an external vent for their surplus production and problems. Also, in contrast to the independent Latin American countries, or semi-autonomous African ones such as Egypt and South Africa, they could not turn the crisis into an opportunity for economic diversification and industrialization. In Brazil, for example, the collapse of agricultural prices reduced the power of the landowners, who had until then dominated the state, and strengthened the hands of those who preferred local industrialization to reliance on agricultural exports. In 1930 a new government came to power and raised tariff protection for local industry. Industrial production rose by about 100% between 1931 and 1938. Industrialization was also given a boost in South Africa and Egypt, both self-governing territories from 1910 and 1922, respectively. By the end of the 1930s Egypt had become self-sufficient in a wide range of consumer goods.

Faced with restive workers at home, frozen foreign investments in Austria and Germany, growing protectionism among the other industrialized countries, rising budget deficits, a falling pound and a run on its gold reserves, Britain abandoned the gold standard, and tightened its hold over the captive colonial markets. The empire became Britain's protectionist bloc. Other regional blocs were also emerging, such as the pact between Italy, Austria and Hungary, and the trade agreements concluded by the United States with a number of Latin American countries. British companies sought to offset losses elsewhere in their imperial backyards of Africa and Asia, just as American companies were extending their tentacles in Canada and some Latin American markets, and German cartels in Central and Eastern Europe.

It was in this context that the first Colonial Development and Welfare Act was born in 1929, which entailed lending money to the colonies for infrastructural development, especially railway construction. The leaders of both the dominant Labour and Liberal parties stressed that colonial development was 'part of the attack on British unemployment... rather than [about] the problem of colonial underdevelopment, although the prevailing belief in the natural harmony of economic interests between metropolis and dependency ensured that those who promoted it could very genuinely believe that it would also 'promote_ (the Colonies') welfare' (Brett 1973:132). An estimated £17.8 million worth of projects were identified in the colonies between 1929 and 1939, for which £8.9 million was actually spent or allocated, of which less than 14% was a free interest grant, and 75% of the funds were spent in Britain. The pattern of attaching thick strings to 'development aid' was established.

The Colonial Development and Welfare Act of 1929, was one of the many instruments the British state developed to deal with, simultaneously, the crisis of legitimacy in the colonial empire and of capitalist rationality at home. It was part of the response to growing anti-colonial resistance in the empire and anti-capitalist revolt in the metropole. It represented, therefore, a defensive, conservative discourse, one that recast the role of the state and framed development as a managed process. State interventionism resonated with as divergent political realities as Nazism in Germany and the New Deal in the United States. It was given theoretical imprimatur by John Maynard Keynes (1961). The Second World War reinforced state intervention, revived capitalism while locking it into a fierce competition with socialism, and bolstered colonial developmentalism, all thanks to the very forces and discourses unleashed by the war in the world at large, as well as in the imperial metropoles and the colonies.

Imperial Drought

One of the most salient features of the post-war era was a shift in the relative balance of power among the metropolitan capitalist countries. The USA emerged as the undisputed dominant power in the capitalist world and the economies of colonial powers like Britain lost their previous predominance (Baran and Sweezy 1966; Mandel 1975; Louis 1977; Magdoff 1978). The emergence of a fundamentally new hierarchy marked by the hegemony of the USA in the political, economic, and military spheres had far-reaching repercussions on the colonial powers and their colonial policies which, in turn, affected the political economies of the colonies themselves.

The roots of the USA's rise to dominance lie in a set of complex factors which we cannot go into here. Suffice it to say that it was during the Second World War that the USA made huge advances in technology, productivity, and labour organization, so that it eclipsed the war-ravaged economies of the Old World. The colonial world could no longer remain sheltered from the spreading tentacles of American capital. The shadows of colonialism, if not its face, could never be the same again. The rise of American capital to a hegemonic position significantly contributed to the reorganization of the home economies of the colonial powers, if only because of their heavy reliance on Marshall Plan Aid in their reconstruction programmes. The inability of British capital to compete with American capital on a global scale led to the reorganization of the British economy; the enhanced interventionist role of the British imperial state at home was to be accompanied by an equally extended interventionist role of the latter in colonial economies. It was, of course, during the Second World War that imperial control over colonial economies was tightened through the imposition of bulk purchasing, currency, tariff, and shipping restrictions (Leubuscher 1956).

Another important feature of the post-war capitalist economy was the growing multi-lateralization of foreign investment. New international agencies such as the World Bank, the International Monetary Fund and the EEC Investment Bank were established, initially to facilitate the reconstruction of war-torn Western capitalist countries before they later extended their operations to the colonial and dependent capitalist countries. Thus, in the long run capitalist investment in the colonies ceased to be an almost exclusive preserve of the metropolitan colonial powers. Paradoxically, it was such diversification in the sources of foreign capital which led to a fuller integration of the colonies into the world capitalist system and laid the roots of neo-colonialism, with the IMF and its twin sister the World Bank later assuming the role of global police for capitalism.

Finally, this period witnessed the growing predominance of multi-national corporations, which was a reflection of the intensifying concentration and centralization of capital on a world scale (Danning 1972; Radice 1975; Widstrand 1975; Kaplinsky 1978; and *Review of African Political Economy*, 1981). The tendencies of international capital export were accordingly altered; in the major capitalist dependencies private capital investment would increasingly be mediated through multi-national corporations, mostly in the form of capital for the establishment and operation of overseas enterprises and branch plants. The entry of multi-national corporations accelerated, in some countries, the trends towards import-substitution industrialization.

In the aftermath of the Second World War, therefore, conditions were propitious for the restructuring of dependency for the colonial world. The more the colonial world became integrated into the international capitalist system, the greater were the pressures, both internal and external, to change colonial policies and relationships. The British government was quite concerned about a possible rush of American investment in the colonies and the adverse impact this could have on the dollar balances of the sterling area. It should be noted that Britain emerged after the 1939-45 war, from being one of the largest creditor nations in the world to being perhaps, the biggest debtor nation, owing £6,000 million (Morgan 1980a:89). The seriousness of the dollar drain was such that in August 1947 there was a balance of payments crisis which forced the British government to adopt drastic dollar conservation policies (Morgan 1980b:4-17; Balogh 1949; Zupnick 1957; Dow 1964; Conan 1966; Gardner 1967). By July 1949, the gravity of the situation prompted the Lord President of the Council to warn that with £12 million worth of gold and dollar reserves vanishing each week, 'they would be down to zero in just over 200 days from now, and in far fewer days than that our position will become untenable unless the fall cannot only be stopped but reversed' (Morgan 1980b:29). For the next few years the situation continued to deteriorate as the balance of payments accounts of Britain and the independent members of the sterling area ran into rapidly increasing deficits in their current accounts both with the dollar area and the rest of the world. American investment in the colonies, if unchecked, would not only acquire 'a dominating position in individual colonial territories', the Secretary of State for the Colonies feared (Gupta 1975:322.), but would also worsen the dollar crisis because such investment 'normally carries with it a dollar liability for remittance of dividends of profits and ultimate liability for repatriation of capital' (Morgan 1980b:106).

As a recipient of Marshall Aid, however, one of the provisions of which gave US nationals access to colonial raw materials, there were limits as to how far the British government could restrict the entry of American capital into their colonial empire (Morgan 1980b:97-108). Thus, it was in order to meet the American challenge and possible challenges from a resurgent Japan and the former West Germany that British firms were encouraged to set up manufacturing industries in the colonies. Colonial import substitution industrialization also promised to cut down colonial imports from America and other dollar areas and increase British exports of capital goods to the colonies which would help to revitalize British industry. In fact, the British government was already putting pressure on the colonies to limit their imports from dollar and other hard currency areas (Morgan 1980b:4-17). However, as a result of the British rearmament programme and the Korean war, British industry was incapable of satisfying colonial needs for items like iron and steel, so that their supplies to the colonies virtually became subject to central allocation from London. Thus, restrictions against imports from the dollar area and the shortage of goods in Britain served to reinforce the centralization of colonial 'development' planning.

Aggravating the dollar deficit and the balance of payments problems was the food crisis. Food stocks, especially of oils and fats, were dangerously low; indeed, the prospects of further cuts and shortages seemed inevitable unless supplies could be increased (Morgan 1980b:Chapter 4). So serious was the situation that the Prime Minister 'proposed the setting up of a Ministerial Committee of World Food Supplies (WFS), under his chairmanship, to keep the situation under review, to coordinate action and to focus on the major issues calling for decision by the Cabinet or international bodies concerned' (Morgan 1980b:177). The WFS held numerous meetings and published reports on the deteriorating food situation. In the end it recommended that the only viable solution lay in increasing colonial agricultural production. Missions were sent to West Africa and other areas to assess their agricultural potential and infrastructural problems.

Britain and the other European colonial powers were only too aware that increased colonial production would not only serve to aid the economic reconstruction of Western Europe, it would also bolster the latter's position in the corridors of world diplomacy and politics. The USA and USSR were now not only the new 'superpowers' but both had, in the words of the British Foreign Secretary, 'tremendous resources...If Western Europe is to achieve its balance of payments and to get world equilibrium, it is essential that colonial resources should be made available' (Padmore 1949:155-156). It is instructive to note that after the war the Colonial Office and its French counterpart, the Ministry of Overseas Territories and before long the Belgians as well, established close organizational contacts. Anglo-French-Belgian collaboration was soon extended to colonial Africa itself where a number of joint conferences were held on economic and labour issues (Padmore 1949:189; Morgan 1980a).

Britain unveiled its new colonial 'development' policy, therefore, amidst a severe economic crisis. Thus hiding behind the rhetorical flourish of the term 'colonial development and social welfare' was hard-nosed British economic self-interest. Gone was faith in *laissez-faire* capitalism at home and in the colonies, and resort to *ad hoc* policies resurrected from the imperial closet each time there was a crisis in some colony. 'In the past,' stated the Economic Policy Committee, 'it had not been necessary for the Colonial Office to exercise a direct control over (colonial) economic investment. As a result of the government's policy of direct participation in the task of colonial development...it was a necessary corollary...that the Colonial Office should be in a position to maintain general control over colonial development' (Morgan 1980a:313). Henceforth, colonial 'development' was to be closely co-ordinated with British reconstruction and development plans. The idea of 'integrated economic development of the colonies and the UK as a dominating factor in development policy' acquired the status of an orthodoxy (Morgan 1980b:80). The eggs from the imperial chickens had become essential for British survival.

It is hardly surprising, therefore, that despite often spirited differences over internal economic and social policies, when it came to the question of colonial development, there was a remarkably high degree of bi-partisan cooperation between the governing Labour Party and the opposition Conservative Party in the first critical years when British post-war colonial policy was being formulated and the machinery to implement it was being created. It is indicative of the prevailing mood that such a stalwart of the Fabian Colonial Bureau and the Labour Movement as Rita Hinden could say categorically that they did not have 'the intention of sacrificing our standard of living for the sake of colonial development, we do not contemplate an evening out of wealth, we know, too, that our development plans are partly inspired by our own needs' (Hinden 1949:89).[6]

There can be little doubt, then, that a consensus emerged in British ruling circles that the dollar and food crises could only be resolved by 'developing' colonial resources (Morgan 1980a:331-33). Some also argued that rapid colonial 'development' would help make Britain independent of the USA. In the words of Ernest Bevin, the Foreign Secretary: 'if we only pushed on and developed Africa, we could have the US dependent on us, and eating out of our hands, in four or five years...[The] US is very barren of essential minerals, and in Africa we've them all' (Gupta, 1975:306).

But ideological and political considerations were no less important. From both sides of the international ideological divide, and for very different reasons, and from the recently formed United Nations and newly independent ex-colonies like India, there came unprecedented and almost universal condemnation of 'old style' colonialism, which rattled the British leadership (Morgan 1980a:Chapter 1). The colonial powers were on the defensive. 'The overwhelming reason', the Secretary of State told the Chancellor of the Exchequer, 'why I feel that these proposals (for colonial development) are essential is the necessity to justify our position as a colonial power' (Morgan 1980c:xxvii). The Chancellor agreed whole-heartedly (Morgan 1980c:xxvi-ii). In short, the policy of colonial development was further elaborated to blunt criticisms against colonialism. Facing the imperial powers was something more lethal than criticism in the international arena. The colonies themselves were on fire.

Colonial Storms

The most sustained pressures for colonial economic transformation came from struggling colonial peoples themselves. The impact of the West Indian riots of 1938 on the Colonial Development and Welfare Act of 1940 has often been noted. This act provided a miserly £5 million a year for development projects and £500,000 for research. The sums provided by the Colonial Development & Welfare Act (CD&W) of 1945 were much larger. It cannot be overemphasized that wartime colonial struggles in Africa had something to do with it.

The Second World War saw the resurgence of some of the gross forms of labour exploitation reminiscent of the earlier phase of primitive colonial accumulation. At the outbreak of the war, colonial states acquired broad powers of coercion over labour. Accordingly, new legislation was passed which gave colonial governors power to order their administrative functionaries to produce quotas of workers for the military, essential services, and sometimes even private agriculture and industry (Zeleza 1982:Chapter 3; Shiroya 1968; Crowder 1981; Olusanya 1973; Holbrook 1978; Killingray 1986). When it is borne in mind that about 200,000 British West African and 98,000 Kenyans, for example, served in the armed forces alone, and that many of them were conscripted, the extent of the conscription drive becomes all to apparent.

Peasant economies were not only drained of labour power, they also became major suppliers of foodstuffs and other raw materials to the metropoles. Colonial governments intervened at the level of both production and exchange. Administrative pressures were either applied for peasants to grow certain crops, or imports and exports were increasingly subject to government demands through the imposition of new systems of quotas, price-controls, and bulk-purchasing. The Marketing Boards which were to become a characteristic feature of the post-war economic scene were born. Such substantial removals of labour power from peasant agriculture and the expansion of production for the war effort produced an explosive combination: by 1942-43 Kenya, for example, was gripped by severe food shortages. Thousands of Africans died from starvation, especially in the Central and Nyanza provinces, and tens of thousands more gauntly watched their health deteriorate. Similar stories were reported from various parts of West Africa. The food shortages were a grim but eloquent testimony to the harsh conditions generated by the war, and an outcome of the cumulative effects of export-oriented agricultural production.

Meanwhile, shortages of food, consumer goods and housing, spiralling inflation, falling real wages, and inadequate social services, afflicted the rapidly growing colonial towns, with their fledgling import-substitution industries, like a plague. Colonial economies were indeed expanding and more people than ever before were entering permanent wage labour, but urban poverty and marginalization persisted, and was even deepening.

The African 'elites' or aspiring 'national bourgeoisie' were not immune either from the deprivations generated by the war. To be sure, unofficial African members were appointed to the Executive Councils of Nigeria and the Gold Coast in 1942, the first African member to the Kenyan Legislative Council was appointed in 1944, and African Representative Councils were established in Northern Rhodesia and Nyasaland two years later (Coleman 1958; Kimble 1963; Roelker 1976; Rotberg 1972). Africans were, however, limited to criticizing colonial policy, not formulating it, even if they formed an elected or unofficial majority as in West Africa. Indeed, state intervention in the economies of the four British West African economies strengthened the large oligopolistic firms to whom the governments gave licenses and other forms of preferential treatment, at the expense of African traders. In Kenya the war gave the settler community a unique opportunity to further consolidate its economic and political position. There was even talk of establishing a 'White Dominion through the whole of East Africa' (Dawe 1942). In British Central Africa this was, of course, no longer mere talk. By 1945 hopes of amalgamating the two Rhodesias and Nyasaland were solidifying into the concrete possibility of a white-supremacist Central African Federation (Creighton 1960; Franck 1960; Keatley 1963). Independence was nowhere on the horizon, notwithstanding the current mythology of 'planned decolonization'.

Superficially, it looked like Europe's 'civilizing mission' in Africa still had a long way to go.

But it was not to be. The 'winds of change' were blowing with increasing nationalist force. Peasants, workers, petty traders, and the 'nationalist bourgeoisie' increasingly became impatient of, and ready to confront, colonial rule. Rural disturbances became widespread. Conscription provoked desertion and flight, forms of labour resistance reminiscent of the early decades of colonial conquest. In parts of the Gold Coast chiefs were chased from collecting market dues. In Kenya, as the position of squatters deteriorated with the expansion of settler capitalist farming, the squatters became more restive than ever before and were poised to become a vanguard in the Mau Mau struggle (Barnett and Njama 1966; Rosberg and Nottingham 1966; Throup 1987; Kanogo 1987; Ogot and Zeleza 1988; Furedi 1989; Lonsdale 1990; Maloba 1993).

The six years of the Second World War also witnessed labour unrest of unprecedented proportions (Davies 1966; Bates 1971; Freund 1981:Chapter 5). On the one hand, there was the re-emergence of desertion as a major form of labour protest, especially among conscripted workers. In Ghana for example, 20% of all men conscripted deserted. On the other hand, workers resorted to strikes on a scale hitherto unknown and wielded the strike weapon with fierce tenacity despite the colonial governments' repressive labour policies. In 1940 the uneasy industrial 'cold war' on the Copperbelt in Northern Rhodesia erupted into a violent strike in which 13 workers were killed and 69 were wounded. In Kenya a strike wave broke loose particularly from 1942 and lasted until the end of the war (Zeleza 1992). In West Africa strike activity peaked in 1942 as well, with 13 major strikes recorded in Sierra Leone, ten in the Gold Coast, and mass protests by railway workers in Nigeria in 1941-42. Hardly had the war ended when there was a successful 37-day national general strike in Nigeria which involved 17 unions representing about 30,000 employees. The seeds of trade unionism sown in the 1920s and 1930s had taken firm root. By 1943 Kenya had ten registered trade unions, and in Nigeria there were 85 unions, with tens of thousands of members, despite the continued repression of trade union activity by the colonial states and capital.

It is indicative of the charged circumstances that the older political organizations shed some of their moderation and timidity, or they were overtaken by new, militant movements. The Youth Movements which had emerged in the turbulent 1930s in British West Africa matured into broad-based and radical nationalist movements, as happened to

Azikiwe's National Council for Nigeria and the Cameroons formed in 1944. In the same year on the other side of the continent the elitist 'native welfare associations' in Nyasaland came together and transformed themselves into the Nyasaland African Congress. And in Kenya, the Kenya African Union, pushed by fiery young activists and the rumblings of Mau Mau, became more rebellious than before.

During the first years after the end of the war these struggles intensified and the sprawling branches, if not the shallow roots, of colonialism were shaken. The East African coastal cities of Dar es Salaam and Mombasa erupted into general strikes in 1947 (Zeleza 1995). In the following year the Gold Coast was rocked by boycotts and riots, and the decade closed with the massacre of 21 striking miners in Enugu, Nigeria (Sandbrook and Cohen 1975; Gutkind et.al., 1978). Trade unions proliferated, officially registered or not. The colonial world was gripped by an upsurge of anti-colonial militancy; mass demonstrations became frequent, the African-run press bitterly denounced colonial exploitation and oppression, and nationalist agitators' publicized their case before the United Nations and other forums, and they forged links with each across the continent and spread their hands to their Asian colleagues. Nkrumah's cry 'Independence now!' was echoed and re-echoed far beyond Ghana's borders. In British Central Africa, the nationalists braced themselves to fight against the imposed Federation of Rhodesia and Nyasaland. And in Kenya the Mau Mau war of national liberation broke out. Colonialism was on the defensive. Reform had become imperative.

Thus, a lot of the pressure for embarking on far-reaching changes in imperial-colonial relationships emanated from the colonies themselves. Cumulatively, the material and social changes during the Second World War influenced and transformed mass political behaviour. The expansion in the number of those who were politically aware, who were conscious of unmet needs, and tired of the material, moral, and mental anguish of the 'colonial situation' resulted in mounting pressures for the transformation of prevailing political practices and the establishment of institutions that would allow for their meaningful participation in political and economic life. Colonial peoples were impatient for rapid economic, political, and social change. The British imperial state could only ignore these pressures at the peril of fuelling the fires of nationalism which might transform the essentially bourgeois reformist nationalist movements into crusaders of more profound struggles for national liberation and social emancipation, and thus 'risk losing goodwill and therefore future economic opportunities' (Fieldhouse 1988:142).

And so post-war colonial 'development and welfare' policy developed. It grew out of an interaction of these complex forces; it was a policy inspired by the imperatives of British economic reconstruction, designed to placate hostile world opinion and colonial critics, and goaded by colonial pressures for political transformation and demands for rapid colonial development.

Tending the Weeds

A number of institutional mechanisms were accordingly created to implement the new colonial 'development' policy. Firstly, there was the CD&W Act of 1945 which together with the amending Acts of 1949 and 1950, provided for a total of £140,000,000 available over the years 1946-56[7]. The colonies were invited to submit ten-year development plans. By June 1948, 17 had been received (Morgan 1980b:Chapter 2). The plans had to take into account money provided under the Act, funds raised from internal resources and on the money markets. Although virtually all the plans related to development in the public sector, the coverage of what constituted such a sector varied from one colony to another, so that it was difficult to apply a single yardstick in making allocations (Morgan 1980b:Chapter 2). For the entire post-war period that the CD&W Act was in operation, Africa received 45% of the funds, with Nigeria ranking as the recipient of the largest grant (£40 million), followed by Kenya (£23 million). The Caribbean area was allotted 22% of the funds, the Mediterranean area (principally Malta) 8%, the Western Pacific 7%, South-East Asia 6%, and the rest went to the small islands and on the services in Britain, mainly survey work and student training (Morgan 1980d:Chapter 1).

The Colonial Development and Welfare programme with its emphasis on the construction of infrastructure and public services such as education, medical and health services, housing and town planning, was useful in paving the way for the eventual entry of private capital on a fairly large scale, but was unlikely to fulfil British economic needs in the short-term. Consequently, the Colonial Office and the Ministry of Food established the Colonial Development Corporation (CDC) and Overseas Food Corporation (OFC), respectively, both of which were to be run on purely commercial lines.

The CDC was formed in 1947 with a capital loan of £100 million from the Treasury, and the right to borrow an additional £15 million from private sources, and later its borrowing powers were increased by £30 million (Morgan 1980d:118-20). Initially, the corporation preferred short-term projects to long-term ones, and those requiring a minimum of capital goods, especially US imports. Eventually it excluded from its sphere of operations loan finance for colonial governments, public works schemes, and any such enterprises that could be financed from other sources. Instead, the corporation invested in, or ran, mining projects, manufacturing and processing industries, civil engineering works, fisheries, agricultural, and forestry projects. In order to facilitate its functions, colonial governments were encouraged to form public corporations which established working relationships with the CDC (Morgan 1980a:Chapter 5; 1980b:Chapter 6).

The financial structure and operational procedure of the OFC was almost similar to that of the CDC, although the former was to have a shorter life-span. The OFC was also established in 1947 with an authorized capital of £50 million, including the £25 million originally advanced to the United African Company (UAC), some of whose operations the OFC took over. The OFC was expected to organize large-scale plantations and food growing projects in the colonies. The disastrous East African Groundnut Scheme, which was originally started by the UAC, was its largest project and the scheme's eventual failure spelled the end of the OFC itself, which was dissolved in 1954 (Morgan 1980b:Chapter 5; 1980d:Chapter 4).

But this investment or 'aid' was only British in name. During the war, the British government had established, as already noted, bulk buying and bulk purchase prices, and the differences between the two prices was supposed 'to constitute a kind of collective post-war credit', according to Harold Macmillan (1967:174-5), a one-time British Prime Minister. Needless to say, the British channelled some of these same commodities on to the world market where the prevailing prices were higher than those of the bulk purchase scheme. In other words, the colonies not only made significant contributions to the British war effort by providing troops, foodstuffs, and raw materials, their huge sterling balances held in London subsidized the British economy at a very critical period, and British sales of their commodities to the dollar area helped in lessening Britain's dollar deficits.

So rapid was the increase in colonial sterling assets that by 1951 they surpassed the £1,000 million mark (Morgan 1980a:311), up from £573 million in 1944 and £805.5 million in 1947 (Morgan 1980c:201). By 1955 sterling balances from the African colonies alone stood as high as £1,446 million, or 'more than half the total gold and dollar reserves of Britain and the Commonwealth, which then stood at £2,120 million' (Rodney 1972:188). The British government sought to freeze the colonial assets, on the one hand, and on the other, to promote colonial 'development'. It was a contradiction which many, especially colonial spokespersons and sympathizers, deplored (Morgan 1980b:46-63). The issue of the colonial sterling balances was a source of fierce interdepartmental infighting between the Treasury, which wanted to have the balances cancelled altogether in view of Britain's economic problems, and the Colonial Office, which feared that such a measure would enrage the colonial world and create serious political problems for Britain. Instead, the latter proposed that the colonies should be asked to 'convert a part of their existing holdings into interest-free loans to His Majesty's Government, which would be repayable only when required to meet certain specified obligations of colonial governments' (Morgan 1980b:58). No wonder that, to queries why colonial sterling balances were not being turned over to the colonies to finance their development, 'it was pointed out that while, from a colonial development point of view, the use of balances to finance development expenditure was likely to accelerate growth, the expedient would be likely to involve increased unrequited exports from the United Kingdom, thus aggravating the balance of payments positions and so should be avoided' (Morgan 1980b:60). The British government's Chief Economic Adviser was, therefore, not exaggerating when in October 1954 he quipped: 'Is it not the case that the increase in colonial sterling balances i.e. short-term lending by the colonies to the United Kingdom, is well in excess of the United Kingdom's long-term lending to the colonies?' (Morgan 1980a:336).

Apart from maintaining huge sterling reserves in Britain, the colonies also played a critical role as dollar earners for the sterling area as the dollar deficit of the United Kingdom and the Dominions grew. The colonial empire was in deficit for dollars in 1945, but by 1948 it was a net dollar earner at the rate of $200 million per annum. In the subsequent years the colonies maintained substantial surpluses, while the dollar deficits of the UK and the Dominions grew larger. In 1951, for instance, the dollar deficits in the UK current account was £472 million and for the Independent Sterling Area, i.e., the Dominions, it was £34 million, or a total of £506 million, whereas the dollar surplus in the colonial current accounts was £235 million (Morgan 1980b:30-1).

Colonial Development and Welfare, therefore, represented a 'recycling' of the fruits of colonial exploitation to generate more exploitation. But this exploitation affected the various social classes in British colonial Africa differently. The funds made available by CD&W and other sources, helped expand the public sector which became increasingly Africanized. The colonial state was consolidating its centrality in the appropriation and distribution of surplus and on its way to becoming perhaps the most important instrument of accumulation for the post-independence bourgeoisie. The incorporation of the 'nationalist bourgeoisie' into the apparatuses of the colonial states in the 1950s reflected, and further shaped, the underlying structural changes in colonial political economies. Colonial governments began improving conditions for, or lifting a number of regulations limiting the ability of the African merchant and agrarian bourgeoisie to accumulate[8]. In West Africa many of the leading commercial firms came to concentrate on wholesale activities and left many of their outlets to African traders. By 1963 the three leading commercial firms in Nigeria accounted for 16% of all imports, down from 49% in 1949. As for Africanization of management positions, the United Africa Company's managerial staff in 1957 was 21% African, up from 7% in 1939 (Hopkins 1973:277). The same process was taking place in the other territories, including that erstwhile 'White Man's Country', Kenya (Leys 1974; Swainson 1980; Kitching 1980). In fact, in Kenya the development of an agrarian bourgeoisie, even in the once exclusive and sacrosanct 'White Highlands', was partially underwritten by the colonial state which undertook land reforms and provided funds for the land purchase schemes (Leo 1984; Ogot and Zeleza 1988; Berman 1990; Berman and Lonsdale 1992; Himbara 1994).

Clearly, the imperial and the colonial states and the more advanced sections of metropolitan capital, impelled by nationalist militancy, led by the long-suffering and highly expectant African bourgeoisie, began to recognize the inevitability of far-reaching social, economic, and political change but wanted to preserve the capitalist system by underpinning these societies on class, not racial stratification, and building up, and aligning with, a mollified and moderate African bourgeoisie. The growing level of state support for African accumulation and the articulation of African commodity production with metropolitan capital led to the development of more pronounced class stratification and began to alter the patterns of social struggles in the colonies. Living conditions in West Africa, for example, for the vast majority of the population, especially peasants and workers, did not rise clearly above the levels of the 1930s until well into the 1950s. Not surprisingly, even Nkrumah's relatively 'progressive' regime in 1954 found itself, like its colonial predecessors, confronted with embittered cocoa farmers demanding higher cocoa prices and the dismantling of some of the state marketing controls. Seven years later a seventeen-day strike among Sekondi-Takoradi workers was harshly suppressed. The point is, although the striking workers were defeated, their strike represented a rejection of 'CPP Socialism' by the most militant and advanced section of the Ghanaian working class (Jeffries 1978:Chapter 5). On the eve of independence Kenya was not only rocked by an unprecedented wave of strikes, many of which were brutally suppressed by the nationalist-led government, there was also the spectre of Mau Mau's resurgence as landless peasants began taking the law into their own hands as they moved into abandoned or undercultivated farms and there was talk of reviving the Land Freedom Army (Kamumchuluh 1975; Lamb 1974).

Mass nationalism had bred political and economic reforms in the colonial system which, in turn, laid the basis for new forms of class exploitation and domination. As the British African colonies attained their independence from 1956 onwards, nationalism, both as a movement and an ideology, was already on the verge of fragmentation as class contradictions, often super-imposed by ethnic, and religious antagonisms, threatened to break loose.

Conclusion

This chapter has tried to show that Colonial Development and Welfare can best be understood in the context of the changed position of the British imperial state in the post-war world capitalist economy, expressed in the immediate aftermath of the war by a severe economic crisis, and, in addition, in response to the political crisis in the colonies which could no longer be contained within the old colonial framework. A new framework had to be found which would not only defuse this political crisis, but also remove obstacles against further accumulation of capital in the colonies. Through the CD&W Acts, the OFC and the CDC, and other less visible institutional measures, the imperial state mobilized material and ideological resources to maximize colonial production and, in conjunction with the colonial states, promoted conditions for, and sought to safeguard the continued successful accumulation of private capital both metropolitan and, increasingly, local.

Thus, on the one hand, the stage was set for the further integration of the soon-to-be independent colonial empire into the British economy, which, in the final analysis, meant the colonies' fuller incorporation into the world capitalist system since, on a world scale, British capital was now subordinate to American capital and new forms of capital organization had emerged. On the other hand, the new conditions of accumulation in the colonies accelerated class differentiation and led to shifts in the arena around which internal struggles began to take place. In this sense, the programmes and ideology of colonial 'development' were an integral part of the process of decolonization, if decolonization is understood to mean a delicate comprise and a restructuring of deepening national and class struggles within and between colonies and colonial powers in a post-war imperialist system characterized by American hegemony, growing multi-lateralization of investment, and the presence of ever-powerful multi-national corporations, and a world, moreover, increasingly polarized between two social systems.

In short, Colonial Development and Welfare was part of a broader process in which conditions of accumulation for the African national bourgeoisie improved noticeably. They were incorporated into the reformed colonial state and gradually integrated intimately into the transnational capitalist economy, while many workers and peasants were being marginalised and pauperized at the same time. In other words, a process of capitalistic expansion was taking place, particularly in the major colonies, by which is meant that there was an increase in the productive forces of social labour and the socialization of that labour, through capital accumulation, rising proletarianization, and intensification of the social division of labour. The main objective in this chapter has been to try and identify the conditions under which the conjunction of struggles in the periphery and structural changes in the centre gave rise to new imperial-colonial relationships, which, in turn, altered the terrain for subsequent social and class struggles, and generated the new discourse of developmentalism that was to dominate the discursive landscape of the post-war world. The expansive or suffocating weeds of developmentalism did not spring naturally from the fertile, whether fiendish or friendly, mind of imperialism, as it appears from idealistic accounts, but they germinated in a world undergoing complex struggles and transformations.

Notes

1. For informative analysis of the development of the term 'development' see Arndt (1978, 1981); also see Hettne (1990) and Marglin (1990).

2. Lewis went on to win the Nobel Prize for Economics, an event duly celebrated among his Caribbean compatriots in *Social and Economic Studies*, 29 (4), 1980.

3. Berman's essay is a vigorous critique of the developmentalist paradigm in Africa. But it also demonstrates the limits of Africanist auto-critiques. While urging us to recognize and embrace the plurality of African histories, cultures, societies, and development trajectories, the West is presented as a homogeneous entity, ruled by instrumental rationality, the fountainhead of all modern institutions and technologies, which unfortunately 'cannot and will not be reproduce anywhere else', least of all Africa, where western development has left behind 'technological chaos' and 'the most spectacular failures of industrial planning of our time'. And almost predictably, the prescriptions roll on: there must be 'maintenance of social diversity with ecological sustainability', the development of 'appropriate technology', and the pursuit of 'ethnodevelopment'. (Berman 1994: 246, 248-9, 253, 254-5). Who will implement this 'new' developmentalist package?

4. Other key participants in the debate are Langdon (1980), Kaplinsky (1980), and Beckman (1980).

5. He concludes: 'In this sense, Eurocentricism mutilates the experience of both the West and the rest, mythologizing the former and caricaturing the latter' (Mamdani 1995: 609).

6. There are numerous references in Padmore (1949), Morgan (1980b), Goldsworthy (1971), and Lee (1967) showing the bi-partisan approach towards colonial policy. For the Paternalistic attitudes of the British labour movement towards colonial labour movements see Zeleza (1982: Chapter 7, and 1987).

7. The original sum was £120 million. For more details, see Morgan (1980c: Chapter 15).

8. The literature on African capitalism and the bourgeoisie has grown steadily, most of it inspired by the need to identify a new agency that can undertake 'development' now that the state seems to have failed. In other words most of these studies are weak on historical analysis and long on functionalist and prescriptive evaluation. One of the few historians to discuss the subject at some length is Illife (1983). Most of the work has been produced by political scientists and development economists, with a sprinkling of sociologists. For earlier studies, see Leys (1974); Osoba (1977); Markovits (1977); Swainson (1980), and Kitching (1980). For some of the more recent ones, see Sender and Smith (1986); Kennedy (1988); Rapley (1993); Berman and Leys (1994). For a review of some of them, see Zeleza (1992).

Chapter Twelve

CULTIVATING HUNGER

Introduction

In 1984, exactly a decade after the devastating famines of the Sahel countries and Ethiopia, the world media rediscovered hunger in Africa. Long the concern of some academics and specialized agencies, the subject once again became a celebrated cause of popular horror, indignation, and despair. Television audiences were treated to harrowing images of suffering and death, drought and desolation in Ethiopia. Newspapers and magazines chronicled the endless conference resolutions on how to resolve the crisis and occasionally screamed with headlines that it was the worst catastrophe in human history, that Africa was a dying continent (*Daily Nation* 30-12-84; *The Standard* 28-5-85). Organizations like FAO (Food and Agricultural Organization) carried such chilling messages in a stream of statistical reports. One hundred and fifty million Africans in 24 countries, later reduced to 21, were facing starvation. The continent was the only major region of the world where food production per capita was declining. Between 1972 and 1983 food imports increased by about two and a half times in volume and five times in value. Despite this, the yawning gap between average food consumption per capita and actual nutritional standards was not closing (FAO 1984; 1985a). It was all overwhelming: donations flooded the coffers of relief agencies, cynical governments promised more 'aid', and, in Africa itself, the euphoria of the 1960s and 'radical' pessimism of the 1970s gave way to sobering disquiet. For the first time the OAU forgot its perennial political squabbles and planned an economic summit (*The Standard* 16-11-84; *Kenya Times* 27-4-85).

Divergent and often contradictory analyses, prognoses, and prescriptions flowed from the pens and presses of academics, 'aid' agencies, specialized global and continental economic institutions and African governments themselves. Some, including many African officials and commentators, put the blame on the weather, mainly the drought that was ravaging vast stretches of the continent. Others, such as the World Bank, attributed the crisis to population 'explosion' and misguided policies pursued by African governments since independence. Then there were those of 'radical' persuasion who saw the prevailing international division of labour as the main culprit and foresaw no lasting solution until it was transformed. In general, the debate and the remedial policies suggested were conducted within the parameters of orthodox development theory and dependency theory, the former with its neoclassical economic assumptions and prescriptions and the latter with its world systems perspectives, sometimes interfaced with Marxist concepts. This chapter seeks to demonstrate the complexity of, and contradictions generated by, the food crisis that gripped some African countries in the mid-1980s, by putting it in its proper historical and global contexts and showing the underlying role of Africa's postcolonial class configurations in reproducing it. First, it challenges the widespread tendency of generalizing both with regard to the dimensions of the agrarian crisis and the prescribed solutions. Then, the 'internalist' and 'externalist' arguments concerning the causality of the crisis are presented and critiqued. Finally, the consequences of the crisis are briefly assessed.

Natives Look Alike

Africa being so large and containing so many countries, 54 in the mid-1980s to be exact, the range of combinations that could be conjured up to represent the entire continent was wide indeed. Authors could choose any number of countries, or even one, to put their personal spin on the narrative of the 'African crisis'. Each writer would outline the sins of commission and omission made by the increasingly despised post-colonial state, which led 'Africa' on the slippery road to economic ruin from which it could only be rescued by, according to the 'men on the right', structural adjustment programmes devised by the international financial institutions, or by social revolution, according to the rapidly dwindling ranks of the 'men on the left'. In either case, the crisis was universalized; it transcended the messy and petty particularities of national history, spatial and social differentiations, state formations and development strategies. Mkandawire vigorously challenged this tendency to overgeneralization. He wrote:

> This general image of agrarian stagnation is misleading since it conceals profound processes of change taking place in African agriculture. It also glosses over vast differences in performances among classes of producers, among commodities, among countries and among regions within the same country. More significantly, the image conceals historical processes of social differentiation induced by and reinforcing the differentiated access to resources that are provided by state and international agencies to African agriculture (Mkandawire 1987:1-2).

Analyzing the food crisis in terms of countries, hid the obvious fact that it was not nations, least of all the 'elites', who suffered hunger, but specific disadvantaged groups, often located in specific areas, and concentrated among the urban and rural poor, among whom women predominated. In short, hunger discriminated according to the intolerant hierarchies of class, location, and gender.

The tendency to conceal the complex social and spatial dimensions of African agricultural performance and food shortages partly arose out of the fetish for statistics in neo-classical economics and its development offspring, in which measurement of national accounts, growth rates, and per capita indices displace questions of differentiation and distribution. It is astonishing how weak the statistical basis of the almost universal view of the African agrarian crisis was. The FAO provided the bulk of the data upon which the conclusions and forecasts of agricultural performance in Africa were based. But, as Raikes (1988:18) has convincingly shown, no less than 75% of all cereal production figures produced by FAO for 1982, for example, were based, wholly or in part, upon 'estimates', which were no better than guesswork. No wonder the frequent revisions that FAO made in its 'estimates' from year to year. Indeed, using those very figures over a long period, the African performance did not vary significantly from the world average. Between 1970 and 1990, for example, African food production, excluding Egypt, Libya, South Africa, and Sudan, rose by 50%, as compared to a world average of about 58%. Thus, contrary to popular view, Maxwell states:

> agricultural and food production have both increased in Africa since 1970... However, unlike the rest of the world, production has risen more slowly than population. The result has been a steady decline in agricultural production per capita, with Africa's per capita food production falling at a comparable rate... Food self-sufficiency has declined to 90 per cent, and import and aid dependence have increased. Things are not all bleak, however. In general, food availability has increased, thanks in part to higher food imports, including food aid. Overall calorie availability in Africa is higher than in 1961, though below the world average and way behind the figure for North America (Maxwell 1992:4).

Therefore, even if we were to accept the comparative global statistics as accurate, the picture they paint is far more complicated than the alarmist reports produced in the media and by the very agencies that produce those statistics. It is certainly inaccurate to date the beginnings of the African agrarian crisis to 1960, the year of African independence, as has become fashionable in revisionist Africanist circles, which smacks of nothing but an apologia for colonialism.

But the global statistics cannot be taken without large calabashes of salt. 'If one took into account official figures of food production and food imports', observes Mkandawire (1987:15), 'and matched those against population growth in Africa, the continued existence of large parts of Africa would simply be inexplicable'. It is difficult to estimate food production in Africa because most of it comes from small peasant farms, the vast majority of which are not covered by any system of registration or reporting. Moreover, in so far as primacy in the food production statistics is given to cereals, reflecting no doubt the consumption patterns in the industrialized countries, while neglecting or underestimating tubers and other crops consumed widely in various parts of Africa, the nature of the African agrarian crisis is overstated and misrepresented. Increased food imports did not necessarily imply a decline in aggregate national production. Rather, Raikes (1988:23) contends, it derived 'primarily from an imbalance between rapidly increasing urban food demand and the capacity of official food purchasing agencies to meet this demand exacerbated in a number of cases by factors making food imports (especially in the form of food aid) more attractive to African governments than local food procurement'. And even on this question, there has been a lot of distortion. The available African food import data for the period 1970-1984 shows that

> by far the largest proportion of cereal imports into Africa does not go to sub-Saharan Africa at all, but to Egypt and the Maghreb, whose imports have also grown more rapidly over the period since 1970. One seldom sees the argument about disparity between population growth and food production capacity put forward in regard to the North African countries, however. This is because it is clear that the reasons lie elsewhere (Raikes 1988:16).

By exaggerating the aggregate production shortfall, Raikes argues, emphasis could be placed on policies aimed at increasing production rapidly through technological change, rather than on the need to generate and increase incomes for poor people and to base production increases on technologies that are environmentally and socially sustainable.

Thus, in discussing food shortages and famine it is crucial to identify where and to whom they occur. It is also important to distinguish the types of food shortages. Raikes (1988:Chapter 4) distinguishes between malnutrition (poorly balanced diet, lacking certain essential nutrients), undernutrition or hunger (not enough even of basic staple food), and starvation. In terms of their incidence over time, he differentiates between constant or long-term food shortages, seasonal food shortages, and exceptional shortfalls or famine. Among the many factors that determine who suffers from food shortage and who has secure access to food three stand out: first, access to land and other productive resources; second, availability and security of employment; and, third, the effectiveness and stability of social networks. Predictably, then, the main victims of food shortages tend to be those who have weak or have lost entitlement to productive resources and employment and the protective umbrella of social networks, as Amartya Sen (1984) has put it in his seminal analysis of the history of global poverty and famines[1]. The evidence is overwhelming that it is the rural and urban poor, among whom women predominate, who bear the brunt of hunger and starvation (Roodkowsky and Leghorn 1977). Certainly that was the case during the famine of the mid-1980s (*The Standard* 20-3-85; 12-10-84). A 1990 survey of food insecurity among the countries that belong to IGADD (Intergovernmental Authority for Drought and Development), namely, Djibouti, Ethiopia, Kenya, Somalia, Sudan and Uganda, found that the rural poor constituted 39.9% of those subject to food insecurity, followed by 34.9% who were affected by war, 12% were the urban poor, 8.9% nomads, 5.3% refugees (Maxwell 1992:4). It can be seen that war is a major cause of food insecurity. In fact, the poorest agricultural performance among African countries in the 1980s was recorded in the war-torn countries.

Starvation, therefore, does not occur only when there is physical shortage of food, but when individuals and social groups lose their entitlement to the food that is available, for whatever reason, from landlessness and joblessness to warfare. Even during periods of actual famine, which is the most dramatic and gruesome expression of hunger, differentiations in entitlement often correspond to differences in the degrees of suffering. Food shortage may be caused by drought, which in turn may be an act of God, but the victims are ordained by society. Hunger is not natural, but political. Natural disasters like drought serve as revelatory mirrors 'in which structural contradictions as well as deteriorating socio-economic conditions are exposed. Paradoxically, however, drought also enables such conditions to be concealed because they can be attributed to the 'crisis' and not to deeper problems and trends' (Solway 1994:471). And these conditions are determined by factors that are both internal and external, structural and conjunctural, economic and political. During moments of 'crisis' the normative order is ruptured and the struggles for, and against, change intensify. This appears to have happened in the countries most affected by the agrarian crisis of the 1980s.

Internal Damnation

To many academics and commentators, the African agrarian crisis of the 1980s was rooted in adverse, internal 'natural' and political conditions. The catalogue of these conditions was long and varied. Singly or collectively, capriciously or cumulatively, natural disasters, population pressures, poor land usage and technological backwardness, with inappropriate government policies superimposed on them, were held to be responsible for Africa's agricultural problems.

Africa it was said, was cursed by the sun. There was always either too much rain or too little. The heat did not help either. The level of micro-organic activity in the soil was so high that the humus in the topsoil tended to be rapidly depleted by the process of microbial decay. Thus, underneath the exuberance of tropical forests, lay soils that were poor in humus and highly susceptible to erosion and leaching. Droughts and floods added their toll. At the outset, therefore, Africa was beset by a harsh environment, which was, in fact, on the verge of collapse (de Vos 1975; Kamarck 1976; Harrison 1982, 1987; Lamb and Seifulaziz 1983; Hall 1984; Timberlake 1985; James 1991, 1993).

According to this view, then, it was the mean weather machine that accounted for poor agricultural productivity in African countries. Drought was singled out for particular retribution. 'Already faced with more than its fair share of political and economic problems', bemoaned *Africa Magazine* (1984:10), 'Africa is now confronted with the catastrophe of continentwide drought'. African political leaders, angered by charges that it was mismanagement not the weather that was responsible for the crisis also sought refuge in the ecological scourge as an explanation. In some environmental circles, controversy began raging as to whether Africa, and the world for that matter, was undergoing permanent climatic change (Darkoh 1989). Alarmist reports that droughts were getting longer, harsher, and more lethal were being churned out (*Daily Nation* 12-4-85; Marei 1976:Chapter 2). The debate of the 1920s and 1930s as to whether the Sahara was expanding or not was revived (Bovill 1920; Stebbing 1935). At a rate of six kilometres a year, UNEP warned, 'the advance of the Sahara could become a major catastrophe by the turn of the century and the loss of earnings caused by the spreading deserts is estimated at around £26,000 million a year' (*The Standard* 11-9-84).

The African people, of course, were regarded as the main culprit in this unfolding ecological tragedy. 'Nineteen eighty-four may be the year when Nature announced it had had enough', the director of Earthscan noted grimly. 'Thousands, perhaps tens of thousands were dying in the highlands of Ethiopia, where growing populations have stripped the land of its ability to feed them' (*The Standard* 24-12-84). The neo-Malthusian scare of the early 1970s, in which the population 'bomb' was thrown into the public consciousness as the chief obstacle to economic development and human progress, was screamed from every available podium (Borgstrom 1969; Marei 1976). A consensus emerged in academic circles and in the popular media blaming Africa's high population growth rates for environmental degradation and declining food supplies. It was said the continent's fragile soils were being eroded due to deforestation for farming and grazing, firewood and timber, so that barren landscapes and deserts grew with a vengeance and droughts became inevitable (Mott and Mott 1980; Lofchie and Commins 1982; Glantz 1987; Goliber 1989; Jolly 1994; Shapiro 1995; *Kenya Times* 1-5-85; 3-10-84; *Daily Nation* 10-5-85; *Sunday Nation* 7-10-84).

Implied in the views that Africa's population growth rate was 'abnormal', indeed, that 'it has many of the characteristics of an epidemic' (*Daily Nation* 14-5-85; 15-5-85; *Sunday Nation* 23-12-84), was the assumption that Africans had as yet to adjust to the technological developments of western bio-medicine which had reduced infant mortality. It was as if in communities without such medicine and artificial contraceptives techniques people had no control over their reproductive processes or that those who opted for high fertility rates were making irrational choices. Neither assertion can, of course, stand up to the evidence. Traditional pharmacopoeias contained a range of biodynamic agents capable of regulating fertility (Farnsworth *et.al.* 1975; Soejarto et.al. 1978). Moreover, as Marxist demographers have shown, in pre-colonial times reproduction was not merely a biological proclivity but was governed by the imperatives of social production (Meillassoux 1981; Guy 1980). As for today, family planning programmes that are based on the assumed irrationality and ignorance of people and take over the social and economic dynamics of fertility are doomed to failure (Sindiga 1985). The linkage between population growth and environmental crisis, Bell (1987) has argued, is actually misconstrued because of misconceptions of African historical demography. At the beginning of the colonial era, Europeans encountered drastically reduced human and livestock populations in Africa due to the slave trade, the rinderpest pandemic of the late nineteenth century, and the effects of colonial conquest itself, so that they formed 'a false impression of the 'natural' relationship between humans and their environments in Africa...the West found in Africa the Garden of Eden of its romantic imagination. The subsequent recovery and development of humans and livestock were therefore seen as unnatural, threatening and ecologically unsound' (Bell 1987:89).

The rapid rates of population and livestock increase, which may, in fact, have been inflated as a result of progressive improvements of census techniques, do not entail a crisis, Bell believes, but mean that Africa's human and livestock populations are not encountering serious constraints and have not yet approached the ecological carrying capacity of the continent. Ignored in all this, was the simple fact that historically, in many societies, population growth by itself had not inevitably led to the deterioration of the land. Indeed, as Boserup (1965) and others have argued[2], population pressure can provide a useful economic stimulus to technical innovation, a general precondition for agricultural progress which, in turn, allows for even higher levels of population concentration. Unless, of course, the population in question is irredeemably backward. Or, and that seems more plausible, the prevailing socio-economic and political conditions in themselves put constraints on technical progress.

It has been suggested, with considerable justification, that the population question is used by richer countries and the middle classes in the poor countries to play down their disproportionate consumption of available resources, including food (Mamdani 1972; Parsons 1977). The average American, for example, consumes as much as three Japanese, 80 Ghanaians, and 150 Bangladeshis. In this sense, the population of the industrialized countries poses a much larger threat to global sustainability than the populations of the so-called Third World, which, therefore, should not be regarded as the cause but a symptom of hunger. Indeed, only broadbased socioeconomic transformation can 'stabilize' population growth. The empirical evidence, in fact, seemed to point to the fact that a demographic transition to lower fertility had actually begun by the 1980s and that some of the high-fertility countries, such as Kenya, Zimbabwe, Botswana, and Nigeria may have entered an era of irreversible fertility decline as result of complex factors, including post-independence material and social development and transformations in gender relations, which resulted in declining infant mortality and death rates, higher life expectancy, and changing attitudes to fertility. The crisis of the 1980s itself accelerated the process because it forced more women to seek off-household incomes and to renegotiate their relations with men to modify their reproductive roles and burdens. In short, existing patriarchal structures and values which favoured high fertility were being undermined from many directions for the rich and poor alike (Ahonsi 1995; Kalipeni 1995, 1996).

Blinded by their gloom, the Afropessimists and ecopessimists neither sought nor saw this evidence. Many of them linked the assumed disastrous consequences of rapid population growth to the supposed backwardness of African agriculture. FAO, for example,

> highlighted the grave impact of population pressure on what was formerly the most widespread of all tropical farming system-shifting cultivation ...increased population densities have led to a drastic reduction in fallow periods. This is causing widespread land degradation and resulting in diminished yields of food (Kenya Times 5-6-85; FAO 1985b).

The harshest criticism was reserved for Africa's picturesque pastoralists who were blamed for laying waste vast tracts of land (Harrison 1982:Chapter 2; Bradburd 1984; Rigby 1992). It is certainly a crude distortion to subsume the wide range of agricultural systems in pre-colonial Africa under the rubric of 'shifting cultivation'. Such generalizations are rooted in the

inability of early European travellers to Africa to understand and
appreciate African agricultural ecology and environmental management,
due in part to evolutionist assumptions about agricultural development, in
which European agricultural techniques and practices stood at the peak
of historical achievement. This encouraged such dubious assertions as
the one made by Goody (1971) that African agriculture was technically
backward in comparison to European agriculture because the plough was
not adopted. In Southern African history it used to be argued that white
agriculture triumphed because of superior white technical knowledge,
entrepreneurship and familiarity with market forces. African agriculture,
on the other hand, deteriorated because it was too frail and backward to
survive the new competition (Macmillan 1930; de Kiewet 1956). Recent
studies have clearly shown that this was not so. African farmers
responded and adapted successfully to commercial farming until they
were later deliberately undermined by the colonial state in order to
promote white capitalist agriculture (Palmer and Parsons 1977; Palmer
1977; Bundy 1980; Beinart et.al. 1986; Beinart and Bundy 1987; Keegan
1987). Similarly, research on West Africa has demonstrated that
'traditional' institutions were not a drag on entrepreneurship. From the
earliest days of commercial farming, farmers used both market and
nonmarket mechanisms to mobilize the working capital needed to
establish their farms and also to insure themselves against economic
failure or personal misfortune. Lineage and ethnic institutions, in
particular, were used to mobilize the capital and information necessary to
set up commercial farms (Hill 1963, 1970, 1977; Berry 1975, 1985; Watts 1983).

For the true neo-classical believers neither nature nor culture, makes a
good culprit. The state is the real nemesis of the invisible hand of the
market, a fiction narrated with elegant simplicity by Lynn Scarlett:

> Ultimately, the poor performance of African agriculture cannot be attributed
> to cultural barriers, entrenched traditional farming techniques, lack of
> demand, physical constraints, or infrastructural defects. Instead, the key to
> Africa's food problem lies in the complex web of interrelationships between
> Africa's agricultural markets and government policies (Scarlett 1981:175).

She argues that African markets in themselves were not disorganized.
Rather, government restrictions impeded their development and,
consequently, undermined agricultural production. Taxation schemes, credit
restrictions, wage and pricing controls, monopolistic state marketing boards
and government directed cooperatives, not only led to gross inequalities in
the distribution of both income and assets, especially between the rural and
urban sectors, they also distorted market operations and proved a disincentive to
farmers. Moreover, by keeping rural income levels low, effective levels of

demand were reduced. The resolution of the rural income problem, however, did not lie in the redistribution of income through 'further tampering with the market economy, but rather in discarding the policies that have significantly distorted income and asset distribution ... (namely) minimum wage policies, food pricing ceilings and the imposition of obligatory hygiene and grading standards' (Scarlett 1981:182).

In the same year that these words were being published, the World Bank weighed in with its omnipotent institutional voice. The Berg Report proclaimed: 'domestic policy issues are at the heart of the crisis in SubSaharan Africa', and not external factors (World Bank 1981:121). In fact, the report goes on, 'for most countries...the terms of trade were favorable or neutral' (World Bank 1981:19). The litany of the asserted policy failures was long, exhaustive, uncompromising. Africa was paying the price for pursuing protectionist trade policies, unrealistic exchange controls, monopoly state marketing, and bloated government expenditure on unproductive parastatals and social welfare. Moreover, governments ignored agriculture in favour of high cost and inefficient import substitution industrialization. These biases were propagated and reinforced by biases towards urban centres where food prices were kept artificially low. The key to accelerated development in future, therefore, lay in the abolition of state marketing and pricing controls, curbing public expenditure, freeing domestic food markets, all underpinned by vigorous private enterprise and expanded agricultural exports. It advised focusing support on the 'progressive' farmers, from whom the benefits would 'trickle-down'. Thus, ended the World Bank's shortlived flirtation with 'basic needs' (World Bank 1975; Hoogvelt 1982:Chapter 2). Aspiring 'aid' recipients were now to pray at the altar of export-oriented production, in the name of structural adjustment programmes. The mask was off: the Bank had, after all, a face as hideous as that of its universally dreaded sister, the IMF (Hayter and Watson 1984; *Daily Nation* 5-5-85; 10-1-85).

Many Africanists dutifully fell into line and began uncovering state 'biases' everywhere. And they found many. To quote Mkandawire:

Failure of African agriculture has been attributed to various kinds of 'biases' - 'commodity bias' (cash versus subsistence crops, exports versus food crops), 'spatial bias' (rural versus urban areas), 'sectoral bias' (industrial versus agriculture, formal versus informal sectors), 'market bias' (external versus domestic markets), 'scale bias' (small-scale versus large-scale), 'gender bias' (men versus women), 'technological bias' (capital-intensive versus labor-intensive, appropriate versus inappropriate technologies), 'class bias' (capitalist versus working classes, feudal class versus peasants), 'fractional bias' (state or bureaucratic bourgeoisie versus agrarian or industrial bourgeoisie (Mkandawire 1987:20).

The literature on each of these binary biases simply exploded, at the centre of which was the heinous post-colonial state (Bates 1981, 1983; Levi and Havinden 1982; Tabatabai 1986; Commins *et.al.* 1986; Nafsziger 1988; Barker 1989; Lofchie 1989; Barkin *et.al.* 1990). The bureaucrats in the international financial institutions could only smile.

It is of course no secret that mismanagement, bureaucratic inefficiencies, and corruption are rampant in many state institutions in Africa. That the peasantry often gets a raw deal at the hands of the marketing boards is again no secret. But it is crass intellectual dishonesty to dismiss the impact on African economies of declining terms of trade, which were deteriorating sharply even as the Berg Report was being drafted. As one critic puts it, the Report

> overestimates what is likely to be achieved by integrating African economies even more firmly into the world system as it currently operates...The Report also greatly understates the importance of external advice and aid generally in the formulation of domestic policy in Africa...the Bank and other aid donors have themselves been responsible for promoting many of the structural characteristics that they find so objectionable (Loxley 1984).

So overwhelming was the psychic presence of the invisible hand, that even those who came to the defense of the maligned state did so in the name of the market. Trying to absolve both the African state and the world system from the failures of market modernization, Hyden (1980, 1983) pleaded for understanding of the state's predicament: it had very little control over the peasantry; all its attempts to 'capture' them had so far failed, thanks to the powerful hold of the peasant mode of production, with its endearing, archaic, and unproductive economy of affection. It was, therefore, futile to try to encourage higher production, as commonly assumed by governments and aid agencies, by offering 'attractive producer prices (because) the market is not a significant factor influencing peasant behavior...in a situation where peasants are only marginally incorporated into the capitalist economy' (Hyden 1980:24). Until capitalism, as a total system, became dominant in Africa and broke 'the hold that the economy of affection has over African society', there was no way out of Africa's low levels of production, especially in agriculture (Hyden 1983:25).

Paradise Lost

But capitalism was already dominant and determinant in contemporary Africa, argued the progressive nationalists and neo-Marxists. So it had been since colonialism. Indeed, to the unapologetic dependistas, the underdevelopment of Africa dated back to the inglorious days of the slave trade (Rodney 1972; Amin 1975; Wallerstein 1976; Inikori 1991).

From that time onwards Africa's economies had continually and progressively been marginalised, subsumed, and incorporated into, the world capitalist system. The agrarian crisis of the mid-1980s was, therefore, a tragic but predictable outcome of this history, and more specifically it was spawned by the prevailing world recession.

Days were, so went the narrative, when Africans primarily produced for their own consumption, and only surpluses were traded. African farming systems, many studies tried to show, were diverse, efficient, and productive. No less than seven different cultivation systems were identified for West Africa alone in the nineteenth century. Shifting cultivation, involving periodic movement of settlements, had long ceased to be common. For centuries most of the settlements were fixed and the chief methods of cultivation were rotational bush fallow, rotational planted fallow, mixed farming, and 'permanent' farming. All these methods entailed careful farming management, the use of annual rotation, manuring, crop mixtures and successions. In addition, tree cultivation was practised in conjunction with other systems, especially rotational bush fallow. Among the tree crops were the oil palm, the kola tree, and the shea tree, all indigenous to West Africa. Finally, floodland and irrigated farming could be found in the floodplains of major rivers, such as the Niger and Senegal rivers (Morgan 1969; Thomas and Whittington 1969; Ruthenberg 1971; Hopkins 1973; Grigg 1974). Research on other parts of Africa similarly showed the existence of diverse methods of cultivation, including extensive irrigation systems and the predominance of intensive and permanent farming (Gray 1963; Miracle 1967; Harlan 1975; Kjekshus 1977; Zeleza 1993).

These systems of cultivation were determined by complex interactions of physical and human geography, including soil conditions, climatic patterns and availability of water resources, population densities, the crop regime, and of course the political economy of land tenure, divisions of labour, development of markets, and the organizations of state and political power. Thus, depending on the specific combination of these factors, there was a continuum of land use, ranging from discontinuous and extensive forms to intensive forms of cultivation. Under the extensive forms output per person hour was extremely high, while permanent cultivation was geared to achieving high returns per acre rather than, or including, per person hour. 'This demonstrates', Hopkins wrote (1973:35), 'the ability of African farmers to adjust to factor proportions in order to achieve optimum results with the resources at their disposal'. Thornton (1990:7-8) concluded: 'African agriculture, even without the plow, was more efficient than that of early modern Europe'.

African land practices, furthermore, reflected the ability of African farmers to manipulate and exploit their environment, but without undermining the environment's capacity for regeneration. Contrary to stereotypical views, Africa has a wide variety of ecosystems each of which demanded different forms of land management. According to Richards, soil moisture conservation and supplementation (irrigation) and erosion control are important in the drier extremes of the tropical zone, while in the equatorial zone maximizing the use of available sunlight and coping with excess soil moisture, apart from erosion control, are the most important land management issues. All these practices assume special significance in the intermediate zone of wetter savannas and drier forests, though at different times during the cultivation cycle. It was here that the widest range of cultivation techniques were to be found, all of which were united by the system of intercropping. Intercropping was the best method of managing the physical properties of African soils. Not only did it minimize the exposure of the soil to erosive rainfall and the spread of pests and diseases, it also maximized the use of available soil moisture, plant nutrients, and sunlight. Moreover, under intercropping yields were extremely high. 'Intercropping, then, is one of the great glories of African science. It is to African agriculture as polyrhythmic drumming is to African music and carving to African art' (Richards 1983:27).

It does not mean, of course, that pre-colonial Africa did not experience food shortages due to weather or pestilence, as well as warfare and other forms of social disruption (Webster 1979; Curtin 1975; Miller 1982). Rather, it was suggested that there existed a wide variety of social mechanisms and ecological responses to prevent, adjust to, recover from, and reduce the impact of food shortages on any one family. To begin with, the organization of the extended household, both as a production and a consumption unit of food and other commodities, reduced the vulnerability of individuals and component nuclear family units. Patterns of redistributive and reciprocal gifts between households, in turn, reinforced the society's ability to withstand a crisis of food shortage. Social insurance against food shortages also extended to the level of food storage and consumption strategies. Elaborate techniques of storage permitted grain to be kept for relatively long periods. Seeds for planting and grains for subsistence were carefully separated. Rituals often guided when granaries could be opened at each point in the agricultural cycle. The tributary appropriation of agricultural surplus further provided a collective insurance against the possibility of famine. Shenton and Watts (1979; Watts 1983) showed that among the Islamised Hausa graduated patron-client relations culminated with the state which used the grain tax stored in central granaries for organized redistribution in times of famine. In addition to these social and political insurance mechanisms, 'unused' land was also kept as a hunting and gathering reserve. Furthermore, the agricultural patterns of intercropping, the selection and ennoblement of drought-resistant strains, and the use of crop combinations which varied with yearly environmental fluctuations minimized the risk of crop failure due to unfavourable climatic conditions, such as drought.

Colonialism drastically altered the social and ecological organization of agriculture and responses to famine and food shortages. The act of conquest itself disrupted agriculture and the established patterns of social relations of production. Famine, food shortages, and migrations followed in the wake of the wars of conquest, with their ruthless scorched earth tactics. First, livestock and lands were seized. Then, millions of men were forcibly uprooted from the rural economy to build the colonial infrastructure, or, as during the First World War, they were conscripted into the colonial armies to serve as porters and canon fodder. The first two decades of colonial rule were simply a demographic disaster for Africa (Crowder 1978; Ranger 1969; Patterson and Hartwig 1978).

An ecological disaster also began to unfold for both humans and animals. There was a sharp decline in African health standards from the late nineteenth century as a result of the introduction of non-African epidemic diseases such as smallpox, venereal disease, influenza, and cholera. Local therapeutic systems were virtually impotent in the face of this intensive onslaught of alien diseases. Livestock was not spared. Rinderpest epidemics swept across the Sahel, and East, Central, and Southern Africa in the 1890s and 1900s. Cattle herds were decimated. The disease apparently entered East Africa via Ethiopia and Sudan from Aden and India on the trails of British and Italian conquest troops. The effects on the pastoral communities and those who practiced mixed farming was catastrophic. Famines broke out. This led to depopulation, accentuated existing social differences, and facilitated European colonization.

> With fewer people to till the fields and fewer cattle and goats to graze the ground and keep the bush at bay, and with imperial laws prohibiting grass-burning and hunting, nature was quick to commence its recovery...wild animals, recovering from the rinderpest, soon moved in to establish grazing grounds in old cultivations. In their wake, the tsetse fly spread to put vast domains of lands beyond the reach of economic activity (Kjekshus 1977:161).

The expansion of the tsetse fly belt weakened immunities among communities who had retreated from old wilderness infection foci. Sleeping sickness spread throughout tropical Africa with abandon (Ford 1971). The expansion of the fly belt also sharply reduced the cattle grazing zones. Quick recovery was hampered by the demarcation of jealously guarded colonial boundaries and, in East Africa, the creation of vast and exclusive game reserves, all of which permanently altered pastoral patterns of transhumance and probably tipped relations between pastoral and agricultural communities away from symbiosis towards antagonism. Hedlund showed in the case of the Kenya Maasai that the alienation of two-thirds of their land for European settlement, the prohibition of the Maasai from using their old trading routes, the imposition of livestock taxation, and the establishment of national parks, where the Maasai were discouraged from grazing their herds, led to the deterioration of the local productive forces as reflected by local overgrazing, soil erosion, and low livestock productivity. Colonial efforts to arrest this by providing permanent sources of water, in the form of boreholes and wells simply led to 'large cattle concentrations form(ing) in areas that were previously not used for dry season grazing', thereby accelerating overgrazing and environmental degradation (Hedlund 1979:28).

Land alienation and the creation of reserves for Africans in the settler colonies and plantation economies had similar effects on productivity and the ecological balance for farming peoples and areas. In the reserves of Southern Africa and Kenya, the precolonial systems of cultivation were put under enormous strain, especially since the reserves were carved out of already fragile agricultural lands. The growth of population and livestock soon led to over-cultivation and over-grazing. Colonial soil conservation measures, which were often ruthlessly enforced, did not improve matters, essentially because they hardly incorporated the time-tested conservation practices of African farmers. They tended to concentrate on preventing soil exhaustion rather than the management of the soil's physical properties; yet 'low plant nutrient status (soil exhaustion) is easier to repair in tropical soils than loss of desirable physical properties' (Richards 1983:26; Greenland and Lal 1977). Faced with deteriorating conditions, peasants had no choice but to trek to the settler farms and expatriate plantations as migrant labourers. That, in fact, was the whole objective of colonial agricultural and labour policy in this type of colony. The reserves were to provide cheap labour and subsidized accumulation in the settler sector. This ruthless appropriation of absolute surplus value was to be paid for in terms of peasants' hunger, ill-health and death, and in environmental deterioration (Bundy 1980; Palmer and Parsons 1977; Brett 1973; Southall 1982; Wylie 1989). But even in the non-settler colonies, patterns of internal labour migration developed. In Ghana, for instance, the routes run from North to South. This was not because the North was 'naturally' poor in resources compared to the South as is often suggested. On the contrary, the North had been at the heart of nineteenth century trade routes and food production. Rather, the North and South were incorporated into the colonial economy differently. Mines and cocoa plantations were centred in the South. The North's role was to provide labour (Plange 1979).

The effect of all this is that the labour exporting regions and countries began to experience a decline in food production. But there was no corresponding increase in food production in the centres of colonial agriculture. The reason for this is very simple: colonial agriculture rested on 'cash' or export crop production. In settler colonies the cash crops were grown almost exclusively by the settlers and in the non-settler colonies by the African peasants themselves, while in colonies dominated by concession companies, as in Mozambique, plantation agriculture went hand in hand with forced peasant cultivation of 'cash' crops (Vail and White 1981). In the mining economies such as Zambia and Zaire agriculture as a whole was underdeveloped in favour of mining (Lanning and Mueller 1979:Chapters 9 and 10)[3].

Apart from naked force, taxation was the most effective weapon with which to whip Africans into 'cash' crop production. Unlike precolonial systems of taxation, the rate of taxation had no bearing on levels of production; drought or no drought, poor harvest or none at all, taxes went up. The monetization of all taxation also dissolved the granaries of precolonial states. The granaries were not replaced, thus increasing the societies' vulnerability to the crises which were made all the more likely now that local food production was falling.

Throughout colonial Africa, then, 'cash' crops ruled. The best lands were given to them; all the available extensions services, research, credit, and marketing facilities were concentrated on the almighty 'cash' crops. Agriculture now meant 'cash' crops. And, for the first time, agriculture was divorced from nourishment. Sisal, cotton, tobacco, coffee, cocoa, groundnuts, tea, sugar, you name them, were not grown because they could feed their producers and fellow citizens, but because they were 'marketable' in international trade. 'Modernization' had indeed arrived with a bang. The central contradiction of African agriculture was born: declining food production in the face of steady, no, phenomenal, increases in export crop production. Exports of cocoa and groundnuts, the leading export from Ghana and British West Africa and from Senegal and French West Africa, respectively, rose by 1,446% and 30,500%, respectively, between the beginning of the century and 1937 (Hopkins 1973:174). Expansion in domestic food production came nowhere remotely close. Food shortages grew, structurally, and the stage was set for famines to begin acquiring epidemic proportions both in their geographical spread and social impact. The imperial hen really came home to roost during the lean years of the 1930s and the Second World War. 'Cash' crop prices collapsed (Helleiner 1966), so that few even had money to buy food. A vicious cycle was created: falling export prices, as well as higher taxes, simply forced peasants to produce more 'cash' crops. The nature of some of these crops, moreover, such as cocoa which takes ten to fifteen years to mature, effectively ruled out quick and flexible responses to market fluctuations.

During the Second World War, as noted briefly in the previous chapter, the colonial powers embarked on policies reminiscent of the first two decades of primitive colonial accumulation. Peasants were coerced to produce record quantities of 'cash' crops for the imperial war effort. In the meantime, the enclaves of hunger became yawning landscapes of famine and starvation in one colony after another (Zeleza 1982; Chapter 3; Holbrook 1978; Crowder 1981:Part VII, Chapter 4). Around the same time, marketing boards were created, thus completing the centralization of colonial marketing (Bates and Lofchie 1980). These boards were ostensibly meant to stabilize producer prices by providing fixed prices, but, in effect, they served as additional instruments of peasant 'over-taxation', for the prices paid to the peasants were far below the prevailing market prices (Leubuscher 1956). The balances were kept in metropolitan coffers to pay off the war debts of the European colonial powers and facilitate their postwar reconstruction on the one hand, and to 'develop' the technical and political infrastructure necessary for the continued expansion of export production in the colonies on the other, through the aegis of Colonial Development and Welfare.

After the Second World War export crop production expanded tremendously, partly thanks to the Colonial Development and Welfare programmes. In Tanzania, 50,000 acres of prime land were given over to the 'Groundnut Scheme' in 1946. In the fledgling settler colony next door, Kenya, African 'progressive farmers' were at last allowed to grow lucrative 'cash' crops. The acreage under African coffee production, for example, rose from 3,702 in 1955 to 133,100 acres twelve years later, a phenomenal increase of 3,595%. In Niger, part of the infamous Sahel region, the area under groundnut cultivation almost trebled from 364,000 acres to 925,000 acres between 1954 to 1961.

Lappe and Collins comment:

> The expansion was at the expense of fallow zones of 'green belts', critical especially during drought years. The cutback of fallow land only compounded the soil depletion caused by the planting of peanuts year after year on the same soil. Peanut cultivation in the 1960s began to spread north, usurping lands traditionally used by pastoralists. This encroachment made the pastoralists and their animals more vulnerable to drought (Lappe and Collins 1977:7677).

It was a hard, cruel legacy. Colonial agriculture, the progressive nationalist and neo-Marxist historians agree, marginalised food production, mined the soil, and degraded the environment. It also introduced patterns of land ownership and social relations of production which gravely undermined coherent social responses to famine or other crises. Indeed, the incorporation of Africa into the world capitalist economy made the continent vulnerable to capitalist crises generated elsewhere, on the one hand, while, on the other, the selective and unequal development of Africa's means and relations of production maintained Africa's vulnerability to locally-rooted crises. The Senegalese peasant in 1960 receiving one-seventh what her grandparents received in 1880 for the same amount of groundnuts (Amin 1973:10) would be forgiven if she envied them, for unlike them, our 'modernized' farmer would most likely be dependent for a growing share of his consumption on imported food, often mislabelled 'food aid'. The African peasant not only came out of colonialism with a hoe, but most likely an imported hoe, and had learned to cultivate hunger.

But dependency accounts of colonial agricultural underdevelopment, while plausible in essential respects, present one-dimensional history. They are so overwhelmed by the structural forces of colonial state pressures against African peasants that, in the process, they lose sight of the peasants themselves, that is, how they struggled against, adapted to, and conditioned the penetration and subsequent development of agrarian capitalism in Africa. For while colonialism may have been a 'one-armed bandit', in Rodney's pregnant phrase (1972:223), peasants certainly had two hands with which to fight back. Numerous examples can be given to show how the process of accumulation in colonial Africa involved the dialectics of domination and the constant reconstitution of that domination through social struggles. But two will suffice.

In West Africa, the marketing boards were not simply dreamt up by the state. They emerged in the aftermath of peasant struggles, most forcefully expressed in the cocoa holdups of 1937, against the unbridled mercantilism of the firms that dominated the import-export trade. The depression of the 1930s had merely exacerbated longstanding peasant grievances against these firms. The colonial state moved in with marketing boards as a way of resolving the conflict and ensuring the continued flow of 'cash' crops: it would buy these crops directly from the peasants, though it would employ the merchant firms to act as its agents in overseas trade. For a while the peasants could celebrate. But this deepened the tentacles of state intervention, which the peasants were later to resent and fight against, and it also set the basis for the postwar transformation of the merchant firms from being horizontally-integrated monopolies to being much more formidable vertically-integrated corporations.

Turning to East Africa, we see that the African 'cash' crop revolution in Kenya was a product of, and a victory for, peasant struggles not merely a policy foisted on reluctant peasants by a Machiavellian state. As I have argued elsewhere (Zeleza 1982) immediately after the war settler hegemony, especially in agriculture, with all that this entailed in terms of monopolizing cash crops production for themselves, was at its height. The colonial state had no intention to change this status quo. But then peasant struggles, reinforced by struggles of workers and the nationalist movement as a whole, all culminating in the Mau Mau war of national liberation, changed the balance of class forces in the country, which, in turn, restructured the social and institutional bases of the colonial state. Only then, from the early fifties onwards, did the colonial state accede to the accumulative demands and interests of the 'rich' peasantry. And that meant allowing them to grow 'cash' crops. Thus, imputing 'economic logic' to historical processes that are essentially born and bred in class relations (struggles and alliances), as dependency writers are wont to do, can obscure our full understanding of the past, the present, and of the forces that might change the future.

Harvests of Hunger

Uhuru was a great political achievement, coming as it did after years of protracted, often bitter, and occasionally violent struggle. But it was not followed by any fundamental restructuring of the economy, either in terms of its internal production structures and relations or external dependency. One does not need conspiratorial 'neocolonialist' theories to explain why this was so. Of course, decolonization did not entail the final defeat of imperialism, it merely represented imperialism's retreat. While nationalism was resolutely anti-colonialist, it was not anti-capitalist, outside of the territories where protracted armed liberation struggles were fought, because the struggle was won before the peasant and working class masses had subsumed the petty-bourgeois led nationalist movements to their social class project, and so neo-colonialism quickly rose from the dying embers of nationalist victories.

'Nation-building' and 'modernization', in short developmentalism, were pursued vigorously, giving the impression of major changes taking place, especially as these goals were often articulated in ideologies clothed in 'socialist' rhetoric. 'Modernization' needed a steady flow of that magic concoction 'foreign exchange', and foreign exchange could not be obtained by producing millet and cassava. It was hoped that before long industrialization, using proceeds from export agriculture or mining, would develop and finally set the stage for these countries' economic 'take off'. In the meantime, 'cash' crop production was intensified, an allout drive for import-substitution industrialization was embarked upon, and the state bureaucracies were expanded accordingly to oversee Africa's race into post-colonial 'modernity'. This was the script followed by most African states regardless of official ideology, among the 'merchant states', such as 'socialist' Tanzania (Msambichaka 1987), 'capitalist' Malawi (Mhone 1987), monarchical Morocco (Bouami and Raki 1987) and Swaziland (Mcfadden 1987), Nasserite and post-Nasserite Egypt (Abdelhakim 1987), the Sahelian Mali (Dembele 1987) and Burkina Faso (Thiombiano 1987), as well as among the 'rentier' states of 'humanist' Zambia (Wood and Shula 1987), 'revolutionary' Algeria (Bedran and Bourenane 1987), and 'petronaira' Nigeria (Titiola 1987). The World Bank and many Africanists cheered on. Agribusiness extended its long helping hands. The cheers were largely of relief that the nationalists would not seek to alter the international division of labour, which extended to agriculture itself. The capital intensive and highly subsidized agriculture in the industrialized regions needed to protect the world market share for its products mostly made up of food.

The military coups of the mid-sixties were the first to puncture this balloon of inflated expectations and euphoria. It was disturbingly clear that everything was not well in the affairs of that much sought-after 'political kingdom' and that the grip of the global economy of affectation was as tight as ever. Shortly after, the bubble burst over the Sahel. It was an accident of nature, many said, the tragic outcome of a vicious cycle of drought. In viewing the Sahel famine as 'natural', the consciences of the liberal mind in the West and the caretakers of the postcolonial state in Africa were saved from guiltridden sleepless nights. The reality, of course, told a different story.

For one thing, the drought seemed to display a strange capacity for discrimination in its impact. The Sudan, for example, was relatively unscathed by the drought and famine that devastated the Sahelian countries and Ethiopia in the late 1960s and early 1970s, while it was not so lucky with the drought of the mid-1980s. What accounts for this dichotomy? O'Brien (1986) believes the answer lies in changes in agricultural policy and conditions of accumulation. Up to the late 1960s agricultural development was based on the expansion of capitalist food production supplying internal markets, while from the early 1970s emphasis was put on export production. This was encouraged by the World Bank and IMF, as well as by Arab capital, which sought to transform the Sudan into the breadbasket of the Arab world. These pressures coincided with the interests of the import-export oriented commercial bourgeoisie whose fortunes began to rise from the 1970s at the expense of the previously dominant agrarian bourgeoisie. The resumption of the civil war in the 1980s simply compounded matters. Peaceful Botswana, on the other hand, escaped serious damage from the droughts of both the 1970s and 1980s. Indeed, it continued to enjoy the highest economic growth rate in the world, thanks to its buoyant mining and beef industries. Although no one died from the drought, it still exposed many of the structural and social contradictions of the existing order (Cliffe and Moorson 1979; Solway 1994). In fact, Southern Africa as a whole 'defeated' the drought of the early 1990s, advertized by the United Nations as 'the worst in living memory', because the countries of the region, through the Southern African Development Community (SADC) were able to anticipate the crisis, monitor it, and mobilize and coordinate local and external resources (Collins 1993).

Perhaps the most inspiring story in Southern Africa in the 1980s came from Zimbabwe. In 1984 gloomy reports lamented that the young republic's hopes of becoming the breadbasket of Southern Africa had died prematurely, thanks to drought. The government set aside $200 million to import maize, the country's staple food. Then the pleasant shock came:

> Zimbabwe's peasant farmers have turned conventional wisdom on its head. Despite the worst drought in living memory they produced a record maize crop last year, proving that it takes more than lack of rainfall to make a famine.... In 1984, the third successive year of drought, Zimbabwe's peasant farmers startled agricultural experts by bringing in more than twice as much as expected their largest ever crop of the country's staple food. And in 1985... peasant farmers are looking forward to delivering a record 600,000 to 800,000 tonnes of maize to the state marketing board, as much as 10 times the maximum ever delivered prior to independence in 1980.... Governments in the rest of Africa are being told by donors like the World Bank to dismantle state marketing boards or much reduce their scope. But western officials make an exception for Zimbabwe's stateowned Grain Marketing Board, which they concede is highly efficient and serves well the interest of both producer and consumer (Daily Nation 23-5-85).

It has been argued that Zimbabwe's relatively well-established transport and industrial infrastructure helped boost agricultural production; the industrial sector provided the agricultural sector with farm machinery and inputs such as fertilizer, while the agricultural sector provided the former with raw materials and foodstuffs. This goes to show that industry and agriculture did not need to be antagonistic as structural adjustment strategies that prioritized agriculture over industry stipulated. Nor did 'cash' crop production have to be abandoned entirely in order to increase food production. But over and above this lay the peasants' greatly improved access to credit, better land, extension services, and marketing, partly reflecting the new independent state's efforts to realize the dreams of the liberation struggle, in which peasants actively participated. The peasants and their supporters were, of course, not entirely satisfied with the slow pace of agrarian reforms, especially over land redistribution. Also, although the 1980s drought was effectively managed, rural differentiation was accelerated (*Moto Magazine* November 1983; Leys 1986; Shopo 1986; Moyo 1995).

But even in the Sahel, drought was quite discerning in its impact according to crop regime and class position. During the drought of the early 1970s, many exports from the Sahelian countries reached record levels. In 1971 alone, the first full year of drought, over 200 million pounds of beef, 56 million pounds of fish, and 32 million pounds of vegetables were exported from the famine-stricken Sahel. Altogether, the total value of agricultural exports from these countries between 1970-74 was a staggering $1.5 billion, which was three times the value of all cereals imported into the region (Lappe and Collins 1977:76). One could go so far as to say the way was paved for the famine. Five years before the drought in Mali there was a significant reduction in the total area dedicated to food grain production, while the acreage devoted to cotton more than doubled. The same thing happened in Chad. Cotton exports reached record levels in both countries. In Niger, the area planted in groundnuts increased by 182,000 acres between 1961 and 1968 to reach a total area of 1,080,000 acres. The country's groundnut exports trebled. The disarticulated and extroverted trunks of colonial capitalism were as robust as ever.

Where did the foreign exchange from the export crops go to? It was used to import food, as well as fertilizers and pesticides in order to grow more 'cash' crops. The rest was whittled away in conspicuous consumption by the fledgling middle classes. In 1971 Niger earned $18 million from cotton and groundnuts, her two major exports, but spent $22 million on imports of clothing, private cars, alcoholic beverages and tobacco products. Similarly, in Senegal, about 30% of the foreign exchange earned in 1974 were just for such items.

In Nigeria booming foreign exchange earnings in the 1970s pushed the country into a 'wheat trap', according to Andrae and Beckman (1986). They show that Nigeria's sharp increase in food imports from the 1970s was partly the result of new patterns of demand following in the wake of rapid oil-induced urbanization, and was partly caused by the inability of the peasant economy to meet the food requirements of the non-agricultural population which was allowed to expand at an excessive rate. The 'wheat trap' was sustained by the congruence of interests of transnational wheat companies and the Nigerian state, in which indigenous agrarian interests were under-represented and those of the commercial bourgeoisie over-represented, and which had to satisfy the demand of the urban masses for whom bread had the advantages of convenience, social attraction, relative availability, and low price. Breaking from the 'wheat trap', they conclude, entails breaking from the structure and process of underdevelopment, a point echoed by Toyo (1986) who argues that it is not simply a question of paying more attention to agriculture, as conventional World Bank wisdom has it, because the Nigerian state has paid a lot of attention to agriculture. The question is what kind of attention, and to whom?

The role of agribusiness in Nigeria's 'wheat trap' was replayed in different ways in the other 'agrarian' traps that emerged to reinforce the disastrous effects of the Sahel drought and famine of the early 1970s. Not only did the famine bring suffering to the inhabitants of the region, leaving thousands of people dead, it also laid the socioeconomic and political basis for still greater suffering in the future. The famine encouraged the further capitalist transformation of the Sahel countries. The debts accumulated by peasants ruined by the drought accelerated the expropriation of their land and its subsequent concentration in the hands of a new class of land owners, mostly bureaucrats and local merchants, some of whom joined hands with international finance capital (Meillassoux 1974; Glantz 1976). The multinational corporations, as usual, presented their investment as impartial 'aid' to fight the spreading of the desert, drought, and future famine. The trail was blazed by Bud Antle, an agribusiness corporation with worldwide interests. It came to Senegal in 1972 and set up large plantations using the latest technology and extensive irrigation to produce green beans, melons, tomatoes, aubergines, strawberries and paprika for the European market. As is the practice with investment by multinational corporations, Bud itself came up with very little of its own capital, however. Most of the capital was provided by the World Bank, the German Development Bank, and the Senegalese government which also helped by removing peasants from the land earmarked for the plantations (Lappe and Collins 1977:199-200). The case of Bud in Senegal demonstrates that agribusiness operates like any other business; it is not into providing nutrition for people. The role of the World Bank in acting as midwife for these corporations in the 'Third World' is critical. The Bank often provides the local government with the loans with which to build the necessary infrastructure - roads, dams, power lines-for the plantations. Even FAO has been turned into an attendant, if a reluctant one, providing technical and political support for the growing global agribusiness (Frank 1980:Chapter 2; George 1976; Feder 1977; Jacoby 1975).

In the 1970s agribusiness penetrated deeper into Africa. In drought-prone Ethiopia, cotton and coffee plantations were expanded into the traditional pasture areas of the pastoralists in the Awash Valley, while in the neighbouring Sudan huge cotton, sugar, and wheat plantations were set up (Franke and Chasin 1980:88-92). In response to falling primary commodity prices and threats of nationalization or labour unrest, agribusiness increasingly adopted new methods of operation, in which they eschewed direct ownership or control of plantations, leaving direct production to outgrowers, and instead concentrated on controlling the processing, trading, transporting, marketing, and distribution of the produce. They saved on wages and benefits and passed on price fluctuations to the 'independent' outgrowers (Dinham and Hines 1983; Zeleza 1988). Often, agribusiness was warmly welcomed by the state and certain factions of local capital. McFadden (1987) shows in the case of Swaziland, for example, that the country's agricultural 'success' was closely tied to agribusiness, whose hegemony in the country's agricultural economy was born out of the collaboration between the state, the traditional monarchy, and international capital and rested on the exploitation and repression of the peasantry.

Thus, in the wake of the Sahel and Ethiopian famines of the early 1970s, Africa saw the establishment of vegetable and fruit plantations and cattle ranches, as well as the expansion of acreage under 'traditional' export staples such as cotton, groundnuts, sugar, coffee, cocoa, bananas, tea, etc. And, yes, Africa discovered that the export of fresh flowers to Europe was a lucrative business! By the 1980s several countries were more than ready for a replay of the 1970s famine, this time on a grander scale. The world professed shock. 'Nature' was again blamed. Others faulted Africa's proverbial backwardness and notoriety for selfproduction. And yet, Africa's cash crop mountain continued to grow even higher. FAO (1983, 1985a) figures, for all that they are worth, showed that between 1974 and 1984 total cereal production fell by almost 11%, while cash crop acreage and production continued to increase. The total acreage for cereals, roots and tubers fell between 1981 and 1984 by 1.8%, while that for cash crops rose by 19.6%. The latter did not even include land used for cattle ranching or for growing vegetables, fruits and flowers. Meat production increased between 1974 and 1983, ranging from 75% for poultry meat, to 29% for goat meat, 28% for mutton and lamb, 24% for pigmeat, and 18% for beef and veal. A lot of this meat found its way to the discriminating palates of Western consumers. And in the early 1980s, between 1981 and 1983, agricultural exports also rose, 28% for cotton lint, 14% for cotton seed, 16% for tea, 4% for sugar, and 1% for coffee. In the meantime, the threat of hunger was looming over the horizon for the poor in several countries.

These aggregate figures do not tell the whole story. It is important to underscore how, as during the Sahel famine of the early 1970s, agricultural exports from the faminehit areas reached record levels in the 1980s. Ethiopia produced 93,000 metric tons of coffee in 1983, about 10% more than in the previous year. Sudan's cotton output grew almost three-and-a-half times from 64,652 metric tons in 1981 to 221,960 metric tons in 1983. Chad and Mali between them produced record cotton harvests in 1983, 23.4% higher than in 1982 (FAO 1983). Burkina Faso's cotton harvest reached 75,572 tonnes in 1982/83, a 30% increase over the previous year (*New African* October 1984:25). Burundi harvested 32,635 tonnes of Arabica coffee in 1983, up 84% over the previous season's 17,758 tonnes (*New African* October 1984:26). Somalia's cash crop productions of bananas, citrus, mangoes, and garden vegetables also grew substantially in 1983, as did sugarcane output at 545,000 tonnes, a record (*African Business* March 1985:22). So did Kenya produce record harvests of coffee and tea in 1982/83 (*African Business* April 1985). All these countries were listed by FAO among the 21 countries critically affected by food shortages in Africa at the time. The list could go on. The conclusion is inescapable: African farmers and peasants were cultivating hunger (Twose, 1984).

It was hunger in another sense as well: from 1979 the bottom fell out of the primary commodity markets as the global capitalist crisis deepened. Prices for cocoa fell by about 44% between 1979 and 1984. Coffee prices fell from a peak of almost 200 US cents per pound at the end of 1979 to a low of 115 US cents, before climbing to just a little over 140 US cents in 1984, still 30% less than in 1979. Sugar prices fluctuated violently from about 9 US cents a pound in 1979 to a high 35 US cents at the beginning of 1981 and down again to about five US cents in 1984. This was even below the cost of production (*African Business* February 1985:9; Zeleza 1988). Given that many African economies depended on one or two crops for their export earnings — in fact nine were dependent on one crop for over 70% of their income — the sharp fall in the terms of trade brought many of them virtually to their knees. Between 1978 and 1981 alone the fall was 36% for Ghana, 30% for Ivory Coast, 27% for Ethiopia, 26% for Madagascar, and so on (World Bank 1982).

As if this was not enough, increasing portions of the dwindling export earnings went into paying debts contracted earlier under usurious interest rates. By 1983, 25% of all export earnings were being spent on debt servicing, up from 7% in the early 1970s (*South Magazine* February 1985:59). By the mid-1980s the outflow of investable surpluses from Africa was more than the expected inflow of foreign investment, loans, and grants. Indeed, some estimates showed that the total loss of foreign exchange earnings to Africa due to falling prices, particularly agricultural prices, from the Second World War to the 1980s had exceeded all foreign funds invested, loaned or granted to the continent during the same period (Lappe and Collins 1977:80). An embittered Nyerere asked: 'must we starve our children to pay our debts'. He called for the formation of a debtors' club, a trade union of the poor (*Sunday Standard* 5-5-85; *Sunday Nation* 19-5-85; *South Magazine* August 1984:35-6; *Africa Magazine* May 1985:38-41).

As the terms of trade fell more countries were forced to expand export crop production, sometimes with the help of agribusiness and multilateral lending institutions. The following press reports amply bear this out:

> Several multinational companies in the UK have shown great interest in Zanzibar to help the twin islands of Zanzibar and Pemba revive their troubled economies...The government is anxious to attract additional investment and to diversify the economy away from the deteriorating cloves sector, on which the island depends for 95% of foreign exchange revenue...In response...Lonrho had said it was looking forward to a massive investment in vegetable and fruit farming on the island (*African Business* February 1985:30).

We meet Lonrho in Zambia as well:

> Zambia looks to agriculture as its biggest export alternative to mining...Lonrho is in the fresh-produce business and airfreights exports of green beans, sweet corn and other vegetables to the UK for the winter market... Other crops with export potential are tobacco and groundnuts (*African Business* February 1985:56-7).

Rwanda was also diversifying:

> Rwanda, it is said, is a country with a thousand hills, and every one of them, it seems, is cultivated from top to bottom...In recent years, a diversification programme, carried out by the government to establish export crops other than coffee, has concentrated its efforts on tea. The country's output is rising steadily as new plantations come into use...(*African Business* April 1985:16).

Tanzania, too, set its eyes on tea:

> Measures have been taken by the Tanzanian government to double the country's tea production to 30,000 tonnes within eight years. Between now and 1989, Shs. 174.7 million will be injected in the crop rehabilitation programmes...Other major cash crops also earmarked for rehabilitation within the next eight years are sisal, cashewnuts, cotton, tobacco, and pyrethrum (*Daily Nation* 9-4-85).

It was cotton for Zimbabwe's peasants:

> Zimbabwe's peasant agriculture is in the grip of a 'white revolution' that promises to overthrow an oppressive system of crippling subsistence farming. Since independence in 1980, agriculturalists have watched happily as peasant production figures for cotton have nearly trebled to reach almost 90,000 tonnes out of a total of 250,000 tonnes harvested this last season...Zimbabwe is now the third largest cotton grower in Africa, headed only by Sudan and Egypt. It is preparing to overtake medium-size producers in Latin America and the Far East (*Daily Nation* 11-1-85).

Some preferred flowers in Kenya:

> Kenya exports about Shs. 9 million worth of flowers to Switzerland every year and there is scope for increasing these foreign exchange earnings...(*The Standard*, 6-6-85).

> Until orchids are added to our export list, Kenya has to do with the good old carnations, blue statice, chingerinchee and astromeria, to roll in the foreign exchange. However, the competition from other countries like Israel, Columbia, South Africa and Sri Lanka is intensifying...Unless something to correct this situation is done speedily, Kenya stands to lose out from the sweet smell of success...(*Sunday Standard* 9-6-85)

It was a free-for-all for coffee:

> Coffee is being seen as providing scope for greater foreign exchange earnings by a wide range of African countries...Now many aid programmes are being designed to expand coffee production. Donors include the World Bank, bilateral agencies and the EEC. The EEC alone is funding coffee programmes in Zimbabwe, Liberia, Tanzania and Ethiopia (*African Business*, January 1985:29).

But Ethiopia's Development Programme had no takers:

Immediate relief is good and we are grateful for it, but we don't want to be beggars all the time. We are a proud nation, and we want to be self-sufficient".

Those are the words of Ethiopian Foreign Minister Goshu Wolde, drawing attention to the fact that no amount of emergency food aid will prevent a repetition of famine which is currently afflicting more than a sixth of the country's 42 million people...It is now evident that since the Ethiopian food crisis erupted nine months ago, attitude in certain quarters, most notably in the US and the UK, have actually hardened, and the PMACs prospects of receiving largescale development funding have become more remote (*African Business* May 1985:31).

Intercropping the Future?

Faced with mounting internal and external pressures, governments began undertaking economic reforms, some of them half-baked or half-hearted, others retrogressive, and a few promising ones. Producer prices for food crops were raised as an incentive for farmers in such countries as Nigeria (*African Business* August 1984:73-5), Burkina Faso (*African Business* October 1984:24-5), Tanzania (*African Business* December 1984:7), Gambia (*African Business* December 1984:42), and Kenya (*Daily Nation* 2-3-85). Pastoralists were implored to grow food (*Daily Nation* 10-5-85). Privatization became fashionable, or so the press claimed, even among avowedly socialist regimes, such as Tanzania (*Daily Nation* 2-5-85), Madagascar (*African Business* February 1985:34), and Mozambique (*African Business* November 1984:40). Parastatals, including marketing boards, were no longer sacred cows; voices were rising in some countries for them to be dismantled or at least streamlined (*Daily Nation* 2-5-85; *Weekly Review* March 8 1985; *African Business* August 1984). On the production front, the 'Green Revolution' was preached fervently and found new converts. Progress was reported on research into new crop strains and crop preservation and storage methods (*African Business* October 1984:73-4; *Daily Nation* 28-11-84). Dams and irrigation networks were being constructed (*African Business* February 1985:19-20; *Sunday Nation* 2-6-85). And plans were announced from several West African countries to increase the local production of fertilizer (*New African* November 1984).

'Economic restructuring', then, was heard with insistent and distressing regularity from the mid-1980s. Some tried to evoke guarded optimism about the future. 'An economic breakthrough is possible in Africa despite the present critical state of the continent's finances', said Professor Adedeji, then Executive Secretary of the United Nations Economic Commission for Africa. He recalled 'that in the fifties and sixties the economic situation in Asia was as desperate as that in Africa today. Many of the countries of Asia have since achieved a significant turn-around' (*Daily Nation* 2-3-85). 'I think', he added, 'there is a new realization among African leaders' (*Sunday Nation* 30-6-85). A point echoed by Colin Legum, a seasoned observer of the African scene. He wrote:

Those who write-off Africa as a basket case, or who think it is a continent doomed to struggle from crisis to crisis, tided over by massive aid from richer nations, simply reflect prejudices or opinions based on selective information. One recalls how, only two decades ago, much the same attitude was being shown towards the Indian subcontinent.... Already, there is evidence that most African leaders have learnt from mistaken policies of the first quarter century of independence. Most now understand the priority needs of agricultural development, support for peasant farmers, and the elimination or, at least, the reform of inefficient parastatal bureaucracies. Many countries are allowing loss-making state firms to go to the wall and handing their activities to the private sector (*New African* March 1985:20-22).

But some underscored the dangers inherent in the solutions that were being proposed and implemented. The World Health Organization (WHO) warned that bilharzia was spreading throughout the Third World, including Africa, as a result of new irrigation schemes and water management projects that 'are illcovered, illplanned' (*Daily Nation* 10-5-85). 'One billion people will starve to death this century', warned the editor of Britain's *Ecologist* magazine, 'and huge projects such as big dams will be one of the main causes... millions will starve because the 'big project' people will have taken away their land, desertified it and destroyed the forests' (*The Standard* 5-12-84). Another source pointed out that 'Africa's dams have also increased the incidence of malaria.(and) the cost of these projects is now so high, up to $10,000 per hectare, that foreign exchange earning export crops must be substituted for staple food crops to pay construction costs. Local people are left hungrier than before' (*Daily Nation* 26-4-85). And there were warnings about the dangers of pesticides dumped by unscrupulous multinational corporations, all in the name of agricultural development, which were poisoning and killing thousands in the Third Word, Africa included (*Kenya Times* 5-6-85; *Daily Nation* 11-1-85; *Sunday Standard* 14-4-85).

In short, caution was expressed about many of the reforms, including the 'green revolution' packages proposed as a panacea by desperate governments, greedy agribusiness, and the pitiless international financial institutions. Critics pointed to the purported revolution's record in Latin America and Asia. Based on the so-called high yielding seed varieties

(HYVs), the 'green revolution' mostly benefited the rich farmers, for the HYVs required lots of irrigation, fertilizer, and pesticides because of their low resistance to disease as compared to locally adapted varieties. Moreover, since they did not remain genetically steady continuously new hybrid seeds needed to be purchased each year in order to maintain high yields. Unable to compete many poorer peasants were either swallowed by big landowners or pushed to marginal land. This led to landlessness and unemployment on the one hand, and accelerated environmental deterioration on the other. And little of the increased food produced found itself on the plates of the growing ranks of the impoverished masses (Palmer 1973; Perelman 1977; Lappe and Collins 1977:Chapters 11-14; Dumont 1980:Chapter 10).

Therefore, without confronting the issue of who controlled and who participated in the production process, technological solutions could actually compound the social problems of underdevelopment. The 'green revolution' strategy rested on the assumption that the 'transfer of technology' was neutral, which was incorrect. According to Amin:

> These are capitalist technologies...that...are controlled by the monopolies. Hence we will be transferring, at the same time as the technology, the underlying capitalist relations of production. Moreover, by this transfer we will not be escaping the domination of imperialist capitalism. On the contrary, we will be extending its scope by integrating the periphery more firmly into the imperialist systems (Amin 1977:172).

Agronomic research had to be organized differently, he stressed, if technology was not only going to increase productivity but also enhance the quality of social life of the direct producers. Such research should not be developed in isolation in laboratories and experimental farms of the monopolies and governments, but should be 'less centralized, established among the producers themselves, which would enable the technical revolution to go hand in hand with the revolution of relations of production and that of culture and ideology' (Amin 1977:175).

Amin's concerns were increasingly echoed by those who, either out of desperation or genuine discovery, came to recognize that lasting agricultural transformation in Africa had to built on indigenous knowledge, on Africa's own agricultural revolution, at whose heart lay intercropping, and not upon imported models based on agricultural experiences in temperate latitudes and inspired by the interest of agribusiness (Igbozurike 1977; Brokensha et.al. 1980; Richards 1983:5058; Dei 1990). The scope, depth, and durability of peasant farming science in Africa had to be acknowledged by scholars, governments, and 'aid agencies' who all too often were mesmerized by the glittering technological gadgets and trinkets of the West. Peasants, moreover, many studies showed, not only produced more per unit of land than the large farmers, they also tended to plough back most of their proceeds into the local economy, unlike the latter whose profits often went into unnecessary consumption and unproductive investment, or as in the case of agribusiness corporations, were repatriated abroad. Enabling the peasants to control the productive process would allow them to keep for their community the gains obtained by technical progress, again unlike the rural bourgeoisie and agribusiness corporations whose technological developments were not pursued to eliminate hunger and raise nutritional levels of the poor, but instead often pushed even more peasants into destitution.

A necessary dimension of empowering the peasantry, some emphasized, was land reform. The mythologies of abundant and communal land could no longer withstand the naked realities of growing land concentration and landlessness, rural differentiation and struggles in many countries, especially the former settler colonies, such as Kenya, Zimbabwe, and Algeria, and the densely populated countries from Rwanda and Burundi, to Malawi and Uganda (Okoth-Ogendo 1993). It was also clear that peasant tenure insecurity had increased even in countries with ostensibly pro-peasant socialist policies, such as Somalia, Ethiopia, and Mozambique, which adopted programmes that initially benefited the peasants, but later assumed the ideological rigidities of excessive bureaucratic intervention or collectivization that weakened peasant agriculture. Moreover, the reforms were implemented with the authoritarian reflexes of the colonial order not the participatory promises of independence (Roth 1993; Rahmato 1993; Bowen 1993).

It became increasingly evident, therefore, that the changes in the relations of production, techniques, culture, and ideology, in short, the reappropriation of the world by peasants, could only take place with the establishment of social, political, and economic democracy at the grassroots. The Economic Commission for Africa (ECA) became the loudest institutional voice for democratic economic reforms in Africa (ECA 1989). This was in direct opposition to the authoritarian, neo-classical package imposed by the World Bank and IMF. The first structural adjustment programme (SAP) loans were approved in 1980, and by September 1983 there were 23, amounting to a total of $3.6 billion (Africa Magazine March 1985:50). They imposed harsh conditionalities, which required massive economic retrenchment and entailed increased political repression and the erosion of some of the social advances made since independence. The term SAP quickly turned into an odious obscenity in Africa and among African intellectuals. Structural adjustment, however, offered a crucial opening for change, not because of its liberalism but because of its authoritarianism: SAPs provoked 'bread riots' and galvanized social movements and democratic forces throughout the continent, peasants among them. Gone were the sterile debates as to whether peasants could be 'captured' or not, and whether they were the 'steam of revolution' or just a 'sack of potatoes', divided and demoralized. They graduated into members of rural 'civil society'. Gone, too, were any bets that the future would look like the present, or the immediate past. Possibilities opened up. And that was a hopeful sign.

Conclusion

In this chapter we have tried to argue that 'nature' — soils, the weather and even population — does not in itself explain the agrarian crisis that afflicted some African countries in the 1970s and 1980s. Neither does the alleged backwardness of African agriculture nor the assumed constraints of the so-called peasant mode of production with its economy of affection. As we have shown, droughts, population growth, and agricultural 'backwardness' did not prevent rapid expansion of export crop production even during the famine years. These explanations fail to provide convincing reasons for the food crisis because they are not historical. The present patterns of ecological abuse have their roots in the colonial period as many progressive nationalist and neo-Marxist historians have demonstrated. So much of what passes as the relics of a moral economy or an economy of affection are not really survivals from the past but are mostly products of the specific historical forms by which such societies were and are enmeshed into a commodity economy dominated by global capitalism. And the alleged failure of the postcolonial state to set a sound agricultural base after independence did not arise because it changed colonial agricultural policies too much, but too little. This was not solely because, as is often argued by the *dependistas*, of neocolonialist schemes and plots. Rather, we have to look at the class basis of the postcolonial state which, in turn, arose from the nature of the anticolonial struggle itself and imperialism's responses to it.

This chapter argues, then, that technological and market solutions held little promise as long as the internal and external power relations were not radically transformed. It has to be remembered that by the 1980s in the core capitalist countries themselves agrarian crises had not been resolved. It was a remarkable testament to the great technical achievements of capitalism and its abysmal social poverty that at the same time when the world headlines were screaming about African hunger, in the incredibly agriculturally productive United States 'just over 20 million people... go hungry for at least several days a month' (*The Standard* 5-5-85; *Daily Nation* 19-3-85; *The Economist,* 13-8-83). Yet, in 1984, the country spent $19 billion in subsidies to farmers, part of it for them to take some of their land out of production. Meanwhile, farm bankruptcies were getting worse (*The International Herald Tribune* 11-6-85).

It was pointed out that the crisis forced many governments to undertake reforms, but the suggestion was made that some of the reforms only papered over the cracks; indeed, they threatened to aggravate the crisis in the longterm, especially those that were either inspired by the neo-classical gospel of the World Bank and IMF, or that were predicated on agribusiness investment. However, therein lay the possibilities of change. The conditions that were being created facilitated the resurgence of social movements committed to the transformation of the status quo, including the apparent 'recolonization' of their countries by the international financial institutions.

The agrarian crisis, therefore, had complex and contradictory causes and consequences on African political economies. It was part of a much larger set of questions and challenges about the conditions and directions of African economies, cultures, and polities, themes that are broached in the following chapters and sections. The analytical challenge always involves examining not only the structural conditioning of social life but also the role of social struggles in transforming those conditions for, after all, the history of capital accumulation and constructions of domination is the history of power and resistance, of political movements, of the affirmation and subversion of ideologies. It is not enough to 'see in poverty nothing but poverty without recognizing in it the revolutionary, subversive side which will overturn the old society' (*Review of African Political Economy* 15/16, 1979:3). That is a privilege of cynical academic tourists, not committed public intellectuals.

Notes

1. Using African evidence de Waal (1990) offers an interesting, but ineffective, critique of Sen, arguing that the theory does not apply in the context of violence and it ignores situations where peasants may choose to suffer in the short-term to preserve their assets for the long-term.

2. Using Boserup's model, Inikori (1982) has argued that, by reducing Africa's population through all the people who were exported or died in the process of acquiring slaves and in transit to the Americas, and through the subsequent reduced fertility rates, Africa lost an important opportunity to make economic progress when other regions, especially Europe and the Americas, were doing so.

3. African colonial economies have been divided into various typologies. Amin (1975) talks of the 'Africa of trade' in West Africa and parts of East Africa, the 'Africa of concession companies' in Central Africa, and the 'Africa of the labor reserves' in Southern Africa and Kenya. Mkandawire (1987) distinguishes between the 'rentier' and 'merchant' states depending on their revenue base. The former depended mostly on mineral rents, while the latter relied heavily on agricultural surpluses. Others simply distinguish between settler and non-settler (or peasant-economies (Biermann and Kössler 1980; Denoon 1983). There are considerable overlaps in these models. For further discussion, see Zeleza (forthcoming).

Chapter Thirteen

CHIEFS AND COMMONERS IN THE GLOBAL VILLAGE

Introduction

The agrarian crisis was a powerful symptom of a much wider crisis confronting many African countries in the 1970s and 1980s. There was little dispute about the scale and seriousness of the crisis. The evidence appeared overwhelming: all the major economic indicators were stubbornly negative agricultural and industrial production had either declined or stagnated; persistent food shortages and recurrent famines were taking their toll on Africans' health, lives, and pride; Africa's terms of trade had virtually collapsed; and the continent was choking under a terrible debt burden. The air was thick with despair. The continent, indeed, seemed to be faced not merely with an economic crisis but an intellectual one as well. Crude, rabidly racist views of Africa, which only a decade or so before were whispered in the dark corners of academia and the lurid pages of the western gutter press, were now proclaimed with open relish even by once 'reputable' Africanists. To some of them Africa's crisis was rooted in the continent's failure to nurture and develop the capitalism that the West had so magnanimously introduced through colonialism. According to Sandbrook (1985), the obstacles to capitalist development lay in Africa's poor natural resources, smallness of local markets, lack of managerial and technical skills, its propensity for generating bitter ethnic and religious schisms, and its singular genius for producing corrupt and dictatorial regimes. In the less abstract hands of the IMF and World Bank, the crisis was placed squarely on the 'policy failures' of African governments (World Bank 1981, 1984). In the face of this onslaught, African scholars rose up in disgust. Many were extremely critical of the African states, but they disputed that these states, or 'internal' conditions, by themselves, 'caused' the crisis. They pointed their indignant fingers at the world economy, specifically the global capitalist crisis, which led to the declining terms of trade for Africa and brought other external shocks. They were particularly dismissive of the economic, political, and moral efficacy of the structural adjustment programmes. The intellectual and ideological gulf between African and Africanist scholarship widened further.

This chapter visits the debates and also tries to provide a reconstruction of the development of the African crisis. It argues that this crisis was, indeed, both a manifestation and an outcome of the world economic crisis that erupted in the early 1970s. Thus, it was not merely dependent capitalism in Africa that was in crisis but capitalism globally. Attempts to treat Africa's crisis as unique are, therefore, theoretically inadequate, empirically misleading, and ideologically suspect. This is not to say that 'internal' conditions in Africa were not 'responsible' for the crisis. Far from it. It is merely to underline the fact that 'internal' and 'external' conditions did not exist in splendid isolation; they reinforced each other in producing and reproducing the differentiated conditions, consequences, and trajectories of the crisis among and within regions, nations, and social classes. The chapter is divided into two major parts. The first examines the roots, nature and dynamics of the world capitalist crisis, and its conflicting analyses by neo-classical and Marxist scholars. The second part deals specifically with the complex evolution and uneven impact of the crisis on African countries, and offers a critique of the economic and political effects of structural adjustment programmes. The implementation of these programmes unleashed powerful political forces: the highly differentiated and sometimes overlapping interests of internal and external actors were pitted against each other, and the structural contradictions and imbalances embedded in the post-colonial social order and patterns of accumulation were prized open. Authoritarian in their conception and execution, opposition to the adjustment programmes was met with increased repression and cooptation, which, in turn, led to more defiance, selective accommodation, and the creation of various coping mechanisms. In short, the terrain on which structural adjustment was imposed, and which it helped refashion, was constantly shifting, so that its effects cannot be deductively assumed either from the fanciful logic of the free-market prescriptions themselves or the rhetoric of staunch nationalist resistance.

When the Chiefs Sneeze

At the turn of the 1970s the world capitalist system entered a prolonged period of economic crisis. The long post-war boom which the industrial market economy countries (IMECs) had enjoyed was suddenly brought to a halt to the bewilderment and consternation of governments and their economic planners. Economic growth rates fell, capacity utilization and investment declined, inflation and unemployment rose simultaneously to unprecedented levels, the international monetary system crumbled, trade and balance of payments deficits skyrocketed, and the ghost of protectionism was resurrected.

From the early 1970s growth rates in the IMECs fell, then collapsed. The average annual growth of the real gross national product per capita in these countries fell from 3.9% in 1960-73, to 1.7% from 1973 to 1980 and to a negative 0.2% in 1980-82 (ILO 1984:34). The growth rates for the IMECs in subsequent years were 1.6% in 1983, 4.1% in 1984, 2.4% in 1985 and 1.9% in 1986 (World Bank 1987:171). The growth rates were lowest in the agricultural sector followed by industry and services. In the periods 1965-73, 1973-80, and 1980-85 agriculture grew at the rates of 1.7%, 0.9% and 1.5%, industry at 5.0%, 2.4% and 2.5%, and services at 4.7%, 3.2% and 2.0%, respectively (World Bank 1987:173).

The fall in economic growth was reflected in increased under-utilization of productive capacity and accompanied by declining investment. According to some estimates, manufacturing industries were operating 16% below capacity in the United States in 1978, 14% in Canada, 19% in Germany, 16% in France, 27% in Italy, 14% in Japan, and 65% in Britain. In all these countries industrial capacity utilization was lower than in the years between 1964 and 1977 (Frank 1980:73-77). Some believed that a much deeper process of decapitalization of industry, or de-industrialization, was underway in many of these countries. Unused excess productive capacity discourages industry from investing to expand capacity still further. So investment tends to fall. The share of investment in GDP among the IMECs dropped from 24.7% in 1973 to 22.8% in 1980 and 19.9% in 1983. It fluctuated between 21.1% in 1984, 20.8% in 1985 and 22.0% in 1986. Thus by 1986 the rate of investment was still lower than in 1973 (World Bank 1987:174).

The IMECs also reeled under high rates of inflation. The average annual rate of inflation in these countries increased from 5.2% in 1965-73 to 7.9% in 1973-84. In some of them the rate exceeded 10%. For example, it was 10.2% in Sweden, 10.4% in Australia, 10.7% in Finland and France, 13.6% in New Zealand, 13.8% in the United Kingdom, 14.4% in Ireland, 16.4% in Spain and 17.2% in Italy (World Bank 1986:181). Inflation was one of the most visible manifestations of the economic crisis. Many western governments cynically dubbed it as 'public enemy number one'. Inflation rose side by side with unemployment, confounding conventional wisdom. The dramatic increase in unemployment was by far the most visible effect of the world economic crisis in the IMECs. According to the ILO (1984:Chapter 2; Chapter 3; 1987:Chapter 3), from 1960 to 1973 unemployment rates in North America varied between 4% and 7%, while they were not higher than 2% to 3% in Europe and Japan. The average for the IMECs in 1973 was 3.5%, then it rose to 5.2% in 1979 and 9% in 1982. By the end of 1982 30 million people were out of work in these countries. In 1983 the number rose to 35 million. The unemployment rate fell slightly to below 8% in 1986 and the number dropped to 31 million. This decline was, however, confined to only three countries, namely, the USA, Canada, and Denmark. Unemployment in many European countries remained at record levels since the 1930s. For example, in 1985 it reached 3.3 million in the U.K., 2.4 million in France, 2.9 million in Spain, 2.3 million in West Germany and 2.5 million in Italy.

These figures hid the true dimensions of the unemployment problem, for they largely did not include workers who were too discouraged to look for work. In other words, official data on unemployment often excluded the long-term unemployed. Workers forced into early retirement, whose numbers increased as the crisis deepened, were also not recorded as unemployed. There is overwhelming evidence that the number of long-term unemployed increased in the 1970s and 1980s in nearly all the IMECs. For example, the proportion of those unemployed over one year in total unemployment between 1981 and 1986 rose from 21% to 27.5% in Australia, 32.5% to 47.8% in France, 19.2% to 32% in West Germany, 13.5% to 17.2% in Japan, 22% to 56.3% in the Netherlands, 43.6% to 56.6% in Spain, and 22% to 41.1% in the UK. Needless to say, unemployment mostly hit certain social groups, especially the youth and the old, women and racial or ethnic minorities.

Apart from generating massive unemployment, the crisis also transformed the structure and opportunities for employment. Industrial employment declined in nearly all the IMECs, particularly in the manufacturing and mining sectors, while employment in the services grew rapidly so that by 1981 it accounted for 63% of total employment in these countries. There was also a marked increase in the number of part-time workers, and in the scale of clandestine employment in the rapidly expanding informal sector. By 1985, for example, the number of part-time workers in the IMECs approximated 45 million. The proportion of part-time workers in the labour force was 28.6% in Norway, 24.6% in Sweden, 23.7% in the Netherlands, 21.2% in the U.K., between 15% and 19% in Austria, the USA and Canada and around 10% in France, West Germany, and Japan. In all these countries the part-time workers were overwhelmingly women. Thus, women bore the brunt of labour casualization brought about by the world economic crisis.

In the meantime, those lucky enough to be employed were faced with declining wages. In all the IMECs governments stepped up intervention in wage fixing. Some of them unilaterally imposed wage freezes or guide-lines. The decline in real-earnings accelerated from the late 1970s. For example, between 1979 and 1983, Norway witnessed a fall of about 5%. In the USA and Canada real earnings also declined, although by a smaller magnitude. There was massive and prolonged state intervention in other spheres traditionally regulated by collective bargaining. Anti-trade union attitudes by employers, which grew, were often reinforced by government measures curbing trade union activities. In the UK, for instance, legislation adopted by Thatcher's government introduced restrictions on the closed shop and direct action, which made it more difficult to call strikes. Thus in many IMECs trade unions became quite weak, partly as a result of draconian government policies, and partly due to falling industrial employment, the traditional source of union membership. Labour was clearly on the defensive. The 'free-market' was for capital, not labour.

Alongside these tribulations, destructive financial storms flared up fuelled by the growing discrepancy between the demands for international liquidity and the use of the US dollar as an international reserve currency. The dollar's decline as a reliable reserve reflected growing US deficits, on the one hand, and the full post-war recovery of the European and Japanese economies, on the other (Nabudere, 1989). The United States, the locomotive of the world capitalist economy, sought to prolong the rapidly decelerating post-war boom through the creation of credit and debt. The federal government began running huge budget deficits. These deficits which normally ranged between $1 and $3 billion in the 1950s, jumped to $23 billion in 1973 and never looked back (Frank 1980:42). By 1980 the budget deficit skyrocketed to $73.8 billion and at the end of 1986 it stood at a staggering $220.7 billion. In the meantime, outstanding federal debt was a mind-boggling $2.4 trillion (Koepp 1987:26-7). And for the first time since the First World War, the USA relinquished its position as the world's largest creditor nation and slid into the ignominy of the world's largest debtor nation. By 1986 it had become the world's largest debtor with its net external debt close to $250 billion and rising by about $150 billion a year (Westlake 1986a:73).

With such a creaking locomotive it is no wonder that the international monetary system was profoundly shaken. This was manifested in violent fluctuations in the exchange rates of the major currencies and in stock markets, as well as in growing balance of payments deficits. The destabilizing impact of the floating exchange rate regime inaugurated from 1971 is now well-documented (Meier 1982; Nabudere 1989). The instability of stock exchange markets culminated in the crash of October 1987, the most devastating since the crash of 1929 which precipitated the Great Depression. The crash spread like wildfire from Wall Street to all the major capitals of IMECs, from London to Bonn, Tokyo to Sydney. Hundreds of billions of dollars vanished in the inferno (*Time* 2-11-87; *Newsweek* 2-11-87).

The deterioration in the balance of payments of many of the IMECs was no less remarkable. In 1970 the USA had a current account surplus of $2.3 billion. By 1985 this had turned into a deficit of $117.8 billion. Eleven other countries experienced a similar sharp decline in their balance of payments. Japan, the emerging colossus of the world capitalist economy, bucked the trend. Its balance of payments improved from about $2 billion in 1970 to $49.2 billion in 1985. Germany, Britain and Spain also enjoyed favourable balance of payments (World Bank 1987:231).

As for the balance of trade, the situation in many of the IMECs began to deteriorate sharply from 1970. UNCTAD (1986:2-3) shows that the trade deficits of all the developed market economies which were, altogether, $3.2 billion in 1960, rose to $11.4 billion in 1970, $31.7 billion in 1975 and by 1980 the deficit had reached $153.9 billion. In 1985 it dropped to $118.9 billion. World Bank (World Bank 1987:221) data on the 19 IMECs indicates that in 1985 these countries collectively had a trade deficit of $137.2 billion. The worst hit was the United States whose trade deficit was $148.5 billion. Ten other countries, including Spain, Italy, New Zealand, Belgium, the UK, Austria, France, Australia, Denmark and Switzerland, also suffered from trade deficits of varying magnitude. In contrast, Japan enjoyed a massive trade surplus of $45.4 billion, a large portion of which was with the USA. By 1986 the American trade deficit had risen to between $150 and $170 billion (Westlake 1986a:73). The American behemoth was clearly in trouble. As late as 1975 the US had enjoyed a trade surplus of over $4 billion (UNCTAD 1986:2-3)

As the balance of trade deteriorated for many of these countries, protectionism reared its ugly head, often in response to pressures from uncompetitive domestic producers. This was manifested in escalating trade wars between Japan, the EEC, and the USA over everything from farm products to electronic goods, and by all three against the imports of manufactured goods from the 'developing' countries. In the course of the 1980s both the EEC and the US imposed import quotas on Japanese products like cars, television sets, and computers. Protectionism took the form of traditional tariff barriers and increasingly non-tariff barriers. The latter included an infinite number of protectionist practices ranging from government subsidies, quantitative restrictions, health and technical standards, to deliberately complex administrative procedures and the so-called 'voluntary export arrangements'. For example, by 1986 the cost of subsidizing farm exports was $25 billion each for the USA and the EEC, and $10 billion for Japan (Westlake and Watkins 1987:10).

In this atmosphere, the efforts of both UNCTAD and GATT to promote fairer world trade balance became increasingly compromised. With lack of interest from the industrialized countries, UNCTAD talks increasingly degenerated into a circus (*Daily Nation* October 1988:3, 10-12). The Eighth Round of GATT talks began in Uruguay in 1986 and dragged on for years because of disputes over the questions of farm subsidies, textiles and the liberalization of trade in services, trade-related investment and intellectual property rights (United Nations 1989). The major industrial nations adopted a product-by-product approach in the negotiations which virtually left 'a large number of developing countries without the possibility of effectively participating in market access negotiations' (Jara 1993:16). The USA, formerly the major guarantor of the multi-lateral system adopted a policy of 'aggressive unilateralism' (Bhagwati 1991), while the EEC's inclination toward managed trade increased. The drift towards regionalism in Europe was replicated in North America with the signing of the Free Trade Agreement between Canada and the United States. Protectionism among the industrialized countries was increasing precisely at a time when the developing countries were being encouraged to liberalize and to adopt export-oriented strategies. In fact, while in the industrialized countries protection continued 'to shelter declining industries... the freedom to protect infant industries in developing countries (was) increasingly restricted' (Glover and Tussie 1993:237).

Dismal Diagnosis

The evidence is therefore compelling that the developed capitalist economies entered a period of profound crisis from the early 1970s. It was not simply an economic crisis but a crisis of conventional economics as well. Neo-classical economists had failed to foresee or 'predict' the crisis. In fact, in the 1950s and 1960s they confidently preached that crises were things of the past banished from the future by the Keynesian revolution. Shonfield (1965) celebrated the emergence of a new kind of capitalism immune from the violent swings of boom to slump, and assured of steady economic growth and social welfare. Galbraith (1966, 1972) discovered the 'affluent society' and the 'new industrial state'.

Development economists assumed the inevitability of modernization for the 'developing' countries once they faithfully followed the tenets of neo-classical developmentalism and prayed at the altar of the market and, with their eyes appropriately closed, received the eucharist of western capital and technology, and then waited to be amply rewarded for their primary products by the god of 'comparative advantage'. In short, neo-classical economists were a cheerful, optimistic lot. To them the term 'crisis' was nothing but Marxist gibberish.

The neo-classical paradigm posits scarcity as the central and universal problem that confronts all societies throughout history. So economics is conceptualized as the study of the allocation of scarce means to unlimited ends. This makes neo-classical analysis inherently ahistorical. This paradigm also identifies and assumes that the three factors of production, namely, land, capital, and labour are all indispensable in production and socially equivalent, so that they are rewarded equally in terms of rent, profits, and wages, respectively. Consequently, exploitation is not regarded as intrinsic to the process of capital accumulation. Furthermore, it is an approach predicated upon a perspective of harmony and equilibrium in the socio-economic system. Conflict is, therefore, regarded as merely transitory and contingent and not structural. Finally, the property relations of capitalism are taken as universal and desirable (Friedman 1953; Lipsey 1971; Livingstone and Ord 1970; Samuelson and Nordhaus 1985).

Thus, neo-classical economics is given to piecemeal approaches and quantitative modelling techniques noted for their sophistry and triviality. It also suffers from compartmentalization and eclecticism. Its abstract analytical apparatus has no room for the analysis of such serious questions as the power structure, conflict, social relations of production, and the historical development of capitalism itself either within a given society or globally. Ideologically, it is apologetic for the capitalist system. By stipulating the existence of universal economic laws, existing capitalist societies are rationalized and legitimized, and the conception of non-capitalist development is dismissed out of hand. In this way, capitalist values and attitudes, development models and institutions are imbued with omnipresence and omnipotence. Forcing them onto the exploited classes and societies becomes, therefore, almost an act of divine intervention.

No wonder the crisis that engulfed the capitalist world from the early 1970s caught neo-classical economists of all persuasions, from the marginalists, to the Keynesians and monetarists, unawares (Roberts 1975:Chapter 2; Gamble and Walton 1976:Chapters 1 and 2; Rousseas

1979; Frank 1980). *Business Week* summed it up neatly when it observed in January 1977 after the meeting of the American Economic Association, that 'economic science is bankrupt' (Frank 1980:101). In explaining the crisis they tended to focus on contingent factors (Roberts 1975:Chapter 3; Gamble and Walton 1976:Chapter 1; McCracken 1977; Frank 1980:Chapter 2), among which the 'oil shocks' of 1973-74 and 1979-80 featured prominently. Alternatively, they mistook symptoms of the crisis for its causes. Inflation in particular, variously blamed on the 'oil shocks', trade unions and high wages, or inflated money supply to finance government expenditure, was accused of triggering the crisis. There was also a tendency to analyze the crisis in each country or region as if it were special or peculiar to it. The circulation process of capital was thereby artificially encrusted in spatial configurations that were dialectically integrated. Celebrations of the emergence of the global village seemed to stop at the door of each nation or region as far as explaining the crisis was concerned.

The neo-classical explanations shared three basic assumptions (Weber 1983). First, that it was a short-run phenomenon. Consequently, short-term restructuring programmes were proposed and implemented. Each temporary recovery was often prematurely hailed as the beginning of another long period of boom. Second, the crisis was believed to have been created in the recent past through incompetent management or 'policy errors' by governments. So it followed, third, that it could only be resolved through better government fiscal management and administrative policies. Both Keynesians and the monetarists, for example, were agreed that it was the government and the state sector that were key to solving the crisis. What they differed on was the exact role the state should play: the Keynesians advocated more government intervention as the solution while the monetarists advocate less. For both schools the assumption was that the 'political' and 'economic' were separate spheres, whose interpenetration, for better or worse, was externally engineered. As the crisis deepened the anti-Keynesians gained the upper hand. Friedman's Chicago Boys scurried the globe converting governments from Keynesian orthodoxy to the monetarist gospel. Government intervention became a dirty world in Washington, London, and Bonn. 'Privatization' became the new incantation of capitalist fetishism. Structural adjustment was born. This is where the Africanist critics of the post-colonial state and proponents of structural adjustment got their ideological inspiration. African countries were being used, once again, merely as empirical test cases.

284 MANUFACTURING AFRICAN STUDIES AND CRISES

Premature Prognosis

Marxist political economists made no such distinction between 'politics' and 'economics'. Moreover, they professed a love for dialectical historical analysis and had no use for universal economic laws. They saw capitalism as one mode of production among several in history, which was subject, like the others before it, to change, decay, and disappearance. To them, therefore, the crisis was not exogenous but intrinsic to the process of capital accumulation. It was spawned by the inner contradictions of the capitalist mode of production, at the root of which was the social nature of production and the private appropriation of the fruits of production (Mandel 1977; Sweezy 1970; Eaton 1966; Onimode 1985).

The early Marxists had predicted that as capitalism developed its periodic crises would intensify, until the entire system collapsed and from its ashes socialism would arise. The Great Depression of the 1930s seemed to provide a grim confirmation of the Marxist analysis of capitalism, and Marxism attracted many new adherents. Then came the Second World War which finally ended the depression and ushered in an unprecedented period of economic boom. The term crisis disappeared from the vocabulary of conventional economics. Bourgeois economists had their field day. Some Marxists lost faith and desperately embraced Keynesianism and found political refuge in social democracy. The 1970s seemed to vindicate Marxist analysis once again. The apparent successes of national liberation movements and OPEC, and the wave of Third World demands for a New International Economic Order and conversion to 'socialism' or 'non-capitalist paths of development' only confirmed the depth of the crisis in the world capitalist system. Marxists brushed off the texts of the old masters seeking explanations[1].

Marxists tend to distinguish between crises associated with the tendency of the rate of profit to fall and those arising from problems of realization, that is from problems of markets. The tendency of the rate of profit to fall springs from the increase in the organic composition of capital in the course of capital accumulation. Realization crises imply overproduction. They arise either from disproportionality between different branches of production in the economy or from underconsumption, that is lack of effective markets. Marxists recognize that there are forces which can counteract these tendencies. The falling rate of profit, for example, can be offset by raising the intensity of exploitation through technological innovation, cutting wages, cheapening the elements of constant capital, expanding foreign trade, and embarking on ruthless exploitation of the 'periphery'. The tendency to overproduction and stagnation can be alleviated by population growth, development of new industries, unproductive consumption, especially of war, and state expenditure. But these counteracting forces cannot postpone the crisis indefinitely. On the contrary, they only succeed in making each subsequent crisis deeper, longer, and more widespread. Overproduction or overaccumulation of capital manifests itself in a number of ways: an overproduction of commodities; underutilization of fixed and money capital; surpluses of labour power expressed as underemployment in production and an expansion of the industrial reserve army; and falling rates of return on capital advanced. Devaluation, the nemesis of overaccumulation, concretely manifests itself in many ways as well: money capital can be devalued by inflation; labour power can be devalued through unemployment and falling real wages; commodities may have to be sold at a loss; and the value embodied in fixed capital may be lost as it lies idle (Harvey 1982:Chapter 7).

Capitalist development, then, is characterized by tendencies towards over-accumulation and devaluation, in short, cyclical progress. Each cycle goes through four phases of crisis, depression, recovery, and advance. In the 19th century, Marx estimated, the crisis cycle tended to be about 10 years (Eaton 1966:157-161). In the 20th century, Menshikov contends, it has been considerably shorter, although after the Second World War the crises were less severe than those of the first half of the century, and they rarely embraced the whole capitalist world, except those of 1957-58 and 1967-71 (Menshikov 1975:Chapters 2 and 3). It would seem that these short business or trade cycles take place within long Kondratieff waves, named after N.D. Kondratieff (1935) who first identified them. Each wave comprises a boom followed by a slump and lasts about 50 years. It has been argued that the uncontrolled growth of the upswing phase sows the seeds for the downswing phase. During the upswing there is rapid expansion which eventually leads to over-production, increased competition, and falling rate of profit. This, in turn, leads to failure of firms, unemployment and wage cuts, in short, stagnation and recession. 'The failure of firms during the downturn helps eliminate competition, the constraints on wages increase profitability, and the mergers that seem to accompany downturns create a large unit of production that is then in a strengthened position for the succeeding upswing' (Bergesen 1983a:16).

Thus, downturns have been characterized by deflation, lower growth rates, cost-reducing innovation, and declining profits. Upswings have correspondingly been marked by increasing prices, production, technological innovations, and rising profits (Mandel 1975; Frank 1980). Each period exhibits certain political and economic transformations that lead, alternately, to recovery and stagnation. During periods of economic expansion one state that exceeds others in commercial, financial, productive and military strength exercises political hegemony. Conversely, downturns are accompanied by competition and rivalry among the core states, which leads to, among other things, global wars, colonial expansion or the tightening up of relations between core and periphery. Long-cycle models of global hegemony have been constructed (Modelski 1978, 1983; Bergesen 1980, 1983b; O'Loughlin 1986). Since the origins of the modern world system in the 16th century Hopkins and Wallerstein (1977, 1979) have identified four hegemonic periods; the Hapsburg (1450-1559), the Netherlands (1620-1650), Great Britain (1850-1873) and the United States (1945-1967). Chase-Dunn (1978, 1979) has identified three hegemonies: the Dutch of the mid-seventeenth century, the British of the mid-nineteenth century, and the American of the mid-twentieth century.

Attempts have been made to periodize the long-waves. Kondratieff identified three: the first rising from the end of the 1780s or the 1790s until 1810-17 and declining until 1844-51; the second rising from 1844-51 until 1870-75 and declining until 1890-96; and the third rising from 1890-96 until 1914-20 and declining from then on (Roberts 1975:34-35). According to Mandel (1975, 1977) and Frank (1980) the first long wave began with an upswing lasting from 1790-1793 to 1819-25 and ended with a downswing until 1847-48. Following the defeat of the revolution of 1848, an upswing of the second long-wave began lasting until 1873 when a down-swing set in until 1893-95. During the latter phase competitive capitalism gave way to monopoly capitalism, Britain began to lose its supremacy in the face of industrial advances made by Germany and the USA, and the 'new imperialism' was launched. Then came the third long-wave with an upswing from 1895 to 1913 and a downswing from 1913-19 to 1940-45, during which two world wars broke out, socialist revolutions triumphed in the Soviet Union and Eastern Europe, and the nationalist movements in Asia and Africa gathered momentum and were poised for victory. Finally, another long-wave began after the Second World War. Its upswing lasted until 1966 followed by a downswing in 1967 which by the 1980s had yet to run its course. Shuman and Rosenau (1972) argue that the upswing began during the war and lasted until 1970 after which the wave slid into a downswing.

These models pointed to the transformations in the processes of global capitalist accumulation and shifts in hegemonies, often obscured in the more narrowly-focused neo-classical analyses of business cycles, in which it is assumed that the market economy itself is essentially stable and the difficulties are basically caused by inappropriate state intervention, politics, or unpredictable changes in mass psychology. But the models tended towards structural determinism, and were underlined by an unflinching belief that the crises were symptoms of capitalism's 'terminal disease', so that the changing institutional contexts of accumulation, brought about by capitalism's very survival of major crises were usually ignored. The links between short-run fluctuations and the long-run or secular tendencies of capital accumulation were not adequately differentiated or theorized.

Out of these concerns and debates the 'regulation approach' emerged in an attempt to offer a more nuanced and historically plausible explanation of capitalist market operations, failures, and institutional mutation. According to this approach, capitalist development, whether in the world economy as a whole or within individual nations, undergoes a series of transformations during which the regime of accumulation, the growth model, and the mode of regulation change. *Regime of accumulation* 'describes the fairly long-term stabilization of social production between consumption and accumulation' (Lipietz 1987:14), while *growth model* 'describes the form of capital accumulation within that particular economy, as well as the nature of its insertion into the world economy as a whole' (Gelb 1991a:11), and the *mode of regulation* refers to the various elements that regulate all aspects of the accumulation process, including the wage relation, structure of demand, competitive interrelations between capitals, and the financial system. Moments of crises, then, are those when there is disequilibrium within the regime of accumulation and incoherence in the mode of regulation, in short, the prevailing growth model disintegrates. In due course a new growth model emerges. The 'regulation approach' eschews advancing a grand theory of capitalist crises, and calls for historical analyses of crises of accumulation as moments of transition spawned by complex processes and pressures, those emanating from the international economy as well as the domestic economy, which are economic, social, and political in nature, for even markets are seen as institutions, as socially produced processes (Aglietta 1970; de Vroey 1984; Noel 1987; Jessop 1989).

Labelled 'Fordism', the post-war growth model of the industrialized countries was based on the correspondence of mass production and mass consumption. This model was sustained by three main factors. First, in the 1950s and 1960s the international monetary system was relatively stable, based on American hegemony. Second, investment and accumulation were profitable because of the massive reconstruction programmes in war-torn Europe and Japan, the appearance of new fields of investment and the cheapening of raw materials and primary products from the underdeveloped countries. Thus, it was boom for the West and bust for the South (Frank 1981:Chapter 1). Third, the state regulation of the economy developed in the capitalist centres during the Second World War was continued after the war. Each state vigorously pursued policies aimed at the management of demand, socialization of the costs of capital accumulation, and the maintenance of social peace and political stability (Shonfield 1965; O'Connor 1973; Gamble and Walton 1976).

By the end of the 1960s the situation had changed significantly. The Bretton Woods international monetary system was breaking down as a result of American hegemonic decline and the emergence of core multi-centricity and instability. The US had abused the position of the dollar as the international reserve currency. In order to finance foreign investment and overseas military expenditure, including the Vietnam War, the US had simply printed more dollars. By the late 1960s American liabilities had grown to more than twice as great as the country's reserves. This could not go on indefinitely, especially now that Western Europe and Japan had recovered. Indeed, America's productive superiority, organizational and technological edge over these countries had been lost (Fitt et.al. 1980). In 1971 the crunch came. The dollar was devalued and before long the system of floating exchange rates was introduced. Currency stability became a thing of the past (Mandel 1972; Meier 1982). Like Britain before it, America had entered the road to hegemonic decline (Goldfrank 1983).

The rate of profit in the major capitalist countries also showed signs of declining. For example, in the USA the rate of net profit declined from 12.7% in 1966 to 5.3% in 1970 and 3.5% in 1975. In Britain the decline was from 2.7% in 1966 to 1.4% in 1970 and 0.9% in 1975 (Frank 1980:33). The 'intensive' regime of capital accumulation that had characterized the developed market economies peaked in the mid-1960s. It had been based on massive increases in productivity, brought about through widespread technological innovations, and sustained by mass consumption, which was generated by rising wages (Aglietta 1970; de Vroey 1984; Thrift 1986). At the end of the 1960s overproduction began to manifest itself. Competition both for markets and raw materials sharpened among the IMECs.

Finally, the unravelling crisis was both a cause and effect of crisis for the state. The state was increasingly confronted with the crises of rationality, legitimation, and motivation. In capitalist society the state has the dual role of promoting and sustaining the capitalist mode of production. As the crisis of accumulation became evident the state in the capitalist centres began to take steps to promote accumulation, without which it would, of course, have nothing to legitimate (Harbemas 1976; Johnston 1986). Social welfare was now anathema. Austerity became the new gospel. One after another western governments drifted to the right. They began waging relentless war against labour in order to reduce real wages. Unemployment was allowed to rise in order to whip organized labour into timidity and maintain the falling wages. The face of liberal democracy wilted under the pressure of increasing political repression. Reactionary ideologies, taking such forms as racism, national chauvinism, religious fundamentalism, and individualism with its escapist banalities, spread with a vengeance (Frank 1980:Chapter 3; O'Connor 1984; Peet 1986; Johnston 1986; Williams 1986; Hammond 1983).

The Global Village

Both neo-classical economists and Marxists agree that the world economy became more globalized after the Second World War, but they differ sharply on its causes and consequences. The former tend to regard the process as essentially benign, if not benevolent, for the 'developing' countries, while conventional Marxist views are much harsher. Marxists have always insisted that capitalism can only be understood as a whole, that is, the course of capital's circulation and reproduction should be examined on the scale of the entire capitalist economy. Marx noted that capitalism is highly dynamic and inevitably expansionary. 'The historical mission of the bourgeoisie', he wrote, 'is accumulation for accumulation's sake, production for production's sake' (Marx 1967:595). He argued that the tendency of the falling rate of profit and the realization problem could be temporarily resolved through external markets, export of capital, and primitive accumulation in the colonies. These ideas were developed further by Hilferding (1978), Luxemburg (1968) Bukharin (1972), and Lenin (1978) who articulated the writings of the early Marxists into his theory of imperialism as the monopoly stage of capitalism.

The notion that the space economy of capitalism is global was fully developed by the dependency writers, whom we examined in some of the preceding chapters. Wallerstein's conceptualization of the world system was probably the most comprehensive and controversial. According to him, the capitalist economy is characterized by the existence of a world market for commodities, a competitive state system, an economic hierarchy of spaces from core through semi-periphery to periphery, economic classes and status groups which interact with one another and with the spatial hierarchy, cyclical long economic waves, and asymptotic secular trends. He sought to overturn the evolutionary model of progress held by both bourgeois scholars and the classical Marxists according to whom capitalism as an historical system was seen as progressive. The myth of progress was created, he believed, by measuring progress in terms laid down by capitalist logic itself. Capitalism, he contended, was not and had never been progressive, certainly not to the world's impoverished majority (Wallerstein 1974, 1979, 1980, 1983, 1984; Taylor 1986).

Building on some of these insights, Harvey presented a persuasive and complex analysis of the spatial configurations of capital accumulation on a global scale. The accumulation of capital and misery, he argued, go hand in hand, concentrated in space. The antagonisms between centre and periphery were not accidental, but 'the coherent product of diverse intersecting forces operating within the overall unity of the circulation process of capital' (Harvey 1982:419). Uneven geographical development under capitalism was based on varied rhythms of accumulation between different regions. In its development capitalism has always been faced with contradictory tendencies. 'On the one hand spatial barriers and regional distinctions must be broken. Yet the means to achieve that end entail the production of new geographical differentiations which form new spatial barriers to be overcome' (Harvey 1982:417). So in the historical geography of capitalism the evolution of productive forces and social relations exist as particular configurations. This explains the regionalization of class and factional history. The circulation of capital is marked by contradictions between fixity and motion, concentration and dispersal, local commitment and global concerns. These contradictions put enormous strain upon the organizational capacities of capitalism. In order to mediate and contain these tensions hierarchical structures of organization or modes of regulation, such as the international financial institutions, have been created to link and mediate the local, national, regional, and the global. Crises therefore become articulated and unfold at various levels of the hierarchical structures in the spheres of finance, production, trade, the state, and so on. Crisis in one region affects the other regions because regional 'boundaries', if they exist at all are highly porous to capital and labour movements. The degree to which over-accumulation problems arising in one place can be relieved by further development or devaluation in another place depends upon the intersection of all manner of diverse and conflicting forces' (Harvey 1982:427).

The impact of crisis varies from one nation and region to another depending on its growth model, which is partly structured by the country's or region's insertion and position in the space economy of capitalism. In times of crisis there are intense social struggles, within and between countries, which exacerbate the disintegration process in the growth model, over who is to bear the burden of devaluation. Classes and regions with massive concentrations of economic and political power seek to load and export the costs of devaluation to the weaker classes and regions. In other words, domestic social antagonisms and imperialism and inter-imperialist rivalries intensify. In the crisis of the 1970s and 1980s, the 'developing' countries, including many in Africa, with their underdeveloped and dependent economies were forced to shoulder most of the costs of devaluation by the developed capitalist economies, who generated the crisis in the first place in so far as it was a global crisis and not specifically a regional one. Once generated, however, the crisis took its own forms in the 'developing' countries, as it rippled through them and anchored itself to domestic conflicts and disharmonies, many of them representing regimes of accumulation and modes of regulation developed out of the temporary resolution of previous crises, and reflecting old patterns of insertion into the international division of labour. In Onimode's words: 'Basically, the African crisis is a composite of underdevelopment and cyclical fluctuations'. Its superimposition on Africa is facilitated by neo-colonialism. Consequently, he concludes, 'since both underdevelopment and neocolonialism are integral aspects of global capitalism, the African crisis is a derivative of the growing crisis of global capitalism' (Onimode 1988:13). In short, the world capitalist system in the 1970s and 1980s had caught a cold and was sneezing furiously. Africa and large parts of the 'Third World' reeled.

The Commoners Catch Pneumonia

After independence African countries pursued a wide range of development strategies, from the unadulterated free enterprise philosophy of Côte d'Ivoire, Kenya, Malawi, Nigeria, and Morocco, and the mild socialism of Algeria, Tanzania and Zambia, to the oscillations of Ghana, Egypt, Guinea and Mali, and the resolute Marxist zeal of post-revolutionary Ethiopia, Mozambique and Angola. These varied ideological permutations sprang from differences in the modalities of accession to independence, colonial legacies, resource endowment, the level of development of the productive forces, the class character of the state, the nature and dynamics of the internal contradictions and struggles, and each country's patterns of integration into the world capitalist system. Their economic performance, therefore, varied, although the variations showed little correlation with official development ideology.

Uneven Exposure

As was observed in the last chapter, the statistics upon which the gloomy assessments of 'Africa's' economic performance and decline are based are problematic. What is true of the agricultural statistics is even more glaring for gross national product statistics. Each of the major institutions that covers Africa, such as the World Bank, and the various United Nations agencies, including the United Nations Conference on Trade and Development (UNCTAD), the United Nations Development Fund (UNDP), and the Economic Commission for Africa (ECA), produces its own statistics, using very divergent methodologies. So there is hardly any unanimity even on the same country. Their definition of what constitutes 'Africa' in their coverage also varies. For the World Bank their 'Africa' is an abbreviation for sub-Saharan Africa, and until recently excluded South Africa and Namibia, as well as North Africa, which is often appended to low and middle income Europe and the Middle East. UNCTAD and the ECA, on the other hand, usually cover the continent as a whole. So generalization about Africa depend on which 'Africa' is being covered.

For example, according to UNCTAD (1988:38; 1989:4), average annual growth rates of real GDP in its Africa were 5.6% in 1961-1972, 2.7% in 1973-85, and 1.5% in 1986, 1.2% in 1987, an estimated 2.6% in 1988, and a forecast 2.8% in 1989 and 2.5% in 1990. In the World Bank's 'Africa' the average annual growth rate was 6.1% in the 1965-73 period, 3.2% in 1973-80, -0.5% in 1980-85, and 3.2% in 1986, -1.3% in 1987 and an estimated 3.1% in 1988 and 3.5% in 1989 (World Bank and UNDP 1989:147, 1990:161). Interestingly, in its 1990 report the World Bank's (1990) estimates for the earlier periods are revised downwards, 5.9% instead of 6.1% for 1965-73, and 2.7% instead of 3.2% for 1973-80. It would not be going too far to argue that the data these organizations produce in their elegant annual reports are nothing more than estimates, some of them very crude, indeed. These statistics exclude the informal sector, which by its very nature is not part of the national accounts. From all indications, this sector grew rapidly in the 1980s partly in response to the crisis of the formal sector itself which was exacerbated by the structural adjustment programmes. Other criticisms of the conventional statistical comparisons is that they do not take into account women's work, which tends to be concentrated in unremunerated or informal sector activities. Also, they do not incorporate differences in the cost of living. Comparisons based on purchasing power parity yield quite different results in national rankings. Does the fact that the per capita income in the United States in 1990 was 5,889% higher than in Kenya mean that the 'average' American lived that much better off than the 'average' Kenyan? Of course not. If the statistics that are bandied around were to be believed, the vast majority of Africans should have died in the 1980s.

The heuristic value of the statistics lies in the trends they depict. There seems to be general agreement that African economies grew reasonably fast up to 1973, when growth decelerated as a result of the world recession. Growth resumed temporarily between 1976 and 1979, after which it plummeted again, thanks to the 1979-82 world recession, the second in a decade. The 1980s were a 'lost decade'. As might be expected, economic performance in Africa since the 1960s has been very uneven between and within countries. But before discussing that let us examine aggregate rates of growth for Africa as a whole in comparison to other regions.

Table 1: Average Annual Growth of Real GDP, 1961-1990 (Percent)

Region	1961-1972	1973-1985	1985-1990
Developed Market Economies	4.9	2.7	3.1
Developing Countries	5.9	4.1	2.3
Of which:			
Africa	5.6	2.7	2.1
Latin America	5.6	3.5	2.3
Asia	6.7	4.7	4.6

Source: UNCTAD (1988b:38; 1989:28).

It can be seen that Africa grew twice as fast from 1961 to 1972 as it did between 1973 and 1985. Except for the period 1973-1985, Africa's performance did not vary significantly from the rest of the developing countries, and was comparable to, or higher than that of the developed market economies for the first two periods. World Bank (1992:221) figures show that in the period 1965-1980 the GDP of sub-Saharan African economies grew at an annual rate of 4.2%, above the world average of 4.0%, and was higher than South Asia's 3.6% and the OECD members' 3.7%. Between 1980 and 1990 sub-Saharan's growth rate dropped to 2.1%, the same as Europe's, and higher than the growth rates of both the Middle East and North Africa, and Latin America and the Caribbean which were 0.5% and 1.6%, respectively.

Table 2 shows annual average growth rates by sector. The table shows that the sectoral performance for sub-Saharan Africa in agriculture was relatively poor, but in industry and services the performance was generally better than the world average, indeed, in the 1965-80 period the industrial growth rate for the region was second only to that of East Asia and the Pacific. The factors behind agriculture's relatively poor performance were analyzed in the last chapter. Suffice it to say, agricultural performance varied enormously between and within countries and sectors and changed over time. Of the 25 countries for which the World Bank had data, in eight agricultural performance between 1965 and 1980 was higher than the world average of 3.7%. The fastest growth rate of 10.7% was recorded in Libya, followed by Botswana with 9.7% and Tunisia at 5.5% (World Bank 1992:220-1).

The same was true of industry. Many of the debates around the agrarian question have also been made with reference to Africa's industrial performance. As Table 2 shows, industry performed much better than agriculture but, as with agriculture, industrial performance varied enormously. In the 1965-1980 period, industrial annual growth rates ranged from 24% for Botswana to -4.3% for Uganda. Of the 24 countries whose data was available, in two the growth rate was over 15%, in three it ranged from 10-14%, in ten 5 - 9.9%, in three 3 - 4.9%, and in four 1 - 2.9%, and in two it fell below zero (World Bank 1992:220-1).

Table 2: Growth of Production by Sector, 1965-1990
Average Annual Growth Rate (Percent)

Region	Agriculture		Industry		Services	
	1965-80	1980-90	1965-80	1980-90	1965-80	1980-90
Sub-Saharan Africa	2.0	2.1	7.2	2.0	7.1	3.6
N. Africa & M. East	4.3	4.3	6.3	0.7	10.9	1.9
South Asia	2.5	3.0	4.3	6.5	4.5	6.3
E. Asia & Pacific	3.2	4.8	10.8	10.2	8.9	8.0
L. America & Caribbean	3.1	1.9	6.6	1.2	6.6	1.7
Europe	..	1.0	..	2.7	..	2.7
OECD Members	..	1.7	2.8	..	4.5	..
World	3.7	4.3	6.0	-1.0	9.6	1.2

Source: World Bank (1992:221).

During the period 1981-90 manufacturing's share in GDP increased in 21 countries, it decreased in 13, and remained unchanged in five. By 1993 manufacturing accounted for about 16% of GDP, and the share of manufacturing in the total exports of the sub-Saharan countries' exports rose from 10% to 15%, although there were wide country variations on both scores (World Bank 1995:47).

There is no question that overall, industry performed much poorer in the 1980s than in the period before. This partly reflected the exhaustion of 'easy' import substitution options. Import substitution was the industrialization strategy that most African countries adopted immediately after independence. It was envisaged that industrial production would move sequentially from consumer goods to intermediate and capital goods. It was also hoped that import substitution industrialization (ISI) would enhance technological development, generate large-scale employment, and improve the balance of payments (Fransman 1982). Before long, however, it became clear that many African countries were stuck at the first stage of consumer goods production, and that technological dependence had increased, relatively few jobs had been created, and the balance of payments constraint had not been alleviated. In short, it appeared ISI had failed to generate a self-sustaining process of industrialization and economic development. The neo-classical critique blamed this on the existence of distorted and inefficient factor and goods markets, mainly caused by government intervention in the economy, especially the excessive promotion of domestic industries behind high tariff barriers (Little et. al. 1970; Killick 1978).

The structuralist or dependency critique attributed the perceived failure of ISI to the inherited distorted productive structure, and fragmented nature of these countries' markets arising from their dependent status in the world capitalist system (Singh 1982; Nixon 1982; Khennas 1992). It was argued that ISI was unsustainable for a number of reasons. First, it generated structural imbalances between agriculture and industry, and between regions and social classes in the distribution of income, and within the financial sphere through the fueling of inflationary pressures. Second, it intensified the balance of payments problems because raw materials for ISI were imported, indeed, the more ISI grew the higher were imports of industrial raw materials and capital goods. Third, ISI encouraged technological imports from the developed capitalist countries and their multi-national corporations (MNCs). In fact, ISI was heavily dependent on foreign capital mediated through MNCs, who promoted the drainage of investable surpluses through repatriation of profits, interest, dividends, royalty payments, management fees, etc. Finally, ISI neither generated sufficient income nor distributed it adequately to justify the full utilization and expansion of industrial capacity.

In the late 1970s, and especially from the 1980s, some countries began to shift to export manufacturing, including the establishment of the so-called Export Processing Zones (EPZs) (Frank 1981:Chapter 3). These were manufacturing enclaves set up to attract foreign capital, often using the incentives of lavish tax holidays and customs exemptions, and cheap, non-unionized labour as bait. EPZs spread particularly rapidly in South-East Asia, the Caribbean and Central America. In Africa six countries; Mauritius, Djibouti, Egypt, Tunisia, Senegal and Liberia had established EPZs by 1985, and more were planning to do so (Westlake and Jayawardena 1985; Bailey 1987). Other countries opted to introduce manufacturing under bond schemes instead. By 1987 Malawi had introduced such a scheme and Kenya announced its intention to start one in order to promote export manufacturing (Ikiara 1987). Working for the world market rather than for the internal market meant that effective demand on the national market was no longer the source of demand for national production. So it would no longer matter if local incomes stagnated or fell.

Both strategies, ISI and export manufacturing, faced severe problems when the world economic crisis broke out. The fall in primary commodity prices meant that many African countries were unable to import industrial raw materials, spare parts and capital goods for their import substitution industries. Thus, industrial production began to fall. For the export-dependent industries there was the added problem of growing protectionism in their main markets, the developed capitalist countries. And with falling domestic wages, the internal market could not absorb the surplus production. The result was declining industrial capacity utilization in many countries.

It is quite evident that the patterns of development were very uneven along national, regional, sectoral and temporal dimensions, so that discussions of the African crisis require careful and differentiated analysis. Table 3 makes even clearer the variations in national performance. Space does not allow for a detailed analysis of each of these countries. But the point has been made that the impact of the crisis on African economies was varied depending on the internal structure of these economies and their insertion into the world economy, as mediated principally by the composition of their exports. Internal political conditions, often tied to complex regional and international forces, also played a major role. The poor performance of Uganda, for example, cannot be divorced from the destabilization brought about by Idi Amin's regime in the 1970s and the country's declining terms of trade (Mamdani 1976, 1983; Nabudere 1980; Kasozi 1994), while Botswana's spectacular performance reflected the buoyancy of its diamond and beef exports as well as political stability fostered by its 'democratic developmentalism' (Holm and Molutsi 1989; Stedman 1983; Tsie 1995). But aggregate national accounts, it cannot be overemphasized, hide a lot of regional, class, and gender differentiations in production, incomes, and living standards.

If the statistics presented above can be believed, then Africa's general economic performance from the 1960s to 1980s compared favourably either with long-term trends or with other countries. During the same period it registered a real per capita growth rate of about 1.6% 'which was the same as for all low-income countries for the same period. Never, probably, in the past has such a performance been achieved by Less Developed Countries', comments Fieldhouse (1988:148-9). He continues: 'Hence, the disappointments and complaints concerning black Africa relate either to particular countries or groups of countries; to different periods within the twenty years; or, finally, to Africa's performance compared with that of more developed countries' (Fieldhouse 1988:149). He believes that the poor assessments of Africa's performance in the 1980s were simply the flip side of the exaggerated hopes of the early 1960s. But his explanation of the economic slowdown in the 1980s leaves a lot to be desired: he blames it all on the fact 'that the tropics are peculiarly difficult terrain for the developer' and 'the social structures and attitudes of Hyden's 'uncaptured peasantry' (Fieldhouse 1988:154-5).

Table 3: Growth of Production, 1965-1990 For Select African Countries

Low Growth Economies	1965-1980	1980-1990
Uganda	0.6	2.8
Zaire	1.9	1.8
Zambia	2.0	0.8
Ethiopia	2.7	1.8
Sierra Leone	2.7	1.5
Medium Growth Economies		
Mali	4.2	4.1
Zimbabwe	5.0	2.9
Cameroon	5.1	2.3
Malawi	5.5	2.9
Morocco	5.7	4.0
High Growth Economies		
Mauritius	5.2	6.0
Congo	6.2	3.6
Kenya	6.8	4.2
Egypt	7.3	5.0
Botswana	13.9	11.3

Source: World Bank (1992:220-1)

This only begs the question: was Africa less tropical between 1960 and 1980, and were the peasants 'captured' then? This is voodoo economic history.

Terms of Tenancy

Neither the weather nor culture had much to do with the economic crisis of the 1980s that faced many African countries. Nor were the alleged 'policy mistakes' for, after all, the same policies had 'produced' growth up till then. The point is not to say 'internal' factors were not involved, but that the crisis was triggered by 'external' factors which exposed and reinforced the 'internal' fragilities of economies that were largely underdeveloped and characterized by a narrow and disarticulated production base, severe sectoral, spatial and social imbalances, fragmented product and factor markets, extreme openness and external dependence with a few commodities accounting for the bulk of total export earnings and government revenue, a situation compounded by an imitative modernism expressed in a 'preference for foreign experts, foreign models, standards and goods' at the expense of 'experimentation, innovation and self-reliant development' (ECA 1989:6-7), and overlaid by weak institutional capability and political authoritarianism and instability. These 'internal' structural bottlenecks and weaknesses, of course, had not been manufactured overnight, but were historical accretions rooted in the colonial economy.

One of the main reasons for Africa's greater vulnerability to the world economic crisis of the 1970s and 1980s was that many of the continent's economies were remarkably 'open'. Export earnings of low income sub-Saharan African countries represented in the 1980s 16% of GDP, nearly twice the 9% average for all low-income developing countries (ILO 1987:66). As a proportion of cash GDP, the export ratio was even greater. Africa was the most commodity-dependent region in the world, with many countries relying for over 90% of their export earnings on only one or two commodities. In short, African economies were basically export-driven. A great deal of the economic contraction in African economies in the 1970s and 1980s was, therefore, directly related to export performance. Table 4 shows the growth of trade between 1960 and 1985. It can be seen that African exports grew faster than the exports of other developing countries only in the period 1960 to 1970. After that African exports grew more slowly. Between 1975 and 1985 Africa's export and import trade grew at a lower rate than the world average. Exports experienced negative growth between 1980 and 1982, as did imports between 1982 and 1986. The decline in imports was a definite response to falling exports (ILO 1987:8). Import compression, in turn, led to reduced utilization of existing capacity and steep declines in further investment and output (Hawkins 1987).

The shrinkage in Africa's export trade can be attributed to the fact that the markets for African exports contracted sharply. It ought to be noted that African exports consist predominantly of primary commodities, which have historically tended to fall, especially in times of economic crisis. In fact, between 1965 and 1985 primary products increased their share in the commodity composition of sub-Saharan Africa's exports from 92% to 94%, as compared to the decline from 80% to 60% for the developing countries as a group (World Bank 1987:222-3). The second notable feature of Africa's export trade is that it is predominantly conducted with the IMECs.

Table 4: Growth of Trade, 1960-1985

Country Group	Exports			Import		
	1960-70	1970-75	1975-85	1960-70	1970-75	1975-85
World	9.2	25.9	8.4	9.1	25.5	8.6
Developed market economies	10.0	23.3	8.2	10.2	24.2	8.4
Developing Countries	7.3	36.0	8.4	6.5	30.0	9.5
Africa	9.2	28.7	5.3	4.9	29.2	6.2

Source: UNCTAD (1986:14-19)

Again, trade between the region and the latter even increased from 78% in 1965 to 80% in 1985, while the average for developing countries as a whole dropped from 67 to 63% over the same period (World Bank 1987:226-7). Thus, Africa was more closely tied to the IMECs than the other 'developing' regions and was therefore more exposed to their economic recession in general, and their depressed demand for primary products in particular.

The result was that the terms of trade for Africa fell more sharply than for other regions in the 'developing' world. Having remained practically constant between 1965-73, they then declined by 23% between 1973 and 1982 (ILO 1987:68). In 1983 they fell by 0.1%, then rose by 1.8% in 1984, then fell again in 1985 by 3.3%, and in 1986 the fall was a staggering 26.3% (World Bank 1987:176). The sharp fall in 1986 alone was translated into a loss of $19 billion. Needless to say, this massive loss of foreign exchange earnings due to falling prices of exports far outstripped all inflows of foreign funds into Africa whether as investment, loans or grants.

The result of the declining terms of trade was growing balance of payments deficits for many African countries. For example, between 1970 and 1985 the deficit of Cameroon rose from $30 million to $165 million, that of Egypt from $145 million to $1,895 million, Kenya's from $49 million to $208 million, Morocco's from $124 million to $889 million and Senegal's from $16 million to $338 million. Over the same period other countries moved from surplus in the balance of payments to deficit. Madagascar, for example, saw its surplus of $106 million turn into a deficit of $151 million, Mauritius' surplus of $8 million dropped into a $30 million deficit and Zambia's $108 million surplus melted into a $98 million deficit. By 1985 the gross international reserves of sub-Saharan African countries stood at $6.3 billion, which represented 4% of the developing countries' total, and were enough for only 1.7 months of import coverage, as compared to the 3.5 months average for the developing countries and 4.2 months for the IMECs (World Bank 1987 230-1). The current account balances of Africa and other regions between 1985 and 1990 are shown in Table 5. The deterioration in the balance of payments forced African countries to begin borrowing heavily from the developed capitalist countries. The debts were thus accumulated to compensate for the worsening external environment rather than to add to productive capacity.

Table 5: Current Account Balances, 1985-1989 ($ billion)

Country Group	1985	1986	1987	1988	1989	1990
	Actual	Actual	Actual	Estimated	Forecasts	Forecasts
Developed market economies	-26.6	13.6	-12.7	12.7	-7.0	5.1
Developing countries	-21.5	-38.8	-9.8	-28.0	-21.8	-20.5
Of which:						
Africa	-8.9*	-14.9	-10.9	-22.9	-24.2	-26.1
West Asia	14.0	-1.5	16.4	10.2	13.3	16.0
Latin America	-9.08	-22.0	-17.0	-17.0	-12.8	-12.7

*Excludes Nigeria

Source: UNCTAD (1988:7; 1989a:21)

African countries and others in the 'developing' world were encouraged by the low interest rates offered during most of the 1970s when the international capital markets were awash with surplus funds from the oil producing countries. Excess liquidity in the markets was also a byproduct of the economic recession in the industrialized countries which was partly caused by a drop in the rate of profit. So capital was looking for greener pastures abroad. Banks zealously competed with each other to lend to desperate 'Third World' governments. They were encouraged by creditor governments 'to whom such voluntary and market-based recycling seemed an efficient and costless way of shifting the oil-exporting countries' surpluses and of further privatizing the international financial system'. The banks 'were able to step up their lending to developing countries by using the technique of lending at variable interest rates and on medium term. This promised to pass the interest-rate risk onto the borrower, and to limit the funding risk' (UNCTAD 1988:92). The Third World debt mountain began to grow higher and higher (Frank 1981:Chapter 4; Castro 1984:Chapter 4).

The bubble burst from the late 1970s as primary commodity prices tumbled to their historic low and, simultaneously, interest rates rose to their historic high. This was a direct result of the adoption by western governments of policies toward combating inflation and the recession of 1979-82. Consequently, the need for debt by the 'developing' countries became more pressing than ever, while its cost escalated. The making of a crisis had begun. For the debtor countries real interest rates rose by about 12% from 1976-78 to 1981-82. Between 1980 and 1982 interest payments increased by 50% in nominal terms and 75% in real terms. The debt indicators worsened accordingly. 'Third World' debts rose at an annual rate of 16.8% between 1978 and 1982, while their economies grew at an annual rate of 3.2% and exports fell by 1.7% during the same period. More devastating was the fact that debt service grew at an annual rate of 23.3%, that is at a greater rate than the increase of the debt itself. This was an indication that the loans were being incurred in order to repay previous loans.

It was usury beyond Shylock's wildest dreams. *South* (November 1987:4) angrily editorialized:

> In 1980 the [Third World's] total debt disbursed and outstanding was US $429.6 billion. Between 1980 and 1986 the developing countries repaid US $658 billion in principal and interest. This means that the total repaid by the developing countries was 53% more than the outstanding debt in 1980. But these repayments did not reduce the Third World's debt liability: that nearly doubled. The amount disbursed and outstanding rose to a staggering US $775 billion in 1986, after lenders had gorged themselves on all the money paid since 1980. This situation has arisen

because the creditors never agreed to charge a fixed rate of interest on their loans and the debtors were exposed to all the vagaries of currency fluctuations and floating interest rates in the lending countries. The debtors could never assess the amount they would be called upon to pay year after year after year. No usurer in history has managed to swing a better deal than this. The burden of inflation in creditor countries was also transferred to the debtors, who were already groaning under extremely unfavorable terms of trade.

Thus, amid all the talk of 'Third World' debt one small fact has gone unnoticed 'that the debt has already been fully repaid...In aggregate terms the principal borrowed by the developing countries has been repaid at least twice'. In fact, the debtor countries had now become net exporters of capital to the creditor countries. In 1987 the net outflow from the 'Third World' debtors reached $38.1 billion. This rose to an estimated $43 billion in 1988 (*Daily Nation* 20-12-88:10). And the debt continued to rise inexorably. By 1988 'Third World' debts totaled a stupefying $1.2 trillion. Never before in history had there been such a drainage of resources from the underdeveloped world to the capitalist centres.

Many African countries were caught in this avalanche. All the achievements they had made since independence, as well as their prospects for future development, were under the threat of erosion by the rising debt flood. Africa's debt more than doubled between 1980 and 1990, and continued to climb in the early 1990s as shown in Table 6. The debt took a heavy toll as many countries spent an ever increasing share of their export earnings and investable surpluses to pay back debts. In the meantime, although direct foreign investment rose, it was too minuscule to offset the rising debt service ratios. It can be seen that in contrast to the mostly commercial bank debt of Latin America and North African countries and Nigeria, the bulk of sub-Saharan Africa's debt was owed to governments or international development banks and agencies. Needless to say, there were considerable country variations. In 1986 external debt as a ratio of GNP ranged from a low of 23-30% for Chad, Rwanda and Lesotho, to a high of 356.5% for Zambia, 234.6% for Mauritania and 178.1% for Congo. The ratio of public debt service to exports was less than 10% for 14 out of 38 countries and over 30% for six. The rest hovered in between (United Nations, 1988:57).

A growing number of countries found themselves unable to service their debts, as indicated in Table 7 by the skyrocketing arrears. The debt burden was becoming unsustainable.

Table 6: Africa's External Debt ($ billion)

Year	1980	1986	1990	1991	1992	1993	1994**
All Africa*	137.8	138.5	291.5	294.8	292.0	296.9	313.0
SS Africa	84.3	231.3	192.2	196.3	195.4	200.4	210.7

* Except Libya and South Africa. ** Projected.

Source: *Africa Recovery* (November, 1995:10).

Table 7: Sub-Saharan Africa's Debt Profile ($ million)

		1980	1990	1993
Debt Stock	Total Debt	84,319	192,202	200,388
	Of which: Long-Term	58,739	156,972	155,890
	Multilateral	7,531	37,120	45,455
	Bilateral	17,038	68,865	71,291
Debt Ratios	Debt/exports (%)	91.5	230.2	253.6
	Debt/GNP (%)	30.7	73.4	73.2
Debt Relief	Total amount rescheduled	n.a	6,555	1,216
	Debt stock rescheduled	n.a	391	1
	Principal rescheduled	n.a	3,984	735
	Interest rescheduled	n.a	1,841	383
	Principal canceled	n.a	1,531	510
	Interest canceled	n.a	81	73
Debt Service	Total paid	9,024	15,149	12,007
	Total due	n.a	25,187	22,004
	Total arrears*	1,364	27,386	49,280
	Of which: official creditors	570	17,214	33,840
Net foreign direct investment		33	854	1,782

*Combines arrears on principal and interest.

Source: *Africa Recovery* (June 1995:11).

Between 1980 and 1987 only 12 of the 44 sub-Saharan African countries were able to service their debts as scheduled, without either running up arrears or resorting to rescheduling agreements with their creditors. Governments, international agencies, and academics spent sleepless nights devising strategies for dealing with the debt crisis. A wide range of prescriptions were offered.

First, the 'radicals' urged debt cancellation. They argued that the debts were simply unpayable (George 1988; Mistry 1988). This was vehemently opposed in many quarters. Such a move, it was said, would destroy the international financial system. For their part, the IMF and World Bank rejected cancellation of their debts, even in part, arguing that to do so would gravely undermine their credit rating in the capital markets and make it impossible for them to continue lending to the debtor countries (*Daily Nation* 24-10-88:6, 10, 12-13). The UN-appointed Advisory Group on Financial Flows for Africa agreed with this assessment and strongly recommended against 'cancellation of even a part of non-concessional loans from...these agencies' (United Nations 1988:26). However, countries like Canada, Scandinavia, and the Netherlands eventually decided to write off some of their African debts. For example, in 1987 Canada canceled $91.8 million owed by seven Francophone African countries, and $103 million owed by five Commonwealth African countries (*South* November 1987:4). These cancellations, often publicized with great fanfare, had a negligible effect, as the above tables indicate.

No African country, nor any country in the Third World for that matter, dared to unilaterally repudiate its debts. Instead some resorted to the second strategy of debt moratorium. This commonly took the form of placing limits on debt service payments. Latin American countries set the trend. In late 1986 Zaire, once the star pupil of the IMF, became the first country in Africa to slap a ceiling on debt payments to 10% of its export earnings and to no more than 20% of the country's annual operating budget (Buren 1986:15; Misser 1987; Islam 1987). But this policy was later abandoned when the government reached a new IMF agreement in May 1987. In January 1986 Nigeria announced that it would allocate only 30% of its export earnings to servicing debts, but it also later abandoned the policy (Duodu 1987). Zambia's defiant revolt against the IMF in May 1987 was followed by the announcement that debt service would be limited to 10% of the balance left after imports of essential goods (Lapper 1987a:30). Also in May 1987 Côte d'Ivoire, for long a model of economic success in Africa, sent shockwaves with its announcement that it was suspending payments on its foreign debt (Bourke 1987).

Thus, on balance, few African countries opted for debt moratorium. The strategy of forming a debtors' cartel in order to enable the debtors to negotiate with the creditor countries from a position of strength met with even less success. Nyerere was one of the most ardent and articulate proponents of this view. 'We', he argued referring to the indebted 'Third World' countries, 'are not completely without power... [we] have a certain kind of power: the power of debt' (*South* August 1984:36). Negotiations between debtors and creditors continued to be conducted on a case-by-case basis between each debtor country acting individually and its various creditors acting in concert.

Other prescriptions that were not followed up included the proposal for the setting up of an international debt redemption agency to buy all the developing countries' external debt, although it was never made clear what the agency would do with the debt thereafter; and the proposal for the cessation of sovereignty over part of the debtor country as debt repayment (Mwarania 1988). But something smacking of the last proposal came to be accepted in the form of debt-conversion schemes, pioneered in Latin America, which included debt-equity swaps, debt buy-backs, and debt securitization. The first strategy involved the exchange of debts for equity in public or private sector enterprises. 'The most common form of swap involves a bank selling a debt owed by a country to a transnational seeking to invest there. The debt is bought at a discount and then exchanged for local currency at the central bank' (Lapper 1987b:95; Westlake 1987). In the debt buy-back arrangement a debtor country would be allowed to buy back, usually at a discount, part of its foreign debt using foreign trade surpluses. A debt-securitization operation 'involves the retiring of existing unsecuritized debt in exchange for new debt that is backed by collateral, either for interest payments or for principal repayment or for both' (African Centre for Monetary Studies, 1992:8). In 1987 the African Development Bank came up with a debt securitization plan to covert official and commercial medium-term and long-term obligations, excluding debt to the IMF and World Bank, into securities with at least 20-year maturity at a fixed below-market interest rate based on the perceived ability of the borrower to pay. By the end of 1988 this plan had yet to win acceptance from the major creditor countries (Lowe 1988; Alagiah and Bourke 1988).

For the 'Third World' as a whole, between 1984 and 1987 about $6 billion of debt was converted into equity. But this represented only 2% of the total commercial debt of the 15 highly indebted developing countries (UNCTAD 1988:106). Debt equity swaps could only be applied to commercial bank debt, so this was not a viable option for most sub-Saharan African countries, whose debt was with governments and international development agencies, except for Nigeria which launched its own debt equity swap programme in the late 1988 (Jason 1988), and the North African countries. The size of the secondary African debt market was $500 million for the 1980-86 period, rising to $800 million for the 1986-89 period (African Centre for Monetary Studies, 1992:28).

The major creditor countries preferred debt rescheduling as the main means of debt relief. Rescheduling of official development assistance (ODA) loans was done at the Paris Club. Between 1980 and 1986, 22 sub-Saharan African countries obtained 55 Paris Club debt rescheduling arrangements, and many more were negotiated in 1987 (United Nations 1988:Chapter 3). Almost invariably, once a country renegotiated its debt, it found itself, sooner or later, repeating the exercise. For example, between 1980 and 1986 the number of multi-lateral debt relief arrangements was as high as nine for Zaire, eight for Sudan, and seven each for Senegal and Madagascar, six for Togo, five for Niger, Liberia and Côte d'Ivoire, and four for Zambia, Sierra Leone and Nigeria (United Nations 1988:56-7). Thus, rescheduling offered only temporary relief. In fact, according to the UN's Advisory Group on Financial Flows for Africa, 'when interest payments are rescheduled, the size of the debt grows rapidly. Over a few years of serial rescheduling, the interest burden increases substantially. For example, such capitalization of interest has added as much to Zaire's debt in the past ten years as net new borrowing' (United Nations 1988:23).

It was partly in response to the growing criticisms against the way the creditor countries were handling the exploding Third World debt that the so-called Baker plan, named after the then US Treasury Secretary, was unveiled. It promised to provide an extra $20 billion over three years from the commercial banks to the 15 biggest debtors, $3 billion a year rise in lending by the World Bank, $2.7 billion from the IMF trust fund for countries with a per capita income less than $550, and a pool of $5 billion for the poorest countries, chiefly in Africa, to be administered by the IMF and the World Bank. But it was met with a cool reaction from commercial banks, donor nations, and the debtor countries themselves who baulked at the implied requirement that all those benefiting from the plan should submit to the strictures of Reaganomics (Westlake 1985; Lycett 1986). The plan floundered and by the end of 1988 Baker had not even secured the additional funding either from the World Bank or the US Congress (Martin and Westlake 1986; Westlake 1988). The US also put up obstacles against improving debt relief measures. At the 1988 Toronto Summit of the G7 some members suggested the extension of repayment periods, the conversion of ODA loans into grants, partial cancellation or waiving of debt repayments. The US only agreed to the extension of repayment periods, a proposal that won grudging acceptance from the IMF (Hodges 1988; Sparks 1988). But extension of payment period meant that the debt, to quote the UN Advisory Group, was merely 'pushed into the future with compound interest. It is tantamount to taking new credit on market-related terms to discharge a maturing liability' (United Nations 1988:23). The Advisory Group itself proposed increased financial resources to Africa, apart from the adoption of better and more comprehensive debt relief measures, as holding the key to Africa's debt crisis. But their stipulated net annual inflow of $5 billion was criticized by African countries as being too inadequate (*Africa Recovery* June 1988:9).

The Clinton Administration showed a little more sympathy and dropped earlier US objections to debt relief for the hardest hit 'developing' countries and proposed its own modest debt relief to 'be costed against the Foreign Aid Budget' (Laishley 1993:3) In the 1990s the most significant debt relief scheme was the Naples initiative agreed to by the G7 at their July 1994 Naples Summit which permitted 'up to 67% reduction in export credit debt stock or debt service (principal and interest) owed to Paris Club creditors'. Ostensibly targeted at the poor, those 'with a debt-to-export ratio of 350% or more (on a net present value basis) or a per capita income of $500 or less', the conditions, set by the IMF were so tough that by the end of 1995 only one country, Uganda, had qualified, but only to discover that 'for all the talk of 67% relief, only 26% of its Paris Club debt was written off' because the Club limited 'relief to debts contracted before the date of a country's first Paris rescheduling, known as the cut-off date'. For Uganda that was in 1981 when its debt was still relatively small (Katsouris 1995:11-12).

Thus, by the mid-1990s no plausible and effective resolution to Africa's debt tragedy had appeared. Rescheduling, with all its shortcomings, remained the most widely used strategy of debt relief. Rescheduling carried another heavy price tag. In order for a country's debts to be rescheduled either at the Paris (official debts) or London Club (commercial debts) that country first had to reach an agreement with the IMF and the World Bank. Thus, the debt crisis forced African countries into the clutches of these agencies. It was a deadly embrace. In exchange for their seal of approval and loans, the IMF and the World Bank exacted their pound of flesh from the emaciated African economies in the form of structural adjustment.

Lethal Medicine

One after another African countries embarked on the rough road to structural adjustment. Seeking to retain some pride, many claimed that these programmes were home-grown. A few did, indeed, try to design their own programmes in a vain effort to keep the World Bank and IMF at bay. And there were mounting domestic pressures for reform, although the visions differed. Some were for a more market-oriented future, others for a more regulated and egalitarian one, and there were those who dreamt of a social democratic developmentalist order. Manufacturers wanted less bureaucratic red-tape and perhaps some deregulation and devaluation, workers aspired to a 'living wage' and a level playing field for collective bargaining, and peasants yearned for fairer prices for their produce. Initially, therefore, several groups welcomed structural adjustment, but some turned against it as the programmes began to bite. The hand of the international financial institutions and western governments behind these programmes became more visible as time went on, jogging uncomfortable memories of colonialism even among the domestic supporters of free-market development.

Increasingly, structural adjustment smacked of surrender of national economic sovereignty to the IMF and World Bank. Zaire was among the first countries in Africa to bow to the IMF. It secured its first IMF standby loan in 1967, forced by economic paralysis after years of civil war and strife, growing authoritarianism and corruption, and falling terms of trade, exacerbated by the sharp decline in the 1970s in the price of copper, the country's main export. The West was keen to prop up Mobutu's regime, which it saw as a bulwark against communism in the region, especially after Angola's independence under the MPLA, and so western 'aid' continued to flow, more IMF standby agreements were signed, and the country sank deeper into debt and structural underdevelopment and dependency (Young and Turner 1985; Islam 1985; Depelchin 1992; Tshishimbi 1994).

Zambia embraced the IMF when its copper goose could no longer lay prosperous eggs. Copper accounted for over 90% of Zambia's export earnings. So when its price collapsed in the 1970s because of reduced demand in the recession-hit industrial markets, the Zambian economy was severely shaken. Between 1974 and 1975, for example, Zambia's terms of trade fell by 46%, her foreign exchange earnings dropped by more than 40%, and the balance of payments deficits grew to 30% of GNP, while government revenue from minerals fell by four-fifths. Initial government responses to the crisis, which included 'imposing a regime of controls in the economy... increased inefficiency and actually contributed to the worsening of the crisis' (Mwanza et.al. 1992:120). The government turned to the IMF and secured standby agreements in 1973, 1976, 1981 and 1986 (Mwanza 1988; Alagiah 1985a). Disappointed with the results, Zambia suspended the IMF/World Bank programme and tried its own brand of structural adjustment, but the pressure was such that it was forced to return to the fold a couple of years later 1989 (Chiposa 1987c, 1988).

Sudan reached the first agreement with the IMF in 1978 after years of resisting the move. But the economy was in deep straits due to a steep fall in the prices of its commodity exports, huge expenditures on state security and administration and on ill-planned development projects. By 1978 Sudan's external reserves had fallen to $28 million, enough for only a few days' imports. The Saudi Arabian government promised Sudan a big loan on condition that the government accepted an IMF agreement. Sudan's tortured relationship with the IMF began (Osman 1985; Hansohm 1986; IFAA 1987).

Uganda, the faded 'pearl of Africa', sought IMF salvation from 1980. Idi Amin's regime had left Uganda devastated. Obote's 'second republic' accepted the IMF diktat, which was assiduously followed by subsequent regimes (Nabudere 1980; Mamdani 1983; Pike 1985a, 1985b). But the IMF was not a client of politically unstable countries only. Malawi, Côte d'Ivoire, and Kenya, once touted as models of free-enterprise development, also sought IMF help to deal with their balance of payments problems. Malawi's 'economic success' which was based on a thin productive base of export estate agriculture, began catching pneumonia from the chilly winds of declining terms of trade which fell at annual rate of 15.5% from 1977, so that by 1980 they were at less than 56% of their 1970 level, and plummeted further in the 1980s, a plight that was compounded by an influx of nearly a million refugees from war-torn Mozambique and the cutting off of Malawi's traditional transport corridor through that country. In late 1979 Malawi embarked on an adjustment programme (Mhone 1987; Sahn and Arulpragasam 1994). Côte d'Ivoire turned to the IMF from 1982 following the drastic fall of its commodity prices, principally coffee and cocoa, which had sustained the 'economic miracle'. The country increasingly found it difficult to repay its debts. In order to reschedule these debts, it needed the IMF's blessings (Schissel 1985; Durufle 1988; Lambert, Schneider and Suwa

1991). Kenya, too, welcomed the IMF's iron grip from the early 1980s when confronted with sagging earnings from coffee and tea and soaring oil prices and rising costs of other imports, which played havoc with its import-substitution industries and increasing population (Coughlin and Ikiara 1988, 1991; Miller and Yeager 1994).

Ghana joined the IMF bandwagon from 1983. The cocoa-export dependent economy was in poor shape. In the preceding decade or so Ghana's terms of trade had fallen drastically and the country had been rocked by coups and counter-coups. Between 1970 and 1982, for example, per capita incomes fell by 30%, cocoa production fell by over half, real export earnings shrank by 52%, import volumes by a third, inflation averaged 44% a year, and the cedi had depreciated sharply. The forced repatriation of hundreds of thousands of workers from Nigeria in 1983 and 1985 only worsened matters. Rawlings' 'radical' regime desperately wanted to reverse the slide and decided to swallow the IMF medicine and proceeded to become the Fund's favourite patient (Hodges 1988; Hutchful 1988a; Essuman-Johnson 1988; Hansen and Ninsin 1989; Gayi 1991; Alderman 1994).

'Socialist' Tanzania fought a long but losing battle against the omnipresent IMF. The external shocks of the 1970s, combined with the internal flaws of the ujamaa experiment, weakened the economy. Up till 1978 Tanzania enjoyed a healthy balance of payments position. In fact, it had accumulated large external reserves. 'The IMF advised Tanzania that the reserves were too large, and she should spend them on imports. She did and by 1979/80 she had a balance of payments crisis and had to seek an IMF loan' (IFAA 1987:12). The 1980 agreement broke down and in the following few years inconclusive talks were held with Tanzania baulking at IMF 'conditionalities', especially devaluation, import liberalization, lifting of price controls, and privatization all of which run counter to the country's official ideology. It tried its own reform programme, but western donors made it clear that relief on its existing debt or new assistance would only be forthcoming when a deal was made with the IMF. Tanzania finally succumbed in June 1986 (Alagiah 1985b; Buckoke 1985; Rake 1986; King 1986; Malima 1986; Biermann and Wagao 1986; Campbell and Stein 1991; Sarris and den Brink 1994).

'Oil rich' Nigeria also put up spirited resistance against the IMF. Nigeria was driven to penury when the price of oil, its main commodity export, fell sharply at the turn of the 1980s, so that the state could no longer finance its ambitious development projects, satisfy the accumulation appetites of its fractionalized bourgeoisie, adequately assuage and manipulate the rising regional, ethnic, and class

312   MANUFACTURING AFRICAN STUDIES AND CRISES

differentiations and struggles, and, of course, repay debts incurred during the glorious years of the oil boom (Bangura 1986; Olukoshi 1995). In 1979 Nigeria boasted of external reserves of $5.1 billion. By 1983 the country had an external debt of about $20 billion. It asked the IMF for a $2 billion loan and the IMF brought its usual package of conditionalities. A national debate ensued and the country resoundingly said 'NO' to an IMF deal (IFAA 1987; Hear 1986a; Jason and Hear 1986). But Babangida's crafty regime brought in the IMF through the backdoor. Nigeria announced a structural adjustment programme with the World Bank, which amounted to an IMF deal without IMF money (Smith 1985). The 1987 budget announced a structural adjustment programme which was said to have started from July 1986. The SAP contained all the IMF prescriptions. This was the minimum required by Nigeria's creditors to reschedule the country's huge debts (Weir 1986). Satisfied with the SAP, they proceeded to do just that (Ankomah 1987; Akinifesi 1988; Adejumobi and Momoh 1995).

Algeria, hit by falling oil prices, was another country that launched a structural adjustment programme with all the austerity measures of an IMF programme but without taking IMF money. The Algerian state 'had worked hard for social and national integration, dishing out incoming export earnings throughout the society', a mission it could no longer discharge especially from the mid-1980s 'following the fall in both oil prices and the dollar exchange rate', which broke open the internal structural imbalances and social tensions long concealed by state populism (Chikhi 1995:321, 323).

And newly liberated Zimbabwe, coming to independence in 1980 when the medicine of structural adjustment was being dispensed with abandon and 'socialism' was in retreat everywhere, found itself buffeted between rising social expenditures to undo a century of racial capitalism, expanding budgetary and balance of payments deficits and foreign exchange shortages because of falling commodity prices and administrative over-regulation inherited from the UDI era, and low levels of investment. No sooner were the independence celebrations over than structural adjustment arrived like a premature hangover (Mandaza 1986; Stoneman 1989; Kadenge 1992).

Altogether, by the end of 1987, 30 countries in sub-Saharan Africa alone were undertaking economic adjustment programmes or were expected to resume them with the IMF and World Bank (United Nations 1988:5). The African economic and political landscape was fundamentally transformed. It was harsh medicine, applied with uncompromising uniformity everywhere, oblivious to the differences of history or the prevailing economic conditions. The policy objectives were pure textbook neo-classical economics: reduction in the size of the public sector, elimination of price distortions in various sectors of the economy, and promotion of trade liberalization and domestic savings. These policies were to be achieved through the deregulation of prices of goods, services and factory inputs to minimize the role of the state in resource allocation; imposition of tight controls of money supply and credit and of high interests to encourage savings; devaluation of currencies to 'realistic' levels to discourage imports and stabilize balance of payments; aggressive pursuit of 'comparative advantage' to increase export earnings, and of an 'open door' policy to attract foreign investment and encourage efficient industrial development; and budget reductions to redress fiscal imbalances. Privatization became the leitmotif of structural adjustment. It was hailed as the panacea of Africa's economic ills. Freed from the suffocating grip of the state and the unproductive cultures of affection, the market would perform its magic and propel African countries from economic crisis to recovery and prosperity.

It was a cruel fiction. Sustained economic growth did not materialize, the rates of investment remained low, indeed, decreased, budget and balance of payments deficits persisted and even widened, the debt mountain continued to grow. In a systematic theoretical and empirical critique, the Economic Commission for Africa (1989) crystallized the African opposition to SAP and outlined an alternative reform strategy which, despite its greater intellectual and political merits, was unfortunately ignored because, unlike the IMF and the World Bank, the ECA could neither back its advice with the carrots of cash nor the sticks of sanctions from international financial markets[2]. Drastic budgetary reductions and cuts in subsidies on social services and essential goods, the ECA argued, led to massive public sector retrenchment and undermined human capital and the enabling environment for future development; the indiscriminate promotion of export production stymieed food production and self-sufficiency and could result in over-supply and falls in prices; across the board credit squeezes contributed to overall economic contraction and capacity underutilization, which accentuated shortages of critical goods and services; excessive devaluation fueled inflation, capital flight, worsened income distribution, and reinforced the production of traditional exports; high interest rates shifted the economy towards speculative and trading activities; total import liberalization reinforced external dependency and threatened infant industries; and excessive dependence on market forces and privatization jeopardized social welfare and human conditions (ECA 1989:37-8).

Elusive Recovery

The evidence, indeed, showed no appreciable difference in the growth rates of the 'strong adjusters', 'weak adjusters', or 'non-adjusters', despite the World Bank's determined statistical acrobatics to manipulate the figures (World Bank and UNDP 1989; Zeleza 1989; Khan 1993)[3]. In countries where the infrastructure was crumbling, thus raising transaction costs for farmers, and where shortages of basic agricultural inputs as well as consumer goods was rampant, thereby offering little opportunity to translate higher prices into incentive goods, raising producer prices had little effect on production, except to increase the food prices for all those who had to purchase food, which did not simply consist of urban dwellers as the thesis of 'urban bias' would have us believe, but included the growing numbers of landless rural workers and the rural poor, many of whom already suffered from high levels of malnutrition (ILO 1987:Chapters 1 and 4). Devaluation also did not have its intended effects. Many African countries devalued their currencies by several hundred percentage points; indeed, Ghana devalued the cedi by 8,227% between April 1983 and October 1988 (Essuman-Johnson 1988:9), and the once mighty Nigerian naira plunged by about 300% between September 1986, on the eve of the introduction of the second tier foreign exchange market, and August 1988 (Anyandike 1988:50). This resulted in more skyrocketing inflation, previously rare in Africa, and falling standards of living than in increased production, for world demand for many African commodities remained stagnant.

Trade liberalization did not provide stimulating competition for domestic industry either. Rather it brought a flood of cheap imported goods, which undercut local manufacturers. In Tanzania, for example, it soon transpired that as foreign goods flooded the market after the adoption of trade liberalization 'locally manufactured products, especially in textiles, began to pile up in shops unsold' (Makaranga 1987:4), while in Nigeria, there were reports that 'closures are rampant, with manufacturers often blaming the difficulties on the reduction of protectionist barriers and on the sharp cost increases for imported raw materials and spare parts brought on by devaluation' (Harsch 198814). It was a refrain heard the length and breadth of Africa. Trade liberalization threatened to replace the difficulties of ISI with de-industrialization and return Africa to the colonial days of unadulterated primary production.

Privatization fever, which gripped many countries, also failed to deliver as much as was promised. Zambia withdrew the monopoly status enjoyed by the National Marketing Board in 1986 (Chiposa 1986) and proceeded to sell several parastatals to foreign companies, some of whom, however, got cold feet after Zambia broke with the IMF in May 1987 (Chiposa 1987a, 1987b). Togo's ambitious privatization scheme started with the leasing of the National Steel Corporation to an American investor (Weir 1985), and eventually encompassed state and mixed-economy companies (Njondo 1988). In Kenya the government signaled a major policy shift towards divestiture and privatization of parastatals with the dissolution of the National Construction Corporation and the flotation of shares of the state-owned Kenya Commercial Bank and National Bank of Kenya in 1988 (Njururi 1988). Also in 1988, Madagascar announced its intention to privatize state-owned banks following reforms already instituted in industry, commerce, and agriculture (Ranaivosoa 1988a, 1988b). Oil-rich Gabon joined the fray with the offer of such flag carriers of state property as Air Gabon and oil concessions to private capital (Karakatounian 1988). Tiny Gambia also entered the road towards privatization with the divestiture of all state shares in banks, trading, insurance, and tourism companies (N'jie 1986, 1987, 1988). And West Africa's economic giant, Nigeria, set in motion the biggest sale of them all, with over a hundred companies and institutions in virtually all fields offered for either full or partial privatization (Hear 1986b; Thompson 1988).

Privatization threatened to lead to economic denationalization in many African countries because the people most able to buy the huge parastatals tended to be foreigners. With some notable exceptions, the indigenous capitalist class was, in the main, too weak to compete effectively with foreign capital in the new and much-vaunted 'open market'. Given the desperation for foreign investment most of the privatized enterprises were fire-sales. The benefits to the ordinary people were not always obvious. In many cases the state monopolies were simply replaced with private monopolies, whose activities were sometimes even more harmful. For example, the withdrawal or drastic reduction of state involvement in agricultural marketing and services had detrimental effects on small-scale farmers, the very people who were supposed to have benefited. In Senegal, privatization of the state agency responsible for irrigation and development in the Senegal River basin, SAED, left many peasants without credit and other services. The result was that during the 1987/88 season thousands of hectares could not be put into cultivation and of those that were, many received insufficient amounts of fertilizer. Similar disruption followed the dissolution of Nigeria's six commodity boards in late 1986. 'The abrupt end to quality control procedures in the purchasing of cocoa in the 1986/87 cocoa season resulted in Nigerian cocoa losing its traditional premium on the world market', which cost the country one fifth of its revenue of that year's crop (Harsch 1988:13; Keen 1988). The damage privatization can cause for industry are no less serious. A UNIDO study warned Côte d'Ivoire 'that if the government were to withdraw too quickly from investment, the private sector might pull out as well. By retaining a stake', UNIDO argued, 'the government could continue to encourage both domestic and foreign investors' (Harsch 1988:13). In other words, untrammeled privatization ignored the naked realities of monopoly capitalism. Perhaps that is why Côte d'Ivoire's privatization drive began to slacken when government authorities became fully aware of its implications, and they now stressed that there was no 'systematic program to denationalize Côte d'Ivoire's state and parastatals corporations' (Bourke 1988:45).

African countries, of course, were not affected by, and did not respond to, the structural adjustment medicine uniformly because of their varied economic, political, ideological, and social constitutions. There were four crucial determining and differentiating factors. First, the fiscal basis of the state, whether it was a 'rentier' or a 'merchant' state (Mkandawire 1987, 1995). In the 'rentier state', in which the state relies on substantial 'external rent' from state-controlled mining production enclaves, the erosion of 'legitimation' expenditures from declining terms of trade and deteriorating economic conditions was likely to have different political and social consequences compared to the 'merchant state', where state revenue is derived largely from domestic taxes and import and export taxes and the state pursues both extractionist and productionist functions. A critical subtext to this was whether or not, in fact, the state had power over its own currency. The Francophone countries belonging to the West African Monetary Union had no power to issue their own currency, unlike their Anglophone counterparts. The state was, therefore, denied 'the possibility of reducing wages by resorting to inflation taxation' through devaluation which, predictably, infuriated the World Bank (Mkandawire 1995:42)[4].

The way the fiscal crisis was played out depended on the social and discursive architecture of the state itself, that is, its administrative capacity, the quality of the leadership, composition of the ruling coalition, the balance between the political and technocratic elements in the state apparatus, and the nature of the dominant discourse (Hutchful

1995). In regimes of accumulation and power where the national bourgeoisie had footholds in both the private and public sectors, as in Kenya and Côte d'Ivoire, the state was less undermined by adjustment than in regimes where the bourgeoisie was concentrated in the public sector as was the case in Tanzania and Mozambique. Also, in the latter countries the hegemonic discourse was far more nationalist, statist, and welfarist than in the more capitalist-oriented countries, so that popular debates and political responses to structural adjustment were far more tortuous and the policy shifts more painful. Moreover, the managerial capacity to undertake the reforms was less developed, with the result that external intervention in the administration of structural adjustment was far more intrusive than in the former group of countries. In other words, in so far as structural adjustment was geared at the dismantling of anti-capitalist development practices and strategies, it was the 'socialist-oriented' states, rather than their capitalist-oriented neighbours, that felt and underwent the most profound transformations.

The density and direction of the changes were predicated, third, on the disposition and differentiation of civil society, which, in turn, was transformed by the timing, tempo, and configuration of the structural adjustment programmes. The implementation of, and responses to, structural adjustment in countries where there were powerful social movements able to defend their access to state resources, and the deflation of the fiscal position of the state occurred suddenly, differed from the patterns in countries where the social movements were relatively weaker and the economic slide occurred more gradually. Hence, the contrasts between Nigeria and Ghana, Zambia and Malawi. In the 'merchant states' of Ghana and Malawi the structural adjustment programme did not provoke the policy twists and turns and immediate political opposition that occurred in the 'rentier states' of Nigeria and Zambia. Not only was civil society in Malawi relatively 'weaker' than in Zambia, the crisis unfolded more gradually than in the latter. Similarly, although civil society was equally strong in Ghana and Nigeria, in the latter the dip in economic fortunes was quite sudden compared to the former. And so the imposition of structural adjustment incited riots in Nigeria and Zambia, but not in Ghana and Malawi.

Finally, the pace of implementation and the nature of the popular reaction to structural adjustment programmes was affected by each country's geopolitical position, which often influenced the design of the programme and the stringency of the conditionalities. France, for example, was quite protective of its 'client states' in the CFA franc-zone, which were shielded from devaluation for many years after it had

become commonplace elsewhere on the continent. What that meant is
that while workers in the Anglophone countries could be paid regularly
their increasingly worthless salaries, since the state could print more
money, and the pacifying illusion of money could be maintained,
workers in the Francophone countries were often not paid, which sparked
off widespread opposition and ensured that civil servants would be in the
forefront of the anti-SAP and pro-democracy movement in Benin and
Mali as compared to Kenya and Tanzania. Geopolitically strategic Egypt
had a powerful patron in the United States, which intervened on its
behalf with the IMF in 1989, while its southern neighbour, Sudan, with
its 'fundamentalist Islamic' military regime was not so 'lucky'. The IMF
and World Bank periodically picked favourite pupils for special
treatment to demonstrate to a skeptical world that the reforms were
working. In the 1970s it was Mobutu's Zaire. In the 1980s Rawlings'
Ghana got the honour and was rewarded with PAMSCAD (Program of
Action to Mitigate the Social Costs of Adjustment). PAMSCAD
represented a grudging acknowledgment that SAP exacted social costs,
as the more 'social democratic' organs of the United Nations, such as
UNICEF, began to point out. In the more sanitized language of the
World Bank and its supporters they talked of 'winners' and 'losers', a
metaphor that burnished the Bank's self-image as a neutral umpire at a
game.

Whatever the differences, structural adjustment programmes failed to
deliver on their promises. Recovery remained as elusive as ever. African
governments increasingly felt betrayed by the ubiquitous international
community on whom they had placed so much faith. They had
swallowed the bitter medicine of structural adjustment, but the palliative
of international financial assistance had failed to materialize. In fact,
capital flows to Africa declined precipitously from 1980. While ODA
disbursements rose in nominal terms, real net flows declined. Official
flows on non-concessional terms also fell as did net flows of official
guaranteed export credits. The latter fell from an annual rate of about $2
billion at the beginning of the decade to $400 million in 1986. The sum
of private lending and trade financing was more than halved by the
middle of the 1980s. Foreign direct investment in sub-Saharan Africa
plummeted from $1.5 billion in 1981 to an annual average of $400 from
1984 (United Nations 1988:11-13). Altogether, in 1986 $18 billion came
into Africa as a whole, of which $16 billion was development assistance
and $2 billion private lending, while $34 billion came out of Africa, $19
billion in losses in potential earnings from the sharp drop in commodity
prices and $15 billion in debt service (*Africa Recovery* 1987:1). This

comes to a net outflow of about $44 million a day. Following the collapse of communism in Central and Eastern Europe at the turn of the 1990s, 'aid' and investment flows to Africa declined either absolutely or relatively. Bilateral ODA declined from $12 billion in 1990 to $10.7 billion in the following year, and sub-Saharan Africa's share of direct foreign investment among the 'developing' countries fell from 13.8% in 1982-86 to 5.3% in 1992-94 (Katsouris 1995:12).

The World Bank and IMF also pocketed the surpluses flowing out of Africa. The IMF, originally created in 1944 at Bretton Woods to provide temporary balance of payments support for the industrialized countries, expanded its activities in Africa from the 1970s. Once its stringent conditionalities were met, it would disburse funds through a series of facilities, including the one year standby facility, the three year extended fund facility, the enlarged access policy, and the compensatory financing facility. In the mid-1980s the Fund created the Enhanced Adjustment Facility ostensibly to help countries undertaking structural adjustment programmes. The World Bank was also established at Bretton Woods to help the reconstruction of devastated post-war Europe. But it soon became involved in the 'Third World', much earlier than the IMF. However, it was not until the 1970s that Africa became a major area of Bank activities, upstaging Asia and Latin America who had featured prominently in the 1960s. The Bank also underwent notable shifts in its project lending policies, moving from emphasizing public sector development in the 1960s and 'basic needs' in the 1970s, to export production at the turn of the 1980s, and before long structural adjustment. With the adoption of structural adjustment the Bank began changing its lending policies, away from loans tied to individual projects to structural adjustment loans (SALs). The first SALs were approved in 1983 and by September 1983 there were 23 of them. In 1985 the Bank introduced its widely hailed Special Facility for Africa, soon followed by the launching of the Africa Project Development Facility to help develop private enterprise on the continent. From about 1982 the Bank had began to include a 'conditionality' aspect to its loans, similar to the conditionalities of its more dreaded sister, the IMF. Thus, the two institutions became increasingly indistinguishable in their policies and practices. The Bank subordinated itself to the Fund whose draconian demand-stabilization agenda imposed on soft-headed 'socialistic' states appealed to the Reaganites and Thatcherites now running the show in the major western capitals. Together the Bank and the Fund set out to construct a global free-market system (Mende 1973; Payer 1974, 1982; Hayter and Watson 1985; Lever and Huhne 1985; Pike 1985b; Browne

1985; Harris 1986; Harrington 1986; Nicholson 1986, Westlake 1986b, 1987; Rake 1988; *Africa Events* 1988; Gibbon 1995).

Neither the Fund nor the Bank lived up to their advanced billing as possible saviors of Africa. On the contrary, they participated in the gory feast of milking Africa dry. In both 1986 and 1987 there was a net transfer of close to $1 billion from sub-Saharan African countries alone to the IMF (United Nations 1988:12). The World Bank also became a drain on the developing countries as a whole. In 1988 it took out about $600 million more in interest and debt repayments than it lent (*Daily Nation* 24-10-88:12; Friedland and Westlake 1986). A growing number of African countries found it harder to service their debts to the IMF and World Bank. By May 1988, five countries, Liberia, Sierra Leone, Somalia, Sudan and Zambia, were ineligible for use of IMF resources because of their arrears to the Fund (*Africa Recovery* August 1988:9). This was a sign that as far as these countries were concerned the IMF policies had failed.

Political Shock Therapy

The major contradiction of the World Bank and IMF structural adjustment programmes was that the state was supposed to be diminished, yet have sufficient capacity to undertake the fundamental restructuring of the economy. It other words, it was expected to roll itself back, to commit political suicide, which only succeeded, wherever this was tried through massive retrenchment of the civil service, in weakening its capacity to implement the structural adjustment reforms themselves (Picard and Garrity 1994). The neo-classical disdain for the post-colonial state was based on a profound misconception of state-civil society relations. Besides representing or being vulnerable to the interests of the so-called 'rent-seekers', the state was not suspended above civil society from which it could evaporate into the clouds of irrelevance. The competing social forces and discourses were inhered within the state itself where they jostled for supremacy. The state, in short, was not merely a bureaucratic outfit amenable to technical remodeling. It was a process, a constellation of complex social contestations and contradictions. Governments that sought to accommodate domestic pressure against various elements of structural adjustment were accused 'of lacking in "political will", the coded language for repressive capacity' (Mkandawire and Olukoshi 1995:3). They were accused of pandering to the discredited 'rent-seekers', despite the fact that any show of 'political will' was often 'directed against groups with no representation in the power structure: workers, the urban poor and the increasingly impoverished middle classes', as the real rent-seekers knew how to 'ingratiate themselves to the political authorities and adjust their business strategies to some of the requirements of the reforms' (Bangura 1995:86). Thus, authoritarianism was a structural necessity for the World Bank's and IMF's structural adjustment. In fact, until political pressures against these programmes could no longer be ignored and the democratic movements gathered momentum, the international financial institutions put all their eggs in the sturdy baskets of authoritarian rule. In fact, one senior Bank official, Deepak Lal, wistfully hailed the military for its capacity to 'ride roughshod over... special' interest groups' (Gibbon 1995:137). From the late 1980s the tune in public changed and the World Bank began talking of democratization and human rights, but the legitimacy of the new dispensation was to be measured by 'good governance', which essentially meant 'the successful implementation of the much-contested adjustment programmes' (Mkandawire and Olukoshi 1995:7). The programmes continued to be authoritarian outfits: designed in secret negotiations, implemented by unelected technocrats, and overseen by fly-by-night experts from London, Paris, and Washington.

It was a raw deal for African countries. In exchange for puny loans, which were subsequently over-repaid, the IMF and World Bank, on behalf of the world capitalist system, accorded themselves the right not only to supervise individual projects, but to manage whole economies entirely, approving their national budgets, foreign exchange budgets and fiscal tariff policies, issuing clearance certificates before these countries could negotiate with other foreign agencies, and even posting representatives to their Central Banks and Ministries of Finance and Trade. It reeked of colonialism. As during the colonial era, it was Africa's masses, the peasants and the workers, and increasingly the professional middle classes, who were paying the price with their sweat, tears, and blood, not only from the deteriorating economic conditions, but increased tyranny as well.

The structural adjustment programmes reinforced the triple crises of legitimation, regulation, and sovereignty for the post-colonial state, which strengthened its authoritarian tentacles, on the one hand, and fueled struggles for fundamental transformation, on the other, culminating in the crusade for the 'second independence', a subject examined in greater detail in the next section. State authoritarianism in post-independent Africa did not spring out of the patrimonial heads of the new rulers. It was partly derived from its colonial progenitor and the nature of the decolonization process, and partly from the imperatives and limitations of accumulation in peripheral societies (Thomas 1984; Iyayi 1986; Young 1988; Goulbourne 1979). The institutional reflexes of the colonial state were authoritarian because the state was an external imposition constructed through force, and it was an agency for the penetration of the capitalist mode of production and guarantor of its development over the articulated corpses of the indigenous modes. From the colonial state, the post-colonial state inherited its schizophrenic tendencies: strong in relation to domestic classes, and weak in relation to imperialism. The overdetermination of domestic social structures and processes by the state and the subsumption of 'national' interests to the centre survived into the post-independence era, for decolonization, while enabling the incipient national bourgeoisie to capture state power, did not lead to autonomous accumulation, which continued to be conditioned and determined by the self-valorization of international capital. In other words, decolonization did not end in the final defeat of imperialism, but merely in its metamorphosis into neo-colonialism.

The intensification of statism after independence was accentuated by the underdeveloped nature of the indigenous capitalist class, and the weak material base of the new rulers. The state became their instrument of accumulation. Being located in neo-colonies themselves, the indigenous bourgeoisie had no colonies or peripheries to loot or plunder, so they depended upon exclusively internal sources for the accumulation of monetary capital. Thus, the rationality of the state rested in promoting and sustaining accumulation for the aspiring indigenous bourgeoisie as well as the ever-present imperialist predators. Its legitimacy lay in providing more schools, hospitals, better jobs, and all those sinecures of colonial privilege to the expectant masses. So after independence the state in Africa was under enormous pressure to mediate between national capital, international capital, and the differentiated masses. It was a juggler's nightmare, and the leviathan tripped many times over. The multiple contradictions and frustrations of the neo-colonial order, with its trappings of independence built on limited sovereignty, political posturing without economic power, and Africanization without genuine indigenization, made the enterprise of state arduous. State intervention in the organization of the economic, social, cultural, and political process intensified as the contradictions deepened and became more open. The monopolization of politics by the state was justified in the wondrous name of development. Economic development became the raison d'être of the state as well as its achilles heel. Developmentalism and development planning attained the sanctity of a religious ritual. But like many such rituals, the plans increasingly lost touch with reality. Almost invariably, hardly was the ink dry on these plans when many of their projections were turned into cruel practical jokes by the capricious chief: the world capitalist economy.

The world capitalist crisis knocked the pedestals of rationality and legitimacy from under the feet of the post-colonial state. Consequently, the state assumed a progressively more precarious and openly repressive character, with frequent coups and rearrangement of ruling cliques, endless constitutional violations and revisions, systematic suppression of political, civil and human rights and basic democratic freedoms, increased militarization and constant civil strife and internal wars. In the 1970s and 1980s Africa took over from Latin America the dubious distinction of being the most coup prone continent in the world (Mazrui 1984; Frank 1981). And the sordid human rights record in many countries finally put to rest the lie of Africans' innate humanism and sense of brotherhood (Hamalengwa 1983; Eze 1984; Hutchful 1988b; Shivji 1989). Militarization as manifested in escalating military expenditures finally exploded the ideology of developmentalism. *South* (January 1985:8) notes sadly that 'while the world military expenditure increased by 30% between 1973 and 1982, military expenditure in Africa more than doubled... public expenditures in Africa in 1980 were estimated at US$26 per capita while expenditure per soldier averaged US$9,449... (and) was as high as US$37,000 for Gabon, US$36,857 for Zambia, US$20,833 for Ivory Coast, US$20,000 for Kenya and US$16,500 for Botswana'[5].

The exceptionally authoritarian posture of the state in the 1980s can be explained by the fact that Africa's capitalist classes were the least able to resolve the crisis on their own being the weakest in the capitalist world system. Structural adjustment reinforced the structural propensities for authoritarianism. In other words, like the colonial state, the post-colonial state had a 'dual mandate' to perform; it straddled 'two levels of contradiction: between the metropole and the [neo-]colony as a whole as well as within the [neo-]colony itself. It therefore bore a dual character: it was at once a subordinate agent in its restructuring of local production to meet metropolitan demand, yet also the local factor of cohesion over the heterogeneous, fragmented and contradictory forces jostling within' (Lonsdale and Berman 1979:490). The embattled post-colonial state extended and tightened its protective umbrella over accumulation to the benefit of capital, both local and foreign, while intensifying the repression of the peasantry, the working classes, and sections of the middle classes, especially students and the intelligentsia.

Conclusion

The same thing, to a different degree of course, was happening in the core capitalist countries themselves, where the repression of labour, ethnic and racial minorities was increasing. This is to suggest that structural adjustment was not confined to the 'developing' countries, let alone Africa. It was a global process, for the capitalist crisis that had generated this package of responses was itself global. These programmes entailed massive retrenchment and increased poverty. Thus, ironically, while the monetarist objective of adjustment was the removal of 'political interference' from economic life, the state had to be reconstituted and made more authoritarian in order for this restructuring to be effected. Labour's repression was deemed necessary to increase the valorization process of capital. So governments of widely differing policy and ideological orientations in the 'developed' and 'developing' countries actively suppressed labour movements and trade union organizations. The World Bank and IMF roamed everywhere actively supervising and regulating the heightened conflicts between regions, nations, and classes in the reorganization of the disintegrating post-war growth model, and in the efforts to construct a new one. But as with previous moments of global restructuring, the weaker and poorer regions, countries, and social classes were being forced to pay a relatively heavy price. Their production, commodities, and services were devalued. In short, their labour power and humanity were devalued. But as before in history, too, they began to rise up and fight back. That, essentially, is the great story of our times: the remaking of global, regional, and national capitalisms, and the struggles by exploited or marginalized nations, classes, ethnic, racial and gender groups for democratic development and governance.

During prolonged periods of crisis in the capitalist world economy far-reaching organizational changes occur which lead to shifts in hegemonic power among the core states and the creation of new or reconstitution of old peripheries. During the crisis of the late 19th century the trust and cartel movement gave way to the giant corporation, a movement that went further and faster in the United States and Germany than it did in Britain, which lost its former hegemonic position, and in the meantime Africa's incorporation into the capitalist world economy was tightened and reconstituted through colonization. In the crisis of the 1970s and 1980s, the integration of large firms with state planning and banking capital was advancing in the most dynamic global capitalist centres of Japan and the Pacific Rim, while a declining Britain and a defensive United States were clinging to, and seeking to revive, antiquated forms of 'free-market' capitalist organization. In being forced to adopt these increasingly backward organizational forms through excessive privatization and doctrinaire economic liberalization, the danger was that African countries and many others in the 'Third World' would be even further removed from the centre of the global village and the reconfigured chief's compound and become more prey to the pneumonia of deepening poverty in future bouts of global capitalist recession and reorganization. That is why the struggles against structural adjustment in Africa were so critical for the continent's future.

Notes

1. Marxist euphoria proved short-lived. The collapse of 'actually existing socialism' in the former Soviet bloc led to a profound crisis of epistemological and political relevance, although some believed new opportunities were opened up for a more open, less economistic, deterministic and deductive Marxism. Indeed, some hoped for a postmodernist post-Marxism! For succinct and fascinating soul-searching essays on post-cold war Marxism in Western Europe and North America, see Callari, Cullenberg and Biewener (1995).

2. The ECA framework built on, and advanced over, the 1980 Lagos Plan of Action and several UN initiatives, including Africa's Priority Programme for Economic Recovery (APPER). It called for adjustment with transformation and identified several major policy directions: enhanced production and efficient resource use; greater and more efficient domestic resource mobilization; improving human resource capacity; strengthening scientific and technological base; vertical and horizontal diversification. Specifically, it called for establishing a pragmatic balance between the public and private sectors; creating an enabling environment for sustainable development; shifting resources from unproductive expenditures and excessive military spending; improvements in the patterns of income distribution among different socio-economic categories of households. African countries were asked to promote food self-sufficiency; lessen import dependence; re-align consumption patterns with production patterns; and establish strong debt management systems. The importance for regional cooperation was also emphasized. It was clearly stated that for these initiatives to be undertaken and bear fruit democratization was imperative.

3. In his study done for the ILO's World Employment Program, Khan shows that out of the 55 adjusting countries world-wide, 'only seven appear to have fulfilled the absolute standard of successful adjustment. Of these seven countries the Republic of Korea is the only unambiguous case of successful adjustment.' The World Bank, he concludes, 'did not adequately appreciate the complexity of the political economy behind structural distortions' (Khan 1993: 37, 46).

4. The Bank tried to circumvent this in its 1981 Structural Adjustment Loan to Senegal by 'demanding a general levy on customs duties and the grant of a subsidy to firms exporting five non-traditional goods. It thought this would have a similar effect to devaluation' (Dieng 1995:106). The Bretton Woods institutions continued putting pressure on both the Francophone countries and France for devaluation of the CFA franc. They finally buckled in 1994 and the CFA franc was devalued by 50 per cent, the first devaluation since 1948. This 'brought immediate shocks as leaping prices, lost purchasing power and grave cash-flow problems for the public and private sector throughout the zone' (Uku 1995:7). As elsewhere in structurally maladjusting Africa, the devaluation had an uneven impact among and within sectors and social groups, depending on their relative production and consumption of the import and export tradeables.

5. The rise in military expenditures in the 1970s and 1980s partly reflected the increase in the number of military governments. According to Epp-Tiessen 1990:1) 'between 1969 and 1978, total African military outlays in current dollars rose from US$2.8 billion to US$7 billion per year, an increase of 250 per cent. Arms expenditures in the same period rose from US$145 million to US$5.25

billion per year, an increase of 3600 per cent'. Militarization was also fueled by outside intervention. It is quite evident that militarization diverted resources that could have been used to promote development. Since most of the arms were imported this represented drainage of valuable foreign exchange. The military generate fewer jobs than other sectors per unit of expenditure because of growing capital intensive military forces. Interestingly, in the 1970s and 1980s military wages in Africa rose relative to civilian wages (Luckham 1980; Sanger 1982; Ghosh 1984; Graham et al 1986; Deger 1986; Porter 1989).

Chapter Fourteen

THE DEVALUATION OF AFRICA'S LABOUR

Introduction

The denigration of labour as a source of value is central to the neo-classical paradigm (Dobb 1979; Dean 1982; Standing and Tokman 1991; Weeks 1991). This perspective assumes various guises in development economics, all of which depict labour, in its constitution, remuneration or agency, as an obstacle to development. At the heart of the structural adjustment regime is the desire to remove the 'distortions' supposedly engendered by urban-based rent-seeking groups who have a disproportionate influence on state policy. Labour is seen as part of the unproductive rent-seeking urban coalition that thrives on the exploitation of the rural peasantry and generates the economic perversions and political rigidities of urban bias (Lofchie 1975, 1989; Lipton 1977; Bates 1981). Deriving its policy prescriptions from these intellectual foundations, the World Bank (1981, 1984) depicted African workers as overpaid, overprotected, and underproductive as compared to their counterparts in Asia, especially in the Newly Industrializing Countries, whose labour was much cheaper, unprotected, and highly productive. Structural adjustment was, therefore, designed, so it was claimed, to 'clear' the labour markets, to discipline Africa's 'lazy' workers and wean them from the deforming comforts of urban bias, the unearned privileges of a 'labour aristocracy'. But the cost was high: unemployment soared, real wages and standards of living plunged.

Until recently the problem of unemployment did not feature prominently in the development literature on Africa. Indeed, there was a school of thought that argued the problem of unemployment in the developing world was exaggerated. Unlike the developed countries where unemployment was cyclical and 'real', in the developing countries unemployment was insignificant, it was contended, because 'opportunities for casual and self-employment are extensive, indeed, such employment is the basis of the economy' (Weeks 1973:62). For Vandemoortele (1987:32) 'the notion of unemployment in African countries is to a large extent invalid because of the absence of a system of unemployment benefits. Most able-bodied persons are engaged in income-earning activities of some kind or another because their very survival depends on it'. What appeared as open unemployment in the developing countries was nothing but 'luxury unemployment' (Udall and Sinclair 1982). Even the ILO tended to view underemployment and poverty, not unemployment, as Africa's main problems (ILO 1972; Jolly 1986; Livingstone 1986). Such an intellectual and policy climate facilitated the assault against labour in the 1980s that was embedded in the structural adjustment programmes. By the end of the decade the situation had become so alarming that the ILO now came to view unemployment as a major problem that 'can no longer be dismissed as an unimportant phenomenon' (ILO/JASPA 1989:ix)[1]. This chapter examines and demonstrates the processes and patterns of African labour's devaluation in the 1970s and 1980s.

The Deregulation Gospel

The World Bank and other proponents of structural adjustment have argued forcefully that labour regulations, standards, and institutions hamper the smooth functioning of the labour market thereby hindering development in general and employment growth in particular. The protective measures are specifically blamed for raising labour costs. Where labour markets are tightly regulated, it is said, there is no relationship between wage levels and productivity. The setting of binding minimum wages is strongly discouraged, and charges for social benefits to cover pensions, health care, disability and compensation are seen as a problem. Administrative restrictions on hiring and firing workers are frowned upon for raising costs and reducing labour mobility. Also deplored are gender-related considerations and restrictions on hours of work and occupations. Public employment services are despised, private ones preferred and, of course, trade unions are abhorred (Horton et al 1991; van Adams et.al. 1992; Hamermesh 1992; Mangum et.al. 1992; World Bank 1992).

Labour law reforms were, therefore, regarded as an essential component of structural adjustment programmes. Labour market conditionalities featured prominently in the third generation of adjustment policies launched towards the end of the 1980s[2]. Countries were forced to review their labour legislation and to reduce government interventions in labour markets and eliminate market price distortions. The Francophone African countries with their strong labour codes came in for particularly severe pressure (Plant 1994:68-71; 167-177). In Senegal, for example, during the third adjustment programme in 1989, government monopoly on recruitment was abolished, followed by amendments to the Labor Code to allow for greater use of temporary contract labour and to the Investment Code to permit enterprises to employ temporary workers for up to five years. Moreover, small and medium enterprises were exempted from government authorization for dismissals, and the government promised to undertake a fundamental review of all existing labour laws and employment practices in order to reduce wage costs and 'free' the labour market from 'rigidities'. In Guinea, one of the conditions for a structural adjustment loan also involved reviewing the existing labour code and devising a new and more flexible one, which entailed, among other things, turning the Labor Office of the Ministry of Labor into an innocuous unit responsible for the collection of labour statistics, rather than a regulatory agency for labour standards.

Côte d'Ivoire came under similar pressures, especially in 1991 during negotiations for the release of the second tranche of the World Bank's structural adjustment loan. The Bank blamed the growing levels of unemployment in the country not, of course, on its own policies and the consequences of previous adjustment programmes, but on protective labour legislation, high formal sector wages, and low labour productivity. The government acceded to intensify its public sector retrenchment programme, which had already cost 127,000 jobs; reduce wages; cut the employers' payroll tax by half immediately and aimed to eliminate it altogether in two years; ease the monopoly of the Public Employment Office and allow recruitment by private employment agencies; and relax restrictions on the recruitment of temporary workers, dismissals, and regulation of overtime. This was only a prelude to a thorough review of labour legislation and institutions.

Anglophone countries, such as Zimbabwe and Ghana, with strong protective labour legislation or labour movements also felt the pressure for deregulation. In 1990 Zimbabwe introduced several amendments in the Labor Relations Act decontrolling the collective bargaining process, allowing more flexibility by employers to hire and fire workers, and reducing government intervention in wage fixing. As a sweetener a National Social Security Act (NSSA) was passed to widen access to social security, and a Social Dimensions Fund (SDF) was set up to help victims of structural adjustment. Both workers and employers welcomed the deregulation of collective bargaining, but in a situation of rising unemployment workers found their power diminishing relative to employers, who also vigorously opposed the NSSA so that years later it was still not fully operational. As for the SDF it provided a meagre Z$20 million. In short, labour lost out to capital (Mhone 1994).

Unlike Zimbabwe, Ghana did not have to be prodded much to retrench workers and reorganize the labour market. Indeed, wages had fallen so drastically before and after the first phase of adjustment that recovery was endangered because of excessively reduced domestic demand. Wage recovery was increasingly seen as vital, although the World Bank insisted that wage rises should be linked to productivity increases and continued to oppose collective wage bargaining settlements which it feared would 'prove inflationary in a context of increasing democratization' (Plant 1994:186). And PAMSCAD, which was meant to cushion the social impact of adjustment for the most vulnerable groups, including retrenched workers, had a rather limited impact: less than 10% of laid-off workers received training or counselling promised under the scheme.

The World Bank's neo-classical paradigm was hostile to trade unions and tripartitism, that is, binding wage negotiations and collective bargaining between government, employers, and employees. According to the ILO (1995:Chapter 4), tripartitism was in retreat everywhere, among the industrialized western countries, the developing countries, and the 'transition' countries of Central and Eastern Europe. Instead, state intervention, often on behalf of capital, actually increased, notwithstanding the deafening rhetoric about 'free markets' and the need to remove the distortions of state interference. The evidence is, in fact, overwhelming that an onslaught was launched against trade unions.

Workers and trade unions encountered mounting problems over a wide range of areas (ILO 1985:Chapters 1, 3 and 4). First, over the question of the establishment of workers' organizations. Although most countries formally recognized the right to organize, legal obstacles increasingly restricted workers' rights. In many African countries public officials were denied the right to organize, or the trade union structure preferred by the state, usually with the connivance of employers, was imposed on the labour movement. Second, public authorities also interfered in the internal administration of trade unions more than ever, mainly in the election and dismissal of trade union officers, the drawing up of constitutions, and financial management. Third, legislation was used to control the activities and programmes of trade unions, so that they were sometimes prohibited from taking part in any kind of political activity, freely engaging in collective bargaining, and going on strike. Finally, attacks on trade unionists' civil liberties became more frequent and systematic. This took various forms, from the arrest and banishment of trade union leaders, especially during strikes, and restrictions on the right to hold meetings or demonstrations, to the seizure of trade union publications, violation of trade union premises, arbitrary suspension and dissolution of trade unions, and discrimination in employment against union activists.

The litany of violation of trade union rights in Africa in the 1980s and early 1990s was a long one, indeed. From Algeria to South Africa, embattled trade unionists reeled under state coercion and the terror of employers. For instance, in 1984 and early 1985, the ILO Committee on Freedom of Association received the following complaints from several African countries (ICFTU 1985:26-27):

Morocco: violent suppression of a 24-hour general strike; arrest of trade union leaders and members of the CDT; dismissal and transfer of trade unionists; occupation and closure of CDT's premises.

Tunisia: non-implementation of wage increases provided for in legislation and collective agreements; prohibition of trade union meetings in enterprises; requisitioning of workers in several sectors to prevent strike action; dismissal of workers and trade union representatives following legal strike action.

Central African Republic: dissolution of UGTC by administrative measures; occupation of trade union premises; freezing of trade union assets; dismissal of trade unions; arrest of the UGTC General Secretary.

Ghana: interference by public authorities in Ghana TUC and affiliated unions; impounding of passports and freezing of bank accounts of trade union leaders; occupation of trade union premises by government-condoned anti-union groups; arrest of trade union leaders.

Press reports from other parts of the continent carried similar stories of governments determined to whip trade unions into submission and tame them for the market. In Nigeria in 1986 the government went even as far as storming 'the conference of the International Transport Workers' Federation (ITWF) in Lagos...(and) whisked away the organizers' (Jason 1986:59) and two years later following a three week strike scores of union leaders were arrested (*African Concord* 17 May 1988). In Zimbabwe as the Government was drafting its restrictive Labor Relations Act, vigorously opposed by the labour movement, '15 left-wing trade unionists and political activists were arrested by security police' (Crisp 1985:68). From Kenya came the news that 'Kenyan managers have been warned...that they should not belong to trade unions as this would cause investors to lose confidence in the country and its future' (Crisp 1984:54). The Kenya Civil Servants' Association was actually deregistered and in late 1988 it was announced, to the consternation of trade unionists, that the country's Central Organization of Trade Unions (COTU) would be affiliated to the national party KANU

(Financial Review October 1988). The Zambian government imposed a strike ban and 'threatened to stop collection of check-off dues for various unions which would cripple the labor movement financially' (Hear 1985:86). Organized African labour was under siege.

The Swelling Reserve Army

In the meantime, the ranks of retrenched and unemployed workers were swelling, real wages falling, and the importance accorded to the informal sector as a labour sponge increasing. Although adequate region-wide data is rather hard to come by, it is possible to discern the general patterns of the employment crisis in Africa by drawing on the statistical coverage that is available on a number of countries. It is quite clear that the growth of wage employment in the modern sector started faltering in many countries from the mid-1970s. According to an ILO/JASPA study (1989:11), 'in the 14 countries for which data are available, the average rate of increase slowed down from 2.8% per annum in the period 1975-80 to 1% between 1980 and 1985'. This decline occurred despite the sharp fall in labour costs.

The decline of modern sector wage employment can be differentiated according to sector, region, gender, and age. It would appear that between the private and the public sector, it is the latter, traditionally the engine of employment growth in post-independence Africa, whose rate of labour absorption was reduced most markedly[3]. The public sector initially grew relatively fast because the post-colonial state was under enormous pressure to Africanize or indigenize and expand the civil service. For the masses *uhuru* meant more social services and jobs. It was in the interest of the new political class to satisfy this hunger for social and economic development and to contain the swelling ranks of the unemployed. And it also served their accumulative interests to expand the state apparatuses. When the employment crisis began rearing its ugly head in the 1970s, the state swung into reflex action. The rate of growth in public sector employment accelerated. It averaged 10% and above in such countries as Ghana, Gambia, Somalia, Sudan, Tanzania, and Congo. Needless to say, this far outstripped the rate of employment growth in the more profit-conscious private sector. For example, public and private sector employment expanded by 6.4% and 0.6%, respectively, in Kenya between 1977 and 1982, and 5.4% and 2.2%, respectively, in Swaziland between 1977 and 1983 (ILO/JASPA, 1989:15).

But the locomotive soon ran out of steam. The fiscal crisis of the state, as well as the structural adjustment programmes, not only curtailed the labour absorption capacity of the public sector, but also, in a growing number of cases, forced the sector to retrench large numbers of workers. A few examples will suffice. The annual growth rate of public sector employment was more than halved in the early 1980s in Somalia, Tanzania, and Malawi. Even more precipitous drops were recorded for the Sudan where the increase in the classified posts in the civil service fell from 10.9% between 1978-79 and 1980-81 to 0.4% between 1980-81 and 1985-86, and the Gambia where the growth rate of established civil service posts fell from 14% in 1975-80 to 2.5% in 1980-85. In Liberia and Senegal the number of civil servants actually dropped by 2% between 1981 and 1985, and 1.4% between 1985 and 1987, respectively (ILO/JASPA, 1989:15).

The decline in wage employment was also precipitated by the slow and dwindling labour absorption capacity of industry. In fact, the growth of employment in industry fell more steeply than in the agricultural and services sectors, from 2.6% in 1975-80 to 0.1% in 1980-85 for the 14 countries on whom JASPA had data[4]. The result is that the share of industrial wage employment in these countries dropped from 25.8% of the total in 1975 to 24.4% in 1985 (ILO/JASPA 1989:12). Part of the explanation for falling growth rates in industrial employment lies in the decline of industrial production itself. Value added in industry for Africa as a whole in 1980-85, 1986 and 1987 grew by -0.6%, 0.5% and -0.3%, respectively. By 1987, industry's share of GDP in Africa had dropped to 31.3% from 39.1% in 1980 (UNDP and World Bank 1989:7). Up to 1980, industry in general and manufacturing in particular had been the fastest growing sectors in Africa. For example, taking the sub-Saharan countries alone, we see that between 1965 and 1980, industry and manufacturing grew at an annual rate of 9.4% and 8.7% respectively, as compared to 1.3% for agriculture and 5.0% for services (World Bank 1990:181).

There is evidence to show that in some countries the growth of industrial employment consistently lagged behind the growth in output. In Ghana, for example, while manufacturing value added grew by 13% per annum between 1962 and 1970, manufacturing employment rose by 10%. Manufacturing employment reached a peak of 89,935 in 1977, and then steadily declined to 66,500 in 1985 (UNIDO 1986a:21, 24). In Kenya's more buoyant economy manufacturing employment rose steadily until 1982 when it peaked at 141,452, before gradually declining to 125,656 by 1985 (UNIDO 1988a:16). Data from Nigeria also shows that not only did most manufacturing enterprises have low employment elasticities, but that these elasticities were significantly lower in the 1980s than they were in the 1970-78 period. Thus, as UNIDO

(1988b:33-5) puts it: 'the impact of the recession on manufacturing employment is seen to be significantly greater than its impact on the generation of value added within the sector'.

By the turn of the 1980s, the size of manufacturing labour force in African countries ranged from small to minuscule, with the notable exceptions of Nigeria and Egypt. By 1984, Nigeria, Africa's most populous country, accounted for nearly 30% of Africa's manufacturing value added (MVA), up from nearly 22% in 1970. But the share of manufacturing in the GDP was still small, 9.3% by 1985. Indeed, MVA per capita in Nigeria was only 70% of the average for the continent, despite the rapid growth of manufacturing during the years of the 'oil boom'. Nigeria boasted the largest manufacturing workforce, numbering 6.6 million people in 1985, or 18.2% of the country's total labour force, up from 16.8% in 1977. Manufacturing employment began showing signs of faltering from 1980. Between 1980 and 1983 it declined at an annual average rate of 7.8%. This rose to a staggering -13.7% in 1986-87 (UNIDO 1988b:33-35). Despite its relatively large size, the manufacturing sector in Nigeria displayed all the structural weaknesses evident elsewhere on the continent. By the 1980s, manufacturing had not moved much beyond the level of 'easy' import substitution, it was monopolistic in character, had weak backward linkages with the rest of the economy, and was heavily import dependent. Manufacturing imported 60% of the raw materials it consumed, but its exports accounted for less than 1% of export earnings. The sector was, therefore, a large consumer of foreign exchange. Not surprisingly, when the economy and foreign exchange earnings went into a tailspin, due to the fall in oil prices, Nigeria's main export, imports of industrial raw material and inputs declined sharply. Moreover, public investment, which was critical and had grown faster than private investment, fell as well. Structural adjustment programmes adopted by the government in the mid-1980s entailed not only reduced state subsidization for industry, but also a massive programme of privatizing hundreds of public enterprises. In the meantime, foreign investment had contracted because it had 'exhausted' the available and most profitable import substitution opportunities. In fact, from 1970 'outflows exceeded net direct, private foreign investment' due to the rise of service payments and profit repatriation (UNIDO 1988b:37, 49). The result was growing capacity underutilization, which for the previously buoyant food, beverage, and tobacco sub-sector fell to 33.6% by 1986, while for the fledgling steel industry it fell to as low as 11% (UNIDO 1988b:37, 49)

Egypt's manufacturing sector was slightly larger than Nigeria's in the mid-1980s, according to UNIDO estimates. Also, Egypt's capital and intermediate goods sectors were more developed. Despite this, it has been maintained that its 'industrial sector (was) still enclaved in the traditional import substitution strategy' (UNIDO 1986b:32). In spite of its larger size, Egypt's manufacturing sector employed far fewer people than Nigerian manufacturing. In 1981-82 their number was about 1.4 million, or about 12.1% of the total labour force, the same as in 1974, despite the rapid growth that the industry had experienced in the second half of the 1970s following Sadat's liberalization policies (UNIDO, 1986b:40). This suggests rising capital intensity in the industry. Egyptian manufacturing was also import-dependent and net a consumer of foreign exchange. In fact, the share of manufactured goods in total exports actually declined from 25.4% in 1973 to 8.1% in 1982. In 1985, for example, imports for industry were US$1.2 billion, twice the value of manufactured exports. And instead of becoming more diversified, the product structure of manufactured exports became more concentrated (UNIDO 1986b:20). Thus the manufacturing sector was becoming more dependent on the foreign exchange capacity of other sectors, while its potential for generating jobs had stagnated.

Generalizations about the African crisis often obscure the fact that there are some economies that did quite well in the 1970s and 1980s. The most outstanding was Botswana, whose GDP grew at an average rate of 14.8% per annum between 1965 and 1973, and 10.7% from 1973 to 1984, making it the fastest growing economy in the world. Its per capita growth rate of 8.4% from 1965 to 1984 was also unmatched (UNIDO 1987:1). This remarkable growth can mainly be attributed to the spectacular expansion of the mining and beef industries. Despite this impressive growth, manufacturing employment remained small. To be sure, it nearly doubled between 1980 and 1985, but from a low base. By 1985 manufacturing workers numbered 10,100, that is, 8.6% of the labour force, up from 6.7% in 1980 (UNIDO 1987:7-8). The industry was dominated by foreign capital. The state avoided investing directly in industry, except in the meat abattoirs. Unlike many African countries, rates of capacity utilization were high because the country did not have foreign exchange problems. But by the same token, 'a comparison with African countries where suitable data exist shows that value added coefficients in Botswana are lower in most sectors. This is hardly surprising when one takes into account the high import content of most branches except meat processing' (UNIDO 1987:20). In short, by the 1980s manufacturing in Botswana was still at very early stages of import substitution, and there was little indication that manufacturing would become the main engine of employment creation, due to its capital intensity and weak linkages with the rest of the economy.

Botswana is classified by the World Bank as a 'middle income' country. Another middle income country that used to enjoy Botswana's reputation for fast economic growth is Côte d'Ivoire. The Ivorian economic miracle' was led by the expansion of export crops. While coffee and cocoa, the main crops, boomed, the manufacturing sector limped along. In fact the share of manufacturing in the GDP declined from 13.2% in 1970 to 11% in 1983. The decline has been attributed to the fact that:

> most industrial and commercial enterprises were under foreign ownership, mostly French, and because the government decided to open up the economy and allow participation by foreign investors, almost a quarter of domestic savings was channeled out of the country in the form of wage transfers (CFAF5.6 billion in 1963, CFAF148 billion in 1983) and dividends (CFAF4.7 billion in 1963, CFAF35 billion in 1978 and CFAF 19 billion in 1983). The government subsequently started to finance public investment by means of foreign loans, rather than from domestic savings, creating the problem of debt (UNIDO 1986c:4-5).

This does not mean, of course, that manufacturing value added did not increase. It did, in fact, quite rapidly in the second half of the 1970s. So did manufacturing employment. But manufacturing was a dependent, not dynamic, sector. As the golden beans of cocoa and coffee lost their lustre on the world market, both public and private investment dropped, and many manufacturing firms began operating well below capacity. MVA growth fell by 6.1% in 1980-81 and 9.9% in 1981-82. Employment followed suit, falling by 3.4% and 2.7%, respectively. The country was paying for its belated debt-financed industrialization and its earlier heavy dependence on external financing, technology and management, which had created an industrial structure that was biased towards capital-intensive techniques and limited its 'capacity to absorb the growing urban labour force' (UNIDO 1986c:35). By 1985 the manufacturing sector employed 81,600 people, a mere 1.8% of the total labour force.

As might be expected, the situation among the 'low income' countries was no better, although, of course, it was quite varied. In Africa's largest country, the Sudan, with its poor infrastructure, acute foreign exchange shortages, and recurrent droughts and civil war, the stagnant manufacturing sector, which was dominated by sugar refining, employed only 144,503 workers in 1981-82, a mere 3.5% of the labour force

(UNIDO 1989:15). Tanzania's manufacturing industry, also beleaguered from the mid-1970s, due to shortages of foreign exchange to import essential industrial raw materials and spare parts, saw its share of GDP fall from 9% in 1978 to 5% in 1984, and its capacity utilization tumble to less than 30%, so that its workforce shrank to 89,000 in 1980, no more than 6% of the total labour force (UNIDO 1986d:x, 19). And there were countries in the mid-1980s where manufacturing employment had fallen to 1% or less of the total labour force. For example, Mali's 20,000 manufacturing workers in 1985 constituted 1% of the country's labour force (UNIDO 1986e:ix, 6-7, 17-20). The small Somali manufacturing sector employed 15,000 workers in 1985, 0.8% of the labour force (UNIDO 1988c:vi, 14, 19). In the Central African Republic the manufacturing sector in 1984 employed only 4,500 people, probably less than 0.5% of the labour force (UNIDO, 1986f:vii-ix, 7). For expansive and resource-rich, but economically depressed, Zaire the importance of manufacturing was insignificant in terms of both GDP and employment (UNIDO 1986g:5-6). The list could go on. But the point has been made that manufacturing employment which was historically low fell sharply in the 1970s and 1980s.

Since public sector and industrial employment were concentrated in the urban areas, their contraction meant that these areas were the hardest hit by rising unemployment. According to some estimates, between 1975 and 1990, the rate of urban unemployment almost doubled from 10% to around 20% (Lachaud 1994:97). The rural areas were, of course, not immune from the crisis. But rural employment does not seem to have fallen as precipitously as in the urban areas. In fact, ILO/JASPA studies suggest that 'the employment prospects in the rural areas have improved recently and look better than those in the urban areas' (ILO/JASPA 1989:ix). In Kenya, for instance, agricultural employment rebounded from 1982, following a decade of steady decline, growing by 3.3% per annum in the 1982-87 period (ILO/JASPA, 1989:23). In Malawi employment in the agricultural, forestry, and fishing industries grew at an annual rate of 15% in 1980-83, its highest since the late 1960s, as compared to 5.7% during 1977-80 (ILO/JASPA 1985a:29).

Evidence from more countries is needed before generalizations about the rise of agricultural employment can be made for Africa as a whole during the 1980s. The ILO/JASPA attributes the trend observed in countries like Malawi and Kenya to policy reforms which reduced 'urban bias'. An 'enabling environment' for agricultural development was created following the liberalization of markets, so that 'agricultural producer prices in particular have increased more rapidly than wages and prices in general so that the long-term decline in the domestic terms of trade has been halted in most countries and reversed in some' (ILO/JASPA 1989:22). For example, in the first half of the 1980s the nominal agricultural output prices outstripped wages by over 70% in the Central African Republic, Senegal, and Zambia. In Kenya and Burundi, agricultural prices began outpacing wages in 1982 and 1983, respectively. Consequently, the rural-urban income gap narrowed. For example, in Côte d'Ivoire the gap fell from a ratio of 3.5:1 in 1980 to 2:1 in 1985 (Addison and Demery 1987). In Burundi by 1986 'the average daily incomes of a farmer approximated the daily minimum wage of a skilled worker in the modern sector' (ILO/JASPA 1989:23). In a 1984 study of 17 countries the ILO/JASPA estimated that by the early 1980s the urban-rural income gap in these countries had been reduced to an average of 4:1[5]. The gap between the average agricultural income and the average wage was even narrower at 1.5:2.5 (ILO/JASPA 1984a:31-38; Jamal and Weeks, 1988). 'The number of countries', concluded Jamal (1986:328), 'where the gap is now so small is a new revelation and certainly requires modification in the usual statement that wage earners are the elite in African countries; they are not in all countries, although they used to be not so long ago'. One result of all this is that in some countries, such as Ghana and Nigeria, according to the World Bank (1990:114), 'urban-rural migration has become significant...three-fifths of all internal migrants in Ghana during 1982-87 came from Accra, the capital'.

However, Ghana's case was far from typical. It would in fact appear that high rates of rural-urban migration persisted despite the falling rural-urban and agricultural-wage differentials. The average annual urban growth rate for the sub-Saharan region actually increased during 1980-88 to 6.2% from 5.8% during 1965-80 (World Bank 1990:239). It ought to be remembered that it was rural-urban migration itself, apart from the greater increase in agricultural prices compared to non-agricultural prices, that facilitated the decline in the rural-urban income gap in the first place (ILO/JASPA 1984a:51-53). It also needs to be pointed out that the rural incomes were catching up with, not rising, but falling urban incomes. Therefore, what was taking place was essentially the equalization of poverty between the rural and urban areas.

The capacity of the rural sector to act as an effective sponge to absorb and retain the rapidly growing population and labour force should not be exaggerated. There is overwhelming evidence that rural poverty was deepening, not lessening, although 'rural development tourists', as Chambers (1983, 1986) calls the rural development experts, continued to under-perceive and mis-perceive it because of various ideological biases. Given the underdeveloped nature of rural industry, rural wage employment was largely provided by agriculture, which, at the most, provided full-time employment to no more than 5% of the total labour force (ILO 1987:71). In fact, formal sector wage employment as a whole had never been large. In 36 sub-Saharan African countries, which represented 95% of the regional labour force, it accounted for only 9.4% of the labour force in 1980 and had probably declined to about 8% by 1988 as a result of the deceleration of wage employment growth in the region (ILO/JASPA 1988, ILO/JASPA 1989:13).

Last Hired, First Fired

In addition to its differential sectoral and spatial impact, the employment crisis also affected households, men and women, youth and adults quite differently. In a detailed study of the effects of structural adjustment on employment and unemployment in the Francophone countries, it was found that 'the incidence of unemployment is between two and three times higher in poor households than in well off ones' (Lachaud 1994:60). This partly reflected the unequal access of women from the poor households to access wage employment. By the 1970s women's wage employment was still relatively low because of various economic, social, cultural, and institutional factors. The marginalization of women was a product of complex combinations of indigenous and imported colonial patriarchal practices and ideologies, which prescribed a rigid gender division of labour that relegated women's work to agricultural and domestic production, and created unequal opportunity structures in education, political roles, and wage employment for women (Bay 1982; Hay and Wright 1982; Robertson and Berger 1986; Zeleza 1988).

Despite these burdens, women's participation in wage employment increased steadily after independence in the 1950s and 1960s. This was facilitated by several factors. First, women's access to education improved, which enhanced their employment prospects. Second, the number of female-headed households increased in many countries, typically to between one-quarter and one-third of all households. This forced many women to seek wage employment for the reproduction of their households, especially as, third, their access to productive resources, such as land, was diminishing as a result of changes in property and social relationships brought about either by the expansion of capitalism or the introduction of ill-conceived socialist experiments (Davison 1988). Finally, the economic crisis of the 1980s itself increasingly made it imperative for more women than ever before to seek off-farm employment or extra income to supplement other sources of household income. By 1975 the proportion of women in total wage employment in the nine countries for which the ILO had comprehensive and comparable data had reached an average of 15.8%, rising to 17.9% in 1980 and 19.1% in 1985 (ILO/JASPA 1989:13; ILO 1986:295-287)[6]. As can be seen, the rate of increase in the 1980-85 period was almost half that of the 1975-80 period. This suggests that the crisis of the 1980s threatened women's access to wage employment. Women's participation in the wage labour force trailed far behind both their crude activity rates and share of the total labour force, which in 1985 stood at 23% and 32%, respectively for Africa as a whole (ILO 1985:7).

Women in Africa, as elsewhere in the world, tended to be concentrated in the lower ranks of the occupational hierarchy. They were grossly under-represented in professional, technical, administrative and managerial positions. For example, according to figures provided by Selassie (1986:106-7) for six African countries, in three of them women occupied less than 2% of these positions, and only in one did their share exceed 5%[7]. My own research on Kenya shows that in 1982 women provided 2.7% of general managers and salaried directors, 7% of middle level executives and department heads, and only 0.3% of architects, engineers, and surveyors. In contrast, about 70% of nurses and other paramedical staff and 93% of secretaries and stenographers were women (Zeleza 1988:177-8). What is remarkable is that in 1968, women made up 72% of secretaries and stenographers. This shows that secretarial work had become more feminized. Indeed, many other service jobs, including sales work and teaching became increasingly feminized. In the meantime, women's employment in manufacturing rose marginally from 5% in 1967 to 7% in 1984 (Zeleza 1988:60-61). Detailed studies by the ILO and JASPA have revealed similar trends in other African countries. In Ethiopia, for example, between 1970 and 1981, women increased their share of managerial and administrative positions from a negligible 0.9% to an insignificant 2.8%, while their share of clerical and service jobs rose sharply from 10.6% to 22.3% and 11.8% to 21.9%, respectively (ILO/JASPA 1986a:24). In Tanzania by the turn of the 1980s women made up about a quarter of the service sector, and only 12% of manufacturing workers (ILO/JASPA 1986b:25). In Zimbabwe two-thirds of all female workers were in services and only 7% in manufacturing (ILO/JASPA 1986c:14).

The increasing feminization of the service industries was facilitated by the perception that these jobs were natural extensions of women's domestic roles. Underlying it all was the conjunction of the interests of male workers to preserve the skilled and higher paying jobs for themselves, and those of the predominantly male capitalists to profit from the labour power which they regarded as particularly cheap, thanks to the prevailing patriarchal practices and ideologies which marginalized women's status in society and saw women's wage employment merely as supplementary to those of 'their' male bread winners. But the crowding of women into the service ghettoes was not simply a product of the prejudices of male workers and employers. It also resulted from female socialization which oriented them towards certain jobs. In schools and colleges women were concentrated in arts subjects and grossly under-represented in the sciences and technical fields, as so many studies have demonstrated (ILO/JASPA 1986d:72-95, 1986e:79-100. Eshiwani 1985). Thus in their pre-employment behaviour (i.e. in their training and job application tendencies) as well as in their on-the-job performance women betrayed their socialization into low-skilled, low-status, and low-paying jobs. In short, sex discrimination in jobs went beyond the work place; it was deeply embedded in the wider cultural and social processes that produced and reproduced the labour force.

Women have traditionally suffered from higher rates of casual employment than men. For example, in Kenya it was found that during the 1972-83 period, '74 per cent of all economically active men were employed as regular workers in comparison to an average of 61 per cent of the female employees. As casual workers the percentage for the men was an average of 13 per cent against 17 per cent for women' (ILO/JASPA 1986d:37). Thus, women who, to begin with, comprised a small segment of the work force were, in addition, much more likely than men to be employed as casual workers. The high incidence of casual labour for women in Africa can be interpreted as partly a response by women to meet their household obligations, and party as a strategy by employers to cheapen female labour power even further. Indeed, it represented the process of feminization of low paying jobs born out of the intensive and contradictory demands placed on women's labour time, and overlaid by capital's propensity for lowering wages in times of economic crisis. The growing rates of female labour casualization amply demonstrated that women faced the sharp edge of the economic and employment crisis in Africa (ECA, 1988a).

The youth were not spared either. The definition of youth differs from country to country, but the internationally accepted definition refers to the age range of 15-24 years. In the mid-1980s youths, thus defined, accounted for about 20% of the population and 30% of the labour force in the sub-Saharan region. As a result of Africa's rapid population growth the youth labour force increased at a phenomenal rate. In 1989 the youth numbered 89 million and their numbers were expected to double to 189 million by the year 2000 (Freedman 1986:81). According to some estimates, in the 1980s some 5.4 million youths joined the labour force every year and the figure was projected to go up to almost 8 million by the end of the 1990s (Vandemoortele 1987:35). Large numbers of these youths seemed destined for the world of unemployment and underemployment. Available data indicates that youths were three times more likely to be unemployed than adults. Of course this was not a scourge peculiar to African youth. Youth unemployment was higher than adult unemployment everywhere, both in the so-called developing and developed countries. The ratio of youth to adult unemployment was 2.8 for Africa, 4.0 for Asia, 3.1 for Latin American and 2.5 for the industrial countries, providing a global average of 3.0 (ILO/JASPA 1989:43).

While there is relatively abundant information on unemployment by age, the data on employment by age is surprisingly rather scanty. Certainly the ILO *Yearbook of Labour Statistics* does not provide it. The 13 studies conducted by the ILO/JASPA in 1986 on youth employment and youth employment programmes in various African countries also give little information on this question. For example, none of the tables in the reports on Zambia, Malawi, and Mauritius, numbering 52, 32, and 41 respectively, provides any data on the patterns of employment according to age (ILO/JASPA 1986f, 1986g, 1986h). The Kenyan report provides only one such table, out of 51, based on a rural labour force survey. It shows that the patterns of employment between male and female youth differ; the latter have fewer wage employment opportunities than the former (ILO/JASPA 1986i:17). The Botswana report offers a slightly fuller picture. In 1981 youths aged 15 to 24 comprised 31.8% of the wage labour force. They were concentrated in the lowest paying branches of the services sector. For example, 30.5% of them worked in personal services, as compared to 21.5% for the adult workers. Comparing the youths and adults along gender lines, the discrepancies are obvious. While 47.4% and 19.5% of female and male youths, respectively, worked in personal services, the equivalent share for adult females and males was 32% and 17.4%, respectively. This also shows that there was considerable occupational differentiation between male and female youths. While nearly half of the female youths were employed in personal services, 41.8% of the male youths were employed in construction and the central government. Altogether the ratio between employed male and female youths was 1.5 (ILO/JASPA 1986j:33-34).

One cannot generalize about the whole of Africa on the basis of this limited data. However, while national differences existed in the employment patterns of youths according to industry, occupation, and gender, it would not be farfetched to argue that youths were concentrated in low-status, low-paying jobs, and that female youths generally fared worse than male youths. What is certainly clear is that youths bore the brunt of the employment crisis. As we noted above, they were more vulnerable to unemployment than adults. Their vulnerability increased quite considerably. For example, in urban Nigeria youth unemployment between 1974 and 1983 increased by 20% as compared to a total unemployment increase of 1%. For Kenya the proportions were 10% and 3%, respectively, in the period 1978-86 (ILO/JASPA 1989:44).

The youth were of course not a homogeneous group. In terms of gender, the rate of female youth unemployment in the mid-1980s in the urban areas of Botswana, Kenya, Ethiopia, and Zambia, was on average at least a third higher than that of male youth unemployment. And for both groups the unemployment rate for adolescents, those aged 15-19, was much higher than for the young adults, those aged 20-24. The ratios for Kenya, Ethiopia, Botswana, Nigeria and Zambia in the early and mid-1980s were 1.24, 1.26, 1.36, 1.69 and 1.80, respectively. The worst of all possible worlds was to be young, female, and adolescent. Data from Seychelles amply bear this out. Between 1983 and 1985 while total unemployment went up from 15.7% to 20.6%, and youth unemployment jumped from 17.7% to 30.9%, the unemployment rate of female adolescents more than trebled from 18.7% to 61%.

Age and gender were not the only variables that affected youth employment and unemployment. Education also had a major bearing. Ironically, unlike the industrialized countries where youth unemployment was often correlated with lack of education and training, in many African countries data shows that the educated youth were more likely to be unemployed than uneducated youths. For example, in Kenya the unemployment rate of uneducated youths between 1977-78 and 1986 increased from 23.2% to 28.7%, while that of secondary school graduates increased from 29.3% to 54.5%. The equivalent figures for Nigeria for 1974-85 were 22.6% to 22.5% for uneducated youths and 24% to 50.3% for those who had completed secondary school. Unemployment appears to have crept up the educational ladder, for in both countries the unemployment rate of primary graduates dropped, 42.4% to 34.2% in Kenya and 53.1% to 23.4% in Nigeria. University graduates began to feel the pinch as well. The rate of graduate unemployment in Nigeria rose from 0.3% to 3.3% during the 1974-85 period (ILO/JASPA

1989:45-48). The labour markets for educated youths deteriorated sharply, partly because the sponge that used to absorb them in the past, the public sector, both directly through guaranteed employment schemes, and indirectly through its high growth rates, had frayed as a result of the fiscal crisis facing the state.

If youths bore the brunt of Africa's growing unemployment women were not very far behind. In Kenya, for instance, the proportion of women in the urban unemployed population more than doubled from 24% to 52% between 1977-78 and 1986. By the 1980s in many countries, such as Kenya, Zambia, Botswana, Ethiopia and Seychelles, women were twice as likely to be unemployed as men (ILO/JASPA 1989:17). There were of course a few exceptions, most notably Mauritius, where registered female unemployment declined from 33.6% in 1979 to 25.5% in 1984. It would seem that the decline was due to growing employment opportunities for women in the rapidly expanding export processing zone industries, where an intensive regime of labour exploitation was instituted (ILO/JASPA 1986h:8-10). The Mauritius case simply proves the rule, that in the absence of economic growth and new employment opportunities, women took a disproportionate share of the ensuing unemployment. It also shows capital's preference for women's labour in the new 'sweat' industries burgeoning in the export processing zones of some so-called Third World countries.

Thus, the faces of women and youths dominated the swelling unemployment lines. There can be little doubt that in the 1980s, open unemployment increasingly became a serious problem in many African countries, where traditionally the problem had been perceived by governments, international agencies like the ILO, and researchers as primarily that of the 'working poor' (i.e. of underemployment and low incomes). To be sure, low-productivity employment continued to be a serious problem. But open unemployment could no longer be ignored either. Many people were being thrown out of the poverty-stricken ranks of the 'working poor' into the destitution of unemployment. Open unemployment became particularly visible in the urban areas, thanks to declining urban wage employment, persistent rural-urban migration, and high population growth rates. Africa's total labour force grew from 152.1 million in 1975 to 170.5 million in 1980 and 191.8 million in 1985, i.e. an annual growth rate of 2.4% (ILO' 1985:6-7). Thus Africa needed to accelerate its job creation capacity, to about 4 million new jobs a year in the sub-Saharan region alone, just to stabilize its already low employment levels.

Let Them Eat Cake

Clearly, this did not happen in the 1980s. In fact, not only did the rate of employment creation decline, as was demonstrated earlier, but there is evidence that the structural adjustment programmes actually led to the loss of some 3.6 to 4 million jobs between 1985 and mid-1987 (ECA 1988b). The tales of woe came from every corner of the continent. From Ghana, the World Bank's star pupil and reputedly enjoying a robust recovery, there were reports that 'as part of the second phase of the economic reform programme now being undertaken, up to 90,000 Ghanaians ... have lost, or will lose their jobs' (Morna 1988:10). In neighbouring Togo it was announced that due to privatization 'it was unfortunately necessary to lose many jobs without compensating the workers or assisting them in relocation' (Njondo 1988:16). In Côte d'Ivoire, too, 'under the watchful eye of the IMF and World Bank the austerity programme ... is resulting in further layoffs... In the construction industry, for example, it has been estimated that 40,000 workers were laid off between 1979 and 1983' (Hear 1984:78). In Senegal, it was revealed in 1988 that 'overall 2,000 jobs will be lost in the parastatal sector (1,500 have gone already) and 7,000 in industry (2,000 so far), while the civil service has been trimmed from 70,000 to 67,000 through voluntary retirement' (*Africa Recovery* August, 1988:24). From Nigeria it was reported in 1984 that 'estimates put the total of public employees sacked since the military government took power at 200,000' (Hear 1984:79), and 'at least 250,000 private sector workers lost their jobs in the first six months of 1984' (Crisp 1984:21). Going east to Kenya the employment news was equally gloomy: 'large scale manufacturing declined 'alarmingly' from 141,452 in 1982 to 125,050 in 1985' (Machua 1988:6). And so it was in the south, too. In Zambia, it was announced in 1986 that 20,000 of the 50,000 employees of Zambia Consolidated Copper Mines, 'the largest employer outside the civil service ... will be declared redundant in the next five years under the World Bank-sponsored five-year production investment plan' (Chiposa 1986:73).

Towns and cities were devastated. Urban unemployment rose at unprecedented annual rates, ranging from 9% in Senegal (1976-80), 11% in Kenya (1978-86) and Togo (1980-85), to 13% in Liberia (1980-84) and by more than 15% in Madagascar (1975-82), Mauritius (1977-84), Niger (1982-86), Burkina Faso (1977-81), and Nigeria (1980-86). By the mid-1980s urban unemployment rates ranged between 15% and 32% in such countries as Somalia, Kenya, Senegal, Mauritius, Liberia, Tanzania, Seychelles, Zambia, Botswana, Ethiopia, and Reunion, in that order (ILO/JASPA 1989:16). For Africa, altogether, recorded unemployment shot up from 5.3% in 1980 to 13% in 1987. Underemployment is also said to have risen from 40% to 50% over the same period (ECA 1988b). For some of the most blighted cities it was being estimated that unemployment 'could be as high as 30-40%, with as many as 70% of the total population living below the poverty line' (Diejomaoh 1987:16). As might be expected, real wages went on a downward spiral as well. The fall in real wages accelerated in the 1980s. Between 1980 and 1985 for example, they fell by an average of 6% per annum for the 18 countries for which comparable data was available to the ILO (ILO/JASPA, 1989:8). Vandemoortele (1991:86) puts the number of countries at 27. This means that by 1985 workers in these countries earned, on average, about a quarter less than they did in 1980. The only exceptions were Burundi and the Seychelles, where real wages rose by an average 1.8% and 2.0% per annum, respectively. In five countries - Zambia, Sudan, Tanzania, Somalia, and Sierra Leone - in that order, the drop in the wages averaged between 10.7% and 20.6% per annum. Wage erosion and compression was more a result of political skullduggery than market forces as shown by the fact that real wages dropped more sharply in the public than the private sector, by 25% compared to 4% between 1977 and 1987 in Kenya, and in Côte d'Ivoire private sector wages rose by 13.4% between 1979 and 1984, whereas public sector wages decreased by 4.4%. Overall, while public sector wages declined by 5% between 1980 and 1987 for the sub-Saharan region, government expenditure rose by 23% (Vandemoortele 1991:89-90). This shows, argues Vandemoortele (1991:90):

real wage levels in Africa have not been rigid in the recent past, neither in absolute nor in relative terms. In some cases real earnings have fallen below their 'efficiency level'. They often need to be supplemented by other income sources[R].

Lachaud (1994:134) simply concludes: 'the pursuit of policies to reduce wages in the public sector is not justified either economically or morally'.

Falling wages brought untold suffering to millions of workers. In Uganda, the minimum wage, which would have bought 1.7 times of an average family's daily food requirements in 1972, could only purchase a quarter in 1984 (ILO 1987:11). In 1987 the minimum wage was Ush 9,000 yet 'one bunch of matoke (green bananas), the staple food for the southerners and westerners, cost an average of Ush 30,000 ... A kilogram of meat cost Ush 10,000 (and) a bottle of beer cost Ush 8,000' (Ojulu 1987). Thus, the minimum wage was almost equivalent to one bottle of beer! In Tanzania, by 1985 the minimum wage would have bought only around 40% of a family's basic foodstuffs (ILO 1987:11). The 'salary of a minimum wage earner who works five days (was enough to) purchase a medium tin of margarine' (Park 1986). Similarly, in Kenya, where the cost of living increased two and a half times as fast as the minimum wage, the standard of living for workers plummeted (Zeleza 1988:Chapter 2). And in 'rapidly recovering' Ghana the minimum wage in 1988 was raised to Ce2,920 per month and that of an upper-middle level civil servant to Ce13,000 'but basic foodstuffs for a family of five was estimated at over Ce37,000 per month' (Morna 1988:10).

Available data on real minimum wages for 28 countries show that, taking 1980 as the base year, in 1970 the wages were much higher than in 1980, in all but three countries, in some cases up to two or three times as high. On average for the 28 countries combined real minimum wages in 1970 were 37% higher than in 1980. By 1986 they were 18% lower than in 1980. Only in six countries, Burundi, Cameroon, Ghana, Malawi, Mali and Zimbabwe, were real minimum wages higher in 1985-86 than they had been in 1980. Ghana led the pack; its minimum wages had risen by 50%, followed by Zimbabwe with 23%. The sharpest decrease occurred in Somalia where the real minimum wage in 1986 was 16% of its 1980 level, followed by Tanzania at 36%, and the Sudan at 45% in 1985 (ILO/JASPA 1989:9).

The decline in real wages led to the compression in the wage structure in several ways. First, income disparities narrowed considerably between the lowest and highest paid workers. For example, in Senegal the disparity declined from 7.8:1 in 1980 to 6.6:1 in 1985. The equivalent figures for Gabon, Mauritania and Kenya were 8.0:1 in 1973 to 3.1:1 in 1983, 4.6:1 in 1974 and 3.6:1 in 1984, and 11.8:1 in 1972 to 7.9:1 in 1982, respectively. In Ghana by 1985 the post-tax wage differential between the highest and lowest paid worker in the public sector had fallen to the incredible level of 1.5:1 (ILO/JASPA, 1989:10).

It is also evident that the ratio of non-agricultural to agricultural wages narrowed. Data for six countries - Ghana, Malawi, Mauritius, Swaziland, Zambia, and Zimbabwe - shows that the average ratio decreased by 6% in the period 1975-80 and as much as 28% between 1980 and 1985. The average ratio decreased from 4.07:1 in 1975 to 3.81:1 in 1980 and 2.75:1 in 1985 (ILO/JASPA 1989:10-11). These figures, if correct, support the proposition made earlier that the urban-rural income gap narrowed quite considerably in many countries in the 1980s.

The data above seems to indicate that incomes for the highest paid workers and for urban workers fell more steeply than for the lowest paid workers and for rural workers. The urban and rural sectors were of course closely interlinked, so that the decline in urban incomes profoundly affected rural incomes. To begin with, as urban wages fell remittances to rural areas also fell. It has to be remembered that many rural households often allocated their labour resources to both urban and rural areas in their income-earning strategies. Moreover, falling urban wages directly reduced the incomes of rural producers, because of declining urban demand for rural produce.

The argument that workers bore the brunt of the economic crisis in Africa can be supported by the fact that in many countries real wages fell more rapidly than per capita incomes. ILO data for 14 countries shows that between 1971 and 1984 growth in per capita incomes outstripped growth in real wages in all but two of the countries. The exceptions were Tunisia and Zimbabwe, where real wages grew faster than per capita incomes by an average rate of 0.6% and 1.8%, respectively. The biggest gap in the growth rates of real wages and per capita incomes was registered in Ghana (-11.9%) and Tanzania (-8.8%) (ILO, 1987:99).

Deciphering the Devaluation

Explaining the rise of open unemployment in Africa in the 1980s might appear at first glance an easy task. It is tempting to attribute it to the economic crisis that was ravaging the continent. The two cannot, of course, be separated, but the relationship between them was more complex than would at first appear. To begin with, the unemployment crisis predated the economic crisis of the 1980s, indeed, even that of the 1970s (Todaro and Harris 1968, 1970; Todaro 1971; Turnham 1970). In Nigeria, for example, unemployment was already a growing problem by the end of the 1950s (Adesina 1994:62), as was the case in Kenya (Zeleza 1982:Chapter 5). So what needs explaining is not simply a cyclical, but a seemingly long-term structural crisis.

The problem of unemployment in Africa, and other so-called Third World regions, has been analyzed in two main ways. First, elegant economic models have been applied, such as the Lewis-Fei-Ranis model, which is concerned with the transfer of labour in a dual economy from the rural to the urban areas, or from the 'subsistence' to the 'modern' sector (Lewis 1954, 1958; Fei and Ranis 1964); the Harrod-Domar model and its variants with their focus on the relationship between the growth of output and employment in the modern sector (Sen 1970); and the Blaug-Loyard-Woodhall model that is noted for its concern with the impact of factor price disequilibrium on resource use and allocation, particularly on employment (Blaug Layard and Woodhall 1969; Drazen 1980, 1982). Undoubtedly these models, separately and collectively, shed some light on the dynamics of employment and unemployment in Africa, but they are often closed and over-formalized analytical systems whose politically and socially disembodied variables can do little to capture and contextualize the complex historical processes of labour formation and labour markets.

The second approach consists of empirical analyses, which tend to be quite eclectic in their choice of explanatory factors behind the different aspects of Africa's employment problems. The empirical studies often differ widely in their analyses of the causes and nature of unemployment for they tend to concentrate on specific case studies covering regions or districts within a country. In short, they are rich in descriptive detail, but weak in explanatory power. It would be wrong of course to suggest that they have no theoretical underpinnings. On the contrary, many of them use the models mentioned above, explicitly or implicitly. But 'a general weakness of many of the empirical analyses', it has been observed, quite correctly, 'even when they are not bound to a particular economic model, is that they have too often been conducted within a limited framework of economic analysis rather than of political economy' (Jolly et al. 1973a:20-21). Few empirical studies systematically analyze the political, social, and class forces that facilitate the generation and regeneration of employment and unemployment in Africa.

As it was demonstrated in the previous section, the unemployment crisis in Africa has many dimensions. A political economy approach is perhaps best suited to unravelling the structural forces that produce and reproduce it. Such an analysis has to be concrete and focus on both the institutional development of the labour markets themselves, and the changes in the organization of African political economies in which the markets are situated and structured. It can be argued that both the African labour markets and political economies, which are reproduced in dependent capitalist formations, exhibit deep structural imbalances that have crippled the continent's capacity to develop satisfactorily and create adequate employment.

African labour markets have been characterized by structural imbalances between labour supply and demand and wide disparities in their aggregate rates of growth. On the supply side what stands out is the rapid growth of the population both in terms of the labour force and the educated population. Africa's demographic explosion began in the 1950s, after many centuries of population decline or stagnation, thanks to the ravages of the slave trade and colonialism (Rodney 1982; Moss and Rathbone 1975; Fyfe and McMaster 1981; Inikori 1982; Cordell et.al.

1987). In 1950 Africa as a whole had 8.9% of the world's population. This figure rose to 11.4% by 1985 (World Resources Institute and IIED 1986:11), still far below Africa's relative size of 20.5% of the world's total land mass. Thus Africa's population rose faster than the world average. The comparable average annual growth rates for Africa for the periods 1960-65, 1970-75, 1980-85 were 2.44%, 1.96%, and 2.74% and the world 2.03%, 3.01%, and 1.67%, respectively (World Resources Institute and IIED 1986:236).

The reasons for Africa's demographic surge are hidden in the intricate web of historical, economic, social, and cultural factors that we cannot go into here. Suffice it to say that after independence living conditions in Africa generally improved so that mortality rates fell from 23.3 per thousand in 1960-65 to 16.5 per thousand in 1980-85, and life expectancy increased by 12.2 years between 1955 and 1985, while fertility rates only fell marginally from 6.54 in 1960-65 to 6.43 in 1980-85 (World Resources Institute and IIED 1986:238). As argued in Chapter 12, some African countries appear to have entered a period of demographic transition from a condition of high mortality-high fertility to one of low mortality-low fertility.

As might be expected, the implications of Africa's rapid population growth on the continent's development and employment were profound. First, African countries were burdened by high and increasing dependency ratios. The proportion of children below 15 rose from 43.6% in 1960 to 45.3% in 1990, compared to world averages of 37.0% and 31.9%, respectively (World Resources Institute and IIED 1988:248). This entailed the expenditure of ever growing slices of national income on subsistence consumption, thereby reducing the rate of savings and hence the potential for future economic growth. Second, the demographic explosion accelerated the growth rate of the labour force, from an annual rate of 2.2% in 1960-70 to 2.5% in 1970-80 and 2.5% in 1980-90, as compared to the world averages of 1.7%, 2.1% and 1.9%, respectively (World Resources Institute and IIED 1988:246). This meant that the continent had to create millions of new jobs each year to accommodate the expanding ranks of potential job seekers.

Not only did Africa's labour force grow rapidly, it also became educated at unprecedented rates. It is worth quoting an ILO/JASPA report at some length. It states that in the Sub-Saharan region:

> the adult literacy rate increased from the very low level of 9 per cent in 1960 to approximately 42 per cent in 1985. In 1960 only one country - Mauritius - had an adult literacy rate of more than 50 per cent. By the mid-1980s, fourteen countries had achieved this benchmark, including

Kenya (54 per cent); Ethiopia (55 per cent); Cameroon (56 per cent); Uganda (57 per cent); Zaire (61 per cent); Gabon (63 per cent); Madagascar (68 per cent); Swaziland (68 per cent); Botswana (71 per cent); Lesotho (74 per cent); Zambia (76 per cent) and Mauritius (83 per cent). Primary school enrollment increased by an average 7.1 per cent between 1960 and 1980 so that the gross ratio more than doubled from 36 to 76. Enrollment in secondary and tertiary education rose even faster, growing at an annual rate of 12.4 per cent and 14.5 per cent respectively...The rapid expansion of the educational system has led to a massive improvement in the African human capital stock. The number of years of formal education attained by the sub-Saharan labour force has doubled from 1.6 to 3.2 between 1970 and 1983 (ILO/JASPA 1989:5).

These improvements had far-reaching effects on African labour. Increased opportunities were created for labour mobility, both spatially and socially, and for raising labour productivity. Expanding education also led to a rise in the aspirations and expectations of job seekers.

There can be little doubt that in most African countries the growth rate of the labour force and the educated population far outstripped demand in the economy. It is not surprising that the youth as the fastest growing segment of the labour force and those most likely to be educated fared worst in the labour market. Some have argued that youth unemployment is a transitory phenomenon, a product of the youths' age, inexperience and immaturity, which makes it difficult for them to compete effectively on the labour markets. According to this view, the 'employment gap' gradually disappears as the successive age cohorts enter the adult range. This explanation assumes an equilibrium between those entering and those leaving the work force. While it is true, as was noted earlier, that unemployment rates in Africa declined with age, the persistent and increasing rates of youth unemployment were not merely transient, but pointed to more a fundamental problem. According to Livingstone (1987:67):

The high unemployment rates in the youth age groups... could be the product of a progressive disequilibrium where the numbers joining the labour force each year are not balanced by departures at the other end and meet with progressive difficulty in being absorbed, as a result of structural factors such as a slowly growing public sector or modern sector or increasing land shortage.

The impact of education on unemployment is quite difficult to decipher. Few would argue that education in itself caused unemployment. But there is evidence, as indicated in the previous section, that unemployment among the educated, particularly educated youths, rose. This has led some scholars to posit the 'mismatch' thesis, according to which the rise in educated unemployment in Africa was a product of the mismatch between the skills, attitudes, and aspirations produced in the education and training system and those required by the labour market. In other words, governments pursued inappropriate education and training policies. In fact, it is said, Africa became bedeviled by the paper qualification syndrome (PQS), whereby employers, both in the public and private sectors, used educational certificates to select people for salaried jobs, despite the little correspondence between the qualifications and skills, training and employment (Oxenham 1986; Eicher 1986). The PQS was reinforced by the examinations system. Examinations played an important role linking formal learning and modern sector employment, but they displayed weak forward links with the income and occupational structure and had an adverse 'backwash' effect on the quality of learning (Little 1986). Ultimately the PQS, it is argued, was sustained by the huge salary differentials between regular modern sector wage workers and informal sector workers and the excessive dependence on pre-career schooling as a screening mechanism for employment. This triggered an insatiable hunger for education, which helped to reproduce and intensify the PQS at ever higher levels, leaving behind a long trail of educated unemployment.

In a series of country studies the ILO/JASPA concluded that the PQS did indeed contribute to unemployment. As one report put it:

> the phenomenon contributes to unemployment as higher educational qualifications produce higher aspirations and in the absence of suitable jobs, create frustrations and unemployment as educated persons are not prepared to accept lower jobs. Geared to producing persons for the modern sector jobs, the education system does not inculcate in them the spirit of self-reliance and creating work for themselves. The attitude of the educated people generally is that since they are formally - or highly qualified for a particular type of salaried job, they should not accept any other kind of employment. This exacerbates open unemployment, prolongs the waiting period, leads to dependence on and living off the earnings of other family members or friends and in turn inhibits the development of opportunities and resources (ILO/JASPA 1982:i-ii).

The PQS undoubtedly aggravated educated unemployment, but it is highly debatable whether it was responsible for it. One only needs to point to the fact that improvements in the education and training system through the introduction of more 'appropriate' educational and training institutions, including trade and technical schools and colleges, and the restructuring of syllabuses and examination systems in some African countries hardly stemmed the tide of educated unemployment in those countries (Livingstone 1987:77). The roots of the PQS, it needs to be underlined, lay in the economic and employment systems, rather than in the schools, colleges, and universities.

Similarly, it is not enough to blame unemployment in Africa, as some labour economists tend to do, on the 'ineffective' functioning of labour markets due to the undeveloped nature of the employment-market information systems, particularly with references to the provision of employment services data. This is not to suggest that such data is not useful. It is indispensable for policy-makers, employers, workers, and the general public including prospective job seekers, for it enables them to plan effectively and improve job opportunities (Nigam 1979; Richter 1986). In a way, then, the inadequate development, dissemination, and use of employment market information in many African countries exacerbated unemployment because it made it difficult to match labour supply and demand. The ILO/JASPA (1989:99) laments that: 'although unemployment and underemployment have been worsening, policy makers and planners in many African countries appear not to have recognized the crucial importance of employment planning in national development plans or as an integral part of a broader socio-economic development planning process'. At best manpower planning in Africa, the ILO/JASPA charged, was preoccupied with outmoded forecasting methods that focused on demand, rather than on reporting and analyzing labour market situations and trends, and the early detecting of imbalances in labour supply and demand in different sectors, occupations, and regions of the economy, including the rural areas (ILO/JASPA 1984b).

But it would surely be going too far to assume that unemployment in Africa in the 1980s would have been significantly curtailed if the employment market information systems had been more developed. It should not be forgotten that in the industrialized countries where labour market information systems were supposedly developed, unemployment reached record levels in the early 1980s. As pointed out in the previous chapter, by 1983, 35 million people in the industrial market economies were out of work, millions more were too discouraged to even look for work, and many others were forced into early retirement. The rise in unemployment in these countries and in Africa had one thing in common: they had little to do either with the PQS or the state of the employment market information systems. They were products, and symptoms, of the global economic crisis.

In addition to the rapid growth of the population, both of the total labour force and the educated labour pool, labour supply pressures in Africa intensified because of profound changes taking place in the agrarian economy. Of critical importance were the changing dynamics of land tenure and distribution. The development literature, inspired by spurious anthropological orthodoxies, long assumed that land in most African countries was an abundant factor, 'communal' tenure systems predominated, and 'subsistence' production prevailed. It was, therefore, concluded that rural inequalities were negligible. As recently as 1987 an ILO report could say that 'there is very little landlessness in Africa' (ILO 1987:9). Historical research shows that this image of what Hopkins (1973:10) calls 'Merrie Africa' is not even correct for the pre-colonial era (Zeleza 1993). Certainly it has very little bearing on reality in post-colonial Africa, except as a mystifying ideology.

After independence the processes of land concentration and privatization and the pauperization of the poor accelerated and deepened in many countries. These processes were particularly pronounced in Mkandawire's (1987) 'merchant states', that is, states dependent on surpluses extracted from agriculture for their revenue. He includes among them Ghana, Burkina Faso, Morocco, Malawi, Mauritius, Tanzania, Swaziland, and Egypt. Kenya and many other countries could be added to the list. The shortage and unequal distribution of land in Mauritius, where the bulk of the land had for decades been under sugar estate production thus leaving virtually no land for smallholders, is well known (Virashawmy 1987). Kenya has attracted a lot of the literature on rural inequality. The studies conducted to date present overwhelming evidence that in the 1960s and 1970s both land concentration and landlessness in the country increased rapidly, and that the population of poor peasants increased. By the end of the 1970s some estimates show, there were at least 2 million absolutely landless people and their numbers were increasing by not less than 1.5% per annum (Okoth-Ogendo 1981:337; Livingstone, 1981; Hunt, 1984; Hazlewood, 1985; ILO/JASPA 1985c; Collier and Lal 1986; Zeleza and Ogot 1988; Zeleza 1991).

In Malawi land inequality also became more marked, so that by the turn of the 1980s many people were either landless or their holdings were too small to sustain them. By 1980-81 two-thirds of the peasant households reportedly held less than 1.5 hectares (Nafsziger 1988:122). Since 36% out of the 38% of the country's cultivable land was already under cultivation, the prospects for poor peasants to expand their production under their current farming practices looked decidedly bleak (ILO/JASPA 1985a:14; Kydd and Christiansen 1982; Mhone 1987). Similar evidence can be gauged from the ILO/JASPA reports on countries with such diverse agrarian systems as Somalia, Liberia, Sierra Leone, Sudan, and Tanzania

(ILO/JASPA 1985d; 1985e; 1985f, 1985g, 1985h) and in general studies of peasant communities in Africa (Barker 1989).

Land concentration and pauperization were not confined to the 'merchant' states, but could also be found in some of the 'rentier' states, such as Zambia, which obtained most of their revenue in the form of rent from the country's mining activities (Wood and Shula 1987). In the various capitalist-oriented 'merchant' and 'rentier' states, the indigenous bourgeoisie used its access to state power to accumulate private land. In the self-declared socialist states, such as Ethiopia and Mozambique, the state extended its control over land. In either case, poor peasants lost out in the land sweepstakes. Their ability to reproduce themselves progressively declined because of their dwindling access to land, agricultural inputs, and sometimes even labour. As a result of their growing reproduction crisis poor peasants increasingly found themselves in need of off-farm income. The pressures for employment intensified.

Structural adjustment programmes contributed directly to these processes. The World Bank pressured captive governments to undertake land reform and registration measures, arguing that private tenure and land titling would increase productivity by improving factor mobility, increasing access to agricultural credit, and encouraging on-farm investments (World Bank 1984, 1989; Feder 1987). In practice, land privatization encouraged land speculation and profiteering, absentee-landlordism and land concentration, and exploitative landlord-tenant relations. Studies on Kenya, where individual land titling had been extensively implemented from the mid-1950s, indicated privatization had not resulted in the expected benefits: expansion of land markets, better access to credit, increased agricultural investment, land improvements and higher productivity (Barrows and Roth 1990; Atwood 1990; Shipton 1992; Place and Hazell 1993). On the contrary, these reforms led to more tenure conflicts within families and deepening tenure insecurity for previously disadvantaged groups, especially the rural poor (Haugerud 1989), women (Mackenzie 1989), and herders (Galaty 1992). Whatever the outcome in terms of production and productivity, land reforms inspired by structural adjustment reinforced the changes in the patterns of accumulation in the rural areas, characterized by the processes of land accumulation, shortage, and dispossession. The growth of the wage-seeking labour force was thereby exacerbated as labour flocked out of landless and poor peasants' households in search of gainful employment.

It can be seen, therefore, that labour supply pressures increased quite substantially in the 1970s and 1980s. But the demand for labour lagged behind. As we noted in the first section of this chapter, the labour absorptive capacity of the public sector and industry either declined or stagnated, while agriculture's ability to generate wage employment remained low and uncertain. Many of the reasons for the incapacity of industry, especially manufacturing, to generate employment were alluded to in the previous section: capital intensity, shortages of foreign exchange, low capacity utilization, weak linkages with the domestic economy, infrastructural constraints and scarcity of entrepreneurs (ILO/JASPA 1985i).

It is generally agreed that import substitution industrialization, the strategy adopted by the new independent states in Africa in the 1960s, failed to generate technological development, large-scale employment and to improve the balance of payments as it was expected to. But, as noted in the previous chapter, the reasons for this are in serious dispute. The structuralist view attributes the 'failure' to the inherited distorted productive structure, and the fragmented nature of these countries' markets arising out of their small size and dependent status in the world capitalist system, while the neo-classical critique blames it on the existence of distorted and inefficient factor and goods markets, mainly created by government intervention in the economy, especially through the excessive protection of domestic industries behind high tariff barriers. The neo-classical view ignores history. Ever since Britain's 'spontaneous' industrialization, in virtually all subsequent industrializations the state has played a central role (Gerschenkron 1962; Sutcliffe 1971; Trebilock 1981; Kemp 1983). Most recently, state intervention facilitated the rise of the much-touted newly industrializing countries, especially South Korea, Taiwan and Singapore (Harris 1986).

Mkandawire (1989) offers a compelling historical and social analysis of Africa's low levels of industrialization and the process of de-industrialization in the 1980s. He argues that throughout this century Africa has been out of sync with the major spurts of industrialization in the developing countries. As colonies, African countries missed the first phase, between 1914 and 1945, when two World Wars and the Depression provided both a stimulus and protection to industrial activities in the politically independent regions of the developing world, especially Latin America. African countries, with the exception of the settler colonies, also missed the second phase, 1945-1960, during which favourable global conditions also allowed countries in Latin America and parts of Asia to extend and deepen import substitution. By the time African countries finally became independent in the 1960s, and free to pursue their own policies of industrialization, the shattered industrial economies had been reconstructed and there was greater opposition to industrialization in the developing countries just entering the race. At the same time the global recession was looming over the horizon. Moreover, the social structure of accumulation that emerged out of Africa's decolonization struggles meant that industrialization would not be subject to the ruthless purposefulness of a class project but would dodder along as an amorphous programme of the nationalist coalition. The debt crisis and the structural adjustment programme of the 1980s simply turned industrial stagnation into de-industrialization for many African countries.

The reduced labour absorption capacity of industry and agriculture led to the rapid growth of the services sector, which was forced to accommodate a high proportion of the new entrants to the labour market. In the 1980s the service sector grew faster than both agriculture and industry, so that its share of the continental GDP rose from 34.1% in 1980 to 38.7% in 1987 (UNDP and World Bank 1989:8). The value added in services in 1987, which stood at $141.0 billion, was almost twice that of agriculture. But the services sector was far from becoming a new reservoir for the rising tide of unemployment. In fact, most of the jobs created in the sector were characterized by low-skills and low-pay.

Informal is Beautiful

As the problems of the labour market intensified, governments, researchers, and the affected individuals themselves increasingly turned to the so-called informal sector as a sponge to absorb the miseries of unemployment away. It is revealing that the informal sector was first discovered in an African country. The term was coined by Hart (1973) in his influential article on informal income opportunities and urban employment in Ghana, and enthusiastically popularized by the ILO (1972) in its famous report on strategies for increasing productive employment in Kenya. The report drew a sharp distinction between the informal and formal sectors. It depicted the latter as the locomotive that would pull Kenya and other Third World countries out of their deepening crisis of employment and unemployment. Informal sector activities, the report enthused, were characterized by ease of entry, reliance on indigenous resources, family ownership of enterprises, small scale of operations, labour-intensive and adapted technology, skills acquired outside the formal school system, and unregulated competitive markets. The characteristics of the formal sector were the obverse of these. Formal sector activities suffered from difficult entry, frequent reliance on overseas resources, corporate ownership, large scale of operation, capital intensive and often imported technology, formally acquired skills, often expatriate, and protected markets (ILO 1972:6). In short, informal sector activities were small, local and beautiful, while those in the formal sector were big, foreign, and ugly. This was populist developmentalism in all its convoluted posturing.

The concept of the informal sector served as a banner to all and sundry. It appealed to, as Peattie (1987:857) has accurately observed:

liberals with an interest in problems of poverty; to economic planners who want their accounting system to represent the actual economy more accurately; to radicals who want to bring into planning analysis a more structuralist view of the economy; and to those who would 'privatize' activities such as housing production, either out of conservative commitment to action by 'the people' or out of a populist commitment to restraint in government welfare expenditures.

International agencies and governments found the concept attractive because it embodied policy implications that appeared to 'offer the possibility of 'helping the poor without any major threat to the rich', a potential compromise between pressures for the redistribution of income and wealth and the desire for stability on the part of economic and political elites' (Bromley 1978:1036).

Writing on the informal sector became an academic growth industry in the 1970s and 1980s (Sethuraman 1981). The concept was, in fact, extended to the advanced capitalist countries where its discovery was prompted in part by the growth in unemployment and the restructuring of work and by feminist critiques of conventional definitions of work (Smith et al., 1984; Redclift and Mingione 1985). It might be noted in passing that:

the way the informal sector is perceived in advance capitalist countries seems to be the exact opposite to its homonym in the developing world...because, even though certain shared characteristics could be identified, the developed and undeveloped informal sectors necessarily fulfil widely differing theoretical and ideological roles, as concepts they respond to different social and political preoccupations (Connolly 1985:59).

There is a large body of literature on the informal sector in African countries, much of it produced by the ILO and its Jobs and Skills Programmes for Africa (JASPA). By 1985 the ILO/JASPA had produced 52 studies covering 21 countries, out of which came an overall synthesis (ILO/JASPA 1985b). These studies were primarily concerned with the employment potential of the informal sector. It was shown that by the mid-1970s the informal sector already employed anything from 40% to

60% of the urban labour force in most African countries and contributed about a quarter to a third of urban incomes and a considerable size of urban value added. There were, of course, significant regional variations. The informal sector dominated the urban economy in West Africa, while in East Africa, where colonial zoning laws and urban planning had been much stricter, the sector was 'emergent'. In the southern African 'labour reservoir economies' it was 'negligible' (ILO/JASPA 1985b:15). The surveys also showed that the sector was highly dependent on family labour, with only 25% of the sectoral labour being paid employees who were not family members. Of this, a third were masters, and the rest 'labourers', a category dominated by apprentices. Informal sector enterprises were small and typically they had between one and three workers, with one-person enterprises comprising about a quarter of the sector. Less than 5% employed more than ten workers. Incomes in the sector varied widely, although around a low average figure. Only the entrepreneurs, and to some extent the skilled workers, earned incomes above or equal to the legal minimum wage. The vast majority of the sector's unskilled workers and apprentices did not. From this the ILO/JASPA concluded that the 'informal sector is a poor man's sector: it is a sector of the poor, for the poor', although of course, the unequal incomes cast 'doubt on any simple belief that the informal sector is a vast undifferentiated body of people all living harmoniously, albeit at low-income levels' (ILO/JASPA 1985b:16, 18).

The ILO/JASPA studies became increasingly cautious in their assessment of the informal sector's potential to create jobs. The 1986 report on Kenya, for example, noted that capital requirements in the sector had risen, thereby making entry more difficult. Also, real incomes in the sector had declined, parallel to the decline in formal sector real wages, as a result of the country's overall economic recession, and more specifically due to the horizontal expansion of informal sector enterprises. The increase in the number of these enterprises stiffened competition among them and increased their attrition rate. Few informal sector enterprises in Kenya managed to survive beyond four or five years. Many of them collapsed or failed to expand because of shortage of capital, lack of equipment, insufficient demand, unfair competition from large firms, inadequate supply of raw materials, scarcity of managerial skills and skilled workers, heavy taxes and licence fees, and lack of access to secure premises for their operations which made them susceptible to official harassment (ILO/JASPA 1986k:88-92). Not surprisingly, official statistics show that the number of jobs created in the sector in the 1980s declined sharply in comparison to the 1970s. In the urban informal sector the number fell from 34,200 in 1981 to 13,200 in 1985 and 19,500 in 1987, while in the rural informal sector only 21,700 jobs were created between 1984 and 1987 (Republic of Kenya 1988:36). Despite these reservations, the ILO/JASPA studies maintain that the informal sector in Africa performed much better than the formal sector in creating jobs. In fact:

the contraction of formal sector activities has created new opportunities for micro-enterprises. Indeed, there is evidence of substitution of informal for formal production, particularly in the field of every-day consumer goods. Moreover, the official attitude toward the informal sector has changed dramatically since the early 1980s. Whereas in the past, the government's position vis-a-vis the sector varied from benign neglect to outright harassment, the informal sector is now being considered as a lead for the creation of new jobs in an increasing number of countries (ILO/JASPA 1989:21).

The report shows that urban informal sector employment in Africa increased by a respectable 6.7% per annum between 1980 and 1985, as compared to 1% for urban formal sector employment. By 1985, the informal sector employed about 60% of the urban labour force, up from 56% in 1980. This meant that the informal sector employed about 15% of the total regional labour force in 1985. Altogether, it is said the urban informal sector created some 6 million jobs between 1980 and 1985, as compared to 0.5 million for the urban formal sector (ILO/JASPA 1989:21).

These statistics have to be taken with extreme caution, for after all, the informal sector derives its very existence from its invisibility in national accounts and statistics. Indeed, by informal sector the ILO/JASPA clearly refers to 'economic activities carried out by individual persons or petty enterprises, which generally go unrecorded in official statistics' (ILO/JASPA 1985b:9). The question such a formulation raises is this: do these activities cease to be informal once they are recorded?[9] Despite its wide usage, the concept of the informal sector has always remained fuzzy, defying definition, except in negative terms, that it is not formal. It is probably this very fuzziness that has made it so attractive; it is an empty vessel that can hold any ideological and theoretical mash. The informal sector has proved difficult to define because of its heterogeneity. Many attempts have been made to classify informal sector enterprises according to the type of activity involved, or the employment characteristics of the participants. House (1981), for example, talks of the 'intermediate sector' of well-off informal sector operators and the rest, the 'community of the poor', while Portes (1984) distinguishes between informal sector entrepreneurs and the informal sector proletariat.

But these classifications and differentiations do not tell us what it is that constitutes the informal sector, or what distinguishes it from the formal sector.

The debate on the nature of the linkages between informal and formal sectors has done little to clarify the distinctiveness of the informal sector, for the distinction between the two sectors is often taken as given. The linkages have been analyzed in three main ways. First, there are those who see the linkages as benign. They include the ILO/JASPA studies which emphasize the autonomy of the informal sector and contend that the informal-formal sector linkages, backward, forward, and technological, are rather limited. 'The only significant linkage', one report states, 'is provided by market demand from formal sector consumers' (ILO/JASPA 1985b:47). Others argue that the linkages are close, but they are complementary rather than exploitative (McGee 1973, 1974). Second, some see the informal sector as subordinate to the formal sector in two senses, either as a marginalized sector denied access to resources of production and product markets (Quijano 1974; Bienefeld 1975; Souza and Tokman 1976), or as a dependent integrated sector whose surpluses are extracted by the formal sector through the mechanisms of unequal exchange of goods and services, subcontracting, and so on (Leys 1973; Gerry 1974, 1978; Bienefeld and Godfrey 1975). Finally, others argue that the linkages between the two sectors are heterogeneous and differentiated, so that they are inherently neither benign nor exploitative. It all depends on the nature of the activity involved and the historical context (Tokman 1978:1071-1073; Traeger 1987:245-46). Although the debate on the nature of informal-formal sector linkages has become more nuanced, we are no nearer defining what the informal sector actually is than we were over 20 years ago when the ILO discovered it.

It has proved difficult to formulate exclusive criteria for the definition and identification of the informal and formal sectors because the two are contrived inventions. The informal-formal dichotomy has a long pedigree in the dualistic conceptualizations of development economics, in which two sets of economic sectors have always been contrasted in the developing countries (Streeten, 1987). The new informal-formal dualism absorbed and replaced other dichotomies, such as 'firm-centered-bazaar' economy (Geertz 1963), 'capitalist-peasant' production system (McGee 1973), the 'upper and lower' circuits (Santos 1979), 'traditional-modern' sector (Weeks 1975; Aboagye and Gozo 1986), 'protected-unprotected' sector (Mazumdar 1975, 1976). Others talked of 'advanced-backward' sector, 'registered-non-registered' sector, 'regulated-unregulated' sector, and 'regular-irregular' sector. Many critics of the informal-formal sector dichotomy have ended up proposing their own dichotomies that bear uncanny resemblance to it. For example, the 'organized-unorganized' dichotomy of Harris (1978) and the 'official-unofficial' dichotomy of Clark (1988).

Many of the activities within the informal sector fall under the rubric of petty commodity production (Moser 1978:1055-62), but to identify the informal sector exclusively with petty commodity production would amount to the substitution of one duality with another, 'petty-full commodity production'. The challenge is not to invent more refined dichotomies but to abandon dualist construction altogether because as Peattie (1987:857-8) has argued, as frameworks 'for considering problems of poverty', they do not work because they 'are factually incorrect and politically obfuscating'. The 'informal sector', however defined, is not necessarily a category within which to locate the poor. There are well-to-do petty entrepreneurs, and underpaid workers in large enterprises. It is politically obfuscating because the use of the category 'informal sector' as central in the analysis of poverty carries, as an implicit theme, the notion that the more highly corporatized the economy, the better will be the position of working women and men. It is well to remember:

> the informal-formal division is inapplicable to many people as they work in both sectors at different stages in their life cycle, times of the year, or even times of the day. The division is even less applicable to households or neighborhoods, as some members may work in the informal sector, while others work in the formal sector (Bromley 1978:1035).

The meshing of informal-formal employment accelerated in the 1980s, as the security of formal sector employment declined and the income gap between them evaporated, so that the contrived distinction between the two became increasingly blurred. It is my contention that the people working in the so-called informal sector did not constitute a separate group, but were an integral part of the working class. What happened in the 1980s, therefore, was not the growth of the 'informal sector', but the emergence of new forms of work and employment in Africa, prompted by the profound changes taking place in the processes of accumulation and the social relations of production. For one thing, the capacity of the fiscally-weakened state to control and oversee the labour process declined. Researchers, armed with dualist baggage and statist conceptions of society, concluded that the 'informal sector' was expanding, indeed, that informalization was the well from which would spring 'an indigenous, technically innovative, and developmental bourgeoisie' (Berman 1994:254), and a revitalized civil society that would foster democracy (Chazan 1988). Individuals and households were simply diversifying their range of remunerative occupations in order to reproduce themselves in the face of an excruciating crisis.

Let Freedom Come

Besides the diversification of income-generating activities by individuals and households, African workers tried to cope with falling wages and declining standards of living in numerous creative ways. Urban dwellers increasingly grew their own food in their backyards, along roadsides and on any patch of available land, or they resuscitated rural linkages and kinship ties and supported rural farming activities. The ruralization of urban residential spaces was one more development that dissolved the sharp rural-urban analytical dichotomy (Bangura 1995:92). Others looked for greener pastures elsewhere, internally, regionally, and internationally. Recent studies show that internal migrations, rural-urban, rural-rural, urban-rural, urban-urban intensified in the 1980s (Toure and Fadayomi 1992; Becker, Hamer, and Morrison 1994; Mafeje and Radwan 1995).

Intra-regional migrations also accelerated. There were the forced migrations of millions of refugees fleeing from natural disasters, such as drought or, more likely, national strife and warfare. These crises and conflicts were, of course, underlined or reinforced by problems of underdevelopment and the poverty-enhancing structural adjustment programmes. Of the fifteen million refugees worldwide at the beginning of the 1990s, five million were in Africa, produced mostly from countries in the Horn - Ethiopia, Sudan, and Somalia; Liberia, Ghana and Chad in West Africa; Rwanda and Burundi in Central Africa; and Angola, Mozambique and South Africa in the southern sub-region (Adelman and Sorenson 1994). Also notable were the increasing migrations of traders operating in trans-border parallel markets, taking advantage of differential national prices, currency values, availability and shortages of goods.

Many more trekked to other African countries as migrant workers. By the mid-1980s about five to six million workers were crossing their border in search of work, mostly to Nigeria, Côte d'Ivoire, Libya, and South Africa. But they were not always welcome. Many were expelled by governments who were at the same time falling over themselves to attract foreign capital. For example, in 1983 and 1985 Nigeria expelled over a million migrant workers from the neighbouring countries. In 1984 Libya expelled about 60,000 migrant workers from other African countries including 30,000 Tunisians. South Africa threatened to expel

1.5 million foreign workers if the international divestment and sanctions campaign against its apartheid regime continued (Hear 1985:75). In 1985 Côte d'Ivoire expelled thousands of Ghanaians 'allegedly in reprisal for football violence against Ivorians in Ghana', and Gabon threatened 'to deport all illegal migrants, adding that all foreign prostitutes would be rounded up and raped by the police force' (Hear 1986:59). In November 1988 Kenya expelled 2,000 Ugandan secondary school teachers. As a result of the dispute between Senegal and Mauritania about 200,000 people were expelled from the two countries (Sy, Ba and Ndiaye 1992:122).

Not surprisingly perhaps, some of Africa's best and brightest were leaving for the western countries. According to UNCTAD (1979), between 1960 and 1975 an estimated 27,000 highly skilled Africans migrated to Western Europe and North America. By 1984 the ECA (1988c) reported the number of African emigrants to the West had risen to 40,000, and by mid-1987 it was up to 70,000, which represented almost 30% of Africa's highly skilled labour stock (Brister 1988). Using UNCTAD's 'imputed value of skills flows' Africa was losing $12.9 billion. In reality, Balogun and Mutahaba (1990:67) contend Africa was losing more through delayed or unimplemented projects and the importation of replacement high-cost expatriate labour. Even taking into account remittances of the emigrants, the costs were high. Needless to say, some countries were more affected than others. By 1984 there were between 1 and 2 million Egyptians and over half a million highly skilled Sudanese were working abroad, mostly in the oil-rich Gulf states. By 1988 there were nearly 600,000 Cape Verdians living overseas, twice the country's domestic population. And there were thousands of Algerians and Francophone West Africans in France, and Ghanaians, Ethiopians, and Somalis scattered in Britain, Italy, Canada, and the United States. And Nigeria, once a destination of skilled labour, saw an exodus of 'top professionals whose salaries have been devalued by thousands of dollars a year... [including] medical specialists, academics, scientists, engineers, airline pilots, sports stars and others' (Oredein 1988:14; Uguru 1988).

It was a brutal devaluation and expatriation of Africa's labour power. It provoked various forms of resistance, both overt and covert, organized and spontaneous, collective and individual, episodic and continuous, some of which coalesced, unevenly and unpredictably, into social movements against the prevailing maladjusted social order, for a new dispensation, for democratization. One of the countries where anti-SAP opposition turned into riots and eventually struggles for democracy was Zambia. As structural adjustment eroded standards of living and the state grew more intolerant, the frayed public nerves finally erupted into the food riots of June 1990 which left several people dead, and a month later the Movement for Multi-Party Democracy (MMD) was born, spearheaded by trade unions and entrepreneurs frustrated by their diminishing opportunities for accumulation. The MMD proceeded to win the elections of 1991, bringing to an end the country's one-party rule (Hamalengwa 1992; Chanda 1995; Isamah 1995). The struggle against structural adjustment and for development and democratization entered a new and complicated phase.

Workers' and popular struggles against SAP and military dictatorship proved particularly difficult in Nigeria. The labour movement, as represented by the Nigerian Labor Congress (NLC), was alternately militant and moderate, confrontational and cautious, and the state tried to crush and coopt it. The NLC campaign, strongly supported by students and other disaffected groups, was galvanized around two highly charged issues: against the removal of the 'petrol subsidy' and for a revision of the minimum wage. The 'petrol subsidy' crusade heated up in 1987 when the government came under pressure from the IMF and World Bank to 'adjust' domestic fuel prices upwards in line with the falling value of the naira. Despite widespread repression and intimidation of labour leaders, including the dissolution of the NLC, the issue had caught fire with the wider public and in April 1988 the country was swept by the 'petrol uprising' in which there were nation-wide demonstrations, riots, and strikes. The state was forced to unban the NLC and back down from implementing a sharp price increase. It then proceeded to refine its oppressive and exploitative machinery and to take advantage of ideological, occupational, ethnic and regional schisms in the labour movement itself and the anti-SAP and pro-democracy social movements in general. In fact, in the subsequent battles the NLC took a back seat to the students and unorganized workers who accused the NLC of having sold out. Reacting to the deterioration in the universities and the aborted aspirations of the middle classes, for which they were training, the students led the Anti-SAP Rising of May 1989, while the NLC focused, some say dissipated, its energies on the formation of a Labor Party following the lifting of the official ban on party politics. Around the same time a campaign for an increased minimum wage and a national conference was launched. Promised a revised minimum wage on the eve of the proposed national conference, the NLC pulled out, and the state banned the conference. The cleavages within the labour movement and the democratic movement were wider than ever while, simultaneously, state authoritarianism escalated and its legitimacy evaporated. Nigeria entered the 1990s on the knife edge of social upheaval (Bangura 1989; Olukoshi 1991; Adesina 1994; Beckman 1995, Albert 1995).

Powerful and contradictory social forces were also unleashed by structural adjustment in the northern corners of the continent. For example, in Tunisia the confrontations between the labour movement, the state, and the Islamic fundamentalists intensified. Led by the Union Générale des Travailleurs Tunisiens (UGTT), Tunisian workers vigorously protested the decline in their purchasing power, while the Islamic fundamentalists sought to fill the ideological space opened up by the fading promises of modernity. Until the late 1970s, the hegemony of the ruling Destour Party was based on an alliance with the trade unions. This alliance was ruptured in the events leading to, and in the aftermath of, the General Strike of January 1978, which was ruthlessly suppressed. At the turn of the 1980s as economic conditions deteriorated and the ruling party's legitimacy waned, leaders of both the labour and religious movements were terrorized by the state through arrests and detentions. Matters came to a head following cuts in food subsidies on 27 December 1983. The price of bread almost doubled overnight. The reaction was immediate. The 'bread riot' of January 1984 erupted, reminiscent of the riots in Egypt in January 1977 and Morocco in June 1981. The riots brought to the public stage women and urban youths formerly marginalized from politics, and 'was the manifestation of the fragility of the civil society and alienation of the state, of the Destour party and even of the trade unions, from the large, unorganized' and differentiated masses (Zghal 1995:127). The price rises were rescinded, but the repression of leaders of the UGTT, the secular opposition, and the Islamic movement increased. Out of the political impasse came the palace coup of November 1987 in which the Prime Minister, Ben Ali, overthrew the increasingly erratic and octogenarian President Bourguiba. The Destour Party changed its name, and the regime sought to reinvent itself through the multi-party elections of April 1989, which it succeeded to win against the demoralized and demobilized secular opposition parties. This demonstrated 'the polarization of political life between a ruling party and a Moslem fundamentalist opposition' (Romdhane 1995a:187; 1995b). Almost sidelined was the UGTT which had been 'the major driving force of the democratic struggle in Tunisia' (Zeghidi 1995:364).

Conclusion

It is quite evident, therefore, that the deterioration in the conditions of living, manifested most poignantly in rising unemployment and falling real wages, which were brought about by the economic crisis and reinforced by the structural adjustment programmes, profoundly altered the political and social terrain in many African countries. State repression and societal resistance increased, at the same time as new accommodations and adjustments were being woven into the quilted tapestry of state-civil society relationships. An attempt has been made in this chapter to outline the dimensions of the unemployment crisis in the 1970s and 1980s, particularly its differentiated impact on various sectors, age groups, and on men and women. It was shown that the public sector experienced the most contraction, thanks to the fiscal crisis of the state, and that industrial employment remained stagnant because of the structural constraints of manufacturing enterprise. The crisis, it was pointed out, was mainly borne by the youth and women, the former because of their rapidly growing numbers at a time of sluggish economic growth, and the latter because of the historical, structural, and institutional biases against them, not only in the labour market, but in all walks of life, which were bolstered by the economic recession and the gender distortions of 'free-market' interventionism.

It was argued that unemployment in Africa in the 1970s and 1980s was not simply caused by labour market distortions. These distortions themselves reflected larger structural imbalances in African political economies. The widening gap between labour supply and demand owed a lot to the rapid growth of both the labour force, itself a product of Africa's post-World War II demographic explosion, and the growth of the educated population, the result of independent Africa's massive investment in the development of its human resources. Changes in the agrarian patterns of accumulation, which released large numbers of people from resource-poor households, also contributed to the labour supply pressures. In the end, it cannot be overemphasized, rising unemployment was engendered and sustained by the domestic and global processes of accumulation, internally-generated and externally-imposed policy responses, principally the structural adjustment programmes. Unemployment represented a huge waste of human resources, and it took an incalculable toll on its immediate victims and their communities. The unemployment problem in Africa cannot be wished away by conjuring up 'economies of affection' in Africa's rural hinterlands, or dynamic 'informal sectors' in the rapidly growing cities as so many writers tend to do. Against this devastating devaluation of their labour power, millions of African men and women increasingly responded by fighting for, and imagining, a new social order, one that was empowering in material and moral terms, one, in short, that was democratic. To that we now turn.

Notes

1. The discrepancies of labor statistics in which, for example, Pakistan has a lower rate of unemployment than the United Kingdom, arise because only a select aspect of unemployment is measured: that of total lack of work and the existence of unemployment insurance and other public relief schemes that can be tracked statistically. In other words, in the developing countries the unemployed are often swept under the carpet of the 'informal sector' and the statistical invisibility of generalized poverty, just as unremunerated household work, mostly performed by women, is not counted in the employment statistics. See the discussion on 'Controversies in labour statistics' in ILO (1995: Chapter 1).

2. The first generation consisted of the initial adjustment policies aimed at economic stabilization; the second comprised comprehensive adjustment policy packages; and the third paid more attention to the social factors, see Cornia, van der Hoeven and Mkandawire (1992); Gibbon, Bangura and Ofstad 1992; and Hoeven and Kraaij 1994.

3. There are those who argue that by the turn of the 1980s the public sector was quite bloated. Indeed, Africa could boast the dubious distinction of having the highest share of public sector workers in non-agricultural wage employment, 54% as compared to 36% for Asia, 27% for Latin America, and 24% for the OECD countries (Heller and Tait 1984).

4. The text talks of 14 countries, but 13 are actually mentioned. They are: Benin, Botswana, Burundi, Côte d'Ivoire, Gambia, Kenya, Malawi, Mauritius, Niger, Seychelles, Swaziland, Zambia, and Zimbabwe.

5. The 17 countries are: Kenya, Sierra Leone, Nigeria, Somalia, Liberia, Ghana, Zambia, Tanzania, Lesotho, Mali, Burkina Faso, Benin, Togo, Congo, Cote d'Ivoire, Senegal and Cameroon.

6. The nine countries are: Botswana, Gambia, Kenya, Liberia, Malawi, Mauritius, Niger, Swaziland and Zimbabwe.

7. The six countries are : Egypt, Malawi, Mali, Seychelles, Tunisia, and Cameroon.

8. He continues, quite wrongly, that 'the steady fusion of the urban labour markets explains a great deal about the absence of any serious social upheavals in the wake of the sharp falls in real wages in African countries during the 1980s.' What were all those 'food riots' and mounting struggles for democratization all about? On the flexibility of wages in Africa also see Colclough (1991).

9. It was not until 1993 at the ILO-sponsored Fifteenth International Conference of Labour Statisticians that a resolution was passed specifying the criteria for a statistical definition of the informal sector and recommending on the design, content and conduct of informal sector surveys, see ILO (1995: 21-3).

Part Four

Imagining Democracy

Chapter Fifteen

POWER IN AFRICA'S PASTS

Introduction

Democracy and democratization have become the new 'flavour of the moment' in African Studies, almost eclipsing development and developmentalism. In the vanguard of this new discourse are political scientists and all those whose intellectual sights are firmly fixed on the murky present, always with an eye to predicting and prescribing the mysterious future. As contestations of conflicting hopes, ideologies, and theories the analyses are vigorous, controversial, and inconclusive, for they are fundamentally debates about African histories, about African pasts and futures, as constructions and reconstructions, prognoses and visions.

The possibilities and limitations of the current democratic projects in Africa are often conceptualised and assessed in the historical contexts of pre-colonial, colonial, and post-colonial traditions and legacies of power and politics. To some, democracy in Africa can only flourish if it is grounded and rooted in the communal and egalitarian values of the pre-colonial past, in indigenous models of governance (Ayittey 1991, 1992; Owusu 1992; Davidson 1992; Landell-Mills 1992). Unrepentant imperialist historians and their acolytes in the western mass media still believe that colonialism bestowed on Africa the gifts of 'good government' and civilization, but unfortunately, it was too brief, so that, quite predictably, anarchy and tyranny returned following decolonization (Austin 1993; also see Duignan and Gann 1967, 1969-1975; Lee 1967; Kirk-Greene ed. 1979; Morris 1980; Morris et al., eds., 1980; Gifford and Louis eds. 1982, 1988). Salvation, it follows, lies in renewed western political tutelage and economic benevolence, in some kind of recolonization[1]. Many Africanist scholars seem to blame the dictatorships, corruptions, and confusions of the post-colonial order on innate cultural traits inherited from the pre-colonial past, conveniently glossing over the structural deformities bestowed by colonialism, and they compete, as suggested in Chapter 7, in coining the worst epithets to describe the post-colonial state in Africa: 'predatory', 'prebendal', 'parasitic', 'precarious', 'patrimonial', 'neopatrimonial', 'lame leviathan', 'swollen', 'collapsed', 'decadent', 'non-developmental', 'kleptocratic', 'greedy', 'crony', 'venal', 'vampire'; it is a state committed to no mission higher than 'belly politics' (Callaghy 1987; Diamond 1987;

Joseph 1987; MacGaffey 1987; Rothchild and Chazan 1988; World Bank 1989; Sandbrook 1990; Fatton 1990, 1992; Jackson and Rosberg 1992; Bayart 1993; Bratton and van de Walle 1994; Zartman 1995). Underlying this scholarship-by-epithets is a profound contempt for Africa and the conviction that there is no alternative to western political modernity[2].

It stands to reason that historians, as students of social processes and change are more than equipped to interrogate, extend, and enrich these debates. They can bring important insights into the dynamics, patterns, and traditions of politics and power over long durations of time, of the connections between state formation and the structured inscriptions of class, community, and gender, between the modes of domination and the modes of resistance, and decipher the conflicting tendencies towards democratization and authoritarianism in Africa's diverse and complex histories. Constructions and reconstructions of the past, indeed, always involve writing contemporary history.

This chapter primarily focuses on the political traditions and legacies of the pre-colonial era, the putative source of Africa's supposedly sick contemporary political order. It is divided into five parts. First, it examines briefly the shifting meanings and content of democracy, followed by an outline of the dominant paradigms that have been used in analysing politics and power in African historiography. Second, it looks at the polities and the structures of power in the state societies. Singled out for a specific critique in this section is Basil Davidson's (1992) recent synthesis, *The Black Man's Burden*. The third part focuses on the so-called stateless and acephalus societies. Finally, the relationship between production and power in the nineteenth century is explored. Finally, the chapter briefly looks at the construction of the colonial states and their complex articulations with the pre-colonial polities.

Claiming Citizenship

Democracy is, of course, a condition everyone professes to know and claims to cherish, but like beauty, its meaning often lies in the eye of the beholder. Particularly powerful and seductive is the Athens-to-Washington narrative of 'democracy' propagated by many western commentators, in which the 'idea' of democracy is said to have been invented in ancient Greece and reached its telos in the contemporary West. This idealistic narrative of democracy as the outcome of concepts not conflicts, insights not instigations, philosophy not practice is, of course, fictitious and ahistorical. Greek society, with its slaves, not to mention the subordination of women, was not democratic[3]. Liberal democracy, as conventionally understood, developed fairly recently following the rise of industrial capitalism in the West. At the same time that this form of democracy was evolving, these countries were perpetrating atrocious violations of human rights in the world at large, including the genocide of the native peoples of the Americas, the Atlantic slave trade, colonial conquests, not to mention the two world wars and numerous imperialist-inspired wars and conflicts and support for tyrannical regimes throughout the world. Indeed, the western countries continue, especially the lone superpower, the United States, to project their laws abroad and readily use force, and to conduct 'economic and sermonic warfare' at will, as Chomsky (1994:16-17) puts it so memorably.

Thus, the development of liberal democracy has been uneven within the economy of global capitalism. In the core capitalist countries themselves, it has been vulnerable periodically, especially during moments of crisis, whether of accumulation or hegemony, as happened in Nazi Germany and McCarthyite America. It also needs to be pointed out that many of the rights associated with liberal democracy, including the right to vote, came quite belatedly to women and racial and ethnic minorities in the western countries, including, in the case of North America, the native peoples. Indeed, democracy in these countries is far from 'complete', regardless of what the Fukuyamas (1992) say about 'the end of history'. If anything, it is in deep crisis, partly because the historic struggles of marginalized groups - women, racial and ethnic minorities - have exposed the limitations and ambiguities of liberal citizenship, and unveiled the disjunction between its formal and substantive aspects. Also, globalization has increasingly weakened the state as the arbiter of citizenship and eroded the certainties underlying claims to citizenship. The exhilarating myth of citizenship as a superordinate identity within the sovereign boundaries of the nation-state, subsuming and coordinating all other identities of class, ethnicity, gender, location, and culture, in pursuit of a common purpose or as a procedural commitment to justice, lost credence as the feminist and civil rights movements and other struggles for difference rejected the universalist and homogenizing claims of liberal citizenship and demanded both the recognition of difference and entitlement to equal opportunity. As the compact of citizenship crumbled, the ritual performances of liberal democracy lost their audience: fewer and fewer people bothered to vote.

Exit from the electoral process is only one of the symptoms of the crisis of liberal democracy. The other is the proliferation of exclusionary practices of citizenship in which class, cultural or ethnic 'outsiders', especially immigrants, are condemned and often criminalized and denied the rights and services of incorporation. These reactionary and racist movements reflect, and are perverse responses to, both the intense internal struggles by formerly marginalized groups, and the destabilizing processes of globalization, which have been rupturing and restructuring rights to reconstituted national spaces. Thus, while the transnationalization of capital has been followed and fostered by the formation of transnational legal regimes that accord corporate citizenship rights, transnationalized labour remains trapped in the exclusionary fictions of nationalism, their rights to the possibilities of citizenship spurned by the stubborn narratives of immigration and otherness[4]. These contradictory claims, concessions, and constraints to citizenship are played out most markedly in the denationalized urban spaces of the megacities, the centres in which international trade and investment are serviced and financed and the transnational companies headquartered, and to which the labour migrants, both the highly skilled professionals and the unskilled crowds, flock. The valorization of transnational capital and devalorization of transnational labour are portrayed, and practised, in the disparities of the glamorous enclaves of business districts and the gloomy zones of squalid neighbourhoods (Burbaker 1989; Dietz 1992; Sassen 1988; 1996; Featherstone 1990; *Index on Censorship* 1994; Holston and Appadurai 1996).

Thus, democracy understood as the access to and exercise of political, economic, social, and cultural rights within the increasingly porous territorial grids of nations, has nowhere yet been achieved. On the contrary, as globalization undermines the articulation between citizenship and nationality, millions of denationalized people, the disempowered legal immigrants and criminalized illegal immigrants, face more exclusions from the rights and services that their labour and humanity should entitle them to. Contemporary western triumphalism in the wake of the collapse of the former Soviet Union hides these inconvenient realities and the fact that liberal democracy as currently constituted in the Western countries is not the culmination of human struggles for empowerment by individuals and collectivities for access to the socially-produced economic and cultural resources, and for control over the determination of their needs and the means of satisfying them; in short, for more popular participation and more open decision making (Lebowitz 1995; Levins 1995; Miliband 1995; Wallerstein 1995).

Historically, the inclusion of economic, social, and cultural rights as part of democratic projects came with the development of Marxism and the triumph of socialist revolutions in Eastern Europe and China, followed by the rise and victories of nationalist movements in Africa and

Asia, and the post-independence developmentalist struggles of these countries, which were reconstituted, together with Latin America, into the 'Third World'. The socialist emphasis on socio-economic rights can be attributed to the fact that socialism emerged out of critiques and struggles against capitalism and tried, with varying degrees of seriousness and success, to transfer real power to the producing classes, the peasants and the workers, who had previously been exploited under capitalism. Similarly, the emphasis on the rights to self-determination and development in Third Worldist circles was spawned by the histories of imperialist colonization and underdevelopment[5].

Given these histories, democracy cannot, and should not, be reduced to the empty shell of competitive party politics, which is often justified on the grounds that it ensures the trinity of 'good governance': efficiency, accountability, and the protection of individual rights. It is not simply for better 'governance' that people all over the world have sacrificed, and continue to sacrifice, their lives: it is for their empowerment in the various spatial economies they occupy, from the local community, to the national, and the global systems. Notwithstanding the universalist pretensions of the dominant western liberal discourses on human rights and democracy, therefore, democracy has neither been the exclusive invention of the western world, nor is it 'finalized', so that all African democratic movements have to do is to import it as a turnkey project from the western political emporium. The meanings and social content of democracy have continued, and will continue, to expand and deepen as subjugated classes, genders, peoples, and nations seek civil and cultural freedoms, political participation, and material development. The crisis of the two dominant historical modes of politics since the nineteenth century, as Wamba-dia-Wamba (1992) calls them, the Westminster model and the Stalinist model, which has virtually collapsed, leaves the door wide open to new constructions, visions, and possibilities of democracy. History indeed continues.

The study of African polities has been dominated by three paradigms: the autonomist, instrumental, and process models (Lonsdale 1981). Underpinning the first paradigm are assumptions of the autonomy of state power and the productivity of political action. The earlier studies written from this tradition conceptualised the state as an independent, active, and creative agency, organized to promote consensus and cohesion. In the later studies, the autonomous state was no longer perceived so benevolently as a symmetrical assemblage of power ensuring social harmonies and equality, but as an asymmetrical concentration of power which generates discords and disparities.

Instrumentalist conceptions of the state took the functional stress further, seeing power in its visible exercise rather than as a many-sided social relation. One version emphasised the centrality of external force, wielded by a minority, as a factor of cohesion in plural, ethnically heterogeneous societies. This analysis was applied to the pre-colonial, colonial, and post-colonial states. The class and neo-colonial versions of state instrumentalism have largely been applied to the colonial and post-colonial states. These states are seen as instruments of bourgeois hegemony, either local or foreign.

The two paradigms have tended to produce idealist histories of politics and power. Analyses inspired by the autonomist paradigm in both its symmetrical and asymmetrical versions have been teleological in that 'they measured past and present change against contemporary hopes for African democracy. In addition, political intentions and values were seen as prime components of power, especially the power to institute change' (Lonsdale 1981:150). The instrumentalist paradigm, on the other hand, produced functionalist studies that did not sufficiently problematize and analyse state formations, and their class compositions, as complex historical processes.

More satisfactory is the third paradigm, that sees the state as a process, whose apparatuses, class composition, ideology, and material base, are historically formed and structured. As historical formations, states manifest themselves in diverse forms, each marking specific articulations between state and nation, state and society, coercion and consent. Consequently they are subject to change as their economic, social, ideological, and cultural configurations and contexts change, and as their legitimacies are ruptured and reorganized.

Many historians recognize this, but it has proven extremely difficult for them to map out the changes in African state and political formations over time, thanks in part to the diversity and complex histories of the continent's polities, cultures, and societies. The focus on the state is itself problematic, for there were many people in various parts of the continent, at certain moments, who did not live together in states. In a situation where there were so many different and changing theatres of politics and power it is difficult to make meaningful generalizations about *African* traditions of democracy and authoritarianism.

Reinventing the Political Kingdom

Awareness of the complexities of the African political pasts has not deterred some students of African political history and contemporary politics from advancing analyses, interpretations, and generalizations about pre-colonial political practices and values, and inferring their legacies for the present and implications for the future. Ever since the de-stooling of the imperialist school in African historiography following the attainment of independence and the consequent institutionalization of African history in the universities in Africa and abroad, pre-colonial Africa ceased to be seen as one long night of savagery and disorder and came to be hailed by the nationalist historians as a golden age of peace and democracy.

This was, of course, not merely an exercise in academic disputation and reconstruction. It was a popular and serious political discourse, one in which the continent's leading 'philosopher kings', as shown in the next section, actively participated. They were intent on finding charters of legitimation and nation-building for their fragile, newly acquired states, and reconfiguring the developmentalist legacies and projects of late colonialism and the nationalist movements. Nationalist historians elaborated on these ideas. Some searched for states to challenge the imperialist myth of an anarchic, stateless Africa, and they found large, powerful kingdoms and empires, with efficient bureaucracies, judicial systems, and military establishments. One of the aims was to show that since Africans had ruled themselves before, they were capable of assuming the mantle of Nkrumah's celebrated 'political kingdom', that they were inheritors of worthy indigenous traditions of statecraft and democracy.

Despite all these celebrations of Africa's pre-colonial political heritage, Davidson (1992:19) laments that Africa's post-colonial ruling classes preferred to adopt political 'models from those very countries or systems that [had] oppressed and despised [them]', instead of modernizing 'from the models of [their] own history, or inventing new models'. This, he believes, is at the root of the continent's contemporary political crisis. The argument deserves closer scrutiny, for Davidson is perhaps the most celebrated Africanist historian advocating the nationalist perspective in African historiography, and his book throws into sharp relief the strengths and weaknesses of this approach.

The reasons for the ill-advised rejection of African models of governance, he argues, are rooted in the modern social history of the educated strata, whose formation began in the nineteenth century in the mission-schools of the 'recaptive' settlements of Sierra Leone and Liberia. This was an elite deeply ambivalent about both Africa and Europe: the former had sent them into slavery and the latter despised them. Consequently, they felt comfortable neither with Africa, to which they belonged physically, nor Europe whose values they were socialised to admire. As an alienated intelligentsia, opposed to both European racism and African 'savagery', it saw its salvation in pursuing the modernization of Africa along European lines in order to liberate the continent and themselves from European control and condescension. It was a mission, Davidson (1992:35) laments, that entailed the erasure of Africa's own history, of the pre-colonial heritage, 'no matter how much they spoke in defence of the virtues of Africa's cultures'.

This alienation and ambivalence among the nineteenth century recaptive intelligentsia has also been observed by other writers, such as Mudimbe (1988:Chapter iv) and Appiah (1992:Chapter One). Needless to say, it was later reinforced by the imposition of colonial rule and the development of colonial schooling which provided, in Rodney's (1982:241) pithy critique, 'education for subordination, exploitation, the creation of mental confusion, and the development of underdevelopment' (also see Mugomba and Nyagah 1977). Some have argued, including myself (see Chapters in Parts I and II above) that cultural alienation and ambivalence among the African intelligentsia has persisted since independence, reproduced through inadequately decolonized educational systems, and the continued domination of a Western epistemological order in African studies, and the externalist orientations of the post-colonial developmentalisms of both the left and the right. It does not necessarily mean, however, that the rejection of pre-colonial models necessarily implies cultural alienation and reactionary politics, for the models themselves may have questionable validity as historical constructions. The histories of state formation in pre-colonial Africa that Davidson outlines demonstrates this quite well, for they are problematic in their simplification and romanticization of African political history.

The first path he traces is the formation of an ethnically distinct nation-state, represented by Asante, and secondly the formation of the ethnically diverse *regna* of the ancient empires of West Africa. Asante rose from a constellation of rival clans into a nation-state, Davidson (1992:59) writes, with all the attributes to justify that label. It had definite boundaries, 'a central government with police and army, a national language and law .... a history of its own state formation', and distinct unifying symbols and rituals of power. Political life in this state, which had by 1750 become a powerful empire-state in 'effective control of the whole of what would become, two centuries later, the republic of Ghana', was organized around three 'principles of good government', based on 'the rule of law, in the diffusion of executive power, and in the encasing of that power within political and legal checks upon its use....'. This was a system that believed in 'participation that must not only work, but must publicly be seen to work... [and] a systemic distrust of power'. The second path he discusses, although it preceded that of Asante historically, is the state formations of the ancient West African empires of Ghana, Mali, Songhay, and Kanem, each of which had a geographically large, ethnically diverse, strong state, which sought to open 'wide regions to the expansion of trade and improvement of trade and production for trade' (Davidson 1992:94), rather than imposing the cultural hegemony and the nationalism of the core peoples of the *regna*. These were, therefore, large, tolerant, federalized political formations.

There are several problems with these reconstructions, some of detail, but largely of interpretation[6]. First, as syntheses they do not incorporate the full range of African political systems spread over time and space, despite Davidson's (1992:60-1) contention that the Asante model was 'characteristic of precolonial political institutions in every African region where stable societies produced one or other form of central government'. Second, there is Davidson's annoying practice of legitimating these models through comparison with European models. Once formed 'the Asante polity', he reassures us, 'proceeded to behave in the best accredited manner of the European nation-state' (Davidson 1992:59). Similarly, he insists, the West African *regna* were comparable to the European *regna*, 'each was able, just like the Normans in the empire of the Franks or the Germans in the Holy Roman Empire to extract from it both tax and tribute'. And, he concludes, 'when the African *regna* in due course fell apart and disappeared from history... their subject peoples - just as in Europe - by no means followed them into the void' (Davidson 1992:93).

After all these years of deconstructing imperialist historiography one would have thought that it was no longer necessary to write African history by imitation, to justify African experiences, institutions, and ideas through the borrowed processes, problematics, and paradigms of European history. It is all the more strange when the argument is on the authenticity of the African political heritage, which the modernized elites have apparently abandoned in favour of European forms. These are the contradictions that the analogous arguments of nationalist historiography inevitably lead to. And in any case who said medieval Europe with its feudalism or the early modern slave-trading European state was democratic? This is a history preoccupied with the forms, not the social content, of structures, not processes.

The third problem in Davidson's typologies of state formation, therefore, is the tendency to search for origins and events, rather than transformations in political processes, and to explain state formation in terms of external forces, whether conquest or the growth of long-distance trade. Also, echoing the equilibrium models of functionalist social anthropology, we are presented with frozen snapshots of these states, idealized moments in time, characterised by little inequality and conflict. While Davidson would be the last person to say so, 'the point of all these hypotheses', Lonsdale (1981:172) correctly points out, 'was that something rather exceptional was needed to explain any concentration of power in a logically tribal Africa. Furthermore, they were all based on plural society assumptions, not on social divisions according to inequalities of access to resources'. Hence, Davidson (1992:11-12) concentrates on 'tribalism', and distinguishes between a pre-colonial 'tribalism' that acted 'as a force for good, a force for creating civil society dependent on laws and the rule of law', and the 'pathological' tribalism-clientelism invented during the colonial period, which persists and sows chaos 'like the economic misery now afflicting much of Africa'. As Mamdani (1993:45) has observed in a trenchant review of the book, this analysis not only misconceives civil society as a concept and its construction as a social process under colonial rule, it is also ahistorical in that it fails to 'appreciate that modern tribalism (like modern religious movements) is a contradictory phenomenon, comprising moments both manipulative and democratic' (for response, see Davidson 1993). Analyses of pre-colonial state formations need more complex periodizations, and conceptions of the state and politics as contested theatres and discourses of power between communities, classes, and gender groups, which were articulated in varied and changing ideological idioms, languages, and cultural traditions. Such analyses, would clearly demonstrate that Africa has many and contradictory political traditions, combining, uneasily, periodically, and tendentiously, practices and values that were democratic and authoritarian, egalitarian and aristocratic, popular and militaristic.

This is, for example, what emerges from Abdoulaye Bathily's (1990; 1994) analysis of the West African state from the earliest times to the post-colonial era. He divides the West African state into five periodizations: three fall into the era before colonial conquest, before the construction of what he calls 'the plunderer state of the colonial regime'. In the first state, called 'the primary state' and subdivided into three forms - the pastoral, agrarian, and artisan states - he argues that the rulers participated in the process of social production and reproduction, and consequently they enjoyed a high level of legitimacy and a considerable degree of democracy prevailed.

During the period of the Trans-Saharan trade there emerged 'the merchant military state', in which Bathily incorporates Davidson's West African *regna*. This state was characterised by the emergence of a predatory ruling class of merchants, Muslim clerics, and a military aristocracy, groups that were constantly engaged in hegemonic rivalries. Despite this, and technological stagnation over the previous phase, the state enjoyed a substantial amount of legitimacy because it provided collective security and freedom to its constituent communities.

The final phase of pre-colonial state formation was from the seventeenth to the nineteenth centuries, the era of the Atlantic slave trade and the subsequent colonial conquest. It witnessed the rise of the 'predatory state'. During this period, in which the region became firmly integrated into Wallerstein's modern world system, militarism and violence increased, and state legitimacy diminished and popular freedoms were curtailed. He concludes that on the eve of colonial conquest, the West African state was parasitic, despotic, and dictatorial, 'almost as alien to [these] societies as the European armies they opposed' (Bathily 1990:15; 1994:56).

One does not have to agree with the specific details, or even the schema, of Bathily's historical model of West African state formation. Indeed, the transition from one to the next form is not always clearly explained, and his characterization of the post-colonial state is as scornful as that of the most contemptuous Africanist. But the analysis makes several important points: that state formation was a slow process, spawned by both internal and external forces, which transformed the social basis of state power and legitimacy; that there were conflicts among the dominant elites, and between them and the general population; and finally, that state institutions, structures, and ideologies also changed. A similar analytical structure can be deduced from the work done on state formations in the histories of Eastern Africa (Salim 1984).

Celebrating Clan Solidarities

Not all people in pre-colonial Africa had the fortune or misfortune to live in state societies. Many lived, at various moments, and in different parts of the continent, in so-called stateless societies. These societies have had their defenders among academic historians and African political leaders. It has been argued that people in such communities did not live in a Hobbesian state of nature, but in ordered, efficient, and just societies, ruled by elders. Nyerere rested his case for *Ujamaa* on the contention that in the traditional setup before the Colonial Fall 'family life was everywhere based on certain practices and attitudes which together mean basic equality, freedom and unity' (Nyerere 1969:10), all had equal access to property, there were no exploiting classes; and there was government by discussion, in which elders 'talk till they agree' (Nyerere 1969:104).

Thus, socialism and democracy existed as two sides of the same coin in good-old communal Africa. It was an attractive, powerful ideology; one that at a stroke, valorised the African pre-colonial past, denounced and dismissed the colonial impact, and affirmed the fondest dreams of the nationalist masses. So infectious was it that even Kenya, Tanzania's avowedly capitalist neighbour, claimed a return to Africa's natural condition of socialism, communal egalitarianism and democracy in its official development policy (Republic of Kenya 1965; Mboya 1963). Earlier, Kenyatta had written unequivocally that 'before the coming of the Europeans, the Gikuyu had a democratic system' (Kenyatta 1938:131). Overseeing this system was the Council of Elders, the *kiama*.

These ideas found echoes in the writings of other prominent African leaders as diverse as Kenneth Kaunda (1967), Sekou Toure (Johnson 1977), and Leopold Sedar Senghor (1964). Academic historians added their weighty thoughts. In the words of Ki-Zerbo, 'these gerontocracies were moderated by democratic principles which assisted the head of family, village or district through an advisory role if not a deliberative one' (Simiyu 1987:51).

The realities of clan power were of course far more complex than these readings might suggest. To begin with, as the histories of Kenya's acephalus peoples such as the Kikuyu, Kamba, Maasai, and Mijikenda in the nineteenth clearly demonstrate, lack of centralized state organization did not entail equality of access to property, community decision making processes, social mobility, or dispute settlement. These were hierarchical societies, in which the exercise of power and the morality of domination were rigidly set, linked to the structured inscriptions of age, gender, status, and wealth (Ogot 1967; Muriuki 1974; Zeleza 1994a, 1994b, 1994c).

Take the example of the Maasai, the quintessential pastoralists of East Africa, and about whom a lot myths have been told, of their communalism, egalitarianism, and aversion to change. Nothing could be further from the truth. In the nineteenth century, this was a society characterised by unequal access to livestock, grazing land, and labour. The rich Maasai not only employed the labour of their poorer neighbours, but also adopted vulnerable outsiders, especially children, to herd their livestock. It was this unequal access to productive resources that sometimes led to conflicts within and between Maasai communities, especially in periods of severe ecological stress, such as drought, as was the case in the last quarter of the nineteenth century. Elderly men generally, and wealthy and powerful individuals in particular, controlled the rituals of power, dispensed justice, and sat at the apex of the male age-set system, through which decisions affecting the entire community were made. While women had their own age-groups and social spaces where they maintained their own networks and exercised authority, no group of women enjoyed the same amount of power wielded by men of their age group.

The same class, age, and gender divisions in access to both productive resources and political power were evident among the Kamba. In the course of the nineteenth century, thanks in part to crises of subsistence caused by recurring droughts, transformations in social structure, and the expansion of coastal trade, groups of powerful men, including traders and those who could mobilise troops in times of conflict, began to concentrate power in their hands as virtual chiefs outside the constrictions of clan structures. This process was halted, then reconstituted, by the colonial conquest state of Kenya.

A different pattern took place for the Mijikenda. There was a dispersal in the spatial and social locus of power as different groups of Mijikenda people dispersed from their original *makaya* settlements in the mid-nineteenth. This was caused and facilitated by agricultural and population expansion, the receding threat from their Maasai and Oromo neighbours, and the changing patterns of regional trade. This led to the diminution in the powers of the elders to control trade, distribute land among the clans, organise collective ritual ceremonies, and regulate social behaviour. Generational conflicts between the elders and young men increased as the latter vainly tried to reassert control. The role of elders' councils were gradually modified and levels of popular participation in matters that affected each community increased, although this did not generally extend to women.

These examples demonstrate that the so-called stateless societies had institutions and structures of power, underpinned and gradually transformed by economic, social, and cultural processes, and ideological mediations. The class, generational, and gender divisions of labour and access to opportunity and authority were often no different in these societies from those with centralized states. Like the latter, democracy in the stateless societies was not a natural condition, ordained by the moral community of clan intimacies, but an ambiguous, tenuous, and partial outcome of conflicts within and between households, clans, communities, and territories, in the process of which the blood solidarities of kinship, both real and fictitious, were often strained, sometimes ruptured, and periodically strengthened.

The Economics of Power

It is evident that the histories of state formation, power, and politics as processes, actions, and authoritative interventions in pre-colonial Africa defy easy generalization. An examination of the material base of African societies, or to use the old Marxian language, the modes of production, further reinforces this point. It is now quite clear that African pre-colonial economies were too diverse and complex to spawn one or two forms of state formation or the circumscribed range of political practices one often reads about (Zeleza 1993).

An analysis of the agrarian relations of production will suffice. Agriculture was the mainstay of African economies in the nineteenth century. There was a remarkable expansion in agricultural production in many parts of the continent. Indeed, it could be said the continent witnessed a kind of agricultural revolution: new food and cash crops were widely adopted, and old indigenous cultigens spread within and across regions. Similarly, new techniques of cultivation were adopted, or old ones spread. But this agricultural revolution exacted heavy political costs. In countries with highly centralized despotic states, such as Egypt, the Northern Sudan, Ethiopia, and the Sokoto Caliphate, peasant surpluses were appropriated through onerous taxation, corvee labour, and sometimes confiscation of produce. These states arose out of revolutionary pressures earlier in the century, which were themselves outcomes of long simmering internal and external forces for change. The intensified regime of economic exploitation in these countries supported different projects, from 'modernization' to social and religious reforms, and state formation itself. But the result was that the peasants probably enjoyed less economic and political freedoms than before. To be sure, the peasants resisted and their struggles escalated.

But even in societies without despotic states, increased commodification of peasant production, led to the expansion of bonded forms of labour, from sharecropping, to indentured, and slave labour. The available literature indicates that the use of slave labour expanded quite considerably in the course of the nineteenth century, especially following the abolition of the Atlantic slave trade. The two were of course connected. Following the abolition of slavery to the Americas the infrastructures developed to supply the slaves remained, and were now used to expand local labour supplies to produce commodities demanded by the European economies. In short, slavery corrupted and terrorised many African societies. As Manning (1990:124) puts it, 'slavery was corruption: it involved theft, bribery, and exercise of brute force as well as ruses. Slavery thus may be seen as one source of modern corruption'.

It is also quite apparent that the agricultural revolution led to a progressive intensification of female labour time in agriculture and subordination of women. The lives of peasant women became harsher, their access to productive resources such as land, less certain, and their control over the disposal of their surplus less assured. This was even true in some of the matrilineal societies, where the position and status of women had historically been higher. Women, and the other exploited and oppressed social groups, did not of course sit idly by, but struggled, in the process of which social relations and the articulation of state power were continually transformed.

This is merely to suggest that there is no simple correlation between economic growth, or development in the late twentieth century lexicon, and democracy, whether in African history or in the history of other regions. Democracy does not fall spontaneously as manna from the heaven of economic progress as the teleological fictions of western modernity proclaim. Some of the most barbaric and horrendous abuses of human rights and freedom have occurred in the twentieth century, the most 'modern' period in world history, and have been perpetrated by the most advanced industrial nations and technologies. Adolph Hitler and Joseph Stalin were not European medieval despots, nor were Idi Amin and Marcias Nguema pre-colonial African chiefs. And none of them relied on pre-industrial technologies to terrorise their nations. All four were products of western modernity in its capitalist, communist, and colonialist garbs. Democracy should entail civil and material freedoms, not just for a few nations and select groups, but for the bulk of humanity. In that sense, democracy is still a long way off globally. If it is eventually achieved, it will not be because of western modernity, but in spite of it, indeed, against it, for western modernity, in the shape of imperialism, is responsible for many of the twentieth century's greatest tragedies of the global commons. One of these was colonialism, whose legacies Africa is still reckoning.

Conclusion: Into the Heart of Darkness

The European colonial conquest of Africa came in the last quarter of the nineteenth with sudden ferocity. African polities crumbled one after the other, with varying durations of resistance, except for Ethiopia, which managed to defeat the Italian invaders and thus kept its ancient imperial state, and Liberia, the state founded by freed American slaves. Thus many of Africa's states were either utterly destroyed or emptied of their powers, broken up, and turned into the native tier of colonial administration. Among these states were some that were in the process of formation at the time of conquest, such as Samori Toure's state in West Africa. Also, snuffed were experiments in 'modern' state construction by the western educated elite in Ghana who formed the Fante Confederation and the Accra Confederation, and in Nigeria where the Egba Union was formed. These experiments were opposed by both the incoming colonial invaders and the traditional ruling elite. And so Europe imposed its imperial will on the continent, and the iron grid of territorial divisions, which, in Mazrui's (1986) apt phrase, separated those who had previously been together, and brought together those who had been separate, thereby sowing the seeds of future conflict in Africa.

The colonial conquest state, as Lonsdale (1989) has called it, involved complex military and political processes, which entailed the ferocious investment of force, as well as the manipulation of local crises and divisions, the accumulation of alliances, and the gradual subordination of the allies and collaborators to the expanding public controls and logic of imperial rule. Colonial state construction was partly facilitated by the political decay and fragility of many African states at the end of the century, partly brought about by their very involvement in the world system, through the slave trade, for example.

The process of colonial state construction reinforced trends towards the political alienation and disempowerment of large sections of the African populations. As Crawford Young (1985) has argued, the scramble for Africa was far more concentrated, intense, and competitive than in other regions. Moreover, the colonial state building venture in Africa included a far more comprehensive cultural project than was the rule in Asia, thanks to the preceding nature of African-European relations. Finally, colonial expansion in Africa occurred when European states were fully developed and consolidated, and therefore less likely to experiment with indigenous political structures, notwithstanding the self-serving fanfare that was made in British colonies of 'Indirect Rule'. The colonial state was an appendage of the imperial state, a ruthless midwife for the construction of the colonial capitalist economy. But it was no mere copy of the latter precisely because of its complex mission: as the mediating agency for imperialism it straddled and articulated, simultaneously, 'local' and 'external' social forces. It was this 'dual mandate' that made the colonial state both fragile and authoritarian (Lonsdale and Berman 1979:90).

In conclusion, it is clear that pre-colonial African history shows complex traditions of authoritarianism, and struggles against such practices, which sometimes led to expansions of political spaces for popular participation and features of democratic rule. But it would be misleading to conclude that democracy as defined earlier in this paper existed in full bloom. But even if it did, we cannot blissfully switch off the last couple of hundred years of imperialism and colonial and post-colonial authoritarianisms, and return back to the Golden Age. Indeed, nowhere in the world has there been a democratic largesse shared equally by all social groups, regardless of class, gender, and ethnicity or race, which democracy-hungry Africa could then import as an appropriate technology. Certainly the West cannot claim to have 'completed' the construction of democracy, for historically and currently the production and reproduction of power in western societies remains unequal along the unyielding hierarchies of class, race, ethnicity, and gender. Africa's democracies will emerge out of the contemporary struggles, whose trajectories will be determined by the continent's complex and contradictory pasts, and the visions being dreamed of, and the practices and capacities being built by, the numerous social and democratic movements across the continent. The only blueprint for democracy in Africa, therefore, lies in the struggles themselves. The future is open to numerous possibilities.

Notes

1. Mazrui (1994) provoked furore among African scholars for suggesting Africa's weak and unstable states may need to be recolonized by the stronger ones and overseen by a benevolent African Security Council composed of five pivotal regional states. Mafeje (1995) reacted ferociously, and Mazrui (1995) responded in kind. For carefully reasoned critique of the debate see Adejumobi (1995) and Bangura (1995). The debate underscores the opprobrium that the term colonization holds in the African imagination.

2. For vigorous African critiques of this 'political sociology of contempt', see Mkandawire (1996) and Eyoh (1995).

3. For a fascinating, if controversial, study on the concept of freedom see Orlando Patterson's (1991) magisterial treatise.

4. This is, of course not restricted to the western countries. Similar trends are evident in Africa, as my discussion of the expatriate African scholars in Chapters 2, and of deportations of migrant workers in the last chapter demonstrate. In early 1996 the parliament of 'democratic' Zambia was preoccupied by an almost surreal debate to change the constitution to bar second generation Zambians from contesting the presidency. The measure was aimed at former president Kaunda who had declared his candidacy. Kaunda, Zambian born, had already ruled for 27 years! In Côte d'Ivoire the incumbent rulers resorted to a similar measure to bar a popular opposition candidate. So much for democratization and Pan-Africanism.

5. For detailed discussions of human rights issues in Africa see O. C. Eze (1984) and I. G. Shivji (1987). For a Eurocentric discussion see R. E. Howard (1986). M. Hamalengwa (1991) has provocatively argued that 'Blacks have contributed to 'political' civilization by overthrowing slave, colonial and apartheid empires - the only race called upon to do so'.

6. As a synthesis Davidson inevitably presents highly summarised overviews. The authoritative study on Asante is Ivor Wilks (1975). For the West African empires, see N. Levitzon (1973) and J. F. A. Ajayi and Michael Crowder, eds. (1976).

Chapter Sixteen

SILENCING STORIES

Introduction

Tyranny demands silence. Colonialism sought to silence African voices and imaginations, to drain and fill them with European fantasies of its superiority and magnificence. It was even denied that Africans had a history, that they could have a future without European tutelage, that they could speak for themselves, to themselves. The successors to the colonial tyrants learned their lessons well. They, too, sought to humiliate and dehumanize their subjects by silencing their stories. As in the colonial days, only the language of oppression could be spoken, the songs of persecution sang. This language, as Toni Morrison poignantly puts it in her Nobel Prize acceptance address:

> does more than represent violence; it is violence: does more than represent the limits of knowledge; it limits knowledge. Whether it is obscuring state language or the faux-language of mindless media: whether it is the proud but calcified language of the academy or the commodity driven language of science: whether it is the malign language of law-without-ethics, or language designed for the estrangement of minorities, hiding its racist plunder in its literary cheek - it must be rejected, altered and exposed. It is the language that drinks blood, laps vulnerabilities, tucks its fascist boots under the crinolines of respectability and patriotism as it moves relentlessly towards the bottom line and the bottomed-out mind (*Index on Censorship* 23 (1/2): 1994:5).

Morrison was talking about the language of oppression in America, the self-proclaimed sweet home of liberty, where the spirit of the Enlightenment resides, or perhaps, is buried. In the post-colonial world the language assumes a different accent, but its grammar remains the same. This chapter is about silence and censorship in one post-colonial state, Malawi, the violence it wrought on the nation's psyche, the lies it bred, the language of dissent it inspired, and the struggles it generated. It is a narrative of the making and unmaking of authoritarian power.

Surreptitious Speech

Banda's Malawi, a thirty-year contraption of totalitarian power, was a land of pervasive fear where words were constantly monitored, manipulated, and mutilated; a country stalked by silence and suspicion; a nation where only the monotonous story of the Ngwazi's achievements could be told and retold; a state of dull uniformity that criminalized difference, ambiguity, and creativity; an omniscient regime with a divine right to nationalise time and thought, history and the popular will. And so it censored memories, stories, and words that contested and mocked its singular authority, banishing and imprisoning numerous opponents, real and imaginary, hunting and murdering exiled 'rebels', and appropriated and dissolved the boundaries between private and public spaces, personal and political spheres, individual and collective lives, so that no one was sure of anyone, not of friends or colleagues, nor relatives, not even of partners and spouses, and even one's careless dreams could be dangerous. All was contaminated by this naked, arbitrary power.

Banda's regime waged an endless war against plurality, against voices that told different stories or sang different songs, stories or songs that did not glorify the everlasting king's infinite wisdom, ululate the miraculous development the country was supposedly undergoing, and wonder at its stability, its enviable peace and calm, law and order in a region wrecked by revolutions, wars, poverty, and decay. Unique, unpredictable stories subverted the four cornerstones of the Party and the Leader: unity, loyalty, obedience, and discipline. They compromised national development, and so they had to be silenced, channelled into praise songs of Malawi's success and efficiency under the Ngwazi's wise and dynamic leadership. He had united the people, so that there were no more Chewas, Tumbukas, Lomwes, Yaos, no regionalism, no poor and no rich, just Malawians, one big homogeneous family, under the guidance of the eternal grand patriarch. An undifferentiated people needed undifferentiated stories. What they wrote, sang, read, thought and dreamt had to be placed under constant surveillance. Censorship was for the public good. It was a protective mantle, for a young nation, a juvenile people.

Under this mantle of silence a whole people were homogenised, infantilised, and demeaned, their tongues burdened with voicing and singing the banalities of what Vaclav Havel (1988) has called totalitarian nihilism, their imaginations denied of dreaming. This totalizing power sought to induce public inertia, to capitulate the popular will and consciousness on the altar of greed and terror for a ruling elite lacking, as Fanon (1963) observed in his searing indictment of the post-colonial order, any historic mission, except its own self-reproduction and mimicry of an imperialist and decadent European bourgeoisie. Censorship becomes an iron veil to hide the lies, deformities, and fantasies of a ruthless, unproductive power. And it begets self-censorship, a numbing collective fear of meaningful social conversation, of public discourse, of openly questioning the way things are and imagining what they ought to be. Censorship silences both the present and the future, defiles authentic memories and forecloses the possibilities of tomorrow. It denies the creative spontaneity of life, the exhilaration and power that lies in unfettered words and stories, which endows us with the humanity that distinguishes us from other living things.

Censorship in Malawi had a universal reach, surveiling and silencing written and oral narratives, intellectual texts and ordinary speech. It could go to grotesque lengths: one lecturer in anatomy was apparently detained for discussing 'the reproductive capacity of old men - deemed disrespectful to the aged Life-President Dr. Kamuzu Banda. There was the teacher who inadvertently referred to Dr. Banda as "the President" instead of "the Life-President". There was the lawyer who criticised the dress code which prohibit[ed] women from wearing trousers' (Carver 1992:14). There were those who dared to party while the President was talking on the radio. And there were many thousands more who languished in jail or enforced exile for refusing to buy the Malawi Congress Party card, failed to donate enough chickens and eggs to the Ngwazi, or to buy their spouses cloths gleaming with the face of a more youthful Banda, or simply made the wrong gesture in public, or perhaps uttered the wrong word in their sleep. 'Exposing injustices', the Catholic Bishops wrote in their historic pastoral letter of 1992, '[is] considered a betrayal; revealing some evils of our society is seen as slandering the country' (Catholic Bishops 1992). The Bishops were vilified and almost killed for daring to urge the establishment of a new and more equitable, just, and democratic order in the country. Fortunately, by then Malawi was feeling the winds of democratic change sweeping across the continent as internal and external pressures were mounting against the dictatorship, which would soon lose the referendum on the introduction of a multi-party system. Till then censorship and incarceration, and sometimes the threat of death, reinforced each other, turning ordinary words, simple appeals to human freedom and decency, into grenades of potential self-destruction.

But censorship in Malawi, as elsewhere in the world, was a complex affair. It is easy to outline its concrete manifestations, the institutional and legal instrumentalities of censorship. It is far more difficult to explain what made this totalitarian nihilism from which the censorship sprang possible. It is tempting, but too simple, to blame it all on a megalomaniac leader, or on an impersonal predatory state, for neither the self-styled Life-President nor the state were suspended above civil society like malevolent clouds. The state and civil society are not binary opposites, embodying exclusive practices and values, but intimately connected structural and moral spaces and spheres of action and reaction that share and appropriate each other's functions and interests. This is to suggest that civil society, and its various constituencies, were not innocent bystanders, but deeply implicated in the construction and reproduction of the 'reign of terror' that turned Malawi into what a journalist once called 'the Land of the Zombies'[1]. Nor can the ubiquitous international community, those 'aid donors' that praised Malawi's 'rapid economic growth' and 'political stability' and cheerfully bankrolled the regime, be absolved. It behoves us to understand the nature of these relationships, the role played by civil society, including the Malawian intelligentsia, in fostering a climate of fear and silence, if we are to prevent the current democratic transition in the country from turning into a brief interlude in a continuous nightmare of tyranny.

The legal framework of censorship in Malawi and other African countries, like most structural features and trappings of the post-colonial state, was inherited from the colonial state, that unmediated authoritarian configuration of power that profoundly altered African political culture and the culture of politics. Working from models provided by the British colonial government and apartheid South Africa, a Censorship and Control of Entertainment Act was passed in 1968 without any serious debate. Its purpose was extensive:

> To regulate and control the making and the exhibition of cinematograph pictures, the importation, production, dissemination and possession of undesirable publications, pictures, statues and records, the performance or presentation of stage plays and public entertainments, the operation of theatres and like places for the performance or presentation of stage plays and public entertainments in the interests of safety, and to provide for matters incidental thereto or connected therewith (Laws of Malawi Vol.IV:2).

The operative terms were broadly defined. For example 'publication' included:

(a) any newspaper, book, periodical, pamphlet, poster, playing card, calendar or other printed matter.

(b) any writing or typescript which has in any manner been duplicated or exhibited or made available to the public or any section of the public (Laws of Malawi Vol IV:3).

The net of prohibited publications was cast wide. Woe to those who imported, printed, published, manufactured, made or produced, distributed, displayed, exhibited or sold or offered or kept for sale any publication, picture, statue or record that was

> indecent or obscene or is offensive or harmful to public morals; or is likely to give offense to the religious convictions or feelings of any section of the public; or bring any member or section of the public into contempt; or harm relations between any sections of the public; or be contrary to the interests of public safety or public order... (Laws of Malawi Vol IV:11).

It was an all-inclusive, unyielding call to silence. In case this was not enough, there were the laws of defamation and libel. Predictably, publications by the President and his ministerial minions were 'absolutely privileged... whether the matter be true or false, and whether it be known or believed to be false, and whether it be or be not published in good faith' (Laws of Malawi Vol.2:76).

Overseeing this regime of public silence and presidential licence was the Censorship Board, whose decisions could not be challenged this side of eternity. The Board discharged its calling with impeccable thoroughness, regularly issuing 'permits' and 'certificates of approval' and declaring numerous publications, pictures, statues and records 'undesirable'. In the first seven and half years of its existence the Board banned over 840 books, more than 100 periodicals, and 16 films. Mercifully, the Board did not always ban publications, but would have parts deemed offensive simply mutilated with impenetrable black ink or sharp razors. The Board was quite generous in its choices of materials to be banned or mutilated: from those containing pornography, works on the communist world and revolutions, to those on the problems of post-independence Africa and of course misguided accounts of Malawi. Malawians had to be protected from the corruption of subversive thinking abroad, their minds kept pure and focused on the momentous tasks of building and developing the nation, that old Cinderella of British Central Africa. And so books, magazines, and records originating outside its borders had to be scrutinised at airports and other entry points for subversive statements and notes. A friend of mine returning from America in the late 1970s had his albums of Donna Summer, the reigning Queen of Disco, confiscated at the airport. The covers looked 'obscene'.

Given their propensity for chronicling the afflictions of post-colonial Africa, books by African writers featured prominently in this hall of infamy. Members of the board were known for sudden seizures of revelation: publications or records that had previously escaped their censure would suddenly be banned. For example, in secondary school in the late 1960s and early 1970s we had read George Orwell's *Animal Farm*, but in 1976 it was discovered that the pigs were in Malawi and the book was banned, together with Achebe's *No Longer at Ease*, Dipoko's *Because of Women*, Lessing's *Grass is Singing*, Nabokov's *Lolita*, Nkrumah's *Dark Days in Ghana*, and Dumont's *False Start in Africa*, just to mention a few. The music of the American soul singer, Percy Sledge, was banned in the mid-1970s after the promoters had audaciously allowed him to use the stadium entrance reserved for the Life-President in an open convertible, his hands gesticulating as if he were the Ngwazi himself shaking his flywhisk of uncontested power. Simon and Garfunkel's 'Cecilia' was banned, for its popular tunes that seemed to parody the depraved domesticity of the presidential power couple:

Cecilia / I'm down on my knees / I'm begging you to please / To come home

The Official Hostess stopped using her compromised first name, Cecilia, and took to using her African middle name, Tamanda, duly prefixed with a daunting matriarchal 'Mama'.

But they were no unschooled minds, these censors. They admired the European classical and medieval writers, from Sophocles to Shakespeare, whose plays were approved without being read, as the chief censor once intimated (Gibbs 1988:19). However, they looked with disdain at the modernist plays of Brecht and Beckett, and were outrightly alarmed by the impudent plays of Africa's own Athol Fugard and Wole Soyinka, whose *Kongi's Harvest* was banned for its politics, *The Trials of Brother Jero* on religious grounds, and *The Lion and the Jewel* was found indecent, the censors taking particular exception to a reference somewhere in the play to 'open breasts'. A passage in Ntwa and Ngcma's play *Woza Albert* about inflation and a salary cut was censored in case the audience read into it the domestic inflation of prices and power, and so was a speech by an ailing Chief Alagba in Kole Omotoso's play, *The Curse*, in which he worries what will happen to his property and women after his death, eerily echoing the lack of obvious heirs to the ailing political leadership.

Publications by Malawi's own literary sons, Legson Kayira, David Rubadiri, and later Jack Mapanje, did not escape the self-righteous wrath of the censors. Kayira had not only written such subversive books as *The Detainee*, he had the gall to suggest in his autobiography, *I Will Try*, that he had walked North in search of an education, appropriating the Life-President's own story that he had walked South in search of an education! Rubadiri had dared to side with the 'Rebel Ministers' in the 1964 Cabinet Crisis and resigned his post as Malawi's first Ambassador

1964 Cabinet Crisis and resigned his post as Malawi's first Ambassador to Washington. When his work was mentioned in a 1976 issue of *Odi*, a bilingual quarterly of Malawian writing run by members of the English Department at the University of Malawi, the entire issue was confiscated. The Editor, Robin Graham, who had deliberately not submitted the offending article to the Censorship Board, was deported, and the English Department came under a heavy cloud of political suspicion. Mapanje (1989:9) tells the awful story of witnessing the head of the English Department, 'Professor James Stewart going through the humiliation of ripping David Rubadiri's poems out of the Heinemann anthology called *Poems from East Africa*, edited by David Cook and David Rubadiri'. And manuscripts already approved could mysteriously be recalled as James Gibbs discovered in 1976 after publishing a collection of plays, *Nine Malawian Plays*, which had just been cleared by the Censorship Board. 'But after the third week or so of publication Gibbs and the publisher were ordered to withdraw all copies of the plays, and the introduction which Gibbs had included after clearing it with the censors was to be removed from the book and another more acceptable one written' (Mapanje 1989:9).

Mapanje himself had a tortured relationship with the censors. He was intimately involved with *Odi* and other literary endeavours in Malawi from the late 1960s, including the formation of the Malawian Writers' Group in 1969. He courageously refused to go into exile, like so many other Malawian writers had done, and continued to publish from Malawi his enigmatic, riddling verse, elevating Malawian poetry to new heights, while simultaneously prodding and pushing the frontiers of political tolerance. When his first award-winning collection of poetry, *Of Chameleons and Gods*, was published by Heinemann in 1981, while he was completing his doctorate in England, the Censorship Board impounded all available copies and trashed them and discouraged school teachers from using the book, although they initially refrained from actually banning it perhaps for fear of turning Mapanje into a literary hero. Upon his return in 1983, he was under constant surveillance, but he refused to be cowered. His luck ran out on 25 September 25 1987, when he was arrested and *Of Chameleons and Gods* (Mapanje 1981) was finally banned.

While no writer could escape the long, harsh hands of the censors, those who wrote in the national language, Chichewa, had to contend with additional demands from the Chichewa Board, and the Life-President himself that 'correct' Chichewa be employed. Never mind that in his infinite wisdom the Life-President never saw it fit to talk to his beloved people in the language. He spoke to them in English through an interpreter. But as the omniscient leader he periodically banished words from Chichewa and imposed new ones from the recesses of his forgotten youth. And the mass media and the nation would purge the 'wrong' words from the vocabulary, and in complying each one of us fell further into silence, surrendering a part of ourselves, of our language, our ability to tell stories to a contemptuous, cynical power.

Manuscripts had to be submitted to the Censorship Board before publication. The manuscripts would return rejected or approved but cleansed in bold strokes of red or black ink of all offensive and inappropriate statements, words, sensibilities. Meeting the Chairman of the Censorship Board to receive the verdict on one's errant manuscript was an occasion to behold. I had such a privilege in 1974 when I went to meet him on a manuscript of a collection of short-stories I had written. I dressed as conservatively as I could, making sure to put on a narrow, dull tie, and trousers that were appropriately tight, for broad gaudy ties then in fashion were too seditiously hippy and flared pants had been banned in the Decency Dress Act of 1973. And I carefully combed my hair and patted it down: the Afro look was too wild. The Life-President hated long hair. Groomed properly, but nervous, I arrived at the Offices of the Censorship Board in Limbe from Chancellor College, about 40 miles away.

The Chairman, a large man with an indifferent voice, looked me sternly in the eye as I sat down. 'Do you want to be like Soyinka?' he asked. I was rather perplexed. Fortunately, he did not wait for my response. 'Do you know where Soyinka is?' he persisted. Again he continued before I could give my bewildered answer. 'He is in jail because he is against his own government. Do you want to spend your life in jail?' My body chilled. 'No', I mumbled, looking down. I was nineteen years old, in my third year, and hungry for life after university, for all those secret joys of adulthood. I knew of lecturers and students who had been detained, who had gone to 'high school', as our registrar used to say, rather gleefully, for reasons that became evident years later: he was an informant for the Special Branch, the Soul Brothers, as we cheekily called them in our hushed conversations. Was this to be my future? Seeing that I was sufficiently shaken he told me the fate of my manuscript: the Board would not allow the publication of six of the stories, for reasons that were explained on three pages of pink paper. One was set in settler Rhodesia and talked about the experiences of a visiting interracial couple from Kenya; it was condemned for contradicting the Life-President's policy of contact and dialogue with the racist settler regimes of Southern Africa. Others were denounced for ridiculing African traditions and for depicting violence and loose morals. He never smiled once. His was the face of unflinching knowing and power. I left frozen with terror. But I was lucky.

Three years later I left the country to pursue graduate studies abroad. I never returned, like so many others before me and after. In the meantime, the noose of tyranny grew tighter and the arrests and detentions more arbitrary. But there was a method in this madness: arbitrary arrests meant that nobody knew why they or anyone was arrested and therefore everyone was afraid. Denied free speech and a credible mass media, Malawians subsisted on whispered rumours or snippets from the BBC and even apartheid Radio South Africa, and the country descended deeper into a dense web of oralised despair. The distinctions between fact and fiction, truth and lies, history and fantasy withered away in the grand spectacles of public celebrations, with their colourful sea of flags, frenzied women's dances and the Life-President's somnolent speeches. Such empty rituals marked public life in Malawi as the carnival of power spread its banality over the length and breadth of this beautiful land. Negotiating through the minefields of expected loyalties, regulations, and prohibitions occupied and drained peoples energies. The stress and tension could be seen in public places on the anxious faces, in the nervous glances, the lowered voices, the stiff body language. And not just in Malawi. Meeting Malawians abroad was a depressing experience. Enquiring about 'how things were back home' always elicited from the new arrival a look of suspicion, stilted conversation, and a hurried departure.

Exile provided little refuge from the terrorising silences of Malawi. Some paid the extreme price. Atati Mpakati was gunned down in the streets of Harare, Mkwapatira Mhango and his family were fire-bombed in Lusaka. Many of the writers found their talents wilting on foreign soil. I did not publish my second book of fiction, a novel entitled *Smouldering Charcoal*, until 1992, sixteen years after my first book was published in Malawi. While studying for an MA and a Ph.D and trying to establish myself as a professional historian took time, my literary silence was largely imposed by the long shadows of Banda's tyrannical regime. I finished writing the manuscript of *Smouldering Charcoal*, an angry attack of the Malawi I knew in the 1970s, in 1982. I was planning to send it for publication when I learned that my young brother then in Malawi was taken for questioning. What happened is that an expatriate Canadian who had taught in Malawi and to whom I had shown a draft of the novel in Halifax, had written someone in Malawi concerning the novel. And the well-honed ears of the security picked it up. So I kept the novel in my drawer, afraid that if it was published I would be endangering the lives of my family and relatives, until 1989 when I met David Rubadiri, who had read it. He convinced me to get it published. By then, also, I was no longer willing to be my own censor. I had recovered my literary voice. Two years after *Smouldering Charcoal* (Zeleza 1992) was published I published a new collection of short-stories, *The Joys of Exile* (Zeleza 1994) marking my first efforts at creative writing in over a decade.

I am sure there are other Malawian writers who can tell similar tales of self-censorship. There were of course compelling reasons for doing this, including concern for one's physical security and that of family, relatives, and friends. But there were also less honourable reasons. Careerism, opportunism, and ideological confusion seduced some to sell their talents to the state for crumbs of tainted silver and favours. For example, one leading writer became Managing Director of the country's biggest publishing house, and another Managing Editor of the country's only daily newspaper, both of which were owned by a conglomerate of companies belonging to the Life-President and his cronies. These writers and intellectuals became sycophantic errand boys of the regime. I remember one of them meeting me in Nairobi in 1987 where I was teaching at the time and trying to convince me to return to Malawi to join the establishment. Between his relatively fat salary, a company car, and the deceptive importance of rubbing shoulders with the Party chiefs in the corridors of power he had forgotten all that anger and contempt we felt and expressed in our student days against the same people he was now busy composing praise songs for. And the censors relied on, as Mapanje later discovered, 'academics, writers, and respectable members of the community', and it was they who happily lent their critical talents to decipher his poetry for the unversed authorities, which resulted in Jack's detention (Mapanje 1989:8).

It is quite evident that in many small ways we as writers and intellectuals not only conceded political space to the state, but sometimes assisted in authenticating its authoritarianism. To quote Mapanje again, one highly placed writer and publisher, who wrote him while he was in London, made the following strange request:

> I have been contemplating a Malawian version of *Of Chameleons and Gods* which should give our literary series a blasting send-off (sic). This is an area I am trying to develop. I need something powerful to start off with. Should you give us your consent to proceed it would be necessary to delete certain titles (poems) which to quote one anonymous analyst

'spoke at wounds that are still raw in Malawian history'. ...Malawians will have the chance to read freely at least 80% of your poems, and I think that both you and the Malawian readership do desire to share experience contained there.

Comments Mapanje sardonically:

The mind still boggles about what the Malawian version of already Malawian verse would be and what the omissions would have been... When I returned home in April 1983, the matter was never raised. We politely avoided mentioning the 'blasting send-off' which they (whoever they were) were probably planning to give me, rather than the poems (Mapanje 1989:8-9).

We have to face up to these ugly truths of our own complicity in tyranny as intellectuals if our criticisms of the Banda regime and others like it in Africa and elsewhere are to be morally credible. As I have argued in Chapters 2 and 3, the state does not bear the sole responsibility for curtailing and circumscribing academic freedom whether in Africa or abroad. The intolerant, hierarchical, arbitrary, exploitative, corrupt, and opportunistic practices and tendencies apparent in the institutions run by the academics themselves, and in the wider society, play a significant role in limiting and undermining academic freedom. A number of simple examples illustrate my point. One evening when I was in my first year of university an Irish film, *Ryan's Daughter*, was being shown on campus by the students union. When it came to a love scene there were groans in the packed audience, not of titillated pleasure as I initially thought in my teenage innocence, but of incredulity and consternation. How could the censors have overlooked this affront to morality? So the film was stopped. In our conformist zeal, we outdid the censors, we extended the boundaries of censorship. We put more bars of silence around ourselves that night regardless of the artistic or intellectual merits of *Ryan's Daughter*.

A few other incidents demonstrate how deeply internalised the culture of submission was in many of us. In 1978 there was a demonstration in London against the detention of Ngugi wa-Thiong'o. I asked some friends of mine from Malawi if we could go. They all refused, saying it was dangerous to attend a political demonstration against a fellow African country. Many Malawian students abroad were even afraid to take part in demonstrations against apartheid South Africa, mindful that the Ngwazi had been feted by the Boers in Pretoria in 1971 and our two countries had diplomatic ties. In typical egocentric style the Ngwazi had declared that he would sup with the devil if it was in Malawi's best interests, as defined by himself, of course. When I arrived in Nairobi in

1979 to do field research for my doctorate I was warned by some Malawian students to avoid meeting David Rubadiri, a 'rebel', then teaching at the University of Nairobi. As far as I knew there was no law in Malawi that said I could not do so. But then Banda's tyranny was beyond the law, and in this case we allowed it to transcend the boundaries of Malawi.

To be sure, there were courageous voices from all walks of life inside and outside of Malawi, including those of academics, students, and writers gathered around the Malawi Writers Group[2], that attacked, subverted, and mocked this totalitarian nihilism. Without them Banda would still be in power today. But he lasted so long also because many people believed in the benefits and necessity of his regime, real and assumed, for themselves or their imagined communities. It was as if in Banda's delusions of grandeur and immortality our young, fragile, and poor nation found a soothing solidity and self-worth. Maja-Pearce (1992:56) is right: tyranny only works 'because we collude in it, because we choose to invest a human being like ourselves with power that no human being can possess over and above our own participation in the exercise of that power'.

The Banda regime was spawned by the specific histories of Malawian nationalism and decolonization, the configuration of social forces and perceived developmentalist imperatives, and its mode of insertion into the sub-regional and global political economies. It tapped into, and reshaped, the existing regional, ethnic, class, religious, and gender dynamics of Malawian society. For example, Banda's projection of himself as the matrilineal uncle, as Nkhoswe Number 1, not only resonated with the cultural traditions of some major ethnic groups, but also exploited deep gender divisions and women's struggles for participation in the male-dominated public arena. This allowed state politics to invade that most private of spaces, the bedroom, for state surveillance to penetrate conjugal relations. Tyranny was totalised. Politics was externalised and refereed to the state in many organizations lacking in institutionalization and dominated by a crass elite desperate for personal accumulation. It would not be farfetched to say, for instance, that many of the lecturers detained at the University of Malawi in the 1970s and 1980s were implicated by their own colleagues anxious for rapid promotion in an institution lacking clear rules of career mobility and a vigorous culture of intellectual production, tolerance, public accountability, and social responsibility.

Conclusion: In Praise of Decency

Thus the culture of authoritarianism, which sustained censorship, was reproduced at many sites, in the home, the school, the work place, in a complex spiral leading to, and reverberating from, the state headed by an infallible Life-President. It is clear, therefore, that while one eye must always maintain a vigilant gaze on state tyranny, the other needs to probe and watch for the totalitarian potentialities of civil society, of tradition, religion, and popular culture, for no truly free, humane, and tolerant society can be created and sustained without the democratization of the state, community life, and other spheres of social existence. Ayittey's (1987) idealistic invocation of some pristine African democracy and customary freedom of speech to which we can magically return to will not do.

The spectre of rising religious and cultural fundamentalisms in some parts of Africa underscores the grave danger that religious and cultural movements, not just the state, can pose to the basic rights of humanity. Often spawned by the failures of a misguided modernity and the discontents of post-colonial tyranny, and inspired by atavistic yearnings for a simpler world, these movements pursue their archaic and intolerant visions with fanaticism, and even self-righteous violence. In many countries groups of Christian extremists inhibit open discussions of many pressing social issues, especially when they touch on morality and sexuality. Islamic extremists have been persecuting and even murdering intellectuals, writers, and journalists in some countries. In Egypt[3] and Algeria (Byrne 1993) those committed to secular and democratic values find themselves crushed between the hammer of state tyranny and the anvil of religious terrorism. When religious fanaticism captivates, as has happened in the Sudan in recent years, 'a political leader or Head of State', Soyinka (1993) writes, 'political power becomes conflated with a sense of divine mission: an internal trail of disaster grows into a blood-stained highway as the convert becomes increasingly obsessed by a mystic mandate to recreate all citizens within his borders in his spiritual image'[4].

The challenges before us, then, and what we must do as intellectuals, writers, activists committed to the establishment of free, humane, and democratic societies in our respective countries, are all too clear. It is not enough to preach tolerance, for tolerance without decency is porous to hate. Contemporary hate groups in the western countries, for example, justify their violent bigotry in the sacred name of 'freedom of speech', yet their ultimate agenda is to eliminate that very freedom and other human rights for those whom they seek to disempower and destroy.

These are difficult questions, troubling moral dilemmas. But we should never tire of fighting the cultures of silence and submission and the languages of oppression whether they are spoken in circuitous official newspeak, clamorous populist voices, or convoluted scholarly discourses, for it is these languages that bred the authoritarian colonial governors and the post-colonial Bandas, and authorise and rationalise contemporary imperialism. If left unchallenged these languages, and the practices of tyranny and the contempt for truth and human life that they embody will continue to thrive and our words and our stories will continue to be banished. And without our stories we forfeit our humanity.

The attainment of the 'second independence' is not so much the end of a chapter as the continuation of an old one. Let Mapanje (1995:86) speak:

On His Excellency's House Arrest
So, now that the febrile lion has accidentally fallen

Into the chasm of his own digging, let us resume
The true fight we abandoned thirty-three years ago
& begin to sing in the native tongues the old guards
Banned under the pretext of building our nation...

Yet today, after the lion has pulped his own cubs dead
Leaving the fragile village tainted in blood & after his
Chums across the valleys & beyond the seas have even
Shelved him; with lethal pythons & scorpions now tame;

Should we pour libation on the streaming ancestral
Stones or shall we perhaps roll up our sleeve for other
More insular & baneful battles, when those old guards,
Not lionised enough by our euphoria, take their revenge?

Those grass huts Mbulaje's clients charred gobbling up
Whatever paracetamols they hacked their way still stand,

Watching his dreaming potholes that'll need our tender
& human crocodiles their wicked amulets conceived
Or the endless cerebral malarias & tuberculoses they
Loved to cast down dressed as AIDS. What chaos, what
Rare sneer won't they raise for our freedom to redress?

M'bulaje jwine, n'lyeje sadaka
M'bulaje jwine, n'lyeje sadaka
M'bulaje jwine, n'lyeje sadaka*
(*Kill another, for you to enjoy the funeral feast)

Notes

1. This phrase was apparently coined by a British journalist, Mike Hall, who was deported from Malawi in February 1990, see A. Maja-Pearce (1991).
2. For a detailed examination of the activities of the group see L. Mphande (1994, 1996). However, based on personal recollections when I was a member of the group from 1972 to 1976 I think Mphande tends to exaggerate the group's political radicalism and effectiveness.
3. See for instance the following articles on Egypt: M. Booth (1989); S. Hetata (1990); J. Napoli (1992); A. Darwish (1992, 1993); and Gavlak (1995). The most recent ones comment on the infamous case of Dr. Nasr Abu-Zaid whose promotion to full professor was blocked by Islamic extremists who also filed a lawsuit seeking to separate him from his wife on the grounds that he is an apostate. And Mahfouz, the 1989 Nobel Laureate, was recently attacked in the streets of Cairo by suspected Islamic extremists. See the sobering report by some of Egypt's leading intellectuals, including Mahfouz, on the 'death of culture on the Nile' (*Index on Censorship*, 24, 1/2: 111-149, 1994).
4. Also see M. Nduru (1989); G. Jones (1990); and J. Wheelwright (1991), who chronicle the systematic harassment of women journalists, medical practionners and other professionals by the Bashir regime.

Chapter Seventeen

AFRICA'S BUMPY ROAD TO DEMOCRACY

Introduction

On three incredible days, April 26 to 28, 1994, the world watched with bated breath as South Africans cast off the yoke of apartheid and voted for a new democratic future, closing the long, sad chapter on 342 years of European colonial rule over that beloved country of rugged beauty and human depravity, lavish natural wealth and intolerable social squalor, white privilege and black poverty, violence and heroism, a country that embodied most poignantly the tragic encounter between Africa and Europe in modern times. Soon after, Mr Nelson Mandela, in an all too familiar transition in recent African history, turned from a once reviled prisoner into a revered president. Relief, indeed euphoria, greeted the transformation, and the fact that the bloodbath predicted by both the foes and friends of apartheid had not materialised. In a world wrecked by ethnic or racial hatreds, gross social inequalities, moral corruption and despair, the demise of apartheid in South Africa marked a rare moment of joy and hope, of faith in the human capacity to transcend the yawning divisions and miseries bequeathed by history.

But the irrational ghosts of history soon reared their ugly heads in Rwanda a month later and the world witnessed a nation devouring itself, with neighbour slaying neighbour, where even churches provided no sanctuary, and the innocence of childhood and the serenity of old age were no salvation; an orgy of violence and carnage that left the green hills littered with bodies, fresh rivers drenched with human remains, from which terrified multitudes fled into the gentler death and miseries of teeming, diseased refugee camps across the borders. Weary of Somalia and Bosnia and all those 'tribal' wars in far away places, the international community initially yawned and did nothing to avert the impending catastrophe, then belatedly despatched relief supplies to the killing fields. Rwanda was an indictment of our collective apathy, one genocide too many in a century filled with genocides and the destructive rages of war, fueled by the illusions of difference, obsessive demonization of otherness, and the pathological will to power.

South Africa and Rwanda. Stories of almost biblical triumph and tragedy, two faces of Africa, one of joyous celebration, another of unrelenting pain, milestones in the continent's bumpy road to democracy, one signalling its successful birth, the other its cruel abortion. In South Africa the transition to democracy was achieved after generations of resistance against organized state terror on behalf of a racial minority, while Rwanda's incipient democratic movement was thwarted by murderous attacks unleashed by a crumbling state against its unarmed citizenry. The South African demos managed to keep its eyes focused steadily on the prize, in Rwanda it was manipulated and diverted into ferocious inter-ethnic violence.

But there were other stories, many more, that did not grace or haunt the world's television rooms, but that were no less momentous for the people involved; heroic and wrenching stories of people determined to reclaim their freedom and dignity, their history and human rights, from the shackles of colonial and post-colonial tyranny. One of the stories unfolded in my own homeland, Malawi, a lovely, little country in Southern Africa, full of tortured memories of British colonial despotism and unhappy experiences with one-party, indeed, one-man, post-colonial dictatorship. The winds of democratic change sweeping across Africa shook the tentacles of Dr. Banda's totalitarian power, and the octogenarian autocrat was forced to concede to multi-party elections in May 1994, which he proceeded to lose, to his utter incomprehension. I, like many other exiled Malawians, returned to witness the elections, to partake in this collective defiance of unproductive power, to affirm a new future for our beloved country. It was my first visit to the country in seventeen years, and for the first time, I understood the exhilaration people in Central and Eastern Europe must have felt when the Berlin War collapsed in October 1989; indeed, what my parent's generation felt in the early 1960s when the imperial sun finally set over Africa.

Nineteen ninety-four was, therefore, for many of us in Malawi our generation's first experience of the sweet taste of freedom, of victory against the forces of oppression, our intoxicating moment of independence. For once the sun did shine brightly, there was music in the air, the trees danced, and the smiles were truly happy. The future looked possible. There have been many Malawis in Africa in the 1990s. This then is the subject I would like to address: the different trajectories of democratic transition in contemporary Africa. What accounts for that, for the ecstasies of South Africa and the agonies of Rwanda, the seeming successes of Malawi and the failures of Zaire, the stalled transitions of Algeria and Nigeria, and the ambiguous transitions of many others, from

Kenya to Zimbabwe. But this begs another set of questions: what led to these transitions to democracy in the first place? What is the content of the democracy that Africans have been fighting for? What does the future hold? Let me hasten to add that as a historian, I am more comfortable deciphering the past than crystal-gazing into the future. Predictions are often no better than projections of wishful thinking. It is hard enough to understand the present. The future is a very long time, always full of surprises.

Prospero and Caliban

Africa's current wave of democratic transitions is so recent, still unfolding, that it is difficult to make many definitive statements about it. At such moments, one wishes one belonged to those social sciences that thrive on making instant analyses of today's events that are proven wrong tomorrow without losing any credibility, at least among one's peers. But even my fellow historians would surely agree, without the advantages of long hindsight and painstaking research into dusty archival records and foggy oral memories, that the changes in Africa's political geography in the last few years have been quite breathtaking. They parallel, in their quantitative scope and qualitative dimension, the changes which occurred at the turn of the 1960s when the majority of African countries gained their independence from colonial rule. No wonder Africans speak of them as constituting Africa's 'second liberation', the 'second independence'.

At the beginning of 1990 the majority of African governments were dictatorships. According to the classification of Freedom House, out of the 52 states it rated at the end of 1989, 34 were 'not free', 15 were 'partly free', and only three were 'free', that is democratic (Diamond 1993a:3). By the end of 1994, the ranks of the 'democratic' states, to use the classification of the African Governance Program at the Carter Center in Atlanta, had swelled and those of the 'authoritarian' ones had fallen sharply. The Center lists 15 as being democratic; 16 pursuing a 'moderate' transition to democracy and six an 'ambiguous' transition; six enjoying 'directed democracy'; three as authoritarian; and eight are characterized by 'contested sovereignty', i.e, civil war[1].

There can be little doubt, therefore, that the late 1980s and early 1990s has been a period of profound change, of questioning the fundamental institutional arrangements of contemporary African states and societies. But it would be a mistake to assume that this 'wind of change', these struggles for the 'second independence' emerged, miraculously, with the onset of the new decade as offshoots of Huntington's (1991) democratic 'third wave'. All too commonly, the growth of the African democratic movements is attributed to external forces, either the demonstration effects of the Eastern European revolutions of 1989, or the diffusion of democratic values, models, and ideologies from the western world (Shaw 1993; Harbeson and Rothchild 1995; Nonneman 1996).

To be sure, the 'Leninist extinction' (Jowitt 1993) in Central and Eastern Europe and the end of the cold war provided a new international context, a new global conjuncture, for Africa's age-old democratic forces and struggles to flourish. The end of superpower rivalry over the continent, in the felicitous phrase of President Museveni of Uganda, 'orphaned' African dictators, who could no longer expect their godfathers in Washington or Moscow to run to their aid when they cried wolf in the face of internal struggles for reform (Mamdani 1990:25).The subsequent collapse of the Soviet empire, Mamdani (1992:312), has argued, 'made it far more difficult for Western governments to explain away — either to people at home or to Africans — pressure for internal reform as a Trojan horse for "Soviet subversion"'[2]. Thus, new spaces were opened up for democratic politics in many African countries. But the consequences were disastrous for a few of them, especially the formerly strategic 'client' states, such as Angola, Zaire, and Somalia. Abandoned by their former superpower godfathers, Angola slid into a ferocious civil war (Funkel 1993; Pereira 1993; Simpson 1993; Hamill 1994; Shiner 1994), Zaire into anarchy (Komisar 1992), and Somalia collapsed (Samatar 1992; Omaar 1993; Weil 1993; Makinda 1993; Doyle 1993).

The demonstration effect of the Eastern European Revolutions should not, however, be exaggerated. As Ali Mazrui (1990: 10-11) has stated, 'the age of political acquiescence in Africa was coming to an end well before the world ever heard much about Mikhail Gorbachev', a point echoed by Colin Legum, a seasoned observer of the African scene[3]. It has been argued that the Soweto eruptions and the West Bank Intifadah had a far greater resonance and a more immediate impact on the growth of the reform movements in sub-Saharan and North Africa, respectively[4]. This is to suggest that the 'demonstration effects' 'have also been evident from within Africa' (Diamond 1993a: 4).

Similarly, the impact of western 'democratic assistance', as Diamond (1993b:54; 1995) facetiously calls it, has not always been examined with the care it deserves. It is true that Western governments and donor agencies, including the mighty World Bank and IMF, began exerting pressure for democratization. The question is why did they embrace the democratic project at the turn of the 1990s, not before? And what type of democracy? Democratization increasingly came to be used as a conditionality for economic assistance. And so the 'economic conditionality' of structural adjustment became tied to the 'political conditionality' of 'good governance' (World Bank 1989). This helped fuel suspicion among some of Africa's ardent nationalists and beleaguered leaders to charge, quite mistakenly, that democratization was the latest in a long line of western conspiracies against Africa, apprehensions that found reinforcement in commentaries in the western media and analyses by Africanist scholars that depicted the West as the source of the trend towards global democratization, a discursive claim that recalled the imperialist historiography of 'planned decolonization', which denied the role played by African nationalism in the decolonization drama (see Chapter 7 above) and reflected the West's post-cold war triumphalism.

But this forced marriage between structural adjustment and democratization was built on a fundamental contradiction. Since structural adjustment programmes entailed the implementation of such draconian and unpopular measures as the devaluation of national currencies, drastic reduction in state expenditure, privatization of state-owned enterprises, and liberalisation of the trade regime, they required more, not less, state coercion, which was bound to be resisted, as demonstrated in the preceding chapters. It would be easy to accuse western governments, therefore, of hypocrisy. After all, they had once actively supported and defended Africa's ruthless dictators - the Mobutus of Zaire and the Mois of Kenya, the Bandas of Malawi, and the Bothas of South Africa. The abandoned dictators, of course, felt betrayed, while the often beleaguered democratic forces welcomed the western powers' new found resolve to isolate their former allies, and tried to exploit it for all that it was worth, which on many an occasion, they found was not much beyond rhetoric, as some of us suspected, and found out[5].

Hypocrisy is, however, too crude and slippery a term to capture the historical forces that have forced western governments and donors to champion democracy in Africa, even if only to neutralise and co-opt the reform movements as Samir Amin (1990) has argued. The fact remains these governments and donors, to quote Mkandawire (1992, 10), 'have been shaken by the realisation that the regimes they have thus far backed are on shaky grounds. To curry favour with the new movements, [they] have had to make sharp turns in their policies'. In short, western governments and donors may have been forced to champion democracy in Africa because of the very strength of the domestic democratic forces in order to neutralize and co-opt them.

Its colonial pedigree aside, the language of 'good governance' resonated with the anti-corruption campaigns of the African reform movements themselves (Harsch 1993), and that of the Economic Commission for Africa, the most eloquent institutional voice against the World Bank-IMF version of structural adjustment in the continent (ECA 1987, 1988, 1989). But the World Bank's 'tropicalised democracy', as Mkandawire (1992:15) ridicules it, which limited accountability to an anti-corruption drive and equated democracy with efficient management, was a poor substitute for the broader democracy advocated by the ECA and many reform movements in Africa. In other words, Africa's social movements wanted far more than the World Bank's governance of efficient management. As in the 1950s during the nationalist struggles against colonialism, their struggles were for a democracy firmly tethered to development. Thus, these were struggles as much for political freedom as they were for economic advancement, for improved conditions of living subverted by the uneven colonial and post-colonial distribution of wealth and the structural adjustment programmes.

The West, then, has not been Prospero to Africa's Caliban in democratic discourse. The African struggles for democracy have in many cases been against the deformities of political culture left behind by colonialism and western economic and political interventions in the post-independence era. Thus attributing the rise of contemporary democratic movements in Africa to western tutelage is to play cat and mouse with history. This does not mean, of course, that these movements have not learned from similar movements in western countries, movements organized around struggles for civil and community rights, economic empowerment, environmental protection, and so on (Hutchful 1992). Western triumphalism hides the fact that liberal democracy, the West's 'dominant historical mode of politics', in Wamba-dia-Wamba's (1992:2) terminology, is in crisis, that democracy in these countries is not 'complete', that beneath the comfortable exterior of what Galbraith (1992) calls the 'culture of contentment' lies profound political alienation, expressed in the progressive withdrawal by large sections of the population from the electoral process. It can be argued, therefore, that social movements in the western countries helped extend the boundaries of participatory democracy and empowerment and put these issues on the international agenda. It seems more than likely that the western governments' and agencies' concern with democracy in Africa reflected the growth and pressures exerted by these movements, as well as by the diasporic communities domiciled in the western countries which provided significant financial, ideological, and political support for the social movements at home.

In short, Africa's democratic transitions during the late 1980s and early 1990s were rooted in domestic struggles against both internal and external forces of oppression and exploitation. As former Zambian President Kaunda once said about the alleged Communist influence behind the struggles against apartheid and for freedom in South Africa: the ordinary South African man, woman and child did not need a communist to tell them that they were oppressed and exploited; they knew it because they experienced it every moment of their lives. Similarly, Africans did not suddenly catch the ennobling virus of democracy from abroad: they had been fighting for freedom for generations. It was not the historical diffusion of democratic values and institutions implanted by the colonialists that inspired them, as Diamond (1993b:52) spuriously asserts, but the memories and traditions of struggle against colonialism. Of course, Africa did not live in splendid isolation and the interaction between external and internal forces and influences behind the resurgence of the democratic movements could not always be disentangled: structural adjustment programmes which fired so many social movements, for example, were externally imposed and implemented by African governments to 'rectify' domestic economic imbalances themselves produced by external shocks and the legacies of colonialism. Thus, the process of democratisation in the continent at the turn of the 1990s was affected by international events and developments. It was occurring in the context of simultaneous, multiple, and contradictory transitions in the global order, characterised by the end of the cold war and the growth of multi-lateralism as well as an international civil society, globalization and regionalization of the capitalist economy, and the re-emergence of nationalist, ethnic and sectarian solidarities and chauvenisms. But Africans were not bit players in other peoples drama. They scripted their own drama.

Seizing Space to Dance

All this is to suggest that the democratic transition in Africa was not, and could not be, simply a turnkey project imported from the West, nor an imitation of the anti-communist revolutions of the East. It had its own history, which went back to all those struggles against the barbarities of slavery, and colonial and post-colonial abuses of power. More immediately, the democratic movements were linked to, and sought to rekindle and transcend, the nationalist struggles of the decolonization era. It should not be forgotten that the struggles for independence were for freedom and development, for participatory democracy long-denied to Africans by the uninvited and unelected colonial authorities. The post-colonial leviathan inherited underdeveloped economies, but strong authoritarian impulses from its colonial progenitor. Initially, it succeeded in delivering the promised fruits of independence: education and employment expanded rapidly, the middle classes grew, life expectancy rose. And so class, social, gender, and regional differentiations became sharper.

In short, civil society became more complex, more vibrant, more restive than ever before. It seized more space to dance. Many of Africa's gerontocratic leaders who started their careers in the colonial days when the ranks of African elites were tiny and the exclusivist ideology of nationalism only permitted the difference between the colonized and the colonizers, us versus them, or the swashbuckling military rulers brought up on a regimen of issuing and following orders, could not understand or tolerate the proliferation of cheeky pluralism. Civil society had outgrown the antiquated and suffocating boundaries of authoritarian power. The challenges of governance escalated into crisis as the state's capacity to discharge its developmentalist mission began to falter with the onset of intermittent recessions from the mid-1970s and the corrosive effects of ill-conceived structural adjustment programmes in the 1980s. As the masses began pointing their fists at the naked emperor, the post-colonial state assumed a progressively more precarious and repressive character. Thus rapid economic growth in the immediate post-independence period changed the social contexts of power and governance, while the economic crises of later years stripped the post-colonial state of its moral and political legitimacy, of its raison d'être.

As its material, coercive, and symbolic resources shrank, the post-colonial state's ability to deliver economic development and political stability, let alone social justice or freedom, and to hold the lid on the boiling steam of internal opposition, diminished. All that remained visible in many countries were its iron claws. Indeed, even these were splintered as the ruling coalitions increasingly lost their cohesion and singular will to power. Pro-democracy movements and discourses grew. The state lost its fearsome seductions. In the immortal words of a Lesotho chief: 'We have two problems - rats and the government' (Ayittey 1992:16). The unproductive political monopolies of the one-party state and military rule began to be vigorously challenged by rejuvenated social movements, once immobilized by reductive nationalist discourse and the independence social contract in which all, the people, the masses, were supposed to pray at the altar of nation-building and development, and the articulation of sectional class, social, community, ethnic, and gender interests was frowned upon as selfish and subversive. Various social movements, both organized and spontaneous, elite and popular, political and cultural, began flexing their muscles in the sprawling urban centres and the rural hinterlands. In the cities the most visible included trade unions, student movements, and professional associations, while in the countryside a wide variety of peasant movements sprang up (Isaacman 1990). Many of the social movements, especially the religious and women's movements and the NGOs, straddled the rural and urban spaces, and were class coalitions of the elites and the poor.

The reform movements were, therefore, numerous and varied, often complex in their composition and contradictory in their agendas. They were organized around the idioms of class, ethnicity, culture, religion and gender. The classes seeking redress were the over-exploited peasantry, underpaid workers, the financially-strapped middle classes, and out-of-power politicians itching to get back into office. Also fighting for democratic participation in the state system or autonomy from it were oppressed ethnic, national and communal groups, and militant religious organizations. In some cases ethnicity represented the spatial and social mobilization of marginalized peasantaries and regions for incorporation, and in other circumstances it provided cultural refuge from the discontents of post-colonial modernization. Women, too, mobilized and sought their empowerment, arguing in the words of the Ghanaian novelist Ama Ata Aidoo (1992:325), that men's political monopoly has lasted long enough. 'If they alone could save us', she wrote, 'they would have done so by now. But instead every decade brings us grimmer realities. It is high time African women moved into center stage' (also see Nzomo 1993, 1995).

The transmission of the messages of popular discontent was facilitated by the spread of mass communications, including the growth of alternative media, both open and clandestine, local and foreign, and Africa's dense networks of oral communication (Diamond 1993a:4; Landell-Mills 1992:552-3; Martin 1992). When the history of this period is written, the role of the photocopier and fax machine will surely merit honourable mention. Intellectuals, who usually wake up to profound movements of the collective consciousness after the fact, abandoned the state-centred focus of their researches and rediscovered the existence of African civil societies and the critical role of social movements as agencies for Africa's transformation. As might be expected, this intellectual homage to the emancipatory potential of civil society first

emerged among the continent's own social scientists in the 1980s, before
it spread to their Africanist colleagues in the West in the 1990s, who
once used to bemoan the 'withdrawal' or 'exit' of social groups from
state-dominated economic and political arenas (Nyong'o 1987; Harbeson
et. al. 1994; Mamdani 1992; 1995; Chole and Ibrahim 1995; Eyoh 1995).
We need to avoid, however, conceiving state-civil society relations in
oppositional, binary terms, seeing the state as the source of all evil, and
civil society as a repository of all that is benign and humane. Almost
always reality is far more complex than such neat models suggest. The
membrane separating the state and civil society in terms of their actions
and visions could be quite thin as so many cases demonstrate: almost
everywhere the struggle was not sharply drawn between the authoritarian
regime, on the one hand, and the democratic masses, on the other, but
pitted the splintered factions within both the state and civil society for
and against each other. The shifts in the alliances were often rapid and
unpredictable. All this calls for careful, detailed empirical research on
Africa's democratic transitions; filtering and forcing them through
preconceived and idealized models of 'democratization' is a futile
intellectual exercise.

It is generally agreed that events in Benin played a crucial
'demonstration effect' in western Africa, especially among the
Francophone countries, as those in South Africa did among the
Anglophone countries in the southern parts of the continent. The Benin
transition wrote the script on using sovereign national conferences as a
mode of democratic transition. In February 1990 representatives drawn
from the political class and the educated elites of civil society gathered
to deliberate the country's future (Robinson 1994a, 1994b). The
conference had been preceded by months of escalating public
demonstrations and strikes, and was agreed to by Benin's long-time
military ruler, Mathieu Kérékou, because of the threat of a general strike.
He and his cronies were shocked by what was to follow. The first thing
the delegates did was to declare the conference sovereign, and they
proceeded to suspend the constitution, dissolve the National Assembly,
adopt plans for multi-party elections, and choose an interim Prime
Minister, who assumed most of the President's powers. A year later
when the elections were held, Kérékou was voted out, and Benin entered
a new political era.

The events in Benin electrified people throughout Francophone Africa,
and the model was followed in one country after another, although the
outcomes varied depending on the strength of the social movements and
the responses of the incumbent regimes. Resting on principles of popular

sovereignty and the right of people to renegotiate the social contract, this model reflected deep popular yearnings for change and tapped into the heritage of the French Revolution, whose bicentennial was in 1989 and was widely marked in the Francophone world. Pro-democracy activists appropriated the moment and incorporated its history into their unfolding culture of politics. Where the ruling parties were able to manipulate and control the process or divide the pro-democracy forces, as happened in Gabon, Côte d'Ivoire, and most sadly Zaire, the ruling parties remained in power. Where they were not able to do so, as was the case in Congo, Mali, and Niger, new governments were voted into office. France, which maintained strong connections with its former colonies and sustained their autocracies, was forced to bow to the pressures as well. At the June 1990 Franco-African summit President Mitterand at last expressed support for democratization in these countries. As late as February 1990, when the Benin Conference was in session, Jacques Chirac, a former Prime Minister and a future President of France, then Mayor of Paris, had declared at a conference of Francophone mayors in Dakar, Senegal, that multi-partyism was a 'political error', a 'luxury', that developing countries could not afford.

In the south it was the troubled regional giant, South Africa, that captured the world's imagination in 1994[6]. South Africa's transition to democracy is a long and complicated affair, which we cannot adequately address here. A few remarks will suffice. Apartheid, declared a 'crime against humanity' by the United Nations, represented a monumental abuse of South Africa's human resources and relations. It led to the development of a highly distorted and eventually stagnant economy, and turned the country into a regional terrorist state and an international pariah, a skunk of the world, as President Mandela put it in his inauguration speech. Following the Soweto uprising of 1976, it became increasingly clear that apartheid was on its way to the dustbin of history after decades of protracted struggles by the liberation movements and other movements coalesced around workers, peasants, women, students, residential and consumer communities, churches and mosques. By then, too, apartheid had become an albatross on the South African economy. The old regime of accumulation and regulation, the growth model of racial capitalism, was disintegrating. As the economic and political crises deepened, the state responded to the emboldened resistance forces with its customary sticks of repression and the untried carrots of reform. Initially attempts were made to co-opt the black middle class, while compartmentalizing the marginalised, unemployed, and unemployable population as 'outsiders' in the Bantustans.

But this was too little too late. Africans wanted the democratization of political power, the equalization of economic opportunities, and the redistribution of the productive resources. The clamp-down of the mid-1980s failed to stem the roaring tide of internal opposition, now reinforced by the chilly winds of international isolation. More reforms were implemented, apartheid laws repealed one after another, but the clamor for a non-racial, inclusive democracy continued unabated. Then in 1990 Mandela and his colleagues were released, the ANC and other organizations unbanned. A national conference was held bringing together political parties and tendencies of all persuasions to hammer out the country's political future. Soon it was all over, the racist fantasy that a white minority could forever dominate their fellow African citizens because of the trivialities of skin color. The fear of large-scale violence that the media kept warning of evaporated soon after the elections were held in 1994 and a coalition government of national unity had been formed. This paralleled what happened in much of Africa during the transition from colonial rule to independence during which ethnic differences were muted by the power of pan-territorial nationalism.

Malawi was one of the countries in Southern Africa that felt the winds of change blowing over South Africa. Banda's regime found itself confronted by a restive population no longer cowered into submission by the vast and ruthless security apparatus of detention camps, networks of informers, and the sadistic storm troopers, known as the Young Pioneers and the Youth League. Besides the momentous changes taking place in South Africa, which had previously bankrolled Banda, the only African leader who had been prepared to establish diplomatic relations with the apartheid state, there was the inspiring democratic transition next-door in Zambia. Internationally, Banda's western allies no longer found his fierce anti-communism useful, and his autocratic rule was becoming an embarrassment in a region and a continent being swept by democratic ferment. And so they suspended aid to pressure the regime to undertake reforms.

There had been protests against Banda's political autocracy before, including a botched guerrilla attack in 1967. The swelling ranks of Malawian exiles in the neighbouring countries formed a myriad of opposition parties, but as is common with exile politics, they often spent as much time attacking each other as they did attacking the Malawi dictatorship. Then in the early 1990s clandestine political parties were formed in the country, workers began flexing their poorly paid muscles, and university students began expressing their youthful intellectual energies on the streets. The end of the Banda era began on a Sunday morning in April 1992, when Catholic priests throughout the country read a carefully crafted pastoral letter calling for justice and fairness, respect for human rights, and wealth redistribution. It was a divine bombshell, for which several bishops were almost killed. But the Rubicon had been crossed. In May the country experienced its worst disturbances. In June 1993 the government lost a referendum on whether or not a multi-party system of government should be introduced. Almost two-thirds voted for a multi-party system. In the period leading up to the elections in May 1994, the army, which had largely been politically-neutral, forcibly disarmed the armed thugs of the ruling Malawi Congress Party, the Young Pioneers. But the two main opposition parties, divided along regional and to a smaller extent ideological lines, refused to unite. Fortunately, the hunger for change in Malawi was so strong that one of them still did manage to win over the 97-year-old Banda. The future had began (Kalipeni 1992, 1995, 1996; Chirwa 1994; Mchombo forthcoming).

The Ghosts of History

But the room which the future seized to dance to the intoxicating rhythms of democracy harbored many ghosts of history; those of poverty, debts, ethnic schisms, and social inequalities, in short, of underdevelopment and dependency. These ghosts threatened to cloud the room, change the music, and even stop the dance in some cases. Relief about the death of apartheid and apprehension about the future hovered over the smiling skies of Pretoria as Mandela took the oath of office in early May, 1994, a poignant scene repeated later the same month in Malawi. History vowed to throw its weight on the new South Africa: simultaneously containing the great expectations for rapid and profound change, as happened elsewhere in post-colonial Africa, and fueling popular demands for meaningful change, for the realization of the dreams of the struggle - democracy, dignity, and development. The new South Africa started off with many social deficits, deep structural imbalances, and a lot of tortured memories, but also with memories of heroic resistance and victory, the assets of an irrepressible civil society, and an economic infrastructure lacking elsewhere in Africa at independence. The last chapter of Africa's sad historical encounter with European colonialism was finally written. A new chapter on a more magnanimous future may have began.

But the journey to the future promised to be a long, bumpy one. By the end of 1994 Nigeria had stalled. It was a strange spectacle: Nigeria and South Africa, Africa's giants, had almost traded places. Until

Mandela left prison for the presidency, Nigeria was in the forefront of the sanctions campaign against apartheid South Africa. Now the future beckoned as brightly in the former racist pariah state as the past stubbornly enveloped Africa's most populous nation in darkness. Freed from the suffocating laager of apartheid, South Africa could begin reinventing itself and realizing its possibilities, while the chains of misrule and mismanagement, despotism and despair, tightened around Nigeria's neck, choking the country of its potential greatness[7]. Out of the 34 years of its independence, less than ten were spent under civilian rule. Successive military dictatorships, interspaced with inept civilian administrations, flirted with then robbed the country of its aspirations for development and democracy, in a relentless pursuit for power and privilege for a decadent political class. Military interventions, usually undertaken during moments of acute political and economic crisis, itself the result of the endless cycle of military oppression and plunder, served to disorganize and dissipate emergences of critical concentrations of popular opposition and were a means of shuffling the deck of accumulation for the competing factions of the military itself and the favoured fractions of the unproductive bourgeoisie.

The Nigerian military has, of course, always claimed a restorative mission to clean up the mess left behind by the structural distortions of previous constitutional arrangements, and the trail of flammable regional and ethnic rivalries, widespread election fraud, economic decline and inefficiency, and a culture of kleptocracy inspired by corrupt competitive party politics. The deceptions and perversions of military rule reached their apotheosis under Babanginda's ruthless, fraudulent charm. The centralization and corruption of power intensified, and uneven regional development and social inequalities deepened, as the state was increasingly privatized and appropriated by dominant domestic and foreign class interests wedded to the structural adjustment programme which wrecked havoc on the livelihoods of peasants, workers, and the professional middle classes. Opponents, real and imaginary, were persecuted, opportunists co-opted, the once fearless press muzzled, civil organizations and political parties were routinely proscribed, extra-judicial murders became more common, and the rule of law was breached by decrees which were issued with the frequency and impunity of machine gun fire. The increasingly restless nation was occasionally bribed with populist rhetoric and dispensations, and above all, by promises of democratic restoration. Accompanying the smoke and mirrors of the elaborate transition programme that was unveiled, was the game of state multiplication, through which the collaborative political elites were rewarded with new bureaucracies, and the rising demands of marginalized regions and ethnic groups for a more equitable political and fiscal federalism were manipulated.

But Babangida's merciless, cynical script lost its flow in the presidential elections of June 1993, which the brutalized nation seized with unexpected alacrity that frightened the bullies in uniform. Judged the freest and fairest elections in Nigeria's post-independence history, an incredulous Babangida annulled them, and Nigeria sank to unprecedented levels of anger and anguish. Abiola was arrested for claiming his rightful victory. Babangida gave way to Abacha's unsmiling face of naked military terror, which the world saw in November 1995 in the brazen judicial murders of nine Ogoni political activists, including the writer, Ken Saro-Wiwa. A shaken Mandela, who had urged cautious diplomacy with the Nigerian military junta, and had risked compromising his moral stature in the process, launched a campaign for sanctions against Nigeria. By the beginning of 1996 Africa was holding its breath as to what would happen next in Nigeria. As with South Africa before it, Nigeria's collapse would be a tragedy too ghastly to contemplate, Rwanda writ large.

By the mid-1990s Algeria's transition had also stalled (Liabes 1995; Chikhi 1995; Sharaway 1995). Following growing popular opposition, culminating in the uprising of 1988 against the highly interventionist government now broke because of the fall in oil prices, its main export, elections were held in 1992. But both the working and entrepreneurial classes, despite their enormous growth since independence in 1962, were weak, or rather, they did not have a distinct democratic and development project substantially different from that of the state that had nurtured them. Only the Islamists did. When the Islamic Salvation Front (FIS) won the first round of the elections and looked poised to win the final round, the military-backed government annulled the elections. This was justified on the grounds that the Islamists were retrogressive fundamentalists, and an FIS victory would foreclose the future of democracy in Algeria. That possibility was of course never tested. But Algeria got a vicious civil war instead. The Algerian Islamic movement, like many other such movements in the Muslim world, constitutes complex coalitions of reactionary interests and traditions and progressive social commitments to the poor. Large segments of the political and middle classes welcomed the cancellation of the elections. The Islamists turned their murderous wrath on them. The irony is that the rise of the Islamists was facilitated by the cultural alienations and the failures of a misguided modernization pursued by the state and supported by those very classes that had been agitating for democracy. The western powers, looking for a new evil empire now that the Soviet Union had collapsed, colluded in the demonization of the Islamists and the miscarriage of the democratic transition (Entelis 1992; Tahi 1992; Hermida 1992). The Algerian story has no angels.

Neither, of course, does that of Rwanda. But there the identity of the devil was clear: the state. Rwanda's conflagration was no simple inter-ethnic warfare, something 'natural', a resurgence of mindless, primordial ethnic antagonisms, as the country's ruthless rulers claimed, and the media made the world believe, but the result of a desperate attempt by a beleaguered authoritarian regime, aided by its regional and international allies, including France, to cling to power. The Tutsi and Hutu share more commonalities than differences: they speak the same language, have similar religious traditions, occupy the same geographical space, and have been governed for centuries by the same aristocracy. In a situation such as this when the real sociological differences among groups are rather tenuous, the groups are pushed towards multiplying their minor differences in an effort to clearly separate friend from foe, an effect Freud called the narcissism of minor difference. The Hutu and Tutsi, thanks in large measure to the divisive politics of Belgian rule, had accumulated hate memories and fears of mutual annihilation, which erupted during moments of political crisis, often exacerbated by foreign powers and neighbouring countries hosting successive waves of refugees. And that is what seems to have happened in the years leading up to the genocide of 1994: the authoritarian Rwandan state, dominated by a narrow-minded ethnic elite, felt threatened by popular insurgent democratic forces, and so it cynically and desperately stoked the memories of inter-ethnic hate and fear, unleashing the pathology of genocidal violence (Ibrahim 1995). Siad Barre's embattled regime in Somalia did the same thing among a people who are ethnically and culturally homogeneous. In this case clans were mobilized and became the deadly markers of difference.

It can be seen that the state was central both as the target and instrument of inter-ethnic conflicts. The existence of ethnic groups was not the problem in and of itself, for if it were all 55 African states would perpetually be at war, instead of the five or so, for all of them by the fiat of colonial map-making are multi-ethnic and multi-cultural. The question is why have ethnic conflicts arisen in some of these countries at the times they did. There is a widely held view that ethnicity is used as an ideological tool by political leaders, much as race is used in multi-racial societies, to circumscribe meaningful political discourse, to frustrate the associational alliances of class and community (Mafeje 1995). The available evidence seems to suggest that most of the major ethnic conflicts have tended to occur when an authoritarian regime is under pressure to democratize, especially when it has some identifiable ethnic character or ethnic beneficiaries who stand to lose in a new democratic dispensation. Whether or not ethnicity plays a role in the transition process appears to depend on the mode of transition (Nnoli 1995).

Ethnicity virtually played no role in countries where sovereign national conferences were held to discuss and work out the modalities of transition. Attended by members representing broad coalitions of civil society, such conferences attained a high level of legality, legitimacy and representativeness, unlike situations where more restrictive constitutional conferences were held, or the transition was effected through constitutional amendments by the incumbent government, or peace agreements were reached among warring factions. Contrary to the self-serving assertions of the defenders of single-party politics against multi-party democracy, it is quite evident that ethnic conflicts tended to occur mostly under authoritarian regimes rather than democratic ones. Clearly, certain modes of transition to democratization can help attenuate, if not neutralize, ethnic conflicts. Others can intensify them, especially if the state is determined to hang on to power, despite the decline in its legitimacy, and the opposition is deeply divided, regardless of the intensity of the popular hunger for democratic change.

That is the Kenyan script. President Moi of Kenya responded to the mounting pressures for change by arguing that multi-party democracy was a foreign import that his developing and multi-ethnic nation did not need. Once regarded as darlings of the West because of their free-enterprise policies in a region filled with socialist experimentation and political instability, Kenyan leaders now found themselves buffeted between rising domestic opposition and open criticism from their western allies, led by the American ambassador. Donor assistance was suspended until they agreed to implement economic and political reforms. Led by the church, lawyers, and women activists, and bolstered by strikes by impoverished workers and demonstrations by ungovernable students and the unemployed youths, the Kenyan pro-democracy movement mobilized various sectors of urban and rural civil society. But it was gravely weakened by the political opportunism and ethnic games played by many of its leaders each of whom desperately wanted to replace Moi in State House. As the election drew nearer, the main opposition party, Forum for the Restoration of Democracy (FORD), split along ethnic and regional lines. Thus fatally divided, Moi could afford to ignore the opposition's calls for a national conference, and managed the process of transition with all the chicanery his regime could muster, which included exploiting rural conflicts over land and fomenting ethnic clashes. The divisions in the ranks of the opposition ensured that although Moi won only 36% of the total votes in the multi-party elections of 1992, he secured more votes than any of his eight or so opponents and retained power. Since then, Kenya has vacillated between openness and repression, the future and the past (Nyong'o 1989; Ajulu 1992; Barkan 1993; Nzomo 1993; Haugerud 1995). But having been awakened, it is not clear the demos can be forced back to the sidelines to watch and cheer their unpopular leaders playing politics with their lives.

Conclusion

And so, as we celebrate the South Africas, Benins and Malawis, we continue waiting for the Algerias and Nigerias, and watching the Hobbesian nightmares of the Rwandas and Liberias (Fleischman 1993; Huband 1993). What, then, does the future hold for Africa's democratic projects? The honest answer is that no one knows. All one can say, with any reasonable degree of confidence, is that Africa's transitions to democracy will continue to take many forms and directions because of the varied constellation of state forms, social movements, class and ethnic forces, regional and international developments facing each country. Visions of what democracy should mean are also going to be different. Democracy is a virtue we all claim to cherish, but its meaning is often in the eye of the beholder. Even tyrants swear by it, often calling their totalitarian contraptions democratic republics. If Africa's new democracies are to be sustainable they must be moulded from Africa's progressive historical and contemporary traditions and visions, those that are rooted in the proliferation and richness of associational life. Pluralism has to mean far more than periodic electoral contests and the badge of citizenship has to entail more than the privilege of casting a vote once every four or five years. It has to be an affirmation of each person's humanity in all its dimensions - political, economic, cultural, social, and moral. And these new democracies will be judged by their ability to deliver economic well-being to their citizens. As in the struggles for the first independence, ordinary Africans are looking for democratic developmental states.

The temptation to draw sweeping predictions from short-term trends is always great, and Africa has perhaps suffered more than most from prescriptive desires masquerading as concrete knowledge. How many times have we read of the continent's death? It is not only the attention span of television addicts that is short: some academics and journalists who make a living channel surfing from one country and one continent to another have already concluded, pointing to Rwanda and Somalia, that the African democratic transitions have failed, ignoring that there are dozens of other countries where progress, usually slow and contradictory, as the historical process usually is, is taking place. Africa is too large, it has too many stories for such glib generalizations. As a historian I may be allowed the indulgence, God willing, of waiting another twenty to thirty years to pore over the written and oral archives of this exceptionally dynamic and complex period in modern African history. Those of us whose lives, whether by choice or by chance, are implicated with Africa, and it should be all of us as human beings, cannot afford the moral luxury of indifference, the intellectual fatalism of Afropessimism, the belief that Africa has no future, or that its salvation lies in benevolent recolonization or the periodic charities of Band Aid. Africa must, and one hopes, will forge its own modes of good governance, its own development models, in short a more generous way of living and being human. Fukuyama (1992) is wrong. History, certainly African history, continues.

Notes

1. For a definition of these terms see *Africa Demos* 3, (2):35, 1995.

2. See the excellent papers in *Review of African Political Economy*, No.50 (March 1991), *Special Issue on Africa in a New World Order*. Many lament Africa's marginalization following the end of the cold war; also see Shaw (1993), Callaghy (1993, 1995). The declining economic and strategic importance of Africa to, say, the United States, may have been exaggerated according to Volman (1993). The narrative of Africa's increasing marginality has uncanny resonances with narratives of the marginality of minorities, especially African Americans, in the United States, and the irrelevance of labour in the brave new neo-classical world. A discourse that glibly writes off a whole continent and measures the relevance and value of its peoples according to their importance to the global market is immoral. It authorises and legitimates global racism. In the United States the discourse of African Americans' irrelevance, indeed, that they are a pathological 'nuisance', reflects and reinforces right-wing racist violence. In Nazi Germany it bred the Holocaust. The comparisons are not as far fetched as they might seem. It is this discourse that allows the world to watch the genocide in Rwanda almost with indifference. Discursive dismissals of races, peoples, societies, countries, social groups, and yes, continents, have often been, history shows, articulated in attempted physical destructions of the objects of vilification.

3. He states categorically (Legum 1992: 205), that the collapse of communism 'was not what has driven Africa to pursue a new political direction. The Second Liberation, in fact, preceded the changes heralded by glasnost and grew out of local experience'.

4. Both Mazrui (1990) and Mamdani (1992) seem to agree on this point.

5. On May 13 1992, the World Bank, on behalf of the Western Consultative Group, announced the suspension of economic assistance to Malawi, except for humanitarian aid, because of the country's poor human rights record. But on 17 June, the Bank approved two loans, one worth $55 million for a power project and another on June 23 of $80 million for balance of payments support. See the World Bank press release, *Meeting of the Consultative Group*, 13 May 1992, and the confidential Office Memorandum, 'Malawi's Macroeconomic Policy Mission Back to Office Report', 12 June 1992, which the Malawi Action Committee, a human rights organization formed by Malawian exiles in North America, to which I belong, obtained. See *MAC Newsletter*, July 1992:3.

6. This section is drawn from a review article of nine books (Zeleza, forthcoming) that deal with the political economy of South Africa's transition from apartheid to a new democratic dispensation: P. G. Moll (1991); P. G. Moll, et. al. (1991); S. Gelb (1991); R. Fines with D. Daved (1991); I. Abedian and B. Standish, eds. (1992); Ben Turok, et. al. (1993); P. H. Baker, et al., eds. (1993); D. Innes, et.al. (1992); African National Congress (1994).

7. This section draws from a review of two books (Zeleza 1995). The books are by T. Olagunju, et. al (1993) and S. Adejumobi and A. Momoh (1995); also see Nnoli (1993) and Olukoshi (1995).

Chapter Eighteen

REWRITING INDEPENDENCE

Introduction

In September 1993, I attended a conference on Malawi's 1964 Cabinet Crisis, a watershed event in the country's post-colonial history, at the Centre for Southern African Studies, University of York in England. Unusual in its composition, organization, and objectives, this conference brought together academics, writers, politicians, and religious leaders, both Malawian and non-Malawian. No formal papers were presented. Instead, we examined the contemporary archival records and interrogated the memories of the participants and observers of the period. It was a serious, open, and sometimes painful enquiry, aimed at re-examining and correcting the historical record, pondering the present, and divining the future. The conference was made possible by the re-emergence of democratic forces in Malawi.

As is common at conferences, some of the most exciting deliberations took place in the corridors and the bars. On a couple of occasions the writers among us, myself and the poets David Rubadiri, Felix Mnthali, Lupenga Mphande, and Jack Mapanje, and the critic Hangson-Mpalive Msiska, sat late into the night, discussing and celebrating the process of political renewal in our beloved homeland, from which we had been exiled for varying lengths of time. We were celebrating, to quote the renowned Nigerian critic Abiola Irele (1992:302), the unmasking of the crisis of political legitimacy in post-colonial Africa, the rupturing of an oppressive system, due to 'a profound movement of the collective consciousness'. Indeed, we were celebrating, as Chinua Achebe (1992:349) would put it, humanity in our country, in our continent - a humanity that has in the past five centuries, faced and triumphed against, the evils of slavery and the obscenities of colonialism, and is now confronting the consequent deformities of the post-colonial order.

But the older ones among us had of course gone through this before, in the early 1960s, during the first wind of change, the first transition to independence, to democracy. And so our celebration was tempered by caution and some trepidation. Indeed, we were only too aware of the potential pitfalls: there was the compromised transition in Kenya, Babangida's farcical manoeuvres in Nigeria, not to mention the descent from tyranny into anarchy in Somalia, and the resumption of a murderous civil war after the elections in Angola.

Lurking beneath the celebratory reflections there were also concerns about our role as writers in the unfolding drama, in the emerging new dispensation. Tyranny had created us, imprisoned and exiled many of us, enraged our consciences, and nourished our imaginations. It had given us the moral inspiration to write, the themes to write about, and often determined the languages, forms and styles of our writing, as well as our audiences and production outlets. Now we were about to be orphaned from this tyranny. We were being challenged to recreate ourselves, our messages, imaginations, and practices.

These, then, are the issues I would like to address. What role have artists, especially writers, played in the struggles for democratization? What does the cultural politics of democratization entail? Most of the analyses, both those diagnosing the nature of the African crisis and prescribing solutions, tend to be political and economic in orientation. The cultural and moral dimensions of the crisis and its rectification have not received the attention they deserve. Writers, however, the chapter will argue, have probed into the tormented state of the modern world more deeply than the academic scholars. This suggests that writers should have a greater voice in discourses about democracy and Africa's future. The chapter begins by briefly discussing the belated academics' and Africanists' concern with issues of democratization. Then, it examines the creative writers' critique of post-colonial Africa, and the argument is made that Africa's democratic regeneration goes beyond the politics of multi-partyism or the economics of development. It is also about cultural and moral renewal.

The Beautiful Ones Are Not Yet Born

Only a few years ago, it was almost an article of faith in Africanist circles that civil society in Africa was fragile, crimped either by primordial allegiances, parochial schisms, and narrow patron-clientelistic loyalties, or by the post-colonial state, which was called various epithets to characterize its allegedly peculiar inconsistencies, incapacities, and irrationalities. Goran Hyden's (1980) thesis of the 'uncaptured peasantry' seemed to confirm that Africans preferred the 'exit' option to pursuing democratic struggles in the political arena dominated by the state, leading to the emergence of what Ayoade (1988) and Riddell (1992) have characterised as 'states without citizens'.

So fascinated did some scholars become with the 'exit' or 'withdrawal' of key social forces within civil society, that they did not even think that structural adjustment would provoke serious political upheaval (Bienen and Gersovits 1986). Part of the problem was conceptual, the tendency by political scientists whether from the Hegelian or Gramscian traditions, to see state-society boundaries in absolute or binary terms, as opposed, rather than interpenetrating, structural arrangements, social and moral spaces (Mersha 1990; Lemarchand 1992; Mamdani 1996). Despite the current struggles, this fascination with the politics of 'exit', which is based on a sharp and simplistic state-society dichotomy, still finds adherents (Fatton 1990).

In hindsight, these analyses, like most analyses of the Soviet empire which assumed its somnolent permanence, appear naive. Many African scholars, including myself, did not anticipate that the changes which have engulfed Africa in the last few years would come so soon. My excuse is that I am an historian, not a social scientist who studies the present or tries to crystal-gaze into the future! But to some, these changes did not come completely by surprise. Mkandawire (1992:4) has noted that in the 1980s while Africanist scholars were busy bemoaning or applauding the 'exit' of peasants and other exploited social classes, 'African social scientists moved in a different direction, casting attention more towards the study of social movements and democracy' (also see, Nyong'o 1987; Mamdani, Mkandawire and Wamba-dia-Wamba 1988; Mamdani and Wamba-dia-Wamba 1995). Mamdani (1992b) notes warily: 'the same circles who used to argue only yesterday that democracy was at best a developmental luxury, today uphold democracy as a developmental necessity!'[2]

Long before these African social scientists had discovered the social movements in the 1980s, African writers, almost from the dawn of independence, knew that the masses were dissatisfied and hungry for meaningful change. The rhetoric of nation-building and development could not fool Ousmane Sembene's ([1960] 1986) restive workers, Ngugi wa Thiongo's (1967) militant peasants, Bessie Head's (1969, 1971) rural exploited women, and Buchi Emecheta's (1979) urban working class women. Achebe (1963, 1966) and Ayi Kwei Armah (1968) showed the economic and cultural hollowness of modernization, before African and Africanist scholars discovered from the 1970s, through the lenses of dependency theory imported from Latin America, that the venerated god had no clothes. And writers as diverse as Alex la Guma (1972) from the dungeons of apartheid to Ama Ata Aidoo (1972, 1979) from the faded Black Star, had been bemoaning exploitation and celebrating the struggles for liberation by the oppressed social classes, peoples, and women, long before academics, variously inspired by Marxist, feminist, and other radical ideologies, began systematically investigating and championing equality and empowerment.

Thus, the social forces behind Africa's present struggles for democracy have been consistently chronicled, dissected, and applauded in African literature in the last three decades. At the heart of the drive for democratization lies the fact that the post-colonial state and the ruling elite have lost legitimacy, as Peter Abrahams (1956) predicted it would in *A Wreath for Udomo* before Fanon (1961) wrote his angry, prophetic, and influential critique of nationalism and the post-colonial order, *The Wretched of the Earth*. The reasons for this and dimensions of the loss are many and varied. They are political, economic, social, cultural, and moral. It is now hard to remember the euphoria that greeted independence, which was won after long and difficult, and sometimes protracted, guerrilla struggles. Independence, it was believed, signalled the end of Africa's exploitation and humiliation and marked the beginning of a new era, the emergence of what Nkrumah (1964, 1965) called the 'African personality' onto the world stage.

We now know, as many writers warned, that the hopes of a new beginning and the beliefs that independence marked a revolutionary conjuncture in Africa were illusory. The weight of Africa's pre-colonial and colonial pasts was heavier than most realised or cared to admit in the intoxicating moment of independence. It cannot be overemphasised that 'contradictions within African societies were not transcended but given new complications by the impact of colonialism' (Irele 1992:297). Far from consolidating the harmonizing 'us' versus 'them' rhetoric of nationalist discourse, independence gave the centrifugal forces of the fragile 'territorial nation' room to manoeuvre. And praying at the altar of developmentalism, of modernization in the western image, Africa's post-independence rulers had no time for the blessings of Africa's own gods and cultures.

The fact that the nationalists inherited, in Basil Davidson's (1973:94) memorable metaphor, an independence 'dish' that 'was old and cracked and little fit for any further use', hardly helped. 'Worse than that,' he continues, 'it was not an empty dish. For it carried the junk and jumble of a century of colonial muddle and 'make do' and this the new... ministers had to accept along with the dish itself. What shone upon its supposedly golden surface was not the reflection of new ideas and ways of liberation, but the shadows of old ideas and ways of servitude'[3].

The nationalists, in other words, inherited the colonial leviathan with all its authoritarian tentacles. The failure to trim these tentacles reflected both the class interests of the new rulers and the weaknesses of the anti-colonial movements, and these countries' continued vulnerability to hostile international forces. The intensification of statism in nationalist discourse after independence was accentuated by the weak material base of the new rulers, so that the state became their instrument of accumulation, and the overwhelming demands of the restive masses for the fruits of *uhuru*. It also reflected the nature and dynamics of nationalist politics. During the struggles for independence, politics was 'essentially referred to the state principle', that is, politics was seen as the state, and the state was politics (Wamba-dia-Wamba 1992:11). The equation of politics with the state was rooted in the homogenizing ambitions of nationalist ideology and the nature of the reforms implemented in the twilight years of colonial rule. The political monopolies of the one-party state and military rule were incubated in nationalist ideas that posited the independence struggle in essentialist and exclusivist terms. After independence, now that the dreaded colonialists, the 'them', had apparently gone, it was time for the 'us', the 'people', to build and develop the nation. Pursuit of separate class, social, cultural or gender interests, could be dismissed as 'sectional' and, therefore, delegitimized[4].

The political reductionism of nationalist discourse was facilitated by the fact that the reforms introduced towards the end of colonial rule 'undid the ties between political and social movements' (Mamdani 1992b:314). The social movements, such as trade unions, peasant, women's, social welfare, and religious organizations increasingly lost their autonomy, for they were required to obtain legal recognition before they could be allowed to operate. The result is that 'accountability to membership became secondary to the organization's accountability to the authorities'. As the social movements atrophied, the 'political movements that once articulated a broad social vision were gradually reshaped by their leaders into vote-gathering machines' (Mamdani 1992b:314).

Thus the 'disintegration' of the links between political and social movements, or what some have referred to as the 'class alliance' of the independence movement (Nyong'o 1992), was already evident at independence. After independence, the compromised autonomy of the social movements and the statism of nationalist discourse deepened, thanks to the new imperative of development. Developmentalism was, in fact, not new. It was an ideology of late colonialism as shown in Chapter 11. Even as radical an ideology as *Ujamaa* had its antecedents in this discourse (Feierman 1990).

The post-colonial leaders worshipped at the altar of economic development in the omnipotent name of modernity either in its capitalist or communist incarnations. It was a daunting mission, but at first the fierce god of modernity seemed to be listening and the leviathan appeared capable of balancing the conflicting demands for rapid development and distribution. Lest we forget in these days of Afropessimism, the puny economic and social infrastructures left by colonialism were greatly expanded. Especially remarkable was the growth of education. For the Sub-Saharan region, excluding South Africa, primary school enrolments rose from 14 million in 1960 to 52 million in 1988, for secondary schools from 1 to 11 million, and for universities from 56,000 to 580,000 (Landell-Mills 1992:554). Health facilities increased. Nutrition levels improved. Life expectancy rose and death rates fell. All this not only led to the rapid growth of population, but also to significant changes in population distribution and social differentiation. Urbanization grew and so did the working classes and the middle classes.

But for many, modernity proved to be a false god, whose power began to fade as the harmattan season of recurrent economic recessions started in the mid-1970s. State intervention in the organization of the economic, social, cultural, and political processes intensified as the contradictions of developmentalism deepened and became more open. The currency of post-colonial state legitimacy, forged in the crucibles of often protracted nationalist struggles and victories of the 1950s and 1960s, began to fade. The SAPs implemented in the 1980s, even the World Bank and IMF now grudgingly concede, only made matters worse[5]. They led to losses of the post-independence gains in welfare, the erosion of populist programmes, and 'compromised the state as the bastion of national sovereignty and... raised the question as to whom the state is accountable', the local society, or external forces, such as the World Bank and the IMF (Mkandawire 1992:7). Various social movements began flexing their muscles in both the urban and rural areas, among men and women, the rich and the poor, to reclaim and renew the tattered independence social contract, to reinvent a more decent and democratic future for their respective constituencies and communities (Mamdani and Wamba-dia-Wamba 1995; Harbeson et. al. 1994). The beautiful ones had to be born. But first the devils on the cross, who cast a spell on them, had to be vanquished.

Devil on the Cross

African writers were among the first to note that the emancipatory potential of independence had been overestimated. Indeed, while many historians and social scientists were busy celebrating the achievements of nationalism or devising models of nation-building and development, African writers had already discovered that the post-colonial emperor was naked (Wastberg 1968). The failure of independence became the overriding theme of African literature in the 1960s. Already in the 1950s, writers such as Abrahams (1956) were warning against exaggerated expectations, indeed, predicting that disillusionment would follow independence.

Potential disillusionment turned into actual disenchantment in Achebe's (1966) *A Man of the People* and Wole Soyinka's (1965) *The Interpreters*, then into despair in Armah's (1968; 1970; 1972) bitter trilogy, *The Beautiful Ones Are Not Yet Born, Fragments,* and *Why Are We So Blest,* and in Kofi Awoonor's (1971) *This Earth, My Brother.* In the 1970s, Wa Thiong'o (1977; 1982) channelled the by now deafening critique of the moral bankruptcy of Africa's post-colonial ruling elite into a furious commitment towards their overthrowal in *Petals of Blood* and *Devil on the Cross.* Thus, the post-colonial novels in Africa, as Appiah (1992:152) states, 'are novels of delegitimation: rejecting the Western imperium, it is true, but also rejecting the nationalist project of the post-colonial national bourgeoisie'[6].

Thus, literature in independent Africa was as deeply political as it had been during the colonial period. The reason for this is simple: the barbarities of colonial and post-colonial rule were too great to allow African writers the indulgence of posing the small questions of post-modernist literature. As the British critic, Peter Lewis (1992:76), recently put it, to many African writers, 'the post-modernist preoccupations of many leading Western writers, indulging in parody and pastiche and playing with words and forms, must seem the nadir of aesthetic decadence'.

Needless to say, not all this literature was good. Commenting on Heinemann's African Writers Series, Soyinka, Africa's first Nobel Laureate in literature, has stated: 'the series, of course, was very uneven; quite a large portion of it was total dross, but a fair amount, quite a good amount was excellent literature... it occurred to me that the series was adopting a policy of anything goes because it's African and therefore must be published' (quoted in Wilkinson, 1992:94)[7]. This view has been echoed in a recent *London Magazine* review of my novel, *Smouldering Charcoal,* in which Michael Kelly (1993:144) contends: 'African writing in English has had its wooden, poorly written, ranting, tedious, pretentious exemplars. The African Writers has at times seemed to be imitating the United Nations' policy of national representation for its own sake'. Mercifully, he had nothing but nice things to say about my novel.

Some African writers and critics are, in fact, uncomfortable with what they regard as the excessive politicisation of African literature. They insist that the writer's responsibility is strictly to the perfection of his/her craft. The Nigerian critic Izevbaye (1977), for example, has long hoped for the development of a literature with a 'suppressed social reference' so that non-sociological criticism can advance. This call for technical and formalistic criticism is supported by many 'Africanist' critics and vigorously opposed by the 'Nationalists', as Jeyifo (1990:37) calls them, who tend to claim, '"natural" proprietary rights in the criticism of African literature' and emphasise 'extra-literary' or 'non-literary' evaluative criteria rooted in anthropological and sociological paradigms[8]. African literary criticism, of course, transcends the 'Africanist-Nationalist' divide, for there are also Marxist and feminist critical perspectives (Gugelberger 1985; Owomoyela 1992; Fran 1984; Jones et al. 1987; Davies and Graves 1986; Davies and Fido 1993; Williams 1992).

The polarized 'sociological' and 'literary' readings of African literature have arisen, because this literature is consumed through a critical treadmill based on European literary traditions with their pseudo-universalism, whereby local European customs are elevated to parables of the human condition, while the evocation of local African customs amounts to mere ethnography. Often in Western universities, African novels are regarded more as academic travelogues, used in social science courses as windows into the African social and cultural worlds, than as literary pieces or aesthetic creations. This devaluation of African literary texts reflects the wider devaluation of African experiences and ideas and resonates with the Eurocentric discourses on democracy discussed in the previous chapter and the well-known neo-colonialist discourses on development. In short, the ideology which assumes that the omnipotent World Bank and IMF, as opposed to Africans, should structure African 'democracy' and 'development' is the same ideology which values western literature as 'art' and African literature for its 'politics' and 'ethnographic' insights.

The question, therefore, is not one of whether or not African literature is political, for all literature is political, but what type of politics it expresses. The intensity, and sometimes ugliness, of the African literary debates, indicate that this literature articulates the politics of liberation, politics that challenges all earlier legitimating narratives in the 'name of the ethical universal; in the name of *humanism*' (Appiah 1992:155), rather than the post-modernist politics of reactionary impotence, whose delegitimation of past narratives is an end in itself.

African literature produced since independence has been political in two senses. First, the theme of resistance has featured prominently. Indeed, this has been a literature of resistance par excellence. The finest of this literature, argues Jeyifo (1992:354), presents 'a sophisticated testamentary tradition that taps the deepest democratic aspirations of the continent and its peoples'. 'The manifest concern', Irele (1990a:xiv) believes, 'of the writers to speak of the immediate issues of social life, to narrate the tensions that traverse their world - to relate their imaginative expression to their particular universe of experience in all its existential concreteness...leave(s) the African critic with hardly any choice but to give precedence to the powerful referential thrust of our literature'[9]. As I write elsewhere: 'the best of this literature is intense and serious, passionately engaging the large questions of human struggle against the dehumanizing evils of oppression and exploitation, materialism and greed, alienation and atomization, as it celebrates memories of struggle and the sanctity of social connectedness and affirms the possibilities of tomorrow, of human survival and transcendence. Often combining realism with myth, poetic vision and fantasy, this is a literature delicately woven from many artistic traditions, fusing oral and written narratives, idioms and conventions into powerful, probing, and poignant tales of the human condition' (Zeleza 1995:87). It is a literature, as Nadine Gordimer (1994:37-38) notes in her appreciation of Soyinka, of multiple languages, and the voices of the gods, who have largely disappeared from other literatures.

African literature has also been political in the way it has been received by both Africa's dictatorial regimes and popular audiences. As is well-known, 'a great number of African writers have had their works banned by these regimes, many have been jailed for long terms, and not a few have been killed or hounded into involuntary exile' (Jeyifo 1992:353). The list of prominent African writers who have suffered detention or been forced into exile is a long and depressing one. From South Africa are Alex La Guma, Dennis Brutus, Breyten Breytenbach, Bessie Head and Njabulo Ndebele. Across the Zambezi, in my own homeland, are Felix Mnthali and Jack Mapanje. Kenya has its Micere Mugos and Ngugi wa Thiong'os, Somalia its Nurridin Farahs, Egypt its Nawal el-Saadawis, and Nigeria its Festus Iyayis and Wole Soyinkas.

This is as good a measure as any of the seriousness with which African literature is taken by African regimes. Achebe (1992:349) recounts an encounter with a Swedish writer at a conference, who said to him and his colleagues: "You fellows are lucky. Your governments put you in prison. Here in Sweden nobody pays any attention to us no matter what we write'. 'We', Achebe says sarcastically, 'apologised profusely to him for our undeserved luck!'. The political marginalization of the arts in the West might be seen as the price European artists have paid for insisting, since the nineteenth century, on 'an intrinsic view of art: art for art's sake' (Schipper 1990:62).

African audiences also take African literature seriously. As is the case in most repressive societies, literature and art become a set of coded messages of protest, resistance, and affirmation, relished privately and collectively by brutalised souls refusing to be numbed into total submission. Ngugi tells the story of how, when his novel *Devil on the Cross* and the play *I Will Marry When I Want*, for which he had been jailed, were first published in 1980, he and the publishers planned to issue

> only a few thousand copies of each, hoping to sell them over a period of two, three, four or more years. But in fact the first editions of each of these works were snapped up within two or three weeks of publication... Now the reception of the novel and the play was really fantastic because they — particularly the novel — were read in buses, in matatus, ordinary taxis; they were read in homes; workers grouped together during the lunch hour or whenever they had their own time to rest and would get one of their literate members to read for them. So in fact the novel was appropriated by the people and made part and parcel of their oral tradition (Wilkinson 1992:129).

Writing in independent Africa, therefore, has been a deadly serious business. What is often at sake, is not merely the possible ridicule of self-righteous critics, but the vengeful wrath of nervous, tyrannical regimes. Consequently, creative writing becomes not merely a bohemian indulgence, a celebration of the exhilarating powers of the human imagination, but a passionate interrogation of the deformed and tortured psyche of modern society. Contrary to widespread misconceptions outside of Africa, most African writers primarily write for their people, and not to titillate western curiosity[10]. Unfortunately, the very tyrannies they criticise and ridicule sometimes make it difficult for them to be read at home.

One conclusion we can draw from all this is that African writers, by calling, choice, and circumstance, have been in the forefront of democratic discourse in Africa, probably longer and more consistently than any other group of intellectual workers. It is clear that this has something to do with the question of audience. Unlike the social scientists with their hermetic discourses, which are often impenetrable to the disciplinary outsiders, creative writers produce for public audiences.

They tend, therefore, to engage the contested public issues, concerns, and visions in a way that academics do not. The creative imagination, moreover, is not encumbered by the strict discursive structures of the academic enterprise.

Like the disappointed masses, it did not take long for African writers to begin lamenting and vigorously protesting the abortion of the first transition to independence. Their deep sense of betrayal was, ironically, a reflection of their class position and ambiguous relationship to the ruling elite. With few exceptions, most of the African writers writing in the European languages, whom I have been discussing in this presentation, belonged to the elite, or the national bourgeoisie, the very class that assumed control of state power. They articulated the same nationalist discourse. In fact, they took it to its logical conclusion. They believed passionately in the unity and strength of the independence struggle, in the emancipatory power of independence. They also hoped independence would lead to the regeneration of African cultures long scarred by European slavery, colonization, and racism.

Many of them tried to commit Cabral's (1969) class suicide, not simply because they misread Fanon or were ideologically confused, as Lazarus (1990) has argued in his fascinating study. There was, in addition, a compelling aesthetic reason. These writers, schooled in western literary traditions, were anxious to forge a distinctive voice for themselves by tapping into the rich artistic reservoirs of their cultures. In this sense, it was not simply class suicide they sought to commit, but also cultural suicide and rebirth, to strip their artistic imaginations of western influences. Many found salvation in the enchanting, haunting and enigmatic myths, folktales, poetry, drama and epics of oral tradition.

Oral tradition seduced even those, especially among the English-speaking writers, who were violently opposed to the negritudist affirmation and romanticization of Africa's supposed idyllic and undynamic primordialities[11]. African poetry shed European idioms and western literary influences and acquired the indigenous accents and rhythms of oral tradition (Okpewho 1988; Ojaide 1992). African theatre reinvigorated itself by incorporating and reconstructing the oral genres, forms and conventions (Jones 1978; Etherton 1982; Barber 1987; James 1990; Wilkinson 1992)[12]. Novelists soon followed suit. They included the old masters, such as Ayi Kwei Armah, whose vision lost its bitter defeatism in the warm creative ambience of orature which infuse his last two novels, *Two Thousand Seasons* and *The Healers* (Lazarus 1992, Chapter 6), and Ngugi, whose devastating attack on modern Kenya found a new intensity in the magic realism of *Devil on the Cross* (Stratton,

1983). Many of the writers who emerged on the literary scene in the 1980s also drew on African mythology and orality, from Chenjerai Hove (1989) in his acclaimed deeply poetic first novel, *Bones*, to Syl Cheney-Coker's (1991) visionary epic, *The Last Harmattan of Alusine Dunbar*, and Ben Okri (1992) whose Booker Prize-winning novel, *The Famished Road*, with its haunting and timeless tales of the *abiku* child, is a penetrating and damning inspection into the tortured psyche of modern Nigeria (see Chapter 20 below)[13].

All too often, literary critics and many African writers themselves see the reappropriation of the genre of oral narrative as a return to authenticity, for it is believed that Africa is ontologically oral, while writing is European. 'Orality and writing are seen not only as exclusive domains but as successive moments' (Julien 1992:21). These essentialist binaries are historically false, for as Gerald (1981) and Scheub (1985) have shown, writing existed in Sub-Saharan Africa, that Africanists' Africa, long before the arrival of the first Europeans. The Ethiopians, for example, were writing before the English had learnt the Roman alphabet. Thus, oral and written forms have coexisted and enriched each other for a long time. The apparent reappropriation of oral genres is not, Eileen Julien (1992:158) has keenly observed, simply a 'search for authenticity', but accountability, 'a desire to bridge a gap... between two populaces, one rich, the other poor, one literate, the other oral'.

Along with building the aesthetic bridges, African writers moved away from the elitist preoccupation with the existential angst of intellectuals and other members of the political elite, and began chronicling the tribulations and celebrating the lives and struggles of ordinary people. The growth in African women's writing expanded and deepened the scope of African literature, for new worlds of experience were opened up - the ironic joys of motherhood in Nwapa's (1966) *Efuru* and Emecheta's (1979) *The Joys of Motherhood*, the feminist impulses and visions of Ama Ata Aidoo's (1972, 1979) *No Sweetness Here* and *Our Sister Killjoy*, and Mariama Ba's (1981) *So Long a Letter*, and the searing tales of struggle in Head's (1974) *A Question of Power*, and Nawal-el-Saadawi's (1983) *Woman at Point Zero*. The self-glorification of the writer 'as teacher', as Achebe (1977) once put it, or as 'the voice of vision in his own time', according to Soyinka (Wastberg 1968:21), gave way to Wa Thiong'o (1981:81) humble injunction that 'African writers must be with the people'[14].

Thus, since independence African writers, far more than the professional academics, have exhibited a commitment to the political cause of the 'masses' and cultural regeneration. While the latter were busy importing theoretical models from the Western and Eastern blocs, African writers were delving deeper into the African condition. For one thing, the writers could not effectively borrow European sensibility and settings in the same way that their academic counterparts could borrow theories manufactured in Europe and use African settings as empirical fodder. It is significant that western literary movements, such as post-modernism, have not found the same resonance among African writers that western paradigms have among academics. Indeed, sometimes African academics seem to judge their sophistication by their capacity to dabble in the terminologies of each intellectual fad that emerges from the intellectual factories of the West. In short, the internal orientation of African writing in inspiration, focus, and audience, contrasts markedly with the dependency of African scholarship on external discourses, as the philosopher Paulin Hountondji (1990), has noted (also see Mkandawire 1989, and the chapters in Part I above).

This triple commitment to the 'masses', at the political and cultural levels and in terms of audience, made, I would argue, many African writers from the early 1960s painfully aware of the disjuncture between the promises and realities of independence. They felt deeply betrayed by the neo-colonial tyrannies and kleptocracies. This sense of betrayal was real but mistaken. If the African ruling elite were guilty at all, this guilt was in a betrayal of the illusory dreams of nationalist discourse, and not the concrete interests of their class position.

Ironically, this failure, or perhaps unwillingness, to separate the rhetoric of nationalist political mobilization and the imperatives of capitalist class accumulation may be what enabled many African writers to offer such powerful indictments of the post-colonial order, and be among the carriers of the democratic discourse forged in the crucible of the struggles for the first independence to this moment, when history seems to be offering us a second chance.

Conclusion: Smouldering Charcoal

At the end of my novel, the narrator, Catherine, says with defiant simplicity: 'The future has began'. But has it? And what kind of future? What type of democracy is going to emerge out of this current transition? Will it be possible to create a social democracy, one in which both the political and economic domains are based on democratic principles as Samir Amin (1990:6) and Bade Onimode (1992) would advocate. Or is it going to be, as Mandaza (1990) and Mafeje (1992) suspect, a 'compradorial democracy', whatever that means. Or will 'multi-partyism' triumph without genuine pluralism, as Mamdani

(1992a:313), Wamba-dia-Wamba (1992:22-23), and Claude Ake fear (1993:240)? Or will it all blow up, and yesterdays despised tyrannies turn into tomorrows ghastly anarchies, as Zolberg (1992) warns? Will new authoritarianisms, perhaps of the Islamic 'fundamentalist' kind, emerge out of the wreckage as Legum (1992:205) and Henze (1992) caution. Or, perhaps might 'the fears of further descent into hell... create opportunities for more egalitarian, popular, and representative systems of governance' as Robert Fatton (1990:456) hopes.

These are all difficult questions, weighty trepidations. The simple answer is that we don't know what the future will bring. Crystal-gazing is an art best left to soothsayers. It is hard enough to comprehend the present. The last thing we need at this moment is more Afropessimism. But we should also avoid exaggerated hope. All one can say, with any reasonable degree of confidence, is that Africa's transitions to democracy will take many forms and directions. Certainly, the future does not belong to democratic models imported from outside, but to those rooted in African traditions. By traditions, I do not mean a return to some mythical 'indigenous roots' suggested by Ayittey (1991, 1992:216-17), or to Owusu's (1992) pristine village democracy. This is the mystical language of nationalist discourse, of Afrocentric essentialism[15].

I refer to traditions of struggle, not false harmonies, traditions that celebrate Africa's diversities, rather than its imaginary uniformities. Governance must be culturally rooted in the current proliferation and richness of associational life, and promote what Davidson and Munslow (1990:11) call 'the increasing decentralizations of power and the increasing regionalizations of power', or in Wa Thiong'o's (1993:xvii) words, moving centres 'between nations and within nations'. Pluralism has to mean far more than periodic electoral contests. I believe the struggles we are witnessing in Africa today are not simply for reforms in the mode of governance or economic development. They are also aimed at the cultural and moral regeneration of our societies.

Therefore, the value of creative work and reflection, deciphering and probing the cultural and moral conditions of our existence in our numerous and diverse countries and communities, affirming our humanity and our possibilities, is inestimable. This undertaking goes beyond the discourse of academic and analytical projects with their objectivist ambitions. As purveyors of the animating powers of the human imagination, but at the same time beholden to wider publics than the exclusive circles of academia, and confronted by the obscenities of material deprivation, cultural confusion, and moral decadence, African writers have used, and will continue to use, their critical consciousness to provoke the consciences of their societies and the world at large. They are thrust into this position by the vocation of creative writing and the weight of African history. They constitute a mirror of what their societies are, and can, and perhaps ought to be.

Thus, as I reflect on the meetings I had with my colleagues in York, England, I realise our tasks are larger than chronicling the political ravages of post-colonial tyranny and the challenges of democratic rebirth, for our responsibility as writers is to imagine and understand the human possibilities from the vantage point of our specific histories and locations in contemporary Africa. We cannot, therefore, afford the indulgence of blind hope or despair. We require sober reflection and renewed commitment to the recreation of our humanity. And that is a profoundly cultural and moral issue that goes beyond conventional academic analyses and discourses.

Notes

1. For an earlier analysis of the 'exit' and 'voice' options see Albert Hirschman (1970).

2. In 1990, Mamdani (1990b) took the political science Africanist establishment in the USA to task for its simplistic analyses and capping to American foreign policy in pushing for ill-conceived 'political conditionality' as a way of bringing democracy to Africa. For an outraged Africanist's response, see Hyden (1990).

3. There are still apologists who deny that the crisis of the post-colonial state has anything to do with the nature of the colonial state itself. See Austin (1992).

4. This is the language that was used for example to delegitimise workers' autonomy and incorporate trade unions into state apparatuses. See Zeleza 1986.

5. The publication of the *Sub-Saharan Africa: From Crisis to Sustainable Growth* by the World Bank in 1989 signalled grudging recognition that SAPs as traditionally conceived were not having the desired effect. Ironically, earlier in 1989 the Bank and the UNDP published a report, *Africa's Adjustment and Growth in the 1980s*, (World Bank and UNDP 1989), lauding the success of SAPs. See my critical review of this report (Zeleza 1989). Stein and Nafziger (1991) have dismissed the shift in the former study as more 'rhetorical' than 'real'. Also see Parfit (1990).

6. Here the term post-colonial is used strictly in the sense of the period after independence, not in the context of the concept of 'postcoloniality'. The concept of 'post-colonial literature' is problematic in the way it privileges colonialism in world history and telescopes crucial geo-political distinctions into invisibility, as McClintock (1992), Shohat (1992), and others have observed. Also see Zeleza 1997.

7. He adds: 'if the series had been run by African intellectuals I would suggest that at least a one-third of what was published would never have been published...'.

8. The most well-known, and in some circles notorious, statement of the 'nationalist' or Afrocentric position is O. J. Chinweizu and I. Madubuike (1980).

9. Irele's (1990b) case for the existence of 'the African imagination' is, however, not compelling.

10. This comes out quite clearly in interviews in Egejuru (1980) of nine African writers, who were all asked who they write for. They reply that they primarily write for their communities, countries, and Africans in general. The nine are Leopold Sedar Senghor, the father of negritude; Chinua Achebe, the doyen of modern African writing in English; Ousmane Sembene, the great Senegalese 'social realist' writer as he calls himself; Mohammed Dib, the most prolific writer of North Africa; Camara Laye, Guinea's most famous writer; Cheik Hamidou Kane from Senegal whose reputation rests on one published book, *Ambiguous Adventure* (London: Heinemann, 1965), one of the great classics of African literature; Ezekiel Mphahlele, the grand old man of South African letters; Pathe Diagne, a less well known figure; and Ngugi wa Thiong'o, Kenya's and indeed East Africa's leading writer.

11. Interestingly to Leopold Senghor, one of the architects of the negritude movement, there was no contradiction between negritude and 'francophonie', according to Kesteloot (1990); also see Kennedy (1989).

12. This is evident in the interviews with African playwrights, such as Wole Soyinka, Ngugi wa Thiong'o, Micere Mugo, and Mohammed ben Abdallah in Wilkinson (1992); and the Tanzanian playwright, Penina Muhando, in James (1990). Also see Jones (1978), Etherton (1982) and Barber (1987).

13. Of course, not all younger writers use oral tradition. For example, Zimbabwe's Tsitsi Dangarembga, author of the fine novel, *Nervous Conditions*, (London: Women's Press, 1988), laments: 'I personally do not have a fund of our cultural tradition or oral history to draw from', see her interview in Wilkinson (1992: 191).

14. There are, of course, some writers who believe writers are a special breed. For example, the Nigerian poet Odia Ofeimun, is 'still quite married to [the] Shelleyan position that writers are actually unacknowledged legislators', while the South African poet, Mazisi Kunene, says he is 'chosen'. See interviews with the two writers in Wilkinson (1992: 67, 144).

15. For a thoughtful critique of the essentialism of Afrocentricity see Appiah (1992).

Chapter Nineteen

VISIONS OF FREEDOM

'Why not leave this country, even Africa, to trial and error?' he said slowly, uncomfortably. This is only my opinion. I don't think I approve of dictatorship in any form, whether for the good of mankind or not. Even if it is painstakingly slow, I prefer a democracy for Africa, come what may', Makhaya in Bessie Head, *When Rain Clouds Gather*, 1968:84.

'In many countries, one hasn't rights; but neither does one really have them in Western Europe or North America although one is made to believe one does... Democracy is the instrument with which the elites whip the masses anywhere; it enables the ruling elite to detain some, impoverish others, and makes them the sole proprietors of power. Who knows what is good for the people? Who knows whom the people love most? Who knows best what the people need?' Zeinab in Nurrudin Farah, *Sardines*, 1981:83.

'Tomorrow it would be the workers and the peasants leading the struggle and seizing power to overturn the system and all its prying bloodthirsty gods and gnomic angels, bringing to an end the reign of the few over the many and the era of drinking blood and feasting on human flesh. Then, only then, would the kingdom of man and woman really begin, they enjoying and loving in creative labour...' Akinyi in Ngugi wa Thiong'o, *Petals of Blood*, 1977:344.

Introduction

As the 1990s draw to a close, Africa has become the most democratically tested continent in the world, so stated the staid *The Economist* (February 3, 1996:17), commenting on the fact that 18 countries in 1996 and another eight in 1997 were expected to go to the polls. It is quite clear from the media, and the analyses presented in the two preceding chapters, that the current struggles and transitions to democracy have been dazzling and messy, their results contradictory and unpredictable, yielding both successes and defeats, concessions to the future and compromises with the past, heroism and tragedy, hope and pessimism. Already, the euphoria that was so evident at the beginning of the decade with the release of Nelson Mandela, the electoral defeats of Benin's Mathieu Kérékou and Zambia's Kenneth Kaunda, and the flight of Ethiopia's Mengistu Haile Mariam and Somalia's Siad Barre, has been tempered by the return of Buyoya in Burundi and the attempted return of Kaunda, not to mention the ambiguous transition of Kenya, the intransigence of Mobutu's regime in Zaire, the terror in Nigeria, the rise of religious fundamentalism in Algeria and the Sudan, and the continuing nightmare of Angola, Liberia, Somalia, and Rwanda.

All these conflicting realities and possibilities have given rise to intense and agonizing debates on the forces that are driving this new 'wind of change', and more importantly, controversies about the social basis of the democratic movements, and the concept of democracy itself, its meanings and trajectories. To some, democracy is coterminous with the pluralism of periodic electoral contests, with efficient governance. To others, democracy must be conceived in broader, more generous and more complex terms, beyond the political domain, beyond the trappings of 'multi-partyism'; it must be seen as a project for collective freedom and empowerment, and Africa's regeneration in all spheres - political, economic, cultural and moral. This chapter seeks to examine the conflicting visions of democracy from the different visions of three writers, namely, Bessie Head, Nurrudin Farah, and Ngugi wa Thiong'o[1]. These writers are products of, and write on, different political formations, so that they offer quite diverse and complex visions of democracy. Head, the South African exile who migrated to Botswana, typifies the Southern African condition, the complex interplay of migration and the constructs of race, nationality, and gender. Also, Botswana is one of the few African countries that has maintained a liberal parliamentary system since independence. Through Head the attributes and limits of this system are thrown into sharp relief. Farah's Somalia is a unique African country in that its people share the same nationality, language, and religion. Its intense internal divisions and, indeed, collapse in the early 1990s, therefore, raises fundamental questions about the articulation of nationhood and statehood in this era of resurgent nationalisms. Moreover, Somalia's official ideology has swung the ideological pendulum of post-colonial Africa from 'scientific socialism', to alliance with the West, and fundamentalist Islam. Finally, Ngugi's Kenya typifies most post-colonial African states: it is multi-ethnic, neo-colonial, and has vigorously pursued an authoritarian developmentalist model. The apparent ambiguity of its current transition to democratic politics also seems ominously typical.

Together, the creative visions of these writers capture the cultural and moral dimensions of freedom and democracy that are sometimes silent in the purely academic discourses. I will begin by briefly examining the African scholarly debate on democracy, against which the visions contained in the novels of the three authors will be read. African intellectuals are generally agreed on the desirability of democracy, but disagree on its form and content. Also, having fed on a diet of developmentalism for so long, they tend to be critical of defining democracy as the exclusive domain of politics, especially the equation between democracy and multi-partyism, but they disagree on the articulations between development and democracy, cultural pluralism and competitive politics.

Visions of Democracy

Like a chameleon, democracy can wear many colours of convenience. Some prefer the gloss of imported liberalism, others the radiance of socialist egalitarianism, or the earthy hue of traditional communalism. The liberal model of competitive party politics is justified on the grounds that it ensures the trinity of 'good governance': efficiency, accountability, and the protection of individual rights (World Bank 1989; Fatton 1990; Harsch 1993). Critics have pointed out that this model has failed in Africa. It primarily serves the interests of the elite and is restrictive for it focuses on political rights at the expense of economic and social rights. And even the political rights, Mamdani argues, are limited and statist in conception, for they are confined to the citizen as a member of the political community defined by the state, and exclude those of resident labour migrants who, thanks to the legacy of colonial capitalism in Africa, are in the tens of millions (Mamdani 1992; Mafeje 1992; Mandaza 1990).

The socialist model has historically put a premium on economic and developmentalist rights. For Onimode (1992) and Ake (1993) political democracy and economic democracy are indissolubly linked[2]. Following the collapse of the Soviet empire, African Marxists have began to revise their visions of socialism. According to Wamba-dia-Wamba (1992:1, 22), neither the Westminster model nor the Stalinist model, the two dominant historical modes of politics since the nineteenth century, 'support a process of human and social emancipation today'. The new emancipative politics must be a politics with 'several sites and a multiplicity of processes'. 'The democratic re-politicization of the people', in Samir Amin's (1990:18) words, 'must be based on reinforcement of their capacity for self-organization, self-development and self-defence'.

To its critics socialism in Africa has only bred statism, repression, corruption and economic stagnation. Its failure in Africa has not only been a manifestation of what Brzezinski (1989) calls the Grand Failure of Communism on a world scale, but also due to the fact that, contrary to what its proponents say, so the argument goes, socialism was an alien European intellectual import which failed to put down deep roots in a foreign cultural and historical terrain. Many of these critics are proponents of the indigenous model which advocates a return to the virtues of village co-operation and consensus politics. As Owusu (1992:387) puts it, all imported democratic models have failed and 'unless the rebirth of democracy in Africa is rooted and grounded in the villages and small communities, the end-product is bound to be elitist and another example of the failed urban-based development' (also see Ayittey 1991, 1992).

The rising waves of concrete struggles and debates about democracy are dissolving the old and predictable ideological lines of confrontation, between revolutionaries and reactionaries, committed internationalist and conservative nationalists, the left and the right. But developmentalism refuses to die. The new debates centre on the developmentalist and distributionist potentialities of the emerging democracies, the possibilities of recasting African states as 'democratic developmental states', to use Mkandawire's (1995) term. Mkandawire, a longstanding and distinguished radical intellectual, believes that out of the current conjuncture will emerge a developmentalism with an abashedly democratic and capitalist face, democratic because of the continuing popular struggles against authoritarianism, and capitalist because capitalism constitutes the political programme of the key actors in the current struggles for democracy and of the dominant forces at the global level, following the collapse of 'actually existing' Soviet socialism. Amin (1995) and others remain unimpressed by the vows of eternal compatibility between capitalism and democracy, and are wary of too close an association between democracy and development, of judging the nascent African democracies on the quicksands of economic performance[3]. An instrumentalist view of democracy cannot be entirely avoided, Olukoshi (1995) has argued, because the struggles for democracy of the 1980s and 1990s were ultimately struggles for material existence, although they were not purely economic. Distinguishing between democratic struggles and democratic consolidation, he contends that the possibilities of democratic consolidation are firmly tied to the demanding ox-cart of development.

We can clearly find echoes of these scholarly discourses on democracy in the works of Bessie Head, Ngugi and Farah as I will demonstrate. As creative works, they present fuller and finer social imaginaries of African worlds, more textured testaments of African problems and possibilities, values and visions. Given considerations of time and space, I will focus only on four visions: what I would call, first, reconstructing the boundaries of home; second, re-imagining the community; third, restructuring development; and fourth, empowering women.

Reconstructing the Boundaries of Home

The post-colonial state in Africa inherited the ill-fitting clothes of liberalism and nationalism. In the liberal tradition, the nation is the bearer of the collective right of self-determination, while the citizen is the bearer of individual rights. This raises vital questions. In a continent where the state is a recent and external construct straddling disparate collections of ethnic groups or nationalities, what constitutes nationality? And doesn't the conception of individual rights in terms of citizenship effectively disenfranchise millions of people given Africa's history of massive labour migration and flows of refugees?

In Ngugi's vision, ethnic nationalism as opposed to pan-territorial nationalism lacks an organic vitality, thanks to the homogenising influence of the anti-colonial struggle, and only thrives because of the class and crass manipulations of the neo-colonial elites. As one of the protagonists, Gatuiria, puts it in *Devil on the Cross* (Wa Thiong'o 1982, henceforth DC):

> We all come from the same womb, the common womb of one Kenya. The blood shed for our freedom has washed away the differences between that clan and this one. Today there is no Luo, Gikuyu, Kamba, Giriama, Luhya, Maasai, Meru, Kalenjin or Turkana. We are all children of one mother. Our mother is Kenya, the mother of all Kenyan people' (DC, p. 234-5).

He plans to compose a 'truly national music for our Kenya, music played by an orchestra made up of the instruments of all the nationalities that make up the Kenyan nation, music that we, the children of Kenya, can sing in one voice rooted in many voices - harmony in polyphony' (DC, p. 60).

These are sentiments echoed by Ngugi's protagonists in the other two novels, such as Wanja and Karega in *Petals of Blood* (henceforth PB) and Matigari in *Matigari*. The heroic Wanja not only knows all the languages of Kenya, but when she establishes her whorehouse she deliberately employs barmaids from different ethnic groups. To Karega, the fiery teacher and later trade union organiser, tribalism is fanned by the corrupt elites in order to divide the workers, demobilise their collective struggles, and facilitate their exploitation. 'A worker', he declares, 'has no particular home...He belongs everywhere and nowhere', for he has no property to bind him to a place, except his labour, which he can carry everywhere with him (PB, p. 291). The archetypal and messianic Matigari represents, and speaks for, all working people in Kenya against the imperialist oppressors and their local compradors.

In Ngugi's nation, therefore, the boundaries of difference are drawn by the polarities of class. It is an unambiguous, dualistic world, of the 'Mercedes family' and the family of the people (PB, p. 98), of the 'clan of parasites' and the 'clan of producers', of people with 'evil hearts' and those with 'good hearts' (DC, p. 53, 54). The overdetermination and simplification of the class dichotomy in *Petals of Blood* deepens as Ngugi's narrative of liberation becomes angrier and more impatient in *Devil on the Cross* and *Matigari*.

This vision of patriots and traitors can be an intolerant one; it certainly has no room for labour migrants and refugees. Thus Karega's statement that 'a worker has no particular home' would seem to apply to labour within the colonial boundaries. Thus contrary to Aizenberg (1991:86), I would argue that Ngugi's narrative does not challenge 'the nation and the identity wrought by independence'. It seeks to constitute it.

Farah questions this grand nationalist discourse. He probes the meaning and boundaries of nationality in a country where the idea of nation appears least problematic. Outwardly, the people of Somalia are one: they speak the same language and adhere to same religion. Yet, In Farah's trilogy, and as the world has seen in recent years, Somalia is a country rent by clan conflicts, tribalism, as Farah calls it, and ruled by 'government of tribal hegemony' (*Sweet and Sour Milk*, Farah 1979: henceforth S&SM, p. 88). In the absence of definable ethnic others, the nation feeds on itself, fracturing into the imagined differences of clan communities.

In *Maps* (Farah 1986) the 'nation's insatiable need', to borrow a phrase from Parker et.al. (1992:5), 'to administer difference through violent acts of segregation, censorship, economic coercion, physical torture, police brutality', is externalised into an aggressive Pan-Somali nationalism of the Ogaden War between Somalia and Ethiopia. The protagonist Askar is born into this disputed territory and is brought up by an Oromo servant woman Misra. The bonding between Askar and Misra, suggestive of the indeterminacy of ethnic and sexual identities, is solely tested by the exclusivist demands of nationalism. Askar is shocked to learn that Misra is excluded on his newly acquired Somali citizenship papers. When he asks what is the essence of Somaliness, he is given a tortured answer by Hilaal:

A Somali... is a man, woman or child whose mother tongue is Somali. Here, mother tongue is very important, very important. Not what one looks like. That is, features have nothing to do with a Somali's Somaliness or no. True, Somalis are easily distinguishable from other people, but one might meet with considerable difficulty in telling an Eritrean, and Ethiopian or a Northern Sudanese apart from a Somali unless one were to consider the cultural difference (*Maps*, p. 166).

He soon adds a caveat, however: although Misra speaks fluent Somali and has lived with Somali communities all her adult life, she does not qualify. Indeed, Somalis living outside the nation-state are inferior because 'they lack what it makes the self strong and whole' (*Maps*, p. 167). And hence, the need to liberate them from their fallen state, to bring them back into the nationalist enclosure. Following Somalia's defeat in the Ogaden war, Misra is accused by the Somali community of betraying them to the Ethiopians, for which she is tortured. She flees to Mogadiscio where Askar, whom she hasn't seen for over a decade, now lives as a member of the Western Somalia Liberation Front. She is bludgeoned to death and Askar is implicated in the murder. Her body is found without the heart, suggesting that he has appropriated important aspects of her character, and the ultimate futility of nationalist and sexual identities constructed around absolutized differences. To Farah nationalism, or nationalist fictions according to Cobham (1991, 1992), is an irredeemably reactionary force. Identities are not fixed, they are provisional, like the notional truth in Askar's maps: 'I identify a truth in the maps I draw', he says to Hilaal (*Maps*, p. 216). This is a celebration of the liberal values of tolerance and pluralism.

The life and work of Bessie Head poignantly captures the dilemmas and incompleteness of the statist liberal conception of individual rights. Bessie Head migrated from apartheid South Africa to Botswana in 1964, where she lived until her death in 1986. She left South Africa because of political repression which she found 'so evil that it was impossible for me to deal with, in creative terms' (Head 1990:67). But her experiences in her newly adopted homeland, Botswana, were traumatic and helped induce a lengthy and debilitating nervous breakdown. Botswana is renowned for its adherence to the liberal principles of parliamentary democracy. But Head painfully discovered that as a refugee she could not enjoy the rights of individual liberty accorded to the citizens. She led a restricted, dependent and humiliating life, as is so evident from her private correspondence. She had to report to the police regularly and even had her mail opened. She was only granted Botswana citizenship in 1979, thanks to the fame she had achieved by then[4].

Head's protagonists, Makhaya in *When Rain Clouds Gather* (henceforth WRCG), Margaret in *Maru* and Elizabeth in *A Question of*

Power (henceforth QP), experience the anguish of migrant and exilic life. Makhaya and Elizabeth are fleeing the oppression and insanity of apartheid South Africa, while Margaret, a Botswana citizen, comes from the outcast Baswara community. The first two expect a new, normal, and fulfilling life in a little corner of the vast continent 'without end'(WRCG, p. 16). But they are haunted by the suffocating insecurities of the refugee condition, a life devoid of the rights and liberties of citizenship, regardless of their productive contributions to their adopted community and country of residence. Makhaya is told in no uncertain terms by the ruthless and greedy chief of Golema Mmidi, Matenge, that he is unwelcome: 'We want you to get out. When are you going?' (WRCG, p. 67). Makhaya lives under the permanent threat of deportation as a 'security risk'. So enraged is he that he comes close to killing the chief.

For Elizabeth the agony of not belonging drives her literally mad. She comes from South Africa on a quest for self-definition and self-actualization, to put together her fragmented life, which began in the tragic circumstances of a mental hospital where her white mother was sent for getting pregnant by a black man. She grew up a marginalised, despised 'Coloured', and was further traumatised by her marriage to a perverse and abusive husband. But in the Botswana village of Motabeng, where she settles, she is detested and rejected as well as a racial inferior, a political alien, a cultural outsider. Her new country begins to feel like South Africa

> Just the other day she had broken down and cried. Her loud wail had only the logic of her inner torment, but it was the same thing: the evils overwhelming her were beginning to sound like South Africa from which she had fled. The reasoning, the viciousness were the same, but this time the faces were black... (QP, p. 57).

Lonely and afraid, she withdraws into herself and descends into a terrifying mental hell filled with witchcraft, evil forces, and misogynist men. While Elizabeth's nightmarish dreams may seem to be rooted in her South African background, the hallucinatory personages, Sello and Dan, who persecute her are both black men, so that her neurosis is an encounter of colonial apartheid and post-colonial pathologies of power[5]. The descriptions of madness as a mental disorder and as a metaphor of political powerlessness and social alienation are hauntingly powerful.

Through Makhaya's and Elizabeth's agonizing struggles against the oppression of exile and marginalization, and their desperate wish to belong, Head articulates a compelling vision for a more humane world, one in which the rights of the individual are inclusive and generous, based on the fact of residence, not merely the fiat of birth, on the universal canvas of labour rather than the narrow shell of citizenship. It is a vision for a truly democratic panAfricanism.

Re-imagining the Community

The community as a formation of cultural practices, a process of associational life, and as a moral landscape, is historically constructed. African writers and intellectuals have grappled with the questions of cultural continuity and change, autonomy and dependency, uniformity and difference, ever since the tragic encounter with an imperialist, intolerant, and universalising Europe. The three writers exhibit divergent understandings of the kinds of traditions and social orders that can accommodate the expansive and humane values of a democratic culture. To Farah and Head 'traditional' values and institutions are largely seen negatively, as authoritarian, patriarchal, and tribalistic, while for Ngugi they offer positive alternatives to the tyrannies, corruptions, and confusions of the neo-colonial era.

In Farah's trilogy the ruthless General is not an arbitrary, superficial presence, but an authentic, organic configuration of Somali life, an embodiment of the articulation between traditional despotism and modern state terror. Indeed, the repression is so vicious precisely because it is woven into an intricate web that links the authoritarian patriarchal family, the tribal oligarchy, and the despotic state.

Farah's vision of traditional culture is bleakest in *Sweet and Sour Milk*, in which the interlocking and mutually reinforcing tyrannies of the family, clan, and the state, are represented by the ruthless, misogynist, and opportunistic Keynaan, an ex-torturer and paid informer of the regime. He forcibly marries Beydan, whose husband he killed, and connives at his son's death in exchange for a police inspector's job and a fat gratuity, part of which he uses to marry his third wife. His son Soyaan, an economic adviser to the president, dies from food poisoning by state agents for his clandestine anti-state activities. But the state cynically wants to honour him as a hero in order to confuse and divide the opposition. As befitting a revolutionary hero his last words are said to have been: 'Labour is Honour and there is no General but our General'.

Keynaan colludes in the lie, at this mockery of Soyaan's memory and negation of his revolutionary activities, despite being challenged by his other son, Loyaan, who was the last person to see Soyaan. To Keynaan, it doesn't matter that he was not there; he is the father, the Grand Patriarch, with power 'to give life and death as I find fit' over his children (S&SM, p. 95). For conducting a quest into his brother's death and disputing the official version of his last words, Loyaan is deported into exile in Eastern Europe.

This omnipresent dictatorship is maintained by the oral mode of communication. The vast oral networks of rumour, gossip and speculation keep everybody guessing, insecure and suspicious. It is an obscurantist tyranny where 'everything is done verbally', and there are no 'traceable... written warrants', death certificates, or imprisonment records (S&SM, p. 136). Once again, traditional culture is mobilised for repression, orality dooms the ordinary people to policing themselves, to sustaining and reproducing state tyranny. The aspiring revolutionaries worship the written word, so Loyaan's quest becomes a search for the secret memorandum his brother wrote. Farah polarizes 'a primitive reactionary orality and a sophisticated Westernized literacy, between oralized despotism and written revolution', as Wright (1990:27) has put it.

This is a nihilistic world in which popular politics and culture are despised, and resistance is reduced to the futile gestures of a tiny intellectual elite. The wholesale condemnation of tradition is tempered somewhat in the two subsequent novels. Through the rebellious activities of the singer Dulman, the official 'Lady of the Revolution', who smuggles tapes of subversive poems in *Sardines* orality is furnished with a more positive face. And in *Close Sesame* (henceforth CS) we get a fascinating portrait of the protagonist, Deeriye, the liberated widowed patriarch who still mourns his wife, and sees himself as a pan-Somalist, a panAfricanist, and as a devout Muslim. He is respected as a hero of the anti-colonial struggle and loved by his children and grandchildren with whom he discusses weighty political and philosophical issues.

Also, we see clan leaders taking a principled stand against arbitrary power, and neighbourhoods and Islamic ideas of brotherhood seem to provide a basis for inter-clan solidarities and friendships. Through Deeriye Farah offers a tantalising vision of transformed gender roles and political values within a harmonious social order. But it is a vision that is overwhelmed by the elitist adventurism and futility of the resistance against the dictatorship, symbolised by the two failed assassination attempts, one of which is made by Deeriye when his son, Mursal, is killed by the security forces for his involvement in the original assassination debacle. All that the two seem to have achieved, according to the women mourning at the end of the novel, is that 'at least neither died an anonymous death - and that was heroic' (CS, p. 207).

Head's Botswana villages are more complex sites of struggle between the forces of good and evil, order and chaos, life and death, tradition and change. Scenes of pastoral beauty vie with harsh images of drought and desolation. Beneath the apparent tranquillity and simplicity of village life lies a world mired in poverty and greed, cruelty and callousness, intrigue and fear, conformity and apathy. It is a world in which women are subjugated by men, ordinary people are terrorised by venal chiefs, and strangers are loathed.

In *Maru* the object of disdain is the Moswara school teacher, Margaret. The abhorrence towards her is far more violent than that shown to the 'Coloured' Elizabeth, as is clear when she meets the headmaster for the first time.

> 'Excuse the question, but are you a Coloured?' he asked. 'No' she replied. 'I am a Moswara'. The shock was so great that he almost jumped into the air... she was no longer a human being...(p. 40).

Margaret encounters racial prejudice as vicious as that accorded to blacks in South Africa. The evils of prejudice and hatred, Head is saying, are not a unique invention of the Europeans, but are deeply embedded in African societies as well. 'They all have their monsters', the narrator laments (*Maru*, p. 11). In Head's vision, therefore, a truly democratic world must be free from the pernicious cancers of racism and tribalism.

There is 'a greater urgency in Head's fiction', unlike Farah's, 'to create 'new worlds' of new men and women' (Brown 1991:160). Out of the confrontation between the forces of good and evil new harmonies can emerge, new communities can be forged, a more humane society can evolve. The solution has to be sought at both the personal and the community levels, for these forces jostle within the individual soul as well as the collective consciousness. In *When Rain Clouds Gather*, Head tilts towards organized communal action, while in *A Question of Power*, it is a harrowingly personal struggle[6].

In Golema Mmidi, the community gradually reconstructs itself through the cooperative and the transformation of the hierarchies of power, symbolised by the mobilization of the women and the suicide of a terrified chief Matenge. The proposed marriage of Makhaya, who speaks to the women 'as an equal' (p. 107), and Paulina, the lonely and fiercely independent leader of the women's farming team, points to the liberating possibilities of a democratised community.

The harmonizing energies of cooperative labour in Motabeng plays a similar role in Elizabeth's recovery from her horrifying madness and alienation. But the healing is also an intensely personal and spiritual one, rooted in a deepening love for her own humanity and the humanity of others, out of the realisation that God is not an absentee Heavenly landlord, but is in 'ordinary, practical, sane people' (QP, p. 31). 'Her realization that God is everything allows for celebration of otherness; power becomes relational rather than a system of domination and submission' (Tucker 1988:175). At the end of the novel, a new beginning beckons: 'As she fell asleep, she placed one soft hand over her land. It was a gesture of belonging' (QP, p. 206).

Such is Head's optimistic vision, or moral idealism, that even Margaret is redeemed from her ostracism, not through a reaffirmation of the conventions of the past, but the invention of new community structures of attitude and reference following her marriage to the protagonist, Maru, the enigmatic and visionary chief-in-waiting, who loves nature, especially flowers, and wants to change the world. He is attracted to Margaret because of her sense of mystery and individuality. Although Maru loses his chieftainship, the marriage marks a new beginning for everyone, including the Baswara themselves who saw 'a door silently opened on the small, dark airless room in which their souls had been shut for a long time' (*Maru*, p. 126).

Head seems to be saying that the liberation and democratization of African societies involves the invention of new traditions, the imagining of new communities, not the pretentious reclamation of some fossilised heritage. In her later writings, though, she speaks in a more affirmative, 'communal voice for the Batswana' (Visel 1990:120)[7].

For Ngugi the past can be retrieved and mobilized for democracy and development. The narrator in *Petals of Blood* tells us that the decrepit and drought-stricken Ilmorog 'had its days of glory: thriving villages with a huge population of sturdy peasants who had tamed nature's forest' (PB, p. 120). In those idyllic days 'when Ilmorog, or all Africa controlled its own earth' (PB, p. 125), 'the land was not for buying' (PB, p. 82), people worked together, and shared 'when a bean fell' (*Matigari*, p. 56). And then the imperialists came, with their trading stations which siphoned off local surpluses abroad; railways, which ate the forests; settlers who stole the land; cities which swallowed the young; Christianity, which imprisoned the people's souls; colonial education which alienated their minds; and chiefs who terrorised their lives.

That glorious past must be remembered and used. The striking students at Siriana, the secondary school in *Petals of Blood*, demand 'to be taught African literature, African history, for we wanted to know ourselves better' (PB, p. 170), and when one of these students, Karega, becomes a teacher, he tries to do just that. Even the more ambivalent Munira, the headmaster of the Ilmorog primary school, resolves to

restore himself 'to my usurped history, my usurped inheritance' in order
to liberate himself (PB, p. 227). Indeed, all of Ngugi's protagonists
wrestle with the past, they seek to reconnect themselves to it.

The very narrative structure of the novels, the intricate interaction of
past, present and future, and the dense allusions to history, in which the
fictional characters often intermingle with historical characters and
events, underscores the importance which Ngugi attaches to history. In
fact, Ngugi has engaged in intense debates with Kenyan historians, most
of whom he criticises for having failed to decolonise the country's
history (Sicherman 1989; Mazrui and Mphande 1993:164-7). They are
collectively represented in *Matigari* in the pathetic figure of the
Permanent Professor of the History of Parrotology.

But Ngugi's past is not a frozen, museum artifact. Indeed, as Karega
matures in his politics he comes to recognize that Africa 'did not have
one but several pasts which were in perpetual struggle' (PB, p. 214), and
that it is not enough to 'talk endlessly about Africa's past glories,
Africa's great feudal cultures', for that would not 'cure one day's pang
of hunger... quench an hour's thirst or to clothe a naked child'. The past
must be studied 'critically, without illusions, and see what lessons we
can draw from it in today's battlefield' (PB, p. 323) 'to right the wrongs
that bring tears to the many and laughter only to a few' (PB, p. 302).

Thus, for Ngugi remembering links the past and the present, the
individual and community; history provides the communal perspective
and the collective myths necessary for the liberation of oppressed
peoples. And so we see that the awareness and militancy of the Ilmorog
community in *Petals of Blood* is heightened along the journey to see the
area's MP in Nairobi during which the members of the group tell each
other their personal histories only to discover how interconnected their
lives are. Even when the revelations are tragic and stressful, they
reinforce the individual and collective determination to fight against the
oppression of their present lives. The murder of the three ruthless
business tycoons who take over Ilmorog occurs in the context of the
unfolding knowledge of Wanja's past and continuing abuse by one of
these men, Kimeria.

In Ngugi we do not meet Farah's alienated, brooding elites, or Head's
lonely, tormented outcasts, but connected individuals, who are always
tapping into each other's memories, and are often found in crowds or
travelling. I have already referred to the purgatorial journey from
Ilmorog to Nairobi in *Petals of Blood*. In *Devil on the Cross* there is the
journey of the protagonists from Nairobi to the Devil's Feast in Ilmorog,

during which the travellers interrogate each other and the corrupt neo-colonial order, and the procession of the angry crowd to the cave where the feast is being held. These journeys are rites of passage, enabling a character like Wariinga to grow from an insecure woman 'into a lucid, decisive woman' (Julien 1992:151).

In *Matigari* there is also the procession and the enraged crowd that gathers at the house where Matigari is expected to make his last Christ-like appearance. The rehabilitation of Christianity in this novel, Brown (1991:177) has argued, reflects Ngugi's 'reassessment of the relationship between culture, religion and politics'. In short, for Ngugi the past and the present, the individual and the community, culture and politics, appropriate each other's states, spaces, and symbols in the struggle for a better tomorrow.

Restructuring Development

Predictably, development in Ngugi's 'literature of combat' is seen in collectivist, socialist terms. Neo-colonial capitalist Kenya is a wretched social terrain that sprouts flowers with petals of blood and roasts under drought, and is controlled by the 'loyalists', those who collaborated with the British colonialists during the Mau Mau war for independence. In *Petals of Blood* Ngugi shows how the coming of 'development', the Trans-African Highway, factories, and modern infrastructure benefits, not the majority of local people, but a few elites and foreign business interests. In fact, the local people lose the little autonomy they had before. Peasants lose their land to speculators and local beer brewers are replaced by multi-national breweries. The New Ilmorog is divided into the two spatial solitudes of affluent suburbs and slum dwellings.

Three of the four protagonists in the novel are destroyed by this 'dependent' development. Abdulla, the proud Mau Mau freedom fighter and owner of the only shop in the old Ilmorog, loses his business and becomes a pitiful alcoholic. When the same fate befalls Wanja, she resorts to prostitution. Munira, the school headmaster, wallows his new found ordinariness in bouts of drunkenness and later embraces the apathetic religiosity of a born-again Christian. Only Karega, who becomes a trade union leader, grows stronger.

The bankruptcy of this development model and the neo-colonial ruling elite is depicted with all the outrage Ngugi can master and the comic vision of grotesque realism, as Berger (1988) calls it, in *Devil on the Cross*, an allegorical story of the competition to choose the seven cleverest thieves and robbers in Ilmorog. The judges are renowned thieves and robbers from seven western countries, 'who are able to roam the whole Earth, grabbing everything - though of course we do leave a few fragments for our friends' (DC, p. 89). The competitors try to outdo each other in displaying their cunning and skills in theft and robbery and ingratiating themselves to the judges, except one who, to the dismay of everybody, issues the nationalist battle cry: 'every robber should go home and rob his own mother! That's true democracy and equality of nations' (DC, p. 171). He is later found dead, which suggests that national accumulation cannot occur and the national bourgeoisie cannot emerge under the present neo-colonial conditions.

This echoes a rather crude dependency perspective. Kenyan capitalism and social class structure is far more complex than is apparent in Ngugi's narrative (Zeleza 1991). Ngugi's critique of Kenyan capitalism is as simplistic and populist as is the process by which he envisages the system will be transformed. He invests messianic faith in what he calls 'the holy trinity' of the masses: peasants, workers, and patriots (DC, p. 230). This voluntaristic vision is based on a romanticization of peasants and workers and assumes the internal homogeneity of each of these classes and an automatic alliance between them. This simplification becomes glaring in *Matigari*[8]. As Gurnah (1991:172) has stated: 'The people, though ritually valorised, are gullible. They remain satisfied with the fantasy of a miraculous redeemer'.

While Ngugi straddles the rural and urban landscapes and powerscapes and deals with the plight of both peasants and workers, Head focuses squarely on rural transformation and farming as a ritual of material growth and moral healing. For his part Farah dwells on the rarefied universe of the professional elites. For Head there is a necessary link between economic, social and spiritual, well-being. Like Ngugi's drought-stricken Ilmorog, Head's villages are landscapes ravaged by the constant threat of drought, which symbolises crisis in the moral and social order. But her conception of development is far more limited than Ngugi's. It is primarily seen in terms of *outside* experts coming to disseminate new techniques to the local people, who lack the capacity to develop by themselves. In *Rain Clouds*, Gilbert, the English agronomist, and Makhaya, are the development experts, while in *A Question of Power* the experts come from Denmark, England, and South Africa, including Elizabeth herself who introduces Cape Gooseberry.

While Head is critical of the insensitivity, arrogance, and naivety displayed by some of the expatriates, like Camilla in *A Question of Power*, she doesn't question their role. Only the greedy chiefs, such as Matenge in *Rain Clouds*, express serious opposition, but since their motives are suspect, their views are discredited. The dedicated, visionary

Gilbert is Head's most endearing development expert. He marries a local woman, the self-possessed Maria, and recruits Makhaya because he has 'the necessary mental and emotional alienation from tribalism to help him accomplish what he had in mind' (WRCG, p. 30). Makhaya eagerly obliges and mobilises the women for the cooperative farming project. Both Gilbert and Makhaya are opposed to socialism because to them 'Africa had a small population, and it might well be that socialism of every kind was an expedient to solve unwieldy population problems' (WRCG, p. 83). Makhaya's vision is shallow and contradictory. He envisions a Golema Mmidi and an Africa

of future millionaires, which would compensate for all the centuries of browbeating, hatred, humiliation, and worldwide derision that had been directed to the person of the African man. And communal systems of development which imposed cooperation and sharing of wealth were much better than the dog-eat-dog policies, take-over bids, and grab-what-you-can of big finance (WRCG, p. 156).

Ngugi, of course despises external dependency. So does Farah. But they have different culprits and visions of their country's economic futures. The capitalist west is the object of Ngugi's wrath, while for Farah it is the Soviet bloc. Soyaan in *Sweet and Sour Milk* notes bitterly in one of his secret memorandums, that East German and Russian aid consists of prisons, not factories, and half-baked experts who come to replace the nation's jailed or exiled 'professionals, intellectuals and technicians' (S&SM, p. 137). Medina, the protagonist in *Sardines*, tells Sandra, the Marxist journalist whose grandfather was once a colonial governor, that the flame of Marxism was lit when the 'light of the civilizing mission of the crucifix...waned' (S&SM, p. 206).

Development in this Scientific-Socialist-Islamic Revolutionary dictatorship comprises of 'showy pieces of tumorous architecture', and new roads and roadsides 'decorated with neon signs illuminating with the brilliance of [the General's] quotes' (S&SM, p. 73, 74). So bankrupt is the dictatorship and the international economic system that famine is big business for the multi-nationals and the General, who also uses it to break the resistance, to 'starve and rule' (*Sardines*, p. 103). There can be no solution to the economic crisis of underdevelopment without dismantling the dictatorship, after which efforts must be made, in the words of the singer Dulman, to belong to the twentieth century and 'to use its technology' (*Sardines*, p. 163).

There is really no blazing vision of a new economic tomorrow from Farah's muddled, elitist intellectuals, as there is from Ngugi's restive masses. Samater meditates with bitter defeatism: 'We the intellectuals are the betrayers... We the intellectuals are the ones who tell our people lies... We are the ones that keep dictators in power' (*Sardines*, p. 72).

Empowering Women

All three writers condemn the oppression of women in African societies and advocate women's liberation as an integral part of the processes of democratization and development. They leave us with unforgettable images of women's subjugation and struggle, marginalization and empowerment. Farah's women struggle against a vicious patriarchal order, in which their voices are silenced, their humanity is denied, and their sexuality is negated through the practice of infibulation. Keynaan expresses disappointment in Loyaan for discussing ideas with women. 'Women', the Grand Patriarch tells his incredulous son, 'are for sleeping with, for giving birth to and bringing up children; they are not good for any other thing' (S&SM, p. 84). Men's domination seems total, except for those women outside the traditional order, such as Margaritta, the half-Somali and half-Italian partner of the murdered Soyaan in *Sweet and Sour Milk* and the educated or foreign women in *Sardines* and *Close Sesame*.

Women are at the centre of the narrative in *Sardines*. Medina, a cosmopolitan journalist, leaves her husband Samater to protect her daughter, Ubax, from her grandmother's threat to 'have her circumcised' (*Sardines*, p. 66-7), and to write a book against the regime which she detests partly because 'the general reminded her of Grandfather who was a [patriarchal] monstrosity' (*Sardines*, p. 16). She is disappointed when Samater accepts a ministerial appointment, ostensibly to protect his clansmen from possible victimization should he refuse. But Farah does not idealise the women. Medina's uncompromising stand increasingly looks self-absolved and obstinate, her single-minded protection of her daughter smacks of emotional and intellectual tyranny. The little girl is rarely allowed to play with other children. As for 'the foreigners Sandra and Atta, respectively the white stooge and black spy of the "revolution" [they] connive at elitism, clan nepotism, and the Islamic subjugation of women in the name of Marxism and Africanity' (Wright 1990:29).

The female privigentsia, as they call themselves, live in the same claustrophobic elitist world as their men, indulging in equally impotent gestures of literary defiance: Medina edits the General's speeches, and anonymous girls write slogans on walls. At the end of the novel Medina is confronted by her sister-in-law, Xaddia, and reminded that her actions contributed to Samater losing his job and others getting arrested. When she is asked: 'What point have you made?' (*Sardines*, p. 246), she has no convincing answer, except to mouth an anguished cliche: 'I say the struggle must go on' (*Sardines*, p. 246). The novel ends with Medina, the just released Samater, and their daughter walking away,

> refusing to play host to the guests who waited to be entertained with explanations, explications and examples. Medina, Samater and Ubax behaved as though they needed one another's company - and no more (Sardines, p. 250).

And so they move away from the oppressive tentacles of the extended family into the liberatory embrace of the nuclear family. Once again, Farah gives us a peculiarly petty-bourgeois vision.

In contrast Ngugi deals mostly with peasant or working class women. Elite women, like the wife of the Minister of Truth and Justice in *Matigari*, who is caught making love with her driver in the back seat of a Mercedes Benz, are portrayed with contempt. He focuses on the lower class women because in his view they are the ones who bear the brunt of class and sexual exploitation. Wanja in *Petals of Blood* and Wariinga in *Devil on the Cross* represent resilient young working class women, while Nyakinyua and Wangari, represent strong older peasant women, respectively. As teenage girls both Wanja and Wariinga are bright, but the chances to develop their potential are dashed when they are made pregnant by wealthy and deceitful old men who deny any responsibility. In desperation Wanja kills her baby, while Wariinga tries to commit suicide. She hates everything about herself, her dark complexion, teeth, and even her manner of walking.

Henceforth, Wanja drifts in and out of prostitution, while Wariinga experiences sexual harassment in every job she applies for or gets. Women working in factories and offices, Wariinga suggests, are treated no differently from prostitutes for their 'arms', 'brains', 'humanity', and 'thighs' are exploited by their male employers. Wanja discovers that even the three men who become her friends in Ilmorog, Munira, Karega, and Abdulla, seek to dominate and control her. The development of the tourist industry simply reinforces the sexual exploitation of women.

Building on the resistance tradition of the older women in their lives or whom they come across who have histories of participation in the independence struggle behind them, the two younger women try to reconstruct their lives on their own terms. Wanja becomes a high-priced prostitute, while Wariinga breaks into the male world of engineering. Men, including her close friends Munira, Karega, and Abdulla, fear and seek to control Wanja's sexuality. The two women regain their self-confidence, which is ultimately augmented by killing the men who sexually exploited and abandoned them when they were young. This signifies not only personal revenge and reaffirmation, but also the force of communal retribution and justice.

Ngugi's women, more so than those in Head's three novels, achieve their gradual sense of empowerment through collective class actions, rather than purely through gender solidarity. The narrator mentions, but only in passing, that 'Wanja and the other women on the ridge had formed' a farming cooperative (PB, p. 200), and 'the women dancers formed themselves into a Tourist Dancers' Union and demanded more money for their art' (PB, p. 305). In the garage cooperative where Wariinga works part-time she is the only female. The women's dependence on male-led class movements becomes troubling in *Matigari*. Guthera, a prostitute reminiscent of Mary Magdalene, follows and submits herself to Matigari, the godly patriarch.

The traditional patriarchal controls that Head's women confront have some similarities to those in Ngugi's and Farah's fiction. They range from economic powerlessness to sexual objectification. A major difference is on the issue of racism. These women are not destroyed, as Katrak (1985) claims. At the beginning of *Rain Clouds*, Gilbert, the English agronomist, tells Makhaya, of the anomaly that it is the women who farm, 'but when it came to programs for improved techniques in agriculture...the lecture rooms were open to men only' (WRCG, p. 34-5). The rest of the novel chronicles the women's cooperative efforts to improve their farming and control the fruits of their labour. When drought strikes and the men have to return from the cattle posts with their emaciated livestock they are 'flushed with pride' to find what the women have been doing in their absence (WRCG, p. 173). Ashamed, they agree to pool their labour and build a long fence for a cattle holding ground. The seeds of change in the power relationship between men and women have been sown.

In *Maru* and *A Question of Power* the links between patriarchal and racial domination, and freedom for women and national liberation, are drawn quite sharply. Both Elizabeth and Margaret suffer from a double sense of powerlessness, as women and as outsiders in a traditional society. Margaret, the artist, survives through sheer endurance and her art that evokes visionary realistic images of ordinary women which loudly proclaim: 'We are the people who have the strength to build a new world!' (*Maru*, p. 108). Elizabeth experiences the full horrors of sexual objectification. She is so insecure that she is initially even attracted to Dan's sadistic and promiscuous masculinity. He is a walking phallus, always parading his women before her, which negates her own sexuality and further silences her, all done to debase and eventually destroy her. But in the end she does heal herself of the hallucinatory Dan and Sello[9]. Both Elizabeth and Margaret are ultimately saved through the solidarity of other women. Head contrasts the warm, sharing and fulfilling friendship of Elizabeth and the peasant woman, Kenosi, and that of Margaret and the royal woman, Dikeledi, with the competitive and manipulative relations between Sello and Dan over Elizabeth, and Maru and Moleka over Margaret. This feminist solidarity not only helps empower the women, but Head also believes, it is central to the transformation of society as a whole, to the dismantling of the hierarchies of power and the dualities of experience that fragment and victimise women and ordinary people.

Conclusion

These readings suggest that the struggles for democracy have to be accompanied by a profound and constant questioning, and in some cases even dismantling, of the old totalising narratives and binary oppositions between nationality and ethnicity, collective and individual rights, and tradition and modernity. What is at stake in this conjuncture of democratic struggle in Africa, and around the world, is the construction of new ways of living and imagining, new visions of being fully human, in this complex, diverse world, where nobody ever realises themselves as individuals outside the affirmative ties of social relations and identities.

Thus these are not narratives that celebrate the ambivalences, contingencies and hybridities of postcoloniality; they are not immersed in the postmodernist fantasies and reactionary fulminations against history and humanist values. As Appiah (1992:66) has observed, their messages of protest against 'the western imperium' and the 'nationalist project of the postcolonial nationalist bourgeoisie... are grounded in an appeal to an ethical universal'. The best of these novels affirm the emancipatory possibilities of the decolonization project; of nationalist struggles firmly tethered to struggles for gender, class, and ethnic equality; of struggles inspired by new and generous visions of community and citizenship, that imagine and seek to create a panAfrican commonwealth that transcends the dangerous and endangered fictions of the nation-state. In much postmodernist and postcolonialist theorizing Third World nationalisms are disparaged and misread as simple derivative discourses of outmoded European nationalism, views that have been strongly rebuked by Ahmad (1995), Lazarus (1994), Parry (1994), and Rosaldo (1994). For many African writers nationalism still matters, their works show that its mission to destabilize imperialism and its neo-colonial configurations in the post-colonial state never ended with the attainment of the 'first independence', nor will it end with the 'second'.

Notes

1. I focus on Bessie Head's three novels, (1968; 1971; 1974); Nurrudin Farah's trilogy (1979; 1981; 983); and Ngugi wa Thiong'o's three post-*A Grain of Wheat* novels (1977; 1982; 1987).

2. Also see the debate between Thandika Mkandawire and Peter Anyang' Nyong'o in *Codesria Bulletin*, Nos.1&2, 1991; and the papers by Nyong'o (1992) and Hamid (1992).

3. For a report of the debates on this and other issues in Africa's current transformations at the Codesria Eighth General Assembly, see Zeleza and Diop (1995), and for an edited collection of the papers see Zeleza and Diop (forthcoming).

4. The first collection of her correspondence contains letters she wrote to Randolph Vigne (Head 1991).

5. My interpretation extends Roger A. Berger's (1990) tantalizing but limited analysis that Elizabeth's madness reflects, in Fanonist terms, her internalization of colonial psychopathology. Oladele Taiwo (1984) also locates her madness in 'the iniquitous system of apartheid', as does Adetokunbo Pearse (1983).

6. I think Cecil Abrahams (1990: 10) is exaggerating when he argues that Head 'opposed vehemently organized class action and sought solutions on the personal level.'

7. For example orality features at the levels of inspiration, composition and narration in her collection of short stories, (Head 1977), see F. Ojo-Ade (1990) and N. Thomas (1990). In *Serowe: Village of the Rain Wind*, (Head 1981) she celebrates the history of this community, and offers an eulogistic portrait of Khama in *A Bewitched Crossroad* (1984).

8. Simon Gikandi (1991), who read the original Kikuyu version, has suggested that the apparent narrative simplicity of the English version of Matigari is a product of the problems in translation.

9. For a fascinating analysis of the psychiatric veracity of Elizabeth's healing process see E. N. Evasdaughter (1989).

Chapter Twenty

CYCLES OF REBIRTH

Introduction

Literature is about memory, it confirms, contests and constructs memories: collective memories of remembered and anticipated histories, places and possibilities, social realities and ruptures, cultural conflicts and conversions, of political and intellectual negotiations. It offers expressive acts of representation, of ordering the competing visions and inscriptions of the past, the present and the future, of imagining the shifting moral economies of being human. And because of its history of violent colonizations, furious resistances, and ambiguous accommodations, Africa is blessed with abundant, conflicting memories, both epic and prosaic, concrete and diffuse in their manifestation through time and space, of nativist struggles for cultural authenticity and reconstruction against narrative silencing and epistemic erasure by the conquering discourses and universalizing fictions of Europe. A complex and constantly changing cultural landscape of antagonistic, fluid, and overlapping memories and moralities of power and being, Africa's aesthetic spaces have spawned a literature that is often intense, impatient, and inspired, a literature that seeks to challenge not only the imported narrative conventions, to slay, as Mudimbe (1994:Chapter 5) has put it, its false western symbolic fathers, but also the imposed conventions of social existence. It is, in Fanon's (1968) insistent revolutionary outcry, a literature of combat. However, the best of this literature does more: while it castigates the monstrosities perpetrated by Europe's seizure of global history - the slave trade, colonialism, racism, the world wars and Nazism - and rebukes Africa for its own failings, it also celebrates the survival and agency of history's victims, thus affirming the deepest yearnings of the human spirit for freedom, for redemption, and intimates possible new memories of being fully human.

Contemporary African societies, with their inherited colonial organizations of despotic power and poverty, and unfulfilled post-colonial ambitions of development and democracy, recall the abiku child, an embodied spirit always migrating between experiences, navigating the memories of different worlds, forever poised on the edge of materiality and spirituality, deprivation and fulfilment, living an incomplete, tentative existence, one that straddles and promises different possibilities of becoming. It is a world of generous contradictions and explosive energies that dissolve the simple dualities of appearance and reality, stasis and change, tradition and modernity, and scatter the boundaries of nation, class, gender and other hierarchies imposed from the intolerant histories of Europe. In the conflictual discursive spaces of these abiku societies, different memories compete to name experiences and dreams, to articulate connections and continuities, to appropriate history and authorise the future.

This chapter discusses the fiction of four West African writers, Ama Ata Aidoo, Ayi Kwe Armah, Buchi Emecheta and Ben Okri. Using different narrative styles, economies of composition, and frames of meanings and images, the four authors poignantly capture the complex clashes, contentions and conversions of cultures in colonial and post-colonial Anglophone West Africa. The differences among them can of course be attributed to their varied social autobiographies, artistic vocabularies, and what Soyinka (1978:115) calls 'secular visions'. Aidoo and Armah are Ghanaian, while Emecheta and Okri are from Nigeria. Their gender identities are obviously critical, too: Aidoo and Emecheta are female and Armah and Okri are male. One could go on signifying the contexts which are crucial to a 'thick' reading of their writing. But this chapter is too short for such an exegesis.

In Search of Redemption

Rooted in a Fanonist discourse, Armah's fiction angrily indicts the decadent, corrupt elite for betraying the emancipatory promises of independence. Armah's neo-colony in his first novel, *The Beautiful Ones Are Not Yet Born* (1968), is a bleak world of infinite solitudes, pitting the opportunistic elite wallowing in debauchery against the suffering masses sinking deeper into squalor. If this novel unravels the politics of neo-colonialism, *Fragments* (1970) dissects its psychology, the elite's pompous mimicry of western alienations and consumption habits, and the consequent subversion and contamination of traditions, relations and expectations. Culture is interrogated, specifically the limits and possibilities of creative, liberatory intellectualism, of cultural democratization in a parasitic neo-colonial social order.

Graphically, cumulatively, we are shown a world where the hopes of independence were aborted. But that is seen as a temporary state in the dialectical march of history, for beneath the decay sprout seeds of rebirth nourished by these very social decompositions. The protagonists, 'the man' in *The Beautiful Ones* and the idealistic journalist, Baako, and sober psychiatrist, Juana, in *Fragments*, refuse to succumb to despair, and search for, in their different ways, affirmative strategies of resistance. Armah seems to suggest that out of the struggles of the conscientized masses and the radical intellectuals real change will come, the beautiful ones will be born to redeem the nation, the fragments will be reassembled, so that a wholesome world will emerge. In this sense, then, these are not novels of hopelessness, as many critics have argued, but of revolutionary faith. In other words, Armah's realist vision walks a tightrope, balancing the pessimism of the intellect with the optimism of the will, to use Lazarus's apt phrase (1990:Chapter 3; also see Fraser, 1980).

But sometimes he trips, as is evident in his third novel, *Why Are We So Blest* (1972). The delicately balanced idealism and realism of the earlier novels gives way to a bitter defeatism, a disavowal of creative intellectualism, of the possibilities of meaningful change. Solo, the narrator, and Modin, the central character, are alienated, impotent intellectuals, their revolutionary energies sapped by their white women and assimilationist fantasies, their will to power paralysed by a manichean understanding of the world. They come to dismiss all national liberation struggles as mystifications contrived by an omnipotent, conspiratorial imperialism. Theirs is a world of unyielding structural dualities and little agency, of absolutes, of the implacable oppositions of whiteness and blackness, as embodied in the essentialized and homogenized terrors of a passive 'Africanity' and a destructive 'Westernity'. It is a nihilistic vision, born out of a messianic view of revolution, an elitist interpretation of history.

In the next two novels, *Two Thousand Seasons* (1973) and *The Healers* (1978), Armah's hope returned. The realistic, harrowing gaze on truncated neo-colonial moments was abandoned for sweeping, visionary reconstructions of the past, as charters for social regeneration. *Two Thousand Seasons* tells the story of an African warrior and a thousand year quest for 'the way', an authentic African social system of being and living. It is narrated in the evocative restorative language of myth, and the affirmative voice of collective consciousness. Focusing on a specific period and a murder case in the 1860s and 1870s when the ancient Ashanti state was collapsing due to encroaching British imperialism, *The Healers*, eschews the conventions of a realist historical novel through constant interruptions and infusions of the narrative idioms of orature, multiplicity of authorial voices, and its celebration of the magnanimous ethics of community.

. It took seventeen years before Armah published his next novel, *Osiris Rising* (1995). The lapse is intriguing. Time does not seem to have enriched his artistic vision. In narrative structure the novel reverts to the realism of his first three novels, and thematically it focuses on the hideous neo-colonial present and the role of intellectuals in the cultural politics of liberation. The protagonists, Ast an African American scholar, reminiscent of Juana in *Fragments*, comes to an African country in search of her roots, a meaningful working life, and to be with the man she loves, Asar, a sensitive, self-assured, committed intellectual, who teaches at a newly established university. She joins him there and together they work to reform the curricula along Afrocentric lines in the face of stiff opposition from insecure Eurocentric expatriates and their African compradors. When the latter find themselves outwitted they conspire with the security forces, and Asar is murdered, with a pregnant Ast helplessly looking on.

This trajectory may suggest the intransigence of totalitarian power, and the futility of intellectual martyrdom, of struggles untethered to the protective mantle of popular movements. But the text suggests, Ast will not capitulate to the misogynist chief of security, Soja, who has unsuccessfully tried to rape her and who has ordered Asar's killing in a desperate bid to defeat and defile her, for she knows, armed with her memories with Asar and the histories of the steady triumphs against slavery, and fortified by spreading networks of subversive knowledge carried by their former students, that 'she was living through the vulnerable beauty of beginnings' (p. 282). Thus, amidst the hardships and harassment, Ast never loses faith in her adopted country, which has given her the chance for rebirth, to retrieve her racial pride and memory from the taunts and violations of America.

This is an Afrocentric tale, an affirmation of racial essentialism, over and above class, gender, nation and other constructions of difference and identity. Fittingly, Ast's doctorate was on ancient Egypt, the cradle of Africanity in Afrocentric discourse, and Asar, too, reads hieroglyphics. But this panAfricanism excludes the predatory overseers of the neo-colonial dungeons, such as Soja, and the fawning assimilationists from America, like Sheldon Tubman, a disgraced former civil rights leader, who comes in search of the imagined glories of a royal past and the misogynist pleasures of polygamy. Not surprisingly, Soja and Tubman find succour in each other's perversions of black power.

In *Osiris Rising*, as in all Armah's fiction, there is an insistent call, not to reform the fraying social fabric, but to discard it and weave one anew from the memories of resistance against imperialism. Essentially, it is a narrative, subsuming all others, of the need for African racial liberation. At its most generous it articulates a panAfricanism that transcends the dangerous fictions of ethnic and statist nationalism, as a precondition for

African liberation and the creation of a truly humanized world. But in its limited version it offers an exclusivist vision, one, moreover, anchored on the quicksand of intellectual elitism and revolutionary voluntarism.

Unconsummated Love

Ama Ata Aidoo, Armah's compatriot, also deals with the experiences and politics of neo-colonialism, but in a radically different, feminist idiom. Versatile, passionate, and often scathingly satirical, Aidoo has written plays, poetry, short-stories, novels and criticism, in which the marginalization and oppression of women is seen as an integral part of the practices and ideologies that affront human dignity, scorn true freedom, and dread genuine equality. Her work depicts a modern world immobilized by capitalism's arrogant hegemony, and an Africa with a tortured psyche desperately trying to reconcile its conflicting memories, to recover its agency ruptured by the barbarities of colonialism and neo-colonialism. The obstacles and challenges are tough and pervasive, but Aidoo's characters, especially the women, are made of stern indomitability. We get a complex vision, one that Odamtten (1994:132) has appropriately called 'pessoptimistic', in which 'its declaration of a positive restructuring of society is tempered by a realistic acknowledgement of doubt'.

Aidoo has written two novels, *Our Sister Killjoy* (OSK) (1979) and *Changes: A Love Story* (1991). In these texts the author demonstrates her formidable talents, craftily combining the conventions of Ghanaian orature and twentieth century modernism, to produce work that distinctively captures the flavours, colours, and conditions of modern African life. The prose narrative is often broken by, and fused with, poetic and dramatic insertions, and there are sharp shifts in sequence and narrative voice. This is particularly true of *Our Sister Killjoy*, whose 'clash of genres', Allan (1991:187) writes, 'styles, tones, and rhythms ... has rendered it unclassifiable. Characterized by critics as a novel, prose poem, and novella, and as "fiction in four episodes" by Aidoo herself, the text is a testimonial to Aidoo's unique creative grammar'.

Our Sister Killjoy is an exploration into the heart of empire, a reappropriation and reversal of the Conradian trip into the imagined darkness of African otherness. The novel opens with a poetic libation to the memories of those Africans and Europeans who have tried to build a more humane and human world, before we are taken 'Into a Bad Dream' of blank whiteness, a metaphorical projection of the Europe that Sissie, the protagonist, will visit. Sissie is sent by an international organization to work in a little Bavarian town, where she encounters the predictable indignities of discrimination and racism. But she also begins to understand the interconnections between colonial and patriarchal abuses of power through her burgeoning friendship with the lonely, insecure housewife, Marija. As the friendship turns into a nascent lesbian affair, the relationship acquires the oppressive power dynamics of a sexist heterosexual one, and Sissie learns, like Modin and Solo in Armah's *Why Are We So Blest*, of the tormented political economy of interracial sexuality, of the infinite capacities of imperialised consciousness to inscribe power to whiteness. In London, on her way home, she witnesses the waste wrought by blank whiteness in the hapless lives of self-exiled 'comatose' African intellectuals (OSK, p. 121) originally seduced by the illusions of finding middle class comfort and cultural universalism.

Changes pursues the complex politics of sexuality and friendship among middle class Africans at home, that is, in their neo-colonial nations and within the domestic sites of patriarchal power. It interrogates the meanings of marriage and love in a society in which professional women are increasingly entering the masculinized public spheres. It chronicles the shifting gender boundaries, the confluence of personal choices and social conventions in the struggles over the redefinitions of gender roles and relations. The changes are complex and slow, the victories partial, the compromises many. It is a story of love awaiting consummation, of liberation unfulfilled.

The three main female characters, Esi, Opokuya, and Fusena are educated, ambitious, strong, and financially independent women. Esi, a statistician, shows no hesitation in naming an unwanted sexual encounter with her insensitive husband as 'marital rape' and she divorces him, to the incomprehension of her mother and grandmother. She avoids wallowing in the self-pity of victimization. Exploiting the charity of her former husband's oppressive extended family, a costly custody battle is avoided, and their daughter is sent to live with her paternal grandmother, freeing Esi of the suffocating joys of motherhood. Incredulously, she falls in love with, and agrees to marry Ali Kondey, a charming entrepreneur, as his second wife. The marriage is mediated through all the formal rituals of 'tradition'. Most remarkably, Esi's concession to polygamy is not accompanied by serious doubts or introspection. Ironically, polygamy seems to give her the space she missed in her monogamous marriage to pursue her career goals. Although remarried she continues to live on her own, contrary to the pattern in the more dependent polygamy of 'tradition'; in fact, she never meets her co-wife, Fusena.

Fusena and Ali were college classmates and good friends before they got married. She lost the cordialities of a friend and acquired the domineering intimacies of a husband, who betrays her by marrying another woman. But she does not leave him. Like Esi, her self-realisation is not tied to her marriage but to her career (she runs a business), although her laconic responses when consulted for her consent to the marriage as demanded by 'tradition', and the textual silence accorded to her afterwards, suggests deep, unspeakable anger. Ali himself is a walking embodiment of the cultural collisions of West Africa. This highly educated Muslim and apparently sensitive man personifies the internationalized West African who remains firmly tied to the patriarchal values of his elders, to the latter's pleasant surprise.

At the end of the novel Esi decides to end her relationship with Ali after she can no longer stand the emotional toll of waiting for him to come, at his convenience, for their increasingly infrequent conjugal trysts. If the novel shows that love between men and women is always implicated with power, it celebrates the permissive joys of female solidarity, textually marked by the friendship of Esi and Opokuya, a spirited midwife who has few illusions about men and marriage. And she is right: her husband, Kubi, nearly assaults Esi. This heterosexual masculinity that objectifies women and threatens female bonding, the novel proclaims, must be eradicated in the construction of a new society.

The Joys of Rebellion

Some of the most powerful images of women's oppression and emancipation can be found in the fiction of Buchi Emecheta, one of Africa's most prolific and widely read writers. She has published a dozen novels, an autobiography, children's stories and television plays. Brilliantly and painstakingly, she has charted the contours of an African literary feminism that explodes and transcends the universalistic pretensions of Western feminism, on the one hand, and on the other, interrogates and subsumes the patriarchal idealizations, distortions and silences of the African male-dominated literary tradition. Emecheta uses realism, Stratton (1994:113) argues, as a 'response to the stereotypical representations of women'.

The novel that confirmed Emecheta's reputation was her fourth, *The Joys of Motherhood* (1979), a powerful, ironic evocation of the ordeals of a womanhood equated with motherhood. Set early in this century when Nigeria was under British rule, it tells the story of Nnu Ego, whose first marriage in Ibuza, her homeland, fails because of her inability to bear children, for which she is stigmatized. She moves to the city of Lagos where she is blessed with a bevy of children that she and her poorly paid second husband can hardly afford to look after. Her miseries accumulate. And she eventually dies alone, unattended by her sons, now educated, individualistic men. Nnu Ego's life is harsher than that of her forebears because she is subject to two forms of oppression rooted in her indigenous past and the colonial present. One demands her fecundity that can no longer be rewarded, the other dictates her economic marginality. Her alter ego and co-wife, Adaku, escapes such a fate by rebelling against both the old and new patriarchal authorities and moralities. Abandoning marriage, she sets up her own business, and is determined to educate her daughters. Nnu Ego gets her revenge in death: her spirit refuses to answer prayers from women who want children.

While *The Joys of Motherhood* offered an intertextual reading of Flora Nwapa's *Efuru* (Andrade, 1990), Emecheta's next novel, *Destination Biafra* (1982), was an intertextual feminist critique of Nigerian war narratives. The novel presents the self-reflexive journal of Debbie, a rebellious, Oxford-educated woman from an upper middle class background, who enlists on the Federal side in the Nigerian Civil War of 1967-70. Like other novels of the war, mostly written by men, Emecheta blames the hostilities on neo-colonial British manoeuvres and the ineffectual leadership of the corrupt Nigerian political class. But unlike them, the gender of this elite is named: it is male, and the war is portrayed as a product of class and patriarchal exploitations of power. Its masculinized heroism is subverted and punctured by portrayals of sexual assaults on women. Debbie herself is raped by fellow soldiers. In finally deciding to care for orphans, Debbie forsakes and overcomes militarised nationalism.

In her subsequent novel, *Double Yoke* (1982), Emecheta's narrative canvas is the gendered nature of narrative itself. The male protagonist, Ete, an aspiring writer, is a student in a creative writing class taught by Miss Bulewao, a famous writer, whom he initially disapproves of because of her gender. As part of his class assignment he writes a story of his relationship with his childhood sweetheart, Nko. Revealing the androcentric values and fantasies of his socialization and imagination, the same story is retold from Nko's perspective by the novel's narrator. We are offered an alternative feminist discourse that valorises gender equality. The dialogic tension of the two interlocked narratives not only seeks to capture different male and female writing styles, but also reflects gender conflicts in the wider society. In the larger literary historiography of modern African fiction, this counterpositioning opens a discursive conversation on gender, while in the textual confines of the novel itself it results in the development of both Eke's character and his talents as a writer.

Emecheta's most recent novel is simply entitled *Kehinde* (1994), after its female protagonist. Told in her customary realistic style, the novel explores many of the themes found in her other novels that we have not examined: the spatial and social contexts of love and marriage, the meanings of home, women's oppression and agency, and the psychic reservoirs of consciousness and well-being. Kehinde and Albert, both from Nigeria, live in London, a typical middle class couple, down to the demographically-correct two children, a daughter and a son, so that when she becomes pregnant again, Albert convinces her to have an abortion. In the social isolation of London, Kehinde, a banker, relates to 'Albert as a friend, a compatriot, a confidant' (*Kehinde*, p. 6). But after eighteen years, Albert is anxious to return to the cultural comforts of home. Kehinde is not. She is left behind alone to sell the house.

To her great surprise, she finds herself gradually ostracised by their former Nigerian friends, including her best friend Moriammo who is banned by her husband from visiting her, fearful that Kehinde's new independence will contaminate the rigid domesticity of his household. Kehinde is forced to return even before selling the house, only to find that Albert has married a second wife, a young university professor. Devastated, jobless, and encouraged by her elder sister and aided by Moriammo, Kehinde returns to London, to live as 'just another black woman' (*Kehinde*, p. 127), now molested by racism. But she refuses to be defeated: slowly she rebuilds her professional and personal life, and when her son returns years later with his manly conceits, she puts him in his place. In choosing to live in London, she is not renouncing Nigeria as home, but affirming her womanhood. She finds racial otherness at the heart of empire less threatening to her integrity as a woman than the patriarchal betrayal of her husband, a decision supported by the spirit of her twin who died at birth. This spirit appears to her at all critical moments, a communion that suggests the erasure of the customary boundaries of belonging constructed by the imagined patriarchal communities of race and nation.

Songs of Enchantment

For Okri, author of ten books, spirits are not ancillary to the narrative, a kind of intertextual cultural gloss, but can constitute the main flow of the narrative itself. Writers of Emecheta's, Armah's, and Aidoo's generation filled their writing with proverbs and folktales, with the idioms of orature, as a means of enlivening and domesticating supposedly alien literary genres, of clearing space for an authentically 'African' aesthetics[1]. In the hands of less capable writers the folkloric infusions stood out as sore thumbs, and even when successfully incorporated, they were often overwhelmed by the realistic mode of narration allegedly imported from Europe. The whole enterprise may indeed have reinforced the reading of African literature in the West as exotic ethnography. Okri's genius in *The Famished Road* (FR) and *Songs of Enchantment* (SE) (1991 and 1993) has been to offer entire texts as exquisitely crafted mythic narratives that boldly re-write and re-interpret the abiku myth so widely known in Nigeria, in a bid to tell large, painful truths about his nation and the human condition in general, and he does so with effortless ease, in incredibly beautiful prose that has the evocative intensity and imagery of poetry, prose that sings and sighs and smells with the enchantments, struggles, and sorrows of modern Nigeria. With Okri the African novel has scaled new heights, finally shedding any lingering apologies to colonial and western sensibilities. Labelling these works post-modern, as some critics seem anxious to do (Hawley 1995), is to trivialise the immensity of Okri's achievement and artistic vision.

The reluctant abiku child, named Azaro, is born to a poor couple in an urban shanty, on the eve of independence, and is a witness to their suffering and sacrifices, occasional laughter and joys, and to the travails and dreams of the wider community, with its petty hierarchies of wealth and power, eternal frictions between men and women, children and adults, its hilarious parties and vicious fights, and its solidarities, jealousies and antagonisms reinforced by divisive independence party politics. The story unfolds slowly, deliberately, movingly, to its compelling conclusion where Azaro's father, who is recuperating from a ferocious boxing match, delivers a haunting prophecy of the post-colonial future of this tortured 'abiku country' (FR, p. 478). It is a vision articulated with striking narrative power, unmatched in recent African fiction. This is serious politics, an artistically sublime interrogation of Africa's pasts and futures, refuting Niven's (1989) hasty prediction that 'Okri and his generation will be more introspective, more personal, less historically ambitious, less radical, than Achebe and his peers'.

Okri's literary project, he declared in one interview, is to blast 'the linear, scientific, imprisoned, tight, mean-spirited, and unsatisfactory description of reality and human beings', and to heal 'the human spirit by giving back to it its full, rich, hidden dimensions' (Ogunsanwo 1995:40). And that's what he gives us in these two novels. The narrative oscillates seamlessly between the mundane and the spiritual, the ordinary and the extraordinary. Azaro lives and apprehends the realities of both worlds simultaneously, is always crossing the thin membrane separating them. We see him communicating with both human beings and spirits, sometimes simultaneously, and living in both the past and the present and peering into the secrets of the future. We are beguiled by his escapades, moved by his battles with his companion spirits who want him to return, and captivated by his love for his sad, troubled mother and frustrated, boisterous father. But Azaro is not the only one who harbours psychic consciousness and powers. His spirit periodically encounters in the interspace the spirits of the indomitable, industrious Madame Koto, owner of the popular bar, and those of the mysterious blind old man, and Azaro's friend, Ade, also an abiku. It is an animated, pantheistic world, marked by constantly transmogrifying reality, endless travel between states of being.

And so the central motif of the novel is the road, an ubiquitous presence inscribed with the incomplete journeys of their lives. It is a road that devours their aspirations, keeps them famished. But their hunger has a regenerative potential: it can 'change the world, make it better, sweeter', Azaro is told by his father. 'All roads', the father continues, 'lead to death, but some roads lead to things which can never be finished. Wonderful things' (FR, p. 498). 'Many people', he concludes, 'reside in us ... many past lives, many future lives. If you listen carefully the air is full of laughter' (FR, p. 499). Inspired, Azaro goes to sleep where he 'found open spaces where I floated without fear. The sky was serene. A good breeze blew over our road, cleaning away the strange excesses in the air... I was not afraid of Time' (FR, p. 500).

Life thus affirmed, he is ready to face new trials in *Song of Enchantment*. Azaro drops out of school, his aging parents temporarily separate, his friend Ade dies while Ade's father is brutally murdered, and Azaro's own father becomes blind. The new 'unborn nation' (SE, p. 90) is ravaged by the political 'wars of the mythologies everywhere' (SE, p. 206), and the melancholies of misguided modernization deepen. But this world has its enchantments, too. Azaro's parents rekindle their love for each other, and they finally acquire the powers to experience the magical realities of the spirit world. Ade occasionally reappears to play with Azaro. And Azaro's father's blindness turns out to be a transitory sojourn into a mesmerising world of new insights and sensations. Deliverance is suggested in his delirious renaming of everything in the universe: inanimate objects and living things, human achievements and afflictions. He even 'named the mosquitoes and praised them for helping to prevent the colonialists from entirely taking over our lands', and he remembered 'our histories, making it necessary for us to invent a science best suited for our continent, making it imperative that we be perpetually creative, constantly inventive, worshippers at shrines of beauty, self-inventors who have to re-dream the world anew because it is always passing away...' (SE, p. 281).

It transpires at the end of the novel that it was Azaro's companion spirits, in their continuing efforts to pressure him to return, who were 'making reality appear more monstrous and grotesque. But so far they had failed. And they had failed because they had forgotten that for the living life is a story and a song, but for the dead life is a dream. I had been living the story, the song, and the dream' (SE, p. 293).

Conclusion

Perhaps that is what great literature is all about, living the story, the song, and the dream of life, a celebration of the infinite mysteries of human existence; it offers imaginative charters for the wretched of the earth, the abiku societies, to rename and repossess the world, remember their pasts and invent new memories of the future. That is what Okri's novels give us, astonishing aesthetic joy about human anguish and struggle, a profound and moving vision of social existence crying out for transformation, for spiritual and moral uplift; a vision that is rooted in the celebration of the extraordinariness of the ordinary, the sanctity of all human life. The other writers examined in this chapter also passionately confront the debasement of social life, the devaluation of human relations in the contemporary world, and suggest in the resilience of their characters the indomitability of the human spirit, the will to survival, to meaning, social and moral.

Even in the most gruesome and grueling conditions, which Africa has had in abundance over the last few centuries, thanks in no small measure to its encounters with the barbarities of western modernity, hope always filters through, the aspirations for new beginnings persist, the possibilities of rebirth, remain. This is the humanist vision that shines through in the finest of African literature; testaments to hope and faith in the human potential for transcendence against the evils of unproductive power and conflicts between nations and nationalities, genders and classes. They offer stories and dreams for a redeemed humanity in a reformed social order, one in which there is democracy and freedom for all.

Notes

1. As argued in Chapter 18 above, the valorization of orality, rooted in the widely held assumption that Africa is ontologically oral and writing is European, is based on an erroneous historiography. Writing has ancient origins in many parts of Africa. Up to the third century, for example, African writers were the most important contributors to the development of Christian thought, see Mudimbe (1994:175-7; Albert Gerald (1981) and Harold Scheub (1985).

Part Five

Towards Panafrican Studies

Chapter Twenty One

THE TRIBULATIONS OF UNDRESSING THE EMPEROR

Edward Said, the Palestinian-American intellectual, is arguably the most exciting, if not the most distinguished, cultural and literary critic now writing in North America. In an age when literary critics seem preoccupied with the petty concerns of post-modernism, Said engages the large questions of comparative historical analyses of literature and cultural politics. His focus is unusually broad and complex, his analysis multi-layered and richly-textured; and when combined with the lucidity of his writing and his immense erudition and political compassion, he compels our attention and respect. Many of the same impressive qualities are evident in his latest book, *Culture and Imperialism* (1993). Nevertheless, this book has some fundamental flaws which many of Said's admirers and critics have ignored.

In several ways the book is a sequel to his influential study, *Orientalism* (1978), whose magisterial sweep, interdisciplinarity, apparent theoretical freshness, and uncompromising indictment of the European invention of the 'Oriental Other', won Said fervent followers and ardent detractors, especially since the book came out at a time when the Left was in retreat and the Right in ascendancy. *Orientalism* contested, but also resonated with, the anti-communisms and post-Marxisms of the Thatcherite and Reaganite era, in which disillusioned leftist intellectuals took refuge in the reactionary anti-humanisms of post-structuralist cultural theories (Ahmad 1992:192-94), while triumphalist conservatives invoked canonical nationalism (Bloom 1988; Hirsch 1988; Cheney 1988; Kimball 1990; D'Souza 1992; Hughes 1993).

In *Culture and Imperialism*, the thesis of *Orientalism* is globalized. Eurocentrism is seen everywhere. It 'penetrated', Said writes, 'to the core of the workers' movement, the women's movement, and the avant-garde arts movement, leaving no one of significance untouched' (1993:222). Its savage, destructive, and all-encompassing energies infused the work of anthropologists as well as historians, Marx, the revolutionary messiah, as well as Kipling, the imperial poet. Imperialism involved more than economic accumulation and territorial acquisition. It was facilitated, sustained, and perhaps even impelled by the cultural affliction of Eurocentricism.

The universalising cultural discourses of Eurocentricism not only constructed and represented, dominated and inferiorised numerous 'others', they also permeated all forms of cultural production in the imperial metropoles themselves, including music and literature. Contrary to what the high-priests of the literary establishment say, these works of art are not autonomous, but implicated, indeed structured in their form, content and aesthetics by the spatial, material, and ideological configurations of empire. The imperial 'structure of attitude and reference', as Said calls it, can be found even in the works of artists rarely associated with imperialism, from Jane Austen to Albert Camus. The latter's novels are still read as parables of the human condition when in fact they 'very precisely distil the traditions, idioms, and discursive strategies of France's appropriation of Algeria' (Said 1993:184).

But *Culture and Imperialism* is no mere sequel to *Orientalism* in its analytical structure. The latter's ahistorical argument, that the seamless web of 'Orientalist Discourse' stretches all the way back to Aeschylus' Greek Antiquity, through Dante's late Middle Ages, to Marx's age of industrial capitalism, and Bernard Lewis's era of American superpower hegemony, has been abandoned for a more focused, and credible, historical analysis. This is to suggest that in *Orientalism* Said essentialised 'Europe' or the 'West' as much as he accused the latter of essentialising the 'Orient'. 'Europe' or the 'West', no more than the 'Orient', or say, 'Asia' or 'Africa', as civilizational entities, are relatively recent historical constructions, not timeless fixtures. In *Orientalism* Said took the Athens-to-Washington narrative of 'Western culture' as a historical reality not for the recent fabrication that it is, as Martin Bernal has so brilliantly demonstrated in his controversial tome, *Black Athena*[1] (1987). Hence, the unresolved analytical tension in the book as to whether 'Orientalist Discourse' embodied 'a system of representations, in the Foucauldian sense, or *mis*representations, in the sense of a realist problematic' (Ahmad 1992:185-86).

In *Culture and Imperialism* Said focuses his critical gaze squarely on the nineteenth and twentieth centuries, the era of modern European and American imperialism. Conrad, not Aeschylus, now gets the dubious distinction of being the precursor

of the Western views of the Third World which one finds in the work of novelists as different as Graham Greene, V. S. Naipaul, and Robert Stone, of theoreticians of imperialism like Hannah Arendt, and of travel writers, filmmakers and polemicists whose speciality is to deliver the non-European world either for analysis and judgement or for satisfying the exotic tastes of European and North American audiences (Said 1993: xvii-xviii).

And he now explicitly recognises that 'partly because of empire all cultures are involved in one another; none is single and pure, all are hybrid, heterogeneous, extraordinarily differentiated, and unmonolithic' (Said 1993: xxv). In fact, he takes to task

> most histories of European aesthetic modernism [which] leave out the massive infusions of non-European cultures into the metropolitan heartland during the early years of this century, despite the patently important influence they had on modernist artists like Picasso, Stravinsky, and Matisse, and on the very fabric of a society that largely believed itself to be homogeneously white and Western (Said 1993: 242).

Said's humane conscience bemoans the fact that despite these processes of cultural globalization the traps of separation and essentialization get tighter: 'Africanizing the African, Orientizing the Oriental, Westernizing the Western, Americanizing the American, for an indefinite time and with no alternative' (Said 1993: 311). This essentialization and tribalization of identities is in fact a by-product of imperialism itself. 'No one today', he states eloquently,

> is purely one thing. Labels like Indian, or woman, or Muslim, or American are no more than starting-points, which if followed into actual experience for only a moment are quickly left behind. Imperialism consolidated the mixture of cultures and identities on a global scale. But its worst and most paradoxical gift was to allow people to believe that they were only, mainly, exclusively, white, or Black, or Western, or Oriental (Said 1993: 336).

Said's indignation is especially compelling in Chapter Four, where he analyses the imperial dynamic of dominating and misrepresenting the so-called Third World societies of Africa, Asia and Latin America. He argues that the crude, reductionist, and coarsely racist representations of the Arab world reproduced in the media helped make the Gulf War popular among Americans. It was an imperial war against the Iraq people, ethnic cleansing and tribalism writ large. As in the old imperial wars of colonial conquest, modern wars of neo-colonial control, are 'prepared for', 'authorised', and 'legitimated' by the narratives produced by artists and intellectuals. Most of these intellectuals have blissfully 'avoided', he contends, 'the major, I would say determining, political horizon of modern Western culture, namely imperialism' (Said 1993:60).

He is particularly harsh on American intellectuals, especially 'the policy-oriented ones, [who] have internalized the norms of the state', and colluded with American imperialism. And by intellectuals he is not simply referring to right-wing ideologues, but also those who consider themselves 'progressive'. The latter, he avers, have been 'defanged' by the American university, with its munificence, utopian sanctuary, and remarkable diversity ... Jargons of an almost unimaginable rebarbativeness dominate their styles. Cults like post-modernism, discourse analysis, New Historicism, deconstruction, neo-pragmatism transport them into the country of the blue; an astonishing sense of weightlessness with regard to the gravity of history and individual responsibility fritters away attention to public matters, and to public discourse. The result is a kind of floundering about that is most dispiriting to witness, even as the society as a whole drifts without direction or coherence. Racism, poverty, ecological ravages, disease, and an appallingly widespread ignorance: these are left to the media and the odd political candidate during an election campaign (Said 1993:303).

Said's fascination with Foucault and Derrida, the latest French intellectual exports to a theory-starved North America, which made his *Orientalism* seem so avant-gardist, seems to have waned. Not surprisingly, *Culture and Imperialism* has not been received with the same critical acclaim that *Orientalism* was. The conservatives have been predictably vitriolic in their condemnation. *The Economist's* reviewer was offended that Jane Austen has been tarnished with the brush of imperialism and finds fault with Said's 'relentlessly turgid jargon... Unhappily', the review concludes, 'this polemic has precious little to do with the relationship of culture to imperialism' ('Novels of Empire': 1993). In a lead article in *The Times Literary Supplement*, the anthropologist Ernest Gellner, mischievously caricatures and dismisses Said's treatise as 'the bogy of orientalism' (1993:3-4)[2].

More sympathetic critics have argued that *Culture and Imperialism* is a potentially great book that fails either because, according to Fred Inglis (1993), it lacks a comprehensive theory of imperialism, or as John M. MacKenzie (1993) has argued, Said simplifies the spatial dimensions of imperialism and constructions of 'otherness' by seeing them purely in terms of relations between Europe and the rest of the world, and not within Europe itself as well. Although Said's conceptualization of imperialism is haphazard, Mackenzie's main criticism is not quite correct. Moreover, his charge that Said misreads Verdi's opera, *Aida*, is unconvincing. To Mackenzie, *Aida* 'is just as anti-imperialist an opera as you can get' (1993:342)[3]. Said succeeds in showing that in its conception, content and production *Aida* was imbued with 'the imperialist structures of attitude and reference' (Said 1993:130).

One of the most effusive reviews of *Culture and Imperialism* is provided by Paul Gilroy, for whom 'Edward Said is a rare commodity these days - a resolutely principled, political intellectual'. 'It would be a great shame', he continues,

> ... if the breadth and imagination of *Culture and Imperialism* were reduced to the status of a riposte to the right. For one thing, Edward Said is a long way from advocating anything like the wholesale transformation of literary scholarship. He makes ritual obeisance to anti-colonial figures like Toni Morrison and Chinua Achebe, but is also keen to reassure his readers that, regardless of the political aspirations of his work, he still appreciates the literary merit of the 19th-century texts he re-reads and re-locates amid the distinctive 'structure of attitude and reference' that characterises imperial power. The idea of aesthetics, like the concept of 'race' survives *Culture and Imperialism* intact (1993a:46).

I find this quite a perceptive reading of the book. For Gilroy, this is its greatest strength. In my view, it is perhaps its central flaw. It underlines the central ambivalence in Said's analysis: his actual reading of the western canonical texts contradicts his anti-imperialist critique contained in the book's opening and closing chapters.

Despite its considerable advances over *Orientalism*, *Culture and Imperialism* is a far less satisfactory work. It is not a matter of finding fault with specific details, or assuming that Said has failed where he might be expected to succeed. The problems are more profound than that. They are methodological, conceptual, and political, some of which are already evident in *Orientalism*. In *Culture and Imperialism* they become glaring because of the very project, the argument Said is trying to advance. The book is, therefore, far less radical and original than many of its detractors and admirers say it is.

Said's objective in *Culture and Imperialism*, as he states at the beginning of the book, is to depict not only the expansion and exactions of western imperialism but also the response and resistance against it in the colonial world which culminated in decolonization. 'These two factors', he states, ' a general world-wide pattern of imperial culture, and a historical experience of resistance against empire inform this book in ways that make it not just a sequel to *Orientalism* but an attempt to do something else' (1993:xii). He seeks to offer, he claims, a 'contrapuntal reading', which 'must take account of both processes, that of imperialism and that of resistance to it' (1993:66). In his view each text or cultural artefact can only be understood in terms of the negative 'other' against which it defines itself and the responses of that 'other'.

However, the way he goes about doing this is problematic and betrays deep ambivalences. To put it simply, he compares apples and mangoes, western canonical texts and Asian and African non-literary texts. The result is to reconfirm the privileged status, the supremacy of the western literary canon and marginalise and dismiss African and Asian literatures, thereby subverting his original thesis. Like a rebellious attendant trying to undress the emperor, he becomes engrossed by the texture of the emperor's robes and forgets his original mission.

All the western canonical texts examined in *Culture and Imperialism* are treated with the kind of hermeneutic engagement, close scrutiny, and informed reading which is not accorded to a single African or Asian literary text, although Said liberally drops names and titles. The story begins with Joseph Conrad, the Polish expatriate, whose *Heart of Darkness* (1925) has an honoured place in imperialist literature[4]. The limitations and ambivalences of Said's 'contrapuntal reading' are quite evident in his assessment of this book. It is as severe as it is forgiving. He observes, quite perceptively and correctly, that *Heart of Darkness* is a narrative of the ravages and the assumed inevitability of empire and the helplessness of the silent 'native'. But he quickly reaffirms Conrad's genius and reminds us that 'Conrad was certainly not a great imperialist entrepreneur like Cecil Rhodes or Frederick Lugard' (1993:24). But who said he was, and so what?

Said is of course only too aware that 'the cultural and ideological evidence that Conrad was wrong in his Eurocentric way is both impressive and rich' (1993:30). For example, he mentions Chinua Achebe's (1989) essay, which criticises Conrad's racist dehumanization of the Africans in *Heart of Darkness*. Said comments: 'Achebe shows he understands how the [novel] form works when, in some of his own novels, he rewrites — painstakingly and with originality — Conrad' (p 76). But that is as far as his analysis of Achebe's *oeuvre* goes. Some 'contrapuntal reading'! In the same paragraph discussing the attacks against Conrad, Said mentions a number of African and Asian writers, such as Ngugi wa Thiong'o and the Sudanese writer Tayeb Salih, who have contested, reclaimed, renamed and reinhabited the imaginative territory Conrad appropriated for imperialism. But none of their works is actually analysed in this chapter. We are promised an analysis in chapter three. I anticipated this chapter only to be disappointed that both Ngugi and Salih are disposed off in a paragraph each (p. 211). Reading Said one would not know that Ngugi has written more novels than *A River*

Between,[5] and that his views on language as represented in *Decolonizing the Mind* (1986), contradict Said's celebration of hybridity. Similarly Salman Rushdie is short-shrifted in one paragraph (Said 1993:216). His troubles with the Ayatollah are analysed at greater length.

More extensively discussed in this chapter on 'Resistance and Opposition' are academic and political texts, including Fanon's prophetic and searing indictment of the post-colonial order *The Wretched of the Earth* (1968), C. L. R. James's classic hortative study of the Haitian Revolution, *The Black Jacobins* ([1938] 1963); George Antonius's classic affirmation of the Arab world, *The Arab Awakening*, ([1938] 1969), and two more recent and rather specialised academic books, Ranajit Guha's *A Rule of Property for Bengal* (1963) and S. H. Alatas's *The Myth of the Lazy Native* (1977). Ironically, pride of place in the literature of decolonization is given to the Irish poet, W. B. Yeats. Said's extended and close reading of Yeats displays both the strengths and weaknesses of his approach. First, contrary to Mackenzie's critique referred to above, it shows that Said does not see imperialism exclusively in terms of interactions between European and non-European nations and societies, for Ireland is the quintessential colonial society. The anguish and tension many Irish writers felt between their nationalism and the English cultural heritage, which both dominated and empowered them, was also experienced by colonial writers. The illuminating parallels that Said draws between Yeats and the negritude poets show the insights that comparative literature can bring. But it would seem that, to Said, Yeats legitimates the narrative of decolonization precisely because he has since been elevated to the hallowed canon.

Said's 'contrapuntal readings' of the other canonical writers in *Culture and Imperialism* are similarly illuminating, limited and ambivalent. After examining Conrad, he focuses on Jane Austen's *Mansfield Park* ([1814] 1966), even though the latter preceded the former. In Said's literary cosmos temporality is often an inconvenience better ignored than interrogated. To be sure, he ably demonstrates that although Austen's allusions to empire are casual compared to Conrad's, the existence of empire, specifically the Betrams' estate in Antigua, is at the centre of Austen's moral geography, framing and sustaining the lives of the characters in the novel. Indeed, Austen's very casualness is a testimony to the acceptance and uncontested reality of empire in British popular culture, a validation of life at 'home' and a devaluation of the 'other worlds'.

The problem is that Austen's gender, and its possible effects on her perceptions of empire, are issues that are never raised. To Said, imperialism, it would seem, is a spatial 'one-armed bandit', to borrow Walter Rodney's damning phrase (1982:205), unsullied and unmediated by the particularities of gender or class. What does he offer for a 'contrapuntal reading' of Austen's novel? A brief mention of Eric Williams's famous book, *Capitalism and Slavery* ([1944] 1964). Whatever its merits or demerits, this book is not a literary work. Thus, once again, Said contests the imperial literary master and the colonial literal subject.

As with Conrad's *Heart of Darkness*, Said's aim is not to remove the imperial robes from Austen's *Mansfield Park*, but to admire their fine texture. 'Yes, Austen belonged to a slave-owning society,' Said states, and then asks rhetorically: 'do we therefore jettison her novels as so many trivial exercises in aesthetic fumpery?' Predictably, he answers: 'Not at all... *Mansfield Park* is a rich work in that its aesthetic intellectual complexity requires that longer and slower analysis that is also required by its geographical problematic...' (Said 1993:96). That is all.

I am not disputing the literary merits of *Mansfield Park*. Rather, I am pointing to Said's tendency 'of alternating between inordinate praise and wholesale rejection', as Ahmad (1992:159) has so acutely observed. This is quite evident in his readings of Kipling's novel *Kim* ([1901] 1941). Kipling, the archetypal imperialist writer, gets the same contradictory, extreme treatment: harsh condemnation for his racist vision and fulsome praise for his 'stylistic' mastery. Kipling's *Kim* is denounced as a narrative of imperialist and misogynist pleasure, but also praised as 'a work of great aesthetic merit' (Said 1993:150). Indeed, to Said it is unfortunate that Kipling, along with Conrad, has been found 'eccentric, often troubling, better treated with circumspection or even avoidance than absorbed into the canon and domesticated along with peers like Dickens and Hardy'(Said 1993:132). A few pages later, the list of Kipling's peers gets longer:

> We have become so used to seeing him alongside Haggard and Buchan that we have forgotten that as an artist he can justifiably be compared with Hardy, Henry James, Meredith, Gissing, and the later George Elliot, George Moore, or Samuel Butler. In France his peers are Flaubert and Zola, even Proust and the early Gide (Said 1993:156).

A page later, Said intones, 'the sheer variousness of [Kipling's] creativity rivals that of Dickens and', guess who, 'Shakespeare' (Said 1993:157). These are weighty claims, indeed.

As Kipling is being elevated to the top of the literary canon, Said reaffirms the permanent inferiority of the '"other" non-European literatures'[6]:

> ... it is a mistake to argue that the 'other' non-European literatures, those with more obviously worldly affiliations to power and politics, can be studied 'respectably', as if they were in actuality as high, autonomous,

aesthetically independent, and satisfying as Western literatures have been made to be. The notion of black skin in a white mask is no more serviceable and dignified in literary study than it is in politics. Emulation and mimicry do not get one very far (Said 1993:316-7).

Kipling himself would not have put it better. Nor would Allan Bloom. Thus, after firing all his seemingly anti-western political salvos, Said comes down to the side of the canonical angels: African and Asian literatures are imitations of the 'real thing'; they are literatures of V. S. Naipaul's (1967) 'mimic men'. 'Naipaul surely,' to quote Ahmad's scathing reaction to the original article containing the above-quoted passage,

> never made a judgement more damning... In direct contrast, we get from the author of *Orientalism*, no less - the characterization of 'French, German and English literatures' as not only 'high' but also 'autonomous', 'aesthetically independent' and 'satisfying'. Now, satisfaction is doubtless a personal matter, but may one ask: 'autonomous' and 'independent' of what? The whole point of *Orientalism*, one would have thought, was that these literatures were *not* autonomous; that they were too complicit in colonialism to be spoken of primarily in terms of 'high' aesthetics(1992:216).

And is this not the same intended message of *Culture and Imperialism*; the same Said who writes on the next page that 'reading and writing texts are never neutral activities: there are interests, powers, passions, pleasures entailed no matter how aesthetic or entertaining the work' (p. 318)?

Said's preoccupation with, and celebration of, western canonical writers is matched by, and in fact may be seen as a corollary to, his growing fixation with exile and 'hybridity' and location in the West, all of which amount to a disavowal of the 'Third World' as spaces of meaningful artistic creativity and political struggle. It betrays a political attitude that is essentially elitist and Eurocentric, if one could dare apply such a term to the author of *Orientalism* himself.

The argument runs like this: the first stage of the ideological and cultural war against imperialism was located in the colonies, from where it spread to the metropolitan world, where it is now concentrated partly because the resurgent nationalisms, despotisms, and ungenerous ideologies of the post-colonial world have betrayed the liberationist struggle, and partly because of migrations from the 'Third World' to these countries, 'voyages in' as Said calls them, not simply of 'nearly forgotten unfortunates', but of intellectuals. Spawned by the ravages of post-colonial and imperial conflicts migration is, in fact, the modern condition. Consequently, liberation as an intellectual mission has now shifted from the settled, established, and domesticated dynamics of culture to its unhoused, decentred, and exilic energies, energies whose incarnation today is the migrant, and whose consciousness is that of the intellectual and artist in exile, the political figure between domains, between forms, between homes, and between languages... There is then [for the exile intellectual] not just the negative advantage of refuge in [his] eccentricity; there is also the positive benefit of challenging the system, describing it in language unavailable to those it has already subdued (Said 1993:332-3).

The era of the 'cosmopolitan' Third World writers and intellectuals, as Brennan (1989) calls them, upon whose hybrid shoulders rest the liberation of both the western and post-colonial worlds, has arrived. This would be laughable if it were not coming from such a renowned scholar.

It is the peculiar arrogance of exiled intellectuals to see themselves in heroic terms, to turn the autobiographies of alienation into fantasies of collective liberation. There was of course a time when exile in the western metropoles often translated into political leadership at home. Many of Africa's first presidents, the Nkrumahs, Kenyattas, Bandas, Houphouet-Boignys, and Senghors flew from metropolitan exile to the colonial governors' mansions, sometimes after a few years of martyrdom in jail. Today, the leadership for the democratic struggles, the 'second independence', is homegrown, and if in exile it is more likely to be in the neighbouring countries than in London, Paris, New York, or Toronto. Said is three decades out of date.

Said's views of the role of exiled intellectuals is problematic for several reasons. To begin with, there are problems of fact. The majority of the migrants and refugees from African and Asian countries move to neighbouring countries or within their own countries, and not to Europe or North America, despite all the political posturing made by the latter countries[7]. Exile is of course nothing new. In any case, for every 'radical' intellectual migrant like Edward Said there is an 'ultra-conservative' Dinesh D'Souza. In fact, my observations are that most recent immigrants are usually too busy 'settling in' and 'trying to make it', often against great odds, to indulge in the world-shattering liberationist politics of which Said dreams.

More troubling in the celebration of the new 'cosmopolitan Third World' intellectuals located in the West is a shocking marginalization of the immense struggles that have been waged, say, in the United States, by generations of African-American intellectuals and activists. This is where Said's 'voyages in' differs from Mazrui's apparently similar concept of 'counter-penetration'. For Mazrui (1978:314-18), raising

Africa's influence in the West through 'counter-penetration' entails, among other things, making alliances with the African diaspora, especially the African-Americans[8]. Said's 'voyage in' is a mission that is singularly unconnected to any social force, confined to lonely textual readings, supplemented by professional contacts and perhaps the conference circuit.

In the documentary film, *Fields of Endless Day* on the history of Africans in what is today Canada from 1604 to the 1930s, an elderly African-Canadian woman says:

> The new people who have just come think that everything that has been accomplished has been accomplished because of them. They forget that we who have been here for a long time built the bridges which they walk on today[9].

This a salutary reminder that the historical and cultural configuration we call North America, or broader still, 'Western' is not purely European as Said himself would be the first to say, and indeed does on several occasions. As Hendrik Hertzberg and Henry Louis Gates note about the African-American presence in the United States:

> those of us whose forebears came here in chains have much deeper roots in American soil, on the average, than those of us whose forebears came here in and for freedom; the vast majority of African-Americans are descended from men and women who arrived before 1776. Except for American Indians, only a shrinking minority of other Americans can say the same (1996:9).

This point is echoed with statistical exactitude by Bohannan and Curtin, who write:

> The median date for the arrival of our African ancestors of Afro-Americans — the date by which half had arrived and half were still to come — is remarkably early, about 1780. The similar median date for the arrival of our European ancestors was remarkably late - about the 1890s. It was not until the 1840s that more Europeans than Africans crossed the Atlantic each year (1988:13).

The Atlantic world, then, as a cultural formation, as a civilizational entity, is no less black than it is white, as much a part of Africa as it is of Europe; built by the blood, tears, and sweat of African slaves and their descendants, and infused and enriched by their struggles, spirit, songs, and speech. But in Said's analysis, this 'Black Atlantic' world, as Gilroy (1993b) calls it, does not exist, and 'western' is often interchangeable with 'European', and oppositional narratives of African-Americans are either ignored or conflated with those of recent immigrant intellectuals, the new vanguard of the global anti-imperialist struggle.

It is idealistic at best to think that the struggles for decolonization are now centred in the North. As Said himself notes at one point, while Americans and the British were sleep-walking under Reagan and Thatcher, there were 'mass uprisings outside the western metropolis [in] Iran, the Philippines, Argentina, Korea, Pakistan, Algeria, China, South Africa, virtually all of eastern Europe' (Said 1993:326). Since the beginning of the 1990s, the convulsions have spread to more countries in Africa, Asia, and Latin America. This is not to deny that important social struggles are being, and have to be waged, in the North, for the political pathologies and social deformities of the modern era are not confined to the so-called Third World, despite what the western media would like us to believe. Rather, it is to point out the absurdity of exiled intellectuals from the South leading these struggles in the North.

The celebration of the hybrid joys of migrancy, ambivalence, and contingency, together with the glorification of globalized, postmodern electronic culture, and the glib dismissal of nationalism, especially in the countries of the 'Third World', constitute the core themes of 'postcoloniality', among whose highpriests are Said, Bhabha (1994), and Spivak (1993). They misread the fact that physical mobilities did not start with the migrations of postcolonial intellectuals like them, nor did cultural cross-fertilizations await the launch of CNN and the Internet; and that as imperialist capital penetrates and transforms all available global spaces, the nation-state form is, simultaneously, proliferating, and progressive nationalism becomes even more crucial to combat the destructive energies of capital against labor, women, and the environment. And they miss the specificity of contemporary intercultural hybridity· that the cultural encounters are not only unequal, but they are occurring among commodified cultures on the imperial market place 'that subordinates cultures, consumers and critics alike to a form of untethering and moral loneliness that wallows in the depthlessness and whimsicality of postmodernism - the cultural logic of Late Capitalism, in Jameson's superb phrase - in a great many guises, including the guises of 'hybridity', contingency, etc.' (Ahmad 1995:17).

Celebratory invocations of cultural hybridity, therefore, are a narrative of privileged migrant elites, not the migrant poor, who 'experience displacement not as cultural plenitude but as torment' and desperately seek 'a *place* from where they may begin anew, with some sense of a stable future' (Ahmad 1995:16). In short, the intellectual exiles Said talks about must show a little more humility and respect both to the people in their native countries struggling for existence and freedom with their lives, not just angry pens in secure North American academies, and the historic and continuing struggles of the descendants of those who built 'the bridges they walk on today' to their secure sinecures in the teaching machine.

How does *Culture and Imperialism* relate to issues and debates in African Studies? Said's critique of Eurocentricism is of course all too familiar to African and Africanist scholars, many of whom have made some of the most vigorous and sustained efforts to deconstruct Eurocentric paradigms and decolonize the Social Sciences and Humanities. These efforts have borne some fruit, although there is still a long way to go as earlier chapters in this book, and the papers in *Africa and the Disciplines* (Bates, Mudimbe and O'Barr 1993 - discussed in next Chapter) clearly show. In the literary field, specifically, the Eurocentricism of canonical and popular authors, including travel writers and journalists, has been amply demonstrated (Pieterse 1992; Pratt 1992; Hawk 1993). Indeed, 'contrapuntal readings' have been made of European and African authors, far more successfully in my view, than Said achieves in this book (see for example, Jones 1984)[10].

Many African and Africanist scholars would, therefore, agree with the general thrust of Said's critique of Eurocentricism. But they would also find his reaffirmation of the primacy of the western canon rather troubling. I have no problem with studying the western canon; in fact, many African intellectuals of my generation were brought up on it. But we also studied African and other literatures. Focusing exclusively on the canon is intellectually stifling. It ought to be clear by now, after all the deconstructionism that has gone on in cultural and literary studies, that the western canon is neither cast in stone nor did it descend from heaven, like Moses's ten commandments. It was constructed. Its boundaries are too narrow, antiquated, and suffocating. There is simply more out there, both within the western tradition itself and outside it. Regardless of what the canonical nationalists say, the 'West' has never been exclusively European, white and male since it was first invented, and neither is it the repository of all the finest artistic achievements of humanity.

Whatever the strengths or shortcomings of the western canon, it is intellectually confining not to expose readers and students in Europe and North America to the literatures of Africa, Asia, and Latin America. Dealing with all these literatures, containing as they do such a wide range of narratives of diverse places and times, which were connected and separated, influenced and transformed in complex and differentiated ways by imperialism, offers incredible opportunities for broadening intellectual horizons, constructing more meaningful and holistic theoretical models, and realizing the grand vision implicit in the name 'university'. The case for a learning atmosphere 'conducive to multicultural inquiry and intercultural literacy', to quote Christopher Miller (1993:219), has never been stronger. This is the opportunity Said missed in his incomplete 'contrapuntal reading' and privileging of the western canonical texts.

Said's failure to accord the literatures of Africa and Asia serious analytical space, recalls the lack of attention and respect on the part of Western scholars, including Africanists, for the critical scholarship of African and Asian intellectuals. It demonstrates, to quote Escobar (1995:16), 'the limits that exist to the Western project of deconstruction and self-critique... deconstruction and other types of critiques do not lead automatically to an 'unproblematic reading of other cultural and discursive systems'. They might be necessary to combat ethnocentricism, 'but they cannot of themselves, unreconstructed, represent otherness' [Bhabha 1990:75]'. Deconstruction, in short, must be accompanied by reconstruction, by the creation of new discursive ways of seeing and acting, of articulating the social and scholarly realities of Africa and Asia, not as the West's 'others', but as themselves. For African studies, this requires the development of genuine and generous Panafrican perspectives and praxis, which recognizes not only the ways in which Africa has been constructed in the last half-millennium through the prism of imperialism, but also taps into the memories of struggle embedded in that encounter, in order to reconstruct and transform Africa's worlds, as part of the process of creating a more benevolent world befitting our collective humanity.

In conclusion, while *Culture and Imperialism* displays much of the lucidity, erudition and perceptiveness we have come to expect from Said, the book has some serious shortcomings. The readings contradict the politics. In fact, Said's political vision seems to have shifted and, some would say, retrogressed. 'The turn from a wholesale denunciation of the West, so uncompromising in *Orientalism*,' Ahmad (1992:211) notes in his review of Said's recent essays, 'to an equally sweeping desire for location in the West, which these latest essays assert, is now complete'. A charitable view might be that this is in keeping with Said's analytical style: the juxtaposition of fierce condemnation and high praise. More probably, it reflects the lack of a clear theoretical focus, rooted in his post-modernist rejection of grand analytical systems and narratives. It leaves him exposed, ultimately unable to transcend the essentialist critique of *Orientalism*, and so he returns to singing praises for the very canon he once denounced so passionately. The emperor still has his clothes on. But the attendant appears naked.

Notes

1. It is indicative of their complicity in the Eurocentric historiographical constructions of both Europe and Africa, the latter being reduced to 'Black' or 'sub-Saharan Africa' that so few Africanists have responded to Bernal, with the notable exception of African-American scholars, as Margaret Johnson-Odim (1993) poignantly observes. Indeed, Bernal has been appropriated, popularised and in many cases over-simplified by Afrocentrists.

2. The review is a typical hatchet job by someone clearly unwilling or unable to confront the book's argument. So he invents a straw argument and concentrates on a small section of the book dealing with Algeria which he proceeds to demolish. Cheekily, he advises those who seem 'to value this free, individualised choice of identity' in the modern world to show 'at least some expression of gratitude to the process which has made such a free choice so much easier', namely, imperialism. I suppose the one million Algerians who died fighting for the independence of their country from obdurate French imperialism, the very French who gave the 'Western' world the slogan of 'liberty, fraternity and equality' should be grateful! For Said's response and other correspondence on the debate, see, *The Times Literary Supplement*, March 19 1993:15 and April 2 1993:17.

3. He continues that the 'composer's essentially liberationist purpose is grotesquely subsumed in an all-embracing myth of cultural imperialism'. MacKenzie is, however, correct in accusing Said of carelessly reading the works of scholars in disciplines such as Geography and History and glibly using terms like 'native' which in the colonial context became offensive and developed racist connotations. I found, for example, Said's reading of Ronald Robinson's (1972), with which he begins section (v) of Chapter 3 simply wrong. Name-dropping and superficial reading erudition does not always make.

4. It is a testament to the enduring power of Conrad's archetypal heroes that Kurtz and Marlow have recently been brought to Toronto in Timothy Findley's, *Headhunter*. Toronto: HarperCollins, 1993.

5. Ngugi has published five novels, several plays, and several collections of essays since then.

6. In the original article he mentions Africa and Asia explicitly, according to the quote in Ahmad (1992: 216). Also, the word 'respectably' is not placed in quotation marks as in the statement in the book reproduced below.

7. The most recent estimates indicate that by the end of 1992 there were of 18.9 million refugees, world-wide, up from 17 million at the beginning of the year. The six most generous host countries in terms of ratio of refugee population to GNP per capita were, Malawi, Pakistan, Ethiopia, Iran, Kenya and Algeria, in that order. See Hella Pick (1993). Many professionals from the South migrate to the North, but many more migrate within the South itself, partly because of immigration restrictions in the Northern countries and the difficulties of getting jobs similar in status to those they left at home. For example, I know of more Malawian academics teaching at the University of Botswana than in Canada and Britain combined.

8. In another context, Mazrui (1995) talks of Africa's counterpenetration or counterconquest of other civilizations, including that of the Arabs, and Africa's impact on France and India. Typical Mazruina: interesting for its speculative boldness, but lacking any theoretical compass.

9. *Fields of Endless Day*, directed by Terence Macartney-Filgate, National Film Board of Canada, 1978.

10. The issue makes comparative readings of such writers as Daniel Defoe and Oluadah Equiano, Graham Greene and Bernard Dadie, Samuel Beckett and Wole Soyinka, W. B. Yeats and J. P. Clark, and Albert Camus and Aimé Césaire. Also see M. Mortimer's (1990: Chapter 1) 'contrapuntal reading' of Joseph Conrad and Camara Laye; and see K. Loeb (1992) who looks at several European and African writers in the last two centuries. Also examined are the African writings of the late Canadian author Margaret Lawrence.

Chapter Twenty Two

AFRICAN STUDIES AND THE DISINTEGRATION OF PARADIGMS

We live in an age of intellectual uncertainty, of paradigmatic disorder. In virtually all the social science disciplines, the master narratives of the past few decades are disintegrating, replaced by various post-isms, from post-modernism to post-coloniality. This turbulence is a product of new currents and ferment in both scholarship and society, of transformations in disciplinary epistemologies and global politics, especially owing to the rise of feminism, the evident crises of both socialism and capitalism in the contending blocs of the old Cold War, and the unravelling of the project of national liberation in the post-colonial world. While no national or regional narrative is immune from fragmentation, it is the decomposition of the dominant 'Western' metascripts that has received most attention, which is frequently credited to the rise of post-modernism in western academies, a creed now being exported to the rest of the world with the missionary zeal of past Eurocentric discourses. Often overlooked are the challenges and confrontations from African, Asian, and Latin American Studies which have played a vital role in the fragmentation, explosion, and deconstruction of the hegemonic western paradigms.

These deconstructions, often articulated in a bewildering array of post-structuralisms, are both desirable and dangerous. Desirable because when they seek to dismantle western discourses from their pedestal of universal claims to knowledge, they strip western modernity of the will to truth, and open up spaces for previously silenced and dissident voices. But all too often, deconstruction becomes an end in itself, an orgy of apolitical theorizing, a mindless celebration of pastiche and eclecticism, valorising *reading* and inter-textual conversations among professional academics who are disengaged from, or even contemptuous of, social movements. Emphasizing the construction of subjective identities and their endless negotiability, much post-structuralist theory hardly pays any attention to political and economic structures that surround, saturate, and signify not only its beloved subjectivities but its very production as theory. By dismissing political economy and debunking all narratives as terrorizing totalizations, the dominant structures and processes of power, within and between nations and among classes and genders, are vaporised. As my friend Dickson Eyoh puts it: 'radical politics is replaced by discursive radicalism'[1]. That is what makes post-structural posturings so politically dangerous[2].

In their different ways, the books by Bates, Mudimbe and O'Barr (1993), Cooper, et.al. (1993), and Appiah (1992a) seek to explore the contributions of research in Africa to the construction, consumption, and dissolution of western social science epistemology and knowledge. To the editors of *Africa and the Disciplines*, 'research in Africa has shaped the disciplines and thereby shaped our convictions as to what may be universally true' (Bates, Mudimbe and O'Barr 1993:xiv). This, they believe, constitutes the ultimate defence for the study of Africa in western universities. Each of the contributing authors to this volume attempts to establish the impact of African research to their specific discipline. Unfortunately, the analyses are uneven and not always compelling.

Predictably, pride of place is given to anthropology, which emerged as the intellectual handmaiden of imperialism and colonialism, to study the 'primitive other', among whom the peoples and societies of the 'Dark Continent' belonged. Moore (1993) traces the development of the discipline from its ignoble beginnings when anthropologists observed and documented the 'customs' of small 'closed', static, 'traditional', 'tribal' communities to the post-colonial, Marxist, structuralist, feminist, and post-modernist preoccupations with economy, symbolic systems, gender, and constructions of identities. Experience in Africa, she states, has played a pivotal role in effecting these changes.

The case for the African influence on economics is less obvious. Collier (1993) laments that Africa has been ignored by economists, although 'during the 1960s a flood of subsequently eminent economists worked in and on Africa'. Despite the limited research, the author enthuses, the African data has helped refine and advance economic models and theories, from the small open economies model and fixed-price theory in macroeconomics, to the microeconomic analysis of factor markets, product markets, and household economics, and it offers wonderful opportunities to study the problems of economic transitions, monetary unions, peasant risk behaviour, and micro-enterprises. In political science, Sklar (1993) avers, Africanists have offered distinctive approaches to political modernization, comparative pluralism, and rational choice theories.

As for African philosophy, Mudimbe and Appiah (1993) contend, it is a field not only connected to American and European philosophy, but it illuminates, and transcends, the divide between analytical and continental philosophy, for it fuses Anglophone and Francophone traditions. It is also interdisciplinary for it combines historical, anthropological, and philosophical analysis, and it raises fundamental questions of the relations between philosophy and culture. Similarly, African art history, according to Blier, exposes the 'fetishes' of western art discourse, including the models of artistic progress or development, and the privileging of things past and the voice of the artist in the determination of artistic meaning, because the nature of the African data has forced researchers to eschew simplistic evolutionary models and appreciate the complex interactions of past and contemporary productions of art and the social constructions of artistic meaning. Also evident from a study of African art are the limitations of post-structuralist methodologies with their 'deconstructionist proclivity to privilege colonial history and western perspective (bias) over all others... where self (the colonizer, the collector, the researcher, the writer) is again accorded the principal, privileged, and exclusive voice' (Blier 1993:157).

Finally, studies of African history and literature have called into question all the master narratives of historical development and processes, and literary production, evaluation, and canonicity. In what is perhaps the best essay in the book, Feierman (1993) demonstrates how the phenomenal growth of African history and the expansion of historical knowledge as a whole has led to the splintering of the discipline's epistemology, methodology, focus, scale, and language, thus making it impossible to write human history as a single clear narrative, as a story of civilization diffusing from the historical heartland of Europe to Africa and the other parts of the world. In another context Fernández-Armesto (1995:31-2) has captured the revisionism going on in historical studies with literary flourish:

> Historians, who formerly tried to crush the facts to fit Procrustean models and schemes, are beginning to enjoy the respectability of uncertainty... Most of the long-term trends and long-term causes conventionally identified by our traditional histories turn out, on close examination, to be composed of brittle links or strung together by conjecture between the gaps...Historians are getting out of the archives into the open air - walking in the woods, strolling in the streets, making inferences from landscapes and cityscapes. The avant-garde are incorporating oral research and personal experience into their work, to the dismay of those trapped in the lanes of a race for objective truth. The best effect of these changes is that there are now again history books that

are works of art as well as scholarship. Great history, like great literature in other genres, is written along the fault-line where experience meets imagination. When well written, it has all the virtues of egghead fiction, plus better plots. Right now, the past has a great future.

The study of African literature, Miller proclaims, is not only good for multiculturalism, the plural nationalism of contemporary western societies, but more importantly, it promotes intercultural literacy and theoretical advance, for it 'demands nothing less than a reconsideration of all the terms of literary analysis, starting with the word "literature" itself, and that such a reconsideration is the best thing that can happen to the field' (Miller 1993:217).

Many of these analyses and encounters of Africanist scholarship with the dominant paradigms of western scholarship are able and fascinating. There can be little doubt that studies of Africa have brought new contexts, methods, insights, and theories that have revolutionised many disciplines, so that the old Eurocentric approaches have lost their paradigmatic prestige and coherence. But the book is still trapped in a western epistemological framework of reference. Its primary aim is to provide a defence for the study of Africa, not on its own terms, but to promote the marketability of Africanists, by demonstrating that their knowledge is relevant and that it has already been successfully incorporated in the traditional academic disciplines. In the words of the editors:

> We therefore abstain from claims for equality of access. For our major point is that, to a degree unacknowledged by either side in these debates, the study of Africa is already lodged in the core of the modern university... Arguments are not privileged by their origins, geographic or cultural; arguments become knowledge when they have been refined by logic and method, and these defences presently fall in the province of the academic disciplines (Bates, Mudimbe, and O'Barr 1993:xii).

This tired appeal to universal logic will not do. Left to logic alone anthropologists would probably still be writing about their beloved 'tribal natives'. Lest we forget, nationalism and decolonization in Africa and the civil rights struggles in America did far more than arcane academic disputes to bring African studies to the segregated corridors of North American universities. As Africanists they should know this: the production of knowledge about Africa has been structured by the social and spatial inscriptions of class, race, nationality, ethnicity, gender, and location, and the contemporary processes of knowledge production and its circuits of circulation and consumption. It is indicative of the perverse conceit of some Africanists that many of the contributors to this book can discuss the developments of their disciplines in Africa without seriously acknowledging, let alone engaging, the work and critiques of African scholars. Moore shoddily dismisses the searing indictment of anthropology made in the 1960s and 1970s as 'drearily conventionalized vituperation' (1993:9). African anthropologists are only mentioned in the concluding remarks, prefaced by the rhetorical question: 'And what will be the involvement of African scholars?' (1993:33). Strange that she seems unaware of such renowned African anthropologists as Mafeje and Magubane whose studies are not even cited in the references.

There is a tendency to assume that Africa's contribution to the disciplines merely lies in its peculiarities and capacity to provide validation to theories already developed elsewhere. Collier and Sklar are quite explicit on this. 'Africa is a gold mine to economists', Collier (1993:58) declares, 'because its economic history has been so extreme: booms, busts, famines, migrations', not to mention its diversity which makes it 'ideally suited to the comparative approach which is the economist's best substitute for the controlled experiment. Until recently this potential has not been realised'. Collier has little to say about political economy, or traditions and writers critical to his neo-classical gospel, whose work has been influential in the analysis of African economic development and change. And predictably, hardly any African economist is mentioned. Sklar (1993:85), in what is possibly the worst chapter in the book, asserts that Africanist scholars could provide more meaningful contributions to political science by examining problems that are specifically or generically African, such as 'parasitic statism, militarism, dictatorship, public corruption, the insufficient accountability of public officials, ineffective political socialization, and differential incorporation of ethnic groups resulting in conflict, among many others'. Surely, these problems are not confined to Africa, are they? If the pathologization of African economic and political behaviour and processes is all that Africanists can offer to economics and political science then they ought to close shop.

Sklar also underscores one of the enduring problems of Africanist scholarship, that of language. He asks:

is linguistic expertise a necessary condition of genuine Afrocentric achievement? I think not; very few political scientists in African studies, apart from those who speak an African language as their mother tongue, have linguistic skills that would be adequate for the purpose of unassisted research in an African language. In this regard, the field of Africa studies is unlike either Asian or Middle eastern studies. In African political studies, research-grade proficiency in an African language is rarely anticipated or attained by individuals who are not native speakers of the language concerned (Sklar 1993:100).

The problem: there are too many African languages! This is like a scholar of Japan or Germany saying he can't master Japanese or German because there are too many Asian or European languages. An American scholar studying Germany without knowing German would not be taken too seriously by experts in the field. The privileging of European languages in African studies is a reflection of relations of dominance of Africa by the West and enhances the capacity of western scholars for intellectual accumulation, appropriation, and domination in African studies. This raises fundamental questions about the authenticity, accuracy, and acceptability of Africanist constructions and representations of Africa.

The contributions in *Confronting Historical Paradigms* (Cooper et.al. 1993) traverse much the same ground. They discuss the rethinking of historical interpretation and paradigm among Africanists and Latin Americanists in the 1980s and early 1990s. Half of the essays are new, while the other half consists of reprints, including Cooper's (1981) and Isaacman's (1990) well-known and comprehensive bibliographic surveys first published in *African Studies Review*.

In Chapter 1 Stern (1993a) echoes many of the points made by Feierman concerning the apparent splintering of historical knowledge due to the confrontations with Third World regional histories, the expansion of the epistemological boundaries of history with the rise of the feminist and social history movements, all of which placed the discipline's old objectivist, androcentric, and Eurocentric paradigms on the defensive. But this picture of extreme fragmentation, Stern correctly argues, has been overstated. Alongside the apparent chaos, trends towards narrowness and specialization

> there also developed a process of "reverberation" - conversations within and across specialized fields and across disciplines; imperfect trackings of historiographical shift and debate; echoes of the tussles with paradigm, method, theory, and grand interpretation in other camps. Much of the "new" scholarship of the 1960s and 1970s self-consciously wrestled with "traditional" paradigms of historical research and interpretation..... Reverberation, an important process of intellectual network and conversation, debate and echo, travel and refraction, does not quite fit the bipolar scheme of unified community versus fragmented tribalism. It mediated, however, imperfectly, the specialized balkanization that is the bane of virtually all fields of contemporary knowledge. It allowed for convergences within a context of difference (Stern 1993a:9)

Historians of Africa and Latin America, he maintains, were not only in the forefront of wrestling with the stereotypes, silences, and misinterpretations of 'traditional' western historiography, they also conversed with each other, which led to fruitful convergences of theme and interpretation, especially on the questions of the capitalist world system, labour processes, and peasantries.

Two of the papers test the utility of the world-system approaches as overarching explanations for historical trends and developments in Latin America and Africa. Stern's (1993b) other paper specifically seeks to probe the promise and limits of Wallerstein's theories on labour relations in Latin America, while Cooper (1993a) offers a broad overview of the paradigms that have dominated the study of African economic history, from modernization, to dependency, and modes of production. Before subjecting Wallerstein's world-system paradigm to critical evaluation, Stern outlines the development and consumption of dependency perspectives in Latin American studies, which antedated Wallerstein's formulation, hence the coolness with which the latter's propositions were received in the region, while earning the author praise and followers in western academies and elsewhere. Using the cases of silver and sugar, he demonstrates that Wallerstein's tripartite division of international labour - free labour in the core, sharecropping in the semi-periphery, and forced labour in the periphery is very misleading on both descriptive and explanatory grounds. A more satisfactory model would have to integrate what he calls the three great motors: 'the European world-system, popular strategies of resistance and survival within the periphery, and the mercantile and elite interests joined to American "centers of gravity"' (Stern 1993b:55).

Cooper's essay, first published in 1981, still makes interesting reading. It painstakingly traces, periodises, and critiques, on empirical, theoretical, and ideological grounds, the three major paradigms in African economic history, beginning with, in the euphoric 1960s, neo-classical economics and its modernization prescriptions, against which followers of the dependence, underdevelopment and world system approaches railed with structuralist and moral outrage in the 'radical' 1970s, before they themselves were found wanting at the turn of the sobering 1980s by Marxists with their articulating modes of production. These critiques, and Cooper's refreshing reconstruction of periodizations, trends, and processes in African economic history from the pre-colonial, to the colonial and post-colonial eras, confirms that none of these paradigms can, on its own, provide an adequate understanding of the evolution of Africa's economies and their places in the world economy. What is needed is creative dialogue between them, and devising comparative approaches that are sensitive to the particularities of time, culture, perception, and struggles. In his postscript, written specifically for this collection, he adds the significance of incorporating the analysis of gender and power. And he cautions against the posturings of 'post-something scholarship', from post-structuralism, post-modernism to post-Marxism (Cooper 1993b:193), a subject on which Appiah (1992a), examined below, has a lot to say.

Despite their differences, the papers by Stern and Cooper are both critical of grand explanations, and offer richly textured histories of the complex and changing interconnections between structure and agency, world system and local process. Only theoretical models and paradigms that integrate these levels of analysis, that combine specific and broader narratives, can describe and explain the historical construction and impact of the capitalist world system, and the development of labour systems and peasantries.

Isaacman (1993) and Roseberry (1993) examine the historiographies of African and Latin American peasantries, respectively. It is incredible to recall, as Isaacman shows, how invisible peasants were in African studies until quite recently, their existence ignored in all the three paradigms discussed by Cooper. Indeed, many Africanists openly doubted the analytical value of the concept of peasant. Only at the turn of the 1980s were African peasants finally admitted to the hallowed theoretical halls of academia, thanks to the deepening agrarian crisis, the collapse of the nationalist project, and the apparent intensification or resurgence of rural struggles. Scholars then began listening to peasant voices, reconstructing their histories and daily lives, and deciphering their organizations of work and struggle against the external forces of state power and exploitative markets, and in response to their own internal divisions and hierarchies of gender, class, and ethnicity. The peasant studies that Isaacman chronicles have assisted in rupturing the narratives of both nationalist and Eurocentric historiographies.

Latin American peasants lost their invisibility much earlier than their African counterparts, partly because of intense debates after the Second World War about agrarian reform and rural unrest among policy makers, intellectuals, and activists. Roseberry argues that the agrarian question was posed simplistically and inappropriately, for the analysis was not conducted on the peasants' own terms, but often through the borrowed terms of Chayanov and Lenin and other writers on the agrarian question in Europe, so that the Latin American peasants often disappeared 'into the structural categories of comfortable, middle, and poor peasants'

(Roseberry 1993:334). The generalised appropriation of European contexts, ideas, and models gave rise to analyses and typologies that were historically and sociologically empty. Only in the 1970s and 1980s, Roseberry tells us, did systematic, detailed, sophisticated, and regionally specific studies of rural life begin to appear. This reflected the expansion of historical research, intellectual fatigue with the transhistorical paradigms of the earlier literature, and the influence of work from other regions, such as James Scott's (1976, 1985) work on Asia which provided a model for understanding peasant consciousness and political activity.

The Latin Americanists working on peasants, argues Mallon (1993) in the last chapter, have much to learn from their Africanist counterparts on how to incorporate gender and ethnicity in their studies, while the Africanists can learn something about reconnecting the specific realities of peasants to the broader cultural, social, and economic contexts in which rural cultivators produce and reproduce their lives. This is only one of the potential reverberations between African and Latin American studies. There have already been many reverberations, including the expositions and critiques of modernization, dependency, and modes of production theories. There are many embodiments of these intellectual ties, the most renowned being perhaps Walter Rodney ([1972] 1982), a Guyanese, who wrote one of the most influential books in African historiography, *How Europe Underdeveloped Africa*. Mallon proposes four areas 'where future dialogue may prove particularly rich: culture and politics, the meaning of nationalism in "Third World" areas, ethnicity and the construction of power relations, and gender and generation' (Mallon 1993:386). These areas, she suggests, will yield more meaningful research than the decadent and depoliticised preoccupations of post-modernism and the other post-isms.

The challenge for Africanists and Latin Americanists, Mallon concludes, is to write complex, meaningful histories without recreating

> the frozen dualisms of the past: before capitalism and after capitalism; before rationality and after rationality; before class struggle and after class struggle. To maintain some sense of narrative line without becoming linear, to maintain a sense of diachronic process and transition without becoming dualistic, to rebuild a sense of explanation and causation without silencing important stories... to make our scripts more flexible and dynamic by rewriting the plots and main cast of characters to include, alongside class, questions of culture, colonialism, politics, ethnicity, and gender and generation. (Mallon 1993:395)

This agenda is very demanding, but essential.

The collection itself demonstrates the considerable advances that have already been made in the studies of Africa and Latin America, which have contributed to the fragmentation of the Eurocentric approaches, while at the same time encouraging new lines of intellectual dialogue, from which broader and more inclusive theories, paradigms, and narratives can be constructed. But the analyses do not always match the declared intent. None of the authors, for example, pays more than lip service to gender analysis. And as in the previous book, this is essentially a conversation *among* Northern academics, recounting *their* generational encounter with Africa and Latin America. To paraphrase Roseberry's critique that the agrarian question in Latin American was posed about, not by, the peasantry, in this book questions are posed about, not by, Africans and Latin Americans. In politics, this might qualify as an aristocratic dispensation of power, not a democratic one. If these texts are any guide, they suggest that Africanist and Latin Americanist studies remain secure in their aristocratic discursive laagers, oblivious to the clamouring of the 'others' they study for equal and meaningful participation in the discourse.

Appiah (1992a) provides an African's voice. His book, *In My Father's House*, deals with many of the same issues examined in the last two books, but in a more satisfying manner. The book combines the engaging readability of a personal memoir and the detached analysis of intellectual trends, at once a reflective essay and an academic treatise, provocative and persuasive, written in prose that is both evocative and precise. But it has conceptual and political flaws. The first two chapters begin with an interesting interrogation of race, noting the different conceptions among Africans and African-Americans, and between the proponents of what he calls intrinsic and extrinsic racism in the world at large. Modern genetics, he points out, has conclusively demonstrated that there are no underlying racial essences among human beings, so that racial discourses and theories are socially-constructed, spurious ideologies. That may be so, but repudiating race in theory does not make it disappear in politics. Race matters, to borrow Cornell West's (1993) phrase, whether in the Americas or globally, despite all the demonstrations that the notion of 'race' has no scientific basis, as much as ethnicity and nation matter despite they, too, being 'artificial' constructions, because it functions as a marker and an anchor to establish and reproduce identity, status, and position in the social or global hierarchy. Races exist because racism does. Race is not merely a social construct that can be neutralized by deconstruction in the academy, but an experienced reality that can be disarmed in political struggles. This is where Appiah's wholly agreeable distaste for racial groupings and consciousness flounders: the politics that constructs race and can deconstruct it is missing.

Appiah's analytical agenda is to demolish essentialist positions not only on race but on Africa. I share many of his concerns and agree with much of what he has to say. He argues, quite correctly in my view, that Africa is not a primordial fixture, but an invented reality, and Africans are not moulded from the same clay of racial and cultural homogeneity. To quote him: 'Whatever Africans share, we do not have a common traditional culture, common languages, a common religious or conceptual vocabulary... we do not even belong to a common race' (Appiah 1992a:26)[3]. Appiah celebrates the diversity, complexity, richness, and contingency of African social and cultural life. This constitutes an essential part of his argument, developed in the next two chapters, against the nativist rhetoric of unproblematised, transhistorical solidarities, which in the African context is articulated most loudly in the essentialist and celebratory fantasies of Afrocentricism. This narrative of African differences, multiple realities, experiences, possibilities, and agencies can be read as a confrontation with the totalizing narratives of both African nationalism and European imperialism, with their dualistic and polarised images of marginality and dominance, subjugation and autonomy. In this context he has much to say about literature and language as markers of nations, ethnicities, cultures, and identities. To go beyond nativist hand waving against western literary discourse, which is often trapped in the epistemological and ideological matrix of that very discourse, it is necessary, he insists, to historicise the analytical terms of 'literature', 'nation', and 'culture' through which the contestation of Eurocentric evaluations and affirmation of Afrocentric authenticity is conducted. As he puts it:

> Inasmuch as the most ardent of Africa's cultural nationalists participates in naturalizing - universalizing - the value-laden categories of "literature" and "culture", the triumph of universalism has, in the face of a silent nolo contendere, already taken place. The Western emperor has ordered the natives to exchange their robes for trousers: their act of defiance is to insist on tailoring them from homespun material. Given their arguments, plainly, the cultural nationalists do not go far enough: they are blind to the fact that their nativist demands inhabit a Western architecture (Appiah 1992a:60).

A compelling critique of the futility of cultural essentialism in an interconnected, imperialized world. But Appiah himself protests too much. His critique of essentialism slips into a utilitarian, promiscuous conception of African identity: 'the African identity is, for its bearers, only one among many' (p. 177), that it is 'a usable identity' (p. 180). This 'epistemological fantasy to becoming multiplicity', to use Bordo's (1990) powerful indictment of the celebrations of identity as a usable, negotiable commodity, ignores the fact that while Africans, indeed, like all other peoples are breathing embodiments of many identities, the identity of being African cannot be worn and discarded with the ease of putting on and taking off clothes. The narrative of identities as social constructions, as inventions, which for historians is self-evident, sometimes ignores the simple fact that being 'inventions' does not mean they lack reality, that they are concretely constituted and historically grounded relational states of being. Perhaps Appiah, with his biracial background, can choose when and whether to use or not to use an African identity, which house of his parents to live in, Ghana or Britain, Africa or Europe. It is a 'privilege' most Africans do not have. Indeed, 'it may not be a coincidence', writes Azoulay (1996:135) 'that many of the voices calling for the deconstruction of essentialist conceptions behind the use of race as a marker for community are situated at the intersection of class and racial boundaries where moving beyond and crossing over are indeed palatable options'.

Appiah's forte is, of course, philosophy. African philosophical traditions, he maintains in Chapter 5, are no more homogeneous than those of any other continent. There is no central body of ideas that Africans have shared across time and space. He agrees with Paulin Hountodji's geographic conception of African philosophy that it is African not 'because it is *about* African concepts or problems, but because ... it is that part of the universal discourse of philosophy that is carried on by Africans' (p. 106), although he insists, following Kwasi Wiredu, that ethnophilosophy must be taken seriously because 'if philosophers are to contribute - at the conceptual level - to the solution of Africa's real problems, then they need to begin with a deep understanding of the traditional worlds the vast majority of their fellow citizens inhabit'. This, however, does not mean African philosophy should be reduced to the pristine wisdom of ethnophilosophy or folk philosophy, which after all, exists in every culture. Thus, the temptation to distil an African conceptual essence, or to impose and impute the philosophical doctrines of some European thinker is unrewarding. Giving the example of trying to find evidence of Cartesian dualism among the Akan, Appiah scoffs:

> I do not myself believe that any of Ghana's Akan peoples were dualist.
> But I do not think that it makes sense to say they are monists either: like
> most Westerners - all Westerners, in fact, without philosophical training -

most simply do not have a view about the issue at all. For, as I have argued already, the examination and systematization of concepts may require us to face questions that, prior to reflection, simply have not been addressed. What the Fanti have is a concept — Okra — ripe for philosophical work. What is needed is someone who does for this concept the sort of work that Descartes did for the concept of the mind, and, in doing this, like Descartes, this Fanti philosopher will be covering new territory (1992a:100).

African philosophers have their work cut out for them. While they may derive conceptual and methodological inspiration from ethnophilosophy and western philosophy, to be meaningful their texts and discourses have to critically engage Africa's contemporary conditions and challenges, an intellectual task that neither ethnophilosophy nor western philosophy is equipped to undertake.

Unfortunately, Appiah's analysis occasionally loses its clarity and relapses into the old, discredited paradigms. The tensions and ambivalences are most evident in Chapter 6. He is only too aware that the binary oppositions of tradition and modernity to describe contemporary society are false, or at least simplistic, as all such dualisms are, but his analytical project, the explication of differences in the cognitive orders of Africa and the West leads him to employ a schema incorporating these very dichotomies. And so he compares the different roles of religion and the modes of thought in the traditional oral cultures of Africa, although the examples are largely drawn from his Asante homeland, and the industrialized literate societies of the West, an evocation of a false dichotomy to begin with, as argued in some of the chapters in the preceding section above. A considerable part of the chapter is in fact taken up by a rather esoteric assessment of the analogies between traditional religion and modern science as explanatory systems and social organizations of inquiry. Why not compare 'traditional' Africa with 'traditional' Europe, or 'modern' Africa with 'modern' Europe? Indeed, why make the comparison at all? In an interesting anecdote about the melange of ceremonies performed at his sister's wedding the point is made effectively that many Africans do indeed lead complex, fluid, and syncretic lives that defy the simple polarities of tradition and modernity.

In the next chapter on post-modernism and post-coloniality Appiah regains his balance. He argues, quite persuasively, that post-modernism's fetishization of fragmentation, subjectivities, and multiplication of distinctions 'flows from the need to clear oneself a space; the need that drives the underlying dynamic of cultural modernity' (Appiah 1992a:145). He elaborates:

> [This], surely, has to do with the sense in which art is increasingly commodified. To sell oneself and one's products as art in the marketplace, it is important above all, to clear a space in which one is distinguished from other producers and products - and one does this by the construction and marking of differences. It is this that accounts for a certain intensification of the long-standing individualism of post-Renaissance art production: in an age of mechanical reproduction, aesthetic individualism - the characterization of the artwork as belonging to the oeuvre of an individual - and the absorption of the artist's life into the conception of the work can be seen precisely as modes of identifying objects for the market (1992a:143).

Oppositionality to the past and the trivialization of current struggles makes post-modernist discourse non-threatening, thus ensuring its marketability. I would add, following Aijaz Ahmad (1992:122-5), that the reactionary anti-humanisms of post-modernism and the other post-isms of contemporary avant-gardist thought in the West reflect a specific political and ideological conjuncture, the global offensive of the Right, and the global retreat of the Left. Thus post-modernism, championed as the end of metanarratives, has itself become a metanarrative. Like the modernisms it denounces, its ambitions are universalistic, as it seeks to force narratives from different societies with alternative times, histories, and causalities into its linear, Eurocentric, and depoliticised tunnel. Some of these societies being catapulted into post-modernity, we need to remember, were until recently seen as pre-modern backwaters!

Post-modernism, of course, claims to thrive on concessions to difference. So for the so-called Third World post-coloniality has become the operative discourse. Emerging in literary theory in the 1980s, post-coloniality carried no memories of the 'radical' political debates of the 1970s about post-colonial states and conditions, for it was created to enable post-modernism's discursive appropriation of the literatures outside Europe and North America. It offered the intellectual gatekeepers of western hegemony a respite from the terrorising terms of 'neo-colonialism' and the 'Third World' (Shohat 1992). It collapsed and homogenized diverse histories and structures, embracing in its generous transhistorical bosom Asian and African countries, the United States and Canada in North America, and Australia and New Zealand in the Pacific. Time and space were dissolved behind mindless contingencies that miraculously left no spatial and historical traces (Ahmad 1995). Thus, while post-colonial theory, like post-modernism, ostensibly 'sought to challenge the grand march of western historicism with its entourage of binaries (self-other, metropolis-colony, center-periphery, etc.), the *term* 'post-colonialism' nonetheless re-orients the globe once more around a single, binary opposition: colonial/post-colonial' (McClintok 1992:85). Post-coloniality recentres global history around Europe, privileging colonialism as the crucible through which the societies and cultures of Africa, Asia, and Latin America were constituted, thereby obfuscating their diverse histories and trajectories, including their very varied experiences with imperialism and colonialism itself. The discourses of post-modernism and post-coloniality arrogantly affirm, as Carole Boyce Davies (1994:2-3) puts it, 'that the West has already invented everything, done everything, experienced everything or thought everything worthwhile, and therefore all we need do is sound the last posts as the world collapses into itself'. In a poignant phrase, she dismisses these totalizing and homogenizing 'posts' as 'nothing but phallic erections without the necessary power to sustain them, and so called for the ways they block new and productive work' (1994:3). Organizing post-colonialism around the binary axis of time, rather than that of power, telescopes into invisibility the continuing ravages of imperialism in the world today, and marginalises the crucial narratives of gender and class in the continued asymmetric configurations of power globally, regionally, and within nations.

Appiah echoes many of these critiques. He argues, echoing Blier (1993), that locating artistic and cultural productions in post-colonial Africa (here the term post-colonial is used in the restricted temporal sense of the period immediately following the end of colonial rule) in the depoliticised matrix of post-coloniality and post-modernism is misleading. Taking the example of African literature, he demonstrates that the novels written after independence

> are novels of delegitimation: rejecting the Western imperium, it is true, but also rejecting the nationalist project of the postcolonial national bourgeoisie. And, so it seems to me, the basis for that project of delegitimation is very much not the postmodernist one: rather, it is grounded in an appeal to an ethical universal; indeed it is based, as intellectual responses to oppression in Africa are largely based, in an appeal to a certain simple respect for human suffering, a fundamental revolt against the endless misery of the last thirty years (Appiah 1992a:152).

I would add the miseries of slavery and colonialism as well. In short, while these writings challenge earlier legitimating narratives, it challenges them 'in the name of humanism... and on that ground it is not an ally for Western postmodernism but an agonist, from which I believe postmodernism may have something to learn' (1992a:155).

The book ends where it began with a reconsideration of African identities in the contemporary world. Appiah argues passionately for the construction of a new panAfricanism, one based on shared popular struggles against imperialist exploitation and visions for a more humane world, a panAfricanism that transcends the superficial entreaties of state relations. 'African unity, African identity', he writes, 'need securer foundations than race' (Appiah 1992a:176), and he believes that 'an African identity is coming into being...that this identity is a new thing; that it is the product of a history' (1992a:174), that in forging it we must not overlook the continent's multifarious communities and cultures, or subsume what Carole Boyce Davies (1994) would call the 'uprising discourses' of gender, ethnicity, and class.

But Appiah overstates his case: so consumed is he in confronting the cancer of racism that he sees racism everywhere, even at the heart of panAfricanism, conveniently forgetting that panAfricanism rose as a counter-discourse to European imperialism and racism, as an emancipatory project for Africans and diaspora Africans, whose realization would entail the extension, not contraction, of the frontiers of human rights and freedoms. Moreover, as Taiwo (1995:44) correctly observes:

contrary to Appiah's contention, Pan-Africanists were much more sophisticated. Yes, race was always present. But it was always problematic and, on a few occasions, it divided Pan-Africanists. It is a testament to their sophistication that they were able to appropriate, when necessary, Indian political heritage, Soviet revolutionary rhetoric, and the like, to build coalitions and enlist non-African support for their cause. Nor should we forget that some of the same Pan-Africanists founded the Organization of African Unity that embraced all of Africa, including Arab Africa!

Thus, historically, panAfricanism has been a house of many mansions. Our challenge, then, is not to erect a new panAfrican house, but to repair, fortify, and improve the old one that has withstood the harsh inclements of European imperialism and racism in order to meet the unrelenting blizzard of contemporary western imperialism and racism, on the one hand, and the reactionary nationalisms and despotisms of our own contemporary African worlds. PanAfricanism, as a political project and a discourse, needs to be re-imagined for the twenty-first century. It is to that question, then, that we now address in conclusion to the book.

Notes

1. Personal communication, May 25, 1996.

2. The literature on post-structuralism and its bevy of 'posts' offspring, such as post-modernism and post-coloniality, is sprouting with the abandon of poisonous wild mushrooms. The classic text on post-modernism is Lyotard (1984). For post-modernisms that seek to retain some fidelity to political economy, see (Harvey (1990) and Jameson (1991); for vigorous criticisms, see (Callinicos (1989) and Ahmad (1992); and for applications and critiques of post-modernism in the Africanist context, see Vaughan (1994), Cooper (1994), Parpart (1995), and Marchand and Parpart (1995).

3. Appiah's arguments have both been praised and attacked. For example, he is applauded by King (1993) for his 'humanistic Africanism', and condemned by Nzegwu (1996) for advancing an imperializing, patriarchal Euro-nativist position to underwrite a neocolonial Africanist career; by Taiwo (1995) for over-simplifying the history of Pan-Africanism; by Owolabi (1995) for his naive universalism and insensitivity to the inequities of Africa's integration into the world system; by Houessou (1995) for misrepresenting Afrocentricity; and by Serequeberham (1996), Azoulay (1996) and Oyegoke (1996) for the philosophical and political weaknesses of his anti-essentialist position. Also see the exchange between Appiah (1992b) and Odia Ofeimun (1992) in *West Africa*; and the review by Nicol (1993).

Chapter Twenty Three

IMAGINING PANAFRICANISM FOR THE TWENTY-FIRST CENTURY

I am going to read excerpts from two stories contained in my short-story collection, *The Joys of Exile* (1994). Initially, that is all I was going to do. But then I thought reading the stories alone did not justify a ticket from the cornfields of Urbana-Champaign to the bright lights of New York, especially on my first visit to this great city. Besides the academic in me could not resist the temptation to write a paper. It would look better on my CV. More seriously, of late I have been having discussions with some friends and colleagues, especially Lynette Jackson and Dickson Eyoh, about panAfricanism, what it means for our generation from the different political places and social spaces that we occupy as people of African descent. So I decided to read the two stories from these conversations, to frame my reflections on panAfricanism around them, to present them as panAfrican narratives.

But this is more than a conceptual convenience, a self-indulgent, selective sampling of meaning from these stories. If you can believe me - a demanding request in this age of post-something sensibility when grand narratives and narrators are supposedly dead and texts will themselves into being - these concerns were central in the very conception of the two stories. The first story I am going to read, *Waiting*, was inspired, indeed, co-authored by my daughter, Natasha Thandile when, as she would say today, she was just a child rather than the infinitely wise and mature 12 year old teenager going on 18 that she is now. I say co-authored because when we step back from the aesthetic conceits of bourgeois individualism, and the fiction of the artist as a lonely genius or a raving lunatic, it is clear that creativity, artistic production, like all intellectual work, in fact, involves complex collaborations and conversations; it represents mediations of shared images, assumptions, and histories, those of the artist and those around her or him, especially those with whom s/he shares the intimacies of co-habitation, friendship, love, passion, and a sense of belonging. Writing, in this sense, is essentially social biography, a revisioning of collective memories.

Ever the prescient child that she was, Thasha, as we call her, would always interrupt our exclusionary adult conversations concerning events that occurred before she was born, inserting her voice, her presence, saying that she was there. Being the doting but impatient adults we were, we would laugh. But our laughter did not impress her, let alone silence her. Her insistence began shaking my paternal, intellectual confidence, mocking my unimaginative, mechanistic views of reality and time as material and lineal; views formed from years of immersion in the illusory certainties of positivist social science, aided of course, by the pretensions and cynicism of advancing years. The more I thought that she might be right, that the membrane separating the material and spiritual worlds may be thin, that reality and time may indeed be nebulous, malleable and open to endless possibilities of becoming, that she may have actually witnessed events that occurred before her birth, the more I grew worried at what she may have seen of my mischief and other unmentionable things that I used to do before she was born.

Such trepidations aside, the story *Waiting* began taking shape. The specific conjuncture of my daughter's birth, as a child of an African father and an African Canadian mother, dictated that the story would be a panAfrican one, a remembering of Africa and the diaspora, a narrative of Africa's dispersal, of the primordial and historical ties that bind, ties of the shared memories of enslavement, colonization, imperialism and racism that are so deeply etched on our collective psyches amidst the differentiating inscriptions of location, class, gender, and the imagined communities of nationality, ethnicity, and religion. The story recounts and reclaims these memories through the eyes of a child desperately waiting for the appropriate auspicious circumstances to be born. Like the *abiku* child of West African fiction, this child connects, lives, and appropriates the past, the present, and the future of the panAfrican condition, weaving in the texture of its very being its formations and possibilities.

The story affirms what all the great panAfricanists have always known and preached, that recalling and reclaiming our histories is a prerequisite to any serious project of emancipation and liberation whether today or tomorrow in the 21st century. This is especially so in our era of wilful historical erasure and amnesia, of easy despair and the arrogant pessimisms of post-humanist scholarship[1], and the resurgent chauvenisms of nationality, ethnicity, race, religion, and masculinity, all fostered by a triumphalist and rapacious global capitalism, a mindless mass media, and reactionary cultural protectionism. It is a narrative plea that the grand narratives of Western modernity's other half - slavery, colonialism, imperialism, racism and fascism, class and gender oppression

- must not be swept away in the hurried march to the shimmering illusions of post-modernity with its loose, unanchored subjectivities. *Waiting*, then, unfolded as a tale of interrogation and affirmation, interrogating the child's possible pasts and futures, and affirming its humanity through its agency and will to survive, its right to be.

These claims may seem rather extravagant, an author's exalted reading of his own work. But it is not a crime to play critic with one's work. Reflecting on this story raises in my mind difficult questions about panAfricanism as a construct and a movement, in its spatial dimensions and cultural content, political and ideological agendas. What is the basis of a panAfrican identity? How important, in fact, is such an identity in the panorama of identities claimed and lived by continental and diaspora Africans - identities of nation, ethnicity, class, gender, sexuality and religion? Is its basis racial as Garvey maintained, or cultural as Du Bois insisted? This debate can, and has been, recast in essentialist and pluralist terms: there are those who see panAfricanism as a narrative of racial identity and solidarity, an evocation and exaltation of pristine African cultural traditions, authenticity and genius, and others who see it as an intercultural and transnational formation, modern, shifting and negotiable, a narrative of the exilic energies of hybridity, multiplicity, and difference masked in black skins. In contemporary panAfricanist discourse the first position is often identified with the Afrocentrists and those who search for African survivals in the 'New World', and the second with the social constructionists, for whom no pure African cultural essence exists either in the diaspora or Africa itself. The latter has been articulated most forcefully by Gilroy (1993) in his book *The Black Atlantic*, and finds strong echoes in the works of Appiah (1992) and Mudimbe (1988, 1994) in their critiques of African cultural essentialism and celebrations of the diversity, complexity, richness, and contingency of African social and cultural life, agencies and possibilities.

Academic discourse, modernist Social Science, thrives on such dichotomies, on binary oppositions. Counterposing essentialist and pluralist conceptions of identity is often an exercise in semantic gymnastics, reflecting different levels of abstraction. Those who emphasize the multiplicity, complexity, and mobility of identity, in short, those who strive to produce more sophisticated accounts of identity, sometimes end up offering more sophisticated essentialisms, for they usually do not deny the existence of the named object or construct from which they derive the disputed or defining practices[2]. Those who essentialise the practices embodied in a particular identity tend to simplify and freeze history. In short, diversity should not preclude articulating a common goal for self-determination. Conversely, a shared agenda can be executed from the different angles of class, gender, location, and other social markers.

This, indeed, has been the history of panAfricanism. PanAfrican identities were forged in the crucibles of slavery in the Americas and colonialism in Africa, and shaped by the 'metalanguage' of race, through which the economic exploitation, political oppression and cultural humiliation of the African communities was articulated, organized, justified and reproduced (Lewis 1995). The braided histories of panAfricanism were woven from the warps and wefts of European racism and resistances against it. Thus panAfricanism emerged as a counter-discourse, both as a repudiation of Western barbarities, hypocrisies, and negative portrayals of Africans, and as an emancipatory project. In so far as the interaction between race, memory, and community formation varied across time, space and the social compositions of class and gender, the messages of panAfricanism took diverse organizational and ideological forms, from the rather scholarly and intermittent Duboisian Congresses and the populist Garveyite Conventions and Back to Africa Movement, to the tentative African student organizations in inter-war Europe and North America and the burgeoning political parties on the continent and in the Caribbean. Predictably, too, they found articulation in the equally diverse languages of nationalism, liberalism, socialism, Marxism, feminism and religion. But there was a unifying theme, an overriding liberatory mission in these discourses: to erase the inhumanity and inferiority stamped on Africans because of the physical constitution of their bodies. In this sense, then, the racial and cultural constructions of panAfricanist consciousness and identity were not contradictory and sequential, but simultaneous and mutually reinforcing. They constituted cumulative, collective social memories, an existential grammar, if you will, mediated by the accents of space and time, the idioms of class and gender, and the expressive specificities of struggle and negotiation[3].

The differentiated accents of space and time in the development of panAfricanism are easy to demonstrate. One example will suffice. In the post-war era struggles for emancipation on the continent and in the Caribbean primarily took the form of struggles for national independence in the colonial territories created by imperialism, while in the United States and Canada the civil rights movement gathered momentum. There were, of course, connections between these movements, intricate networks of political conversation, strategy and tactics, hopes and dreams which allowed for convergences within a context of difference. The reverberations among the African, Caribbean and North American political and social movements occurred in a rapidly changing global context, itself partly brought about by the struggles of these very movements. Their relative successes, in turn, changed the contexts and complexion of panAfricanism. Africans turned inward towards the politics of nation-building, the economics of development, and the aspirations of continental unity, while in North America many turned their eyes steadily on the prize of civil rights: political enfranchisement, economic opportunity, and racial equality. The panAfricanism of Du Bois and Garvey, Nkrumah and Azikiwe receded into memory. Almost.

This growing drift is explored in the second story I am going to read, *Homecoming*, about an African American artist visiting Africa for the first time, seeking the nurturing affirmation and inspiration of his roots. In his encounters and experiences he finds his American and panAfrican identities swinging dangerously along the slippery poles of collision and conflict, convergence and conversion. As his cultural journey proceeds, he gradually finds some solace, not in the suffocating grip of a contrived homogeneity, but in the warm embrace of transcultural friendship. While this story did not emerge out of the animating interventions of my daughter's rich imagination, it can be seen as a narrative of her possible future encounters, of what faces many of those who seek to quench diasporic alienations, real and imagined, with the supposedly pristine waters of Africa, only to discover that these waters are distilled from many sources, including the salty springs of western modernity. If there is an identifiable source for the story at all, then perhaps it lies in the African American exchange students that I used to meet and teach in Kenya. And perhaps it is also a translation of my wife's experiences. Both of us were foreigners in Kenya, but the way we were perceived and treated and the cultural negotiations we made differed, reflecting no doubt our different personalities and gender, as well as our different points of origin on the panAfrican spatial and social map. This story, I would say if pressed, is an argument against the nativist rhetoric of unproblematized, transhistorical solidarities, and a celebration of the transformative power of critical engagement and mutual respect across the numerous boundaries that cover the panAfrican world[4].

Where are we now in this complex and diverse world? The picture is very mixed. A couple of years ago we witnessed the closure of the long, sad history of European savagery in Africa with the demise of the apartheid regime, which grew out of, and embodied, all the crimes committed against African peoples in modern times - slavery, colonialism, and genocidal racism. But our joys have been tempered by the severe challenges that confront many African countries and the

African diaspora in North America. From the 1980s, the independence and civil rights projects in Africa and North America, respectively, began to unravel in the face of structural adjustment programmes and the shift to the right in global politics and economic management, which saw determined assaults against the developmentalist state in the South and the welfare state in the North, the rolling back of economic sovereignty in Africa and of affirmative action in North America. This new global conjuncture, the resurgence of a virulent imperialism, with its familiar entourage of racism, sexism, and class exploitation, which threatens the hardwon achievements of the independence and civil rights movements, demands panAfricanist responses and struggles. The intense struggles for democratization, for the 'second independence', that have rocked Africa in recent years, and the mounting restlessness of the diaspora African middle classes and the ungovernability of the inner cities in North America, represent some of the responses. Our challenge as panAfrican intellectuals and activists is to decipher and develop new reverberations among these movements.

As we reflect on contemporary panAfricanism and its future directions, we may fruitfully focus our critical attention on several arenas, as I call them, by which I mean organizations of discourse and action. The arenas proposed can serve both as paradigms of analysis and praxis. I identify five: the arenas of state power, social movements, intellectual production, popular culture, and interpersonal relations. These arenas are of course interpenetrated, but each marks a discursive space of signifying practices, structures of attitude and reference, and configurations of power relations. Since 1945 African nationalisms have succeeded in capturing state power in Africa and the Caribbean and in swelling the ranks of African American elected officials at all levels of government in the United States. At one level, therefore, panAfricanism has to confront the realities of state power, it has to be articulated in the context of inter-state relations, wear the face of foreign policy. In so far as foreign policy is often an extension of domestic policy, except for client states and dictatorships, the mobilization of domestic constituencies that serve the national, regional and global interests of African peoples become extremely crucial.

The agency and productivity of state power is currently being eroded thanks to the globalization and regionalization of the world economy. What should be the panAfrican response to these developments? Regional integration schemes are being pursued with renewed determination and sometimes desperation in parts of Africa and the Caribbean. Can these schemes be extended, should they be extended, beyond their predictable geographical confines into the more imaginative panAfrican triangular linkages of a South Atlantic region perhaps anchored on South Africa and Brazil, a reconfigured North Atlantic region incorporating Western Africa, and an Indian Ocean region connecting Northeastern Africa to the Arabian peninsula and beyond? Such futures require the dreams and pockets of the African and diaspora African bourgeoisies. As Garvey and many others recognized (West 1993), and more recently as Moshood Abiola, the incarcerated Nigerian tycoon and political leader, and the late US Commerce Secretary, Ron Brown, understood, panAfricanism needs an economic agenda.

The manufacturing and distribution of state power is conditioned by the constitution of civil society, the richness and dynamism of associational life, by the organizational capacities of social movements. In an increasingly globalized corporate world of powerful multi-national corporations, and state-controlled international agencies from the United Nations and the practically defunct Non-Aligned Movement, to the Organization of African Unity and the Organization of American States and their numerous sub-regional equivalents, how thick are the associational ties between the social movements of the civil societies of the panAfrican world, both the old ones, including the churches, trade unions, political parties, and the newer ones, such as NGOs, environmental and women's movements?

All these social movements are gendered in their construction, composition, and conceptions of the future. For example, in most African countries, as Aubrey (1995, 1996) has demonstrated, women represent the overwhelming majority of NGO members, and many African American women interact with Africa through NGOs. The role of NGOs as conduits for a more generous and gendered panAfricanism have not been adequately explored and exploited. The question of gender has, indeed, not received the attention it deserves in panAfrican discourses. Frances White (1990) has correctly observed that the political memories and futures invoked by some nationalists either ignore gender completely, or collapse it into the false complementarities of an invented African traditional egalitarianism. As with most nationalisms, panAfricanist narratives have been loudly masculine, in which men march to redeem 'Mother Africa' 'raped' by Europe[5].

These are all serious organizational and intellectual challenges. The intellectual arena has its own specific demands. Apart from the extremely uneven provision of infrastructural and institutional resources for intellectual production and reproduction within and between countries, and differentiated access to the available resources because of the discriminatory idioms and filters of class, gender, religion, ethnicity, and location, scholarly discourses are often enclosed in impenetrable and intolerant national, regional, and ethno-cultural geographies. The cultural wars of the American academy, for example, and the reactionary babble of political correctness are often exported to terrorize others in regions with their own intellectual battles and concerns. Development, for instance, has been the dominant discourse in African studies in the last three decades, while diaspora studies have concentrated on questions of racial and cultural identities. There is need for scholarly engagement between the two, to free diaspora studies from the dangers of cultural relativism by incorporating political economy, and to rescue development studies from the pitfalls of economic reductionism by including issues of race and culture, and for both to be gendered (Robinson 1995; Aubrey 1995, 1996; Zack-Williams 1995).

Our challenge as panAfricanists, I believe, is not only to push for the democratization of access to our respective institutions, but also for the internationalization, gendering and diversification of the gate-keeping functions of academia: the development and composition of courses and conferences, and designing transparent criterial systems for reviewing publications and jobs. Also, there is the question of balancing the demands of a narrow academic careerism and the social responsibilities of public intellectual intervention, of promoting collective insurgency against the injustices of our various worlds. In so far as the perception of Africa is the source of our marginalization, we need to constantly intervene in the contested discursive terrain on Africa, and vigorously challenge and transcend the terms of the dominant discourses, through which our oppression and exploitation have been articulated historically and continue to be reproduced today. This requires us to consciously cross the various boundaries of scholarly production and communication, to develop vibrant, critical, serious and sustainable collaborative transatlantic intellectual networks and conversations that consciously engage the burning issues and questions of our day. As part of this agenda, there is need for establishing African diaspora studies centres in African universities[6].

The arena of popular culture points to the immense possibilities of transatlantic communication. There are intense, intricate connections and cross-fertilizations of music and dance, sartorial fashions, hair styles, and youth body languages, of expressive conversations that revitalize and recreate in new voices and styles and performances the vernacular cultures of the panAfrican world, invoking old cultural continuities and imagining new ones. And there has been what Irele (1995) calls the 'cycle of reciprocities' between the literary movements of Africa and the diaspora, most significantly the Harlem Renaissance and the Negritude movement whose echoes found resonances in the Black Arts and the Black Aesthetics movements of the United States in the 1960s and 1970s (Warren 1990; Ogren 1994; Harper 1995). But the reverberations of popular culture are increasingly transmitted through the homogenizing and banalizing channels of multi-national media companies, in the process of which they are sometimes sterilized of their subversive potential, stripped of their oppositional power. And there are many other arts that have yet to make their imprint in the popular arena: the theatrical and cinematic arts, for example, television and radio, not to mention the much vaunted Internet.

PanAfricanism was born in the hideous voyages of the Middle Passage, and it has been sustained in each subsequent generation by the continuous redemptive travels of ideas, images, individuals, travels across the vast spaces linking Africa, Europe and the Americas, and in the local melting pots of port cities and industrial towns, plantations and mining centres, churches and colleges. This is to suggest that panAfricanism can also be located in the intimate intersections of interpersonal relations. In large cities like New York, London, Lagos, or Johannesburg you encounter African peoples in all their splendid diversities of origins, classes, genders, sexualities, religions, sensibilities, obsessions, limitations, possibilities, and yes, cultures. One, then, can travel within, locally, or even as with my daughter, in our own homes, to experience and celebrate, not only our survival through the hells of slavery, colonialism and apartheid, but our humanity in all its richness and mysteries. *Thank you. Now it is story time.*

Waiting[7]

This is my last chance, the very last chance. I have had two chances before, but I refused to take them. One is only given three chances. This is my third. If I do not take it this time I will never have another chance. I will never live. I will forever remain a possibility that never was.

My first chance came many years ago, in fact centuries ago, in the days when the land and the mountains were still green and the rivers and lakes sparkled with sweet water and fish. Most people lived off the land. Their lives were a lot simpler, and maybe a little happier. My parents were a young couple, full of vigour and hope. That is why I chose them. They had grown up together. They were actually distant cousins. I was going to be their first child. I was not sure whether I wanted to be a boy or a girl. Each sex presented its own advantages and disadvantages.

Besides farming, my father was a fisherman. He had grown up by the Zambezi river and had fished ever since he was a little boy. There were times when I wanted to be born a boy so that he could take me fishing. He was also a remarkable swimmer and a canoeist, indeed, the fastest in the village. He had many prizes and trophies to show for it. At the end of the harvesting season the village organized all sorts of competitive games and events. He won all the swimming and canoeing contests from the time he was fifteen until the age of twenty-two when he decided to get married.

My mother was the most beautiful woman in the village. Every young man wanted to marry her. But she chose my father, or rather, their two families, who could see that the young couple were in love, encouraged their relationship. I was happy for both of them and for me. My mother was renowned for more than her beauty, however. She was one of the finest potters and weavers in the village, a trade she had learned from her mother who, in turn, had learned it from her mother, and so on, going all the way back to the misty past. At times I wanted to be born a girl so that she could teach me the fine arts of pottery and weaving.

With such parents I expected to lead a secure and comfortable life. Why should I want anything less? That is why I had waited for so long to find them. I had seen what had happened to some of my more impatient friends, who ended up with parents who were either too poor or too cruel to look after them properly. And so after a few miserable years, sometimes even months, weeks, or days, they would return dead and heartbroken. I was determined to avoid such a fate. I wanted a long, satisfying life.

My spirit friends were envious when I told them of my parents. This only made me wish the customary nine-month waiting period was telescoped to a few weeks, or better still, days, for I was more than ready to assume my new life.

If we were really given a choice none of us would have wanted to be born, to go into the terrifying exile of human life, leaving behind the charmed life we led in the spirit world. We spent our time playing and feasting and we knew only love and none of the pain, loneliness, heartlessness, and cruelty that humans seem to love inflicting on one another. But we had to be born at least once in half a millennium if we wanted to continue enjoying the eternal joys of the spirit world. That was the pact each one of us made with the Master of Eternity when, in his and her infinite wisdom, for the Master was both male and female, or rather neither, summoned us from nothingness. It was a small price to pay: human life is short, made shorter by all that suffering, brutality, and unhappiness. Many of us in fact preferred early death. It was easier in the old days when infant deaths were so common and the span of human life was so much shorter than it is now.

All was going well until the fourth month of the waiting period. One evening my father went to catch some fish for supper, and also to relax, for he had spent most of the day plastering the inside of their newly built coral stone house. He never came back. Dusk turned into night without any sign of him. My mother grew worried, then alarmed. She went to the banks of the river and called his name until she was almost hoarse. But all she could hear were echoes of her own voice, mingled with the occasional noises of croaking frogs and chirping birds and the enchanting roar of the river. Neighbours came out of their houses and started talking among themselves, wondering what was going on.

Scared, she ran to her parent's house, which was on the anthill a few hundred yards away.

'What's the matter?' her mother asked.

'It's Abu', she said. 'He's gone'.

'Gone?' her father repeated. 'Gone where?'

She pointed in the direction of the river.

Her mother put both hands on her mouth, as if in prayer, and cried, 'Oh, no!'

'Wait here', her father said. He mobilized the men of the village and went to the river. There was a full moon, which helped considerably. But there was no sign of Abu anywhere, not even of his fishing gear. The strong swimmers dived into the cold water, just in case he had drowned. They were there until the cocks began crowing, but there was still no trace of Abu. The search was called off at the break of dawn. By noon news had spread that there were other men from the neighbouring villages who had also disappeared.

The whole village mourned, for everyone knew Abu would never return. He was not dead, but he was gone forever. In a sense, then, he had died. He had been taken to the coast to be packed on a ship for a journey to the end of the world across the vast, furious oceans.

My mother wept for weeks. How could they do that to her? How could her husband be captured, sold, and bought as if he were a chicken? How could human beings do that to one another?

My spirit friends asked me what I was going to do now. I was so confused. I loved both of my parents. I could not imagine growing up without my father. I knew my mother would love me and do the best she could to raise me properly. But I was afraid, afraid of disappearing like my father, and like so many other people since this curse had fallen upon the land. What else could it be called, if not a curse, this cheapening of human life, turning human beings into commodities exchangeable for worthless trinkets and guns? So it was that a few weeks after my father disappeared I left my mother. She was devastated. I felt sorry for her, but there was nothing I could do. I don't know whether she ever forgave me.

This business of selling and buying human beings went on for many generations. I decided to remain where I was until it was over, which took longer than I thought. Four centuries is a long time. In the meantime, I watched my friends going and coming. But some never returned. I suspect they, too, disappeared beyond the oceans, stolen people to build stolen lands.

It was a great relief when this sordid business came to an end. I did not care why or who finally stopped it. I was just glad that it was over. At last I could be born and taste all those agonizing mysteries of human life my spirit friends had always told me about.

The search for parents took longer than I would have liked, for I was really looking for people who were replicas of my first parents. It was only gradually that it dawned on me that this was a futile search, the world had changed, and I could not find any two people like them. The sordid business had left widespread devastation, including new diseases from the distant lands, diseases of both the body and the spirit. That is one reason why I chose my second parents.

My father was a *sing'anga* and my mother was the guardian of the shrine where the people of the region came to pray and make offerings to Chauta. Both had inherited their positions from their clans. Unlike almost all other families in the region, all ten of their children were alive. I needed that kind of assurance that I would survive my infancy and grow into a healthy man or woman until ripe old age before coming back to the spirit world.

I was not worried that both my parents were advancing in years. My father was half bald, while my mother's hair had gone almost completely grey. I knew they would be very old by the time I was initiated. That was another attraction. As their last child, they would spoil me. Last children, like single children, are always spoiled. I wanted to be spoiled. Moreover, I would have all those big brothers and sisters to spoil me, too. Three of my brothers and sisters were in fact already married and had their own families.

My spirit friends were happy for me, including those who had come to think of me as a coward, and those who had grown tired of feeling sorry for me, or of my questions about human life. Now I would have a chance to experience it for myself. I was more than prepared, or so I thought. Given my mother's advanced age I feared she would be reluctant to accept me. But she was more than thrilled, and so was my father. It was a sign, they told all their relatives and friends, that Chauta still had work for them to do. They agreed that they would call me Madalitso, the blessed one. I was blessed because I was coming towards the sunset of their lives, so I would keep them company and they would shower me with all their love. I was also blessed because I need not grow up worrying, like my elder brothers and sisters did when growing up, of being snatched by strangers and sold to the pink people at the coast and taken to the lands of no return. I liked this name, Madalitso, and so, too, did my spirit friends.

My mother carried me with ease. She liked to talk to me, telling me stories of her clan and how she became the guardian of Chauta's shrine, and that if I decided to come out as a girl, I would, as the last born child, take over from her. My father would gently admonish her, saying he was sure I would be a boy and that he would teach me about all the plants and how to cure all the diseases that afflicted people in the region, including those brought recently by the pink people.

I was in the sixth month of the waiting when tragedy struck. Rumours had been circulating for months that the pink people at the coast were moving inland, deposing chiefs, taking over land, and giving it to their followers. The chief and elders of the region gathered and decided to send a team to the coast to investigate. The investigative team came back and confirmed everything. The pink people had indeed taken over the land at the coast and were moving inland, assisted by recruits of local shiftless young men and short-sighted leaders keen on settling scores with their neighbours. Entire villages were burnt, livestock seized, and young women captured and turned into concubines for the marauding armies.

The chief and the elders made preparations to defend the region from the invaders. Weapons, including guns, were manufactured, repaired or bought. Underground pits were dug to hide foodstuffs, and temporary shelters were built in the surrounding mountains and forests, to which the elderly, women, and children were evacuated.

The invaders found empty villages and towns. They were so overconfident that they came in broad daylight. They put up a flag in the courtyard of the chief's compound after which they went about looting. But there was not much to loot. To their delight, in some of the houses they found drums filled with beer. They happily sat down to celebrate their victory. Many did not wake up from their drunken stupor, for the beer had been 'treated'.

When the news of the disaster reached the coast, a larger, angrier, and more ferocious army was mobilised. They arrived at night and torched the villages and towns. The next day they made for the surrounding forests and mountains where, a local informer had told them, people were hiding. The local soldiers were more than ready for them. The fighting lasted longer than the invaders thought it would. In fact, it went on for years in spurts of pitched battles, little skirmishes, and lulls of inactivity. Many of the region's inhabitants moved to more inaccessible places beyond the mountains.

But my mother was not one of them. As the guardian of Chauta's shrine, she could not leave. She did not want to leave. Leaving would mean abandoning Chauta, who could not be abandoned any more than society could be abandoned. Chauta was everything, the overseer of their past and their future, their ancestors and their unborn. Chauta was their Creator, the Supreme God. How could she then leave his house unattended?

The invaders found my parents inside the shrine. They were both on their knees praying. The pink soldiers started laughing. But the native soldiers, some of whom were from the other side of the river, looked scared.

'What are you afraid of?' one of them asked, roaring with laughter as he tried to push some of them in.

My parents continued praying without looking up.

'This is idolatry!' another added as he kicked my parents and started throwing the holy objects around.

'You pagans! Up on your feet!'

My parents did not respond. The pink soldiers began beating my father. Before long, they were joined by the native soldiers. When my father passed out, they grabbed my mother.

'I will start', the commander announced as he motioned the others to go outside. They waited in a queue.

'Let's see whether your god will come to help you', he snorted.

My mother fell to the floor and her body shook in violent spasms as her lips mumbled, almost of their own volition, some strange voices. The commander's laughter and grunts became louder. One after another they went in.

I was not born. I did not want to, nor could I be.....

Homecoming[8]

As the plane touched down, the passengers clapped and cheered. Some tried to peek through the misty windows to catch their first glimpse of African soil. A few could be seen combing their hair or putting on make-up. An air steward welcomed them to the airport and the city and thanked them for flying the air line.

'The local time is 5.30 a.m.' she said. 'Have a pleasant stay'.

The luggage compartments swung open and people scrambled to get their hand baggage. Everyone looked suddenly alert and alive, relieved that the eight hour ordeal was over and anxious to meet awaiting relatives, friends, tour guides, business associates, conference organisers, or government officials.

The brightest smiles were worn by young men and women returning home with hard-earned degrees and diplomas, ready to join the ranks of the elite. Also grinning broadly were those who had saved for years for a dream holiday in the African sun, for a taste of the exotic and the wild, a chance to tan and boast to envious neighbours and friends that they had escaped the tyranny of winter.

Probably nobody in the plane was as relieved, anxious, and excited as Richard. He had dreamed about this trip for as long as he could remember. At last he was coming home. Africa. So far away and yet so near, his mysterious, ancient homeland. Yes, he had come to see the real Africa, not the Africa of the evening television news and World Vision infomercials, that Africa of starving women and children, marauding armies and mindless violence, but an Africa which pulsated with enchanting rhythms, simplicity, and communal warmth. A land blessed with nature's wonders and cultures, uncorrupted by the superficiality and materialism of America.

'Welcome home', a steward said as he approached the exit door. Richard bowed and muttered '*Asante sana*'. He felt as if she recognized him.

'Are you an American?' she asked.

'An African', he said emphatically. 'Why?'

'Your accent. Welcome to Africa', she smiled more broadly.

He flinched for some unknown reason, then closed his eyes for a fleeting moment, wondering whether it was all really happening. Africa!

There were black faces everywhere. All the crew, the customs and immigration officials, and the sales people in the duty free shops were black. He had never seen anything like it before. The whites were no longer dominant as they were in the plane. As they walked towards the arrivals hall, their voices were lower, their eyes slightly more vigilant, their walk a little awkward. Richard suddenly felt normal, safe, and secure. The burden and ambiguity of his blackness wilted before Africa's embrace.

Waiting outside the arrivals hall were more people. Richard eagerly searched the crowd, although he was not looking for anybody in particular. Some faces had a hazy familiarity as if he had seen them somewhere before. It was their smiles, the gestures, the laughter.

He was soon mobbed by taxi drivers. He chose to go with an elderly looking man, figuring that he would be more reliable and informative. He was hungry for the African experience.

It felt quite chilly outside. As they drove away he was astonished by how big and busy the airport was. The road was lined with jacaranda trees, interspersed with hedges of luxuriant tropical flowers and cactus plants. Richard could not resist taking photographs.

'Is this your first visit to this country?' the taxi driver asked.

'Yes, it is my first visit to Africa', Richard said, beaming.

'Oh, really, I thought you were from Ethiopia. You must be from America then'.

'Why Ethiopia?'

'You look like an Ethiopian. How long are you here for?'

'I am not sure. There is so much to see. How are things here?' Richard spoke with ill-concealed enthusiasm.

'It depends on what you mean', the taxi driver said wryly. 'Politics stinks and the economy is in bad shape. We small people are suffering. The big people eat everything'.

Richard did not pursue the discussion. He was engrossed by the view of the flat savanna grassland spotted with trees and shrubs. As the airport receded from sight, factory sheds with familiar names such as Firestone, General Motors, and Coca Cola began to appear. Further on, housing estates and low apartment blocks came into view, while shimmering in the distance was an array of towering office blocs, hotels, and conference centres. The traffic on the highway became heavier and the crowds at bus stops grew larger. It all looked so familiar, and yet so strange. He had not expected Africa to look like this.

When he got to his hotel, he took a shower and had some breakfast. Although he needed sleep, he decided to walk around the city, camera in hand. There were crowds everywhere, cars hooted endlessly, street vendors screamed on top of their voices to advertise their wares, shoeshine boys beckoned, and some beggars accosted him. The city was noisy, vibrant, alive. But it all looked so unreal.

He walked aimlessly through shops and supermarkets, across wide boulevards and narrow streets, to the city market with its stalls of fresh food, fruits, vegetables, and curios. In the nearby park, which was deserted at that time of day except for a team of school kids and isolated men basking in the sun, he sat for a while and watched the flamingoes peck one another by the banks of a river.

He could not find his way back to the hotel. It was lunchtime, and he was beginning to feel hungry, so he decided to stop by any restaurant and eat some real African food. It turned out to be a long search. He went into smart restaurants and open cafes patronised by the elegant elite where food he could get in any Chicago restaurant was served. The cheaper restaurants served nothing but hamburgers and fries, or chips, as they called them here.

'Excuse me', he said. 'Where can I get a good African restaurant?'

'What do you mean?' the man looked bewildered. 'There are restaurants all over the place'.

'I mean, where I can eat African food'.

'But all these restaurants serve African food', The man looked perplexed.

'No, I mean what ordinary people eat'.

'Which ordinary people? For their lunch many of them eat chips and soda, sometimes with sausages, a hamburger, or a piece of chicken if they can afford it'.

'But what do they eat at home in the evenings?'

'Different things. Some live on *ugali* or *matoke*, others on *chapati* or rice. Take your pick'.

'Can you take me to a restaurant that serves such food?'

'Sure', He chuckled. Tourists!

'Oh, by the way, my name is Richard Gates'. He stretched out his hand.

'I am Maina Kamau. Are you an American?'

'Yes, an African-American. It's the accent, right?'

Maina nodded.

The restaurant Maina took Richard to was on the outskirts of the city. As they left the centre of town, the streets and buildings became smaller and more decrepit, the crowds thicker and more weary-looking, and the street vendors more numerous. Music blared from the shops, street kiosks, and overloaded *matatu* vans. Here and there rapt crowds listened to preachers, comedians, and musicians screaming into loudspeakers. Others gathered in circles to watch some enterprising young men try their luck at acrobatics, magic, or snake charming.

By the time they got to the restaurant, the place was almost half-empty, for the lunch hour was virtually over. It was a dingy little place with bare tables packed together. Maina noticed the expression of surprise on Richard's face. 'You said you wanted a place where ordinary people eat. Not so?' he insisted defensively.

'Yes, it's all right, really'. Richard tried to sound unconcerned as he pulled up a chair. A plump waiter came and asked them what they wanted. 'What should I have? You mentioned some dishes earlier'.

'Well, I don't know. What do you have?' Maina asked, turning to the waitress.

She frowned as she wiped her sweating brow.

'*Ugali* and *chapati* are finished. Only *matoke* and chips are remaining'.

'Then you try *matoke*,' Maina told Richard, 'I'll have chips with chicken'.

'What's *matoke*?' Richard asked.

'Boiled and mashed bananas. Oh, what will you drink?' But Maina did not wait for Richard to respond before calling back the waiter. 'Bring us some drinks. Coke for me and passion fruit for my friend here'.

Richard found the food far more delicious than he expected, given the surroundings. The *matoke* was served with dry fish cooked in groundnut stew, and the drink had a pleasantly sweet and sour tingle.

'Aren't you going to be late for work?' Richard looked at his watch. It was a little after two. Maina shrugged. 'I have my own business,' he said with an air of self-importance.

'What kind of business?'

'I do all sorts of things. What about you?' Maina asked, quickly changing the subject.

'I am an artist. I paint and do sculpture'.

'What? Can you make enough to live on that?'

'Sometimes. I love it. I wouldn't trade it for any other job in the world.'

'Maybe in America. Here you can't survive on that. It's mostly tourists who buy that kind of stuff. What are your plans for the afternoon?' he asked, changing the subject again.

'Why, I was thinking of going back to the hotel to rest for a while'.

'Well, I thought you might be interested in being shown around'.

'That would be nice. All right, where would you like to take me?'

'Anywhere. But first, we need to get a car'.

That afternoon Maina drove Richard all over the city. It was much larger than he had thought. They visited the exclusive western suburbs with their lavish homes, lush lawns, bolted gates, and barking dogs. They toured the middle class estates in the south where handsome bungalows, townhouses, and flats competed for prominence. And they took in the crowded, shabby working class neighbourhoods in the east. Finally they drove past the desolate slums that stretched as far as the eye could see - a disparate collection of dwellings made of anything handy, including mud, cardboard, metal sheets, grass, and polythene bags. The place smelled like a dump.

It was a long, exhausting tour. Richard did not know what to make of it all. The images were so overwhelming and confusing. Maina appeared to know the city like the back of his hand. He obligingly explained everything Richard wanted to know, even the history and scandals of each neighbourhood. They did not get to the hotel until six, just as dusk was falling and the afternoon heat began to give way to a cool, refreshing breeze.

'You look a little tired', Maina remarked.

Richard nodded.

'Well, you haven't seen nothing yet. Go get some rest, then I will come and pick you up later'.

'No, not tonight', Richard protested feebly. Why was Maina being so friendly, spending his precious time keeping him company? he wondered. Then he reproached himself for thinking ill of Maina; the brother was simply being hospitable. So when Maina asked him if he could change his money into local currency on the 'parallel' market where the exchange rate was double the official one offered by the banks, Richard agreed and gave him a couple hundred dollars. Maina promised to bring the money and pick him up at nine.

When Richard got to his room, he asked the reception to wake him up at nine. The telephone rang promptly at that hour, but he was reluctant to get up, then remembered that Maina was on his way. After he took a quick bath and put on fresh clothes, Maina had still not showed up, even though it was nine-thirty. Switching on the TV, he was surprised that an old episode of *Good Times* was just ending. A few moments later *The Cosby Show* came on. He skimmed through the TV listings. At ten would be the main evening news, followed by *Dallas*. The whole night seemed to be filled with American programs, except for the news. Where were the African plays, productions, voices, images?

By ten Maina had still not come. How could he have been so naive to trust his money with someone he had just met? He would not have done that in Chicago.

He decided to go to the bar downstairs where he ordered a beer and watched the scene with fascination. Before long a young woman asked if she could join him. She was very pretty but barely twenty.

'Can I get you a beer?' he asked.

'If you want', she replied in a husky voice older than her years. Her name was Maria. She was inquisitive about America. She seemed to know the names and lives of the major African-American singers, actors, sports personalities, and politicians.

'Where do you get all that information?'

'American magazines, *Ebony*, *Essence*, and all that stuff. I would like to go to America and live there. Life here is miserable'.

'America is also miserable. There is poverty there, too, lots of it, and violence'.

But she looked unconvinced and bombarded him with more questions.

Suddenly, he heard Maina's voice...

Notes

1. The post-something constructions — postmodernism, post-structuralism, post-coloniality — are so deeply marked by moral nihilism that they are best described as anti-humanist. They have borrowed, appropriated, and ultimately subverted the critiques, deconstructions and explosions of western modernity first advanced by anti-Western African and other Thirdworldist intellectuals and activists. For a compelling critique of postmodernist and poststructuralist literary theorizing see Ahmad (1992).

2. This point is developed persuasively in the African diasporic context by Michaels (1995).

3. There is a vast literature on all of these subjects. On the history of Panafricanism see Legum (1965), Langley (1973), Adakunle (1973), Weisbord (1973) Geiss (1974), Drachler (1975), Abraham (1982), Esedebe (1982), Harris (1982), Ofuatey-Kodjoe (1986), Magubane (1987), and Lemelle and Kelley (1994). The question of identity formation is a staple of cultural studies (During 1993; Appiah and Gates 1995), the new cultural geography (Keith and Pile, 1993; Duncan and Ley, 1993; Bird, et.al., 1993; Lash and Urry, 1994; Godlewska and Smith, 1994), cultural politics (Jordan and Weedon 1995); and of course diaspora studies (Chow 1993).

4. For an exemplary study of the complexities of transnational identities see Lake's (1995) research on diaspora African-American repatriates in Ghana.

5. For stimulating studies of the relationships between nationalism, gender, sexuality and narratives of identity see Parker (1992).

6. Conversations with Thandika Mkandawire and Dianne Pinderhughes, February 24, 1996, Champaign, Illinois.

7. Excerpt is pp. 1-10 of 'Waiting' in *The Joys of Exile*, pp. 1-20.

8. Excerpt is pp.133-142 of 'Homecoming' in *The Joys of Exile*, pp. 133-158.

References

AAS & AAAS, 1992, *Electronic Networking in Africa: Advancing Science and Technology for Development*, Nairobi, African Academy of Sciences and the American Association for the Advancement of Science.

Abdelhakim, T., 1987, 'Agriculture and State Policies in Egypt', in Mkandawire and Bourenane, ed.

Abedian, I. and B. Standish, eds., 1992, *Economic Growth in South Africa: Selected Policy Issues*, Cape Town, Oxford University Press.

Abid, A., 1992, 'Improving Access to Scientific Literature in Developing Countries: A UNESCO Programme Review', *IFLA Journal*, Vol. 18/4, pp. 315-24.

Aboagye, A. A. and K. M. Gozo, 1986, 'The Informal Sector: A Critical Appraisal of the Concept', in JASPA.

Abraham, K., 1982, *From Race to Class: Links and Parallels in African and Black American Protest Expression*, London, Grassroots Publishers.

Abrahams, C., ed., 1990, *The Tragic Life: Bessie Head and Literature in Southern Africa*, Trenton, NJ, Africa World Press.

Abrahams, P., 1956, *A Wreath for Udomo*, London, Faber and Faber.

Abun-Nasr, J. M., 1975, *A History of the Maghrib*, 2nd ed., Cambridge, Cambridge University Press.

Abun-Nasr, J. M., 1987, *A History of the Maghrib in the Islamic Period*, Cambridge, Cambridge University Press.

Achebe, C., 1963, *No Longer at Ease*, London, Heinemann.

Achebe, C., 1966, *A Man of the People*, London, Heinemann.

Achebe, C., 1977, 'The Novelist as Teacher', in Morning Yet on Creation Day, London, Heinemann.

Achebe, C., 1989, 'An Image of Africa: Racism in Conrad's Heart of Darkness', in Hopes and Impediments: Selected Essays, New York, Doubleday, Anchor.

Achebe, C., 1992, 'African Literature as Celebration', *Dissent*, Vol. 39, pp. 344-9.

Adakunle, A., 1973, *Pan-Africanism: Evolution, Progress and Prospects*, London Andre Deutsch.

Addison, T. and L. Demery, 1987, 'Alleviating Poverty Under Structural Adjustment', *Finance and Development*, 24.

Adejumobi, S. and A. Momoh, eds., 1995, *The Political Economy of Nigeria Under Military Rule: 1984-1993*, Harare, Sapes Books.

Adejumobi, S., 1995, 'Ali Mazrui and his African Dream', *CODESRIA Bulletin*, n° 4, pp. 16-18.

Adeoye, A. O., 1992, 'Understanding the Crisis in modern Nigerian Historiography', History in Africa, n° 19, pp. 1-11.

Adepoju, A., 1983, 'Patterns of Migration by Sex', in Oppong, ed.

Adesina, J., 1994, *Labour in the Explanation of the African Crisis*, Dakar, CODESRIA Book Series.

Afary, J. and A. Lavrin, 1989, 'Some Reflections on Third World Feminist Historiography', *Journal of Women's History*, Vol. 1, p. 147-152.

Afigbo, E. A. et al., 1986, *The Making of Modern Africa*, vol. 1, The Nineteenth Century; Vol. 2, The Twentieth Century, London, Longman.

Afonja, S., 1981, 'Changing Modes of Production and the Sexual Division of Labour Among the Yoruba', *Signs*, Vol. 7, pp. 299-313.

Afonja, S., 1986, 'Land Control: A Critical Factor in Yoruba Gender Stratification', in Robertson and Berger, eds.

Africa Demos, 1995, Vol. 3, n° 2.

Africa Events, 1988, London, November.

Africa Magazine, 1984-1985, London, UK.

Africa Recovery, 1987-1995, New York, United Nations Department of Public Information.

Africa Watch, 1990, 'African Universities: Case Studies of Abuses of Academic Freedom', CODESRIA Symposium on Academic Freedom, Research and the Social Responsibility of the Intellectual in Africa', Kampala, Uganda, November.

African Business, 1984-1985, London, IC Publications.

African Center for Monetary Studies, 1992, Debt Conversion Schemes in Africa: Lessons from the Experience of Developing Countries, London, James Currey.

African Concord, 1988, London, UK.

African National Congress, 1994, *The Reconstruction and Development Programme*, Johannesburg, Umanyano Publications.

Afshar, H. ed., 1987, *Women, State and Ideology: Studies from Africa and Asia*, Albany, State University.

Agarwala, A. A. and S. P. Singh, eds., 1958, *The Economics of Development*, New York, Oxford University Press.

Aglietta, M., 1970, *A Theory of Capitalist Regulation: The US Experience*, London, New Left Books.

Ahmad, A., 1992, *In Theory, Classes, Nations, Literatures*, London and New York, Verso.

Ahmad, A., 1995, 'The Politics of Literary Postcoloniality', *Race & Class*, Vol. 36, pp. 1-20.

Ahmed, L., 1992, *Women and Gender in Islam: Historical Roots of a Modern Debate*, New Haven, Yale University Press.

Ahonsi, B., 1995, 'Economic Crisis, Gender Relations and Emergent Transitions in Africa's Demography', CODESRIA Eighth General Assembly, Dakar, June.

Aidoo, A. A., 1972, *No Sweetness Here*, New York, Doubleday.

Aidoo, A. A., 1979, *Our Sister Killjoy: Or, Reflections from a Black-Eyed Squint*, New York, Nok.

Aidoo, A. A., 1981, 'Asante Queen Mothers in Government and Politics in the Nineteenth Century', in Steady, ed.

Aidoo, A. A., 1991, Changes: A Love Story. New York, The Feminist Press.

Aidoo, A. A., 1992, 'The African Woman Today', *Dissent*, Vol. 39, pp. 319-25.

Aina, T. A., 1995, 'Library Acquisitions of African Books: An Academic Publisher's Viewpoint', APNET Open Forum: Library Acquisition of African Books, Harare, August 2.

Aiyepeku, W. O., 1983, *International Socio-economic Information Systems: An Evaluation of DEVIS-type Programs*, Ottawa, International Development Research Centre.

Aizenberg, E., 1990, 'The Untruths of the Nation: Petals of Blood and Fuente's The Death of Artemio Cruz', *Research in African Literatures*, Vol. 21, pp. 85-103.

Ajayi, J. F. A. and M. Crowder, eds., 1978, *History of West Africa*, Vol. 2, London, Longman.

Ajayi, J. F. A. and Michael Crowder, eds., 1976, *History of West Africa*, Vol. 1, 2nd ed., London, Longman.

Ajayi, J. F. A., 1989, 'Historical Education in Africa', paper presented to Meeting of Experts on Textbooks and Teaching of History in African Schools, UNESCO, Nairobi, March.

Ajayi, J. F. A., L. K. H., Goma, and G. A., Johnson, 1996, *The African Experience with Higher Education*, Accra: The Association of African Universities.

Ajulu, R., 1992, 'Kenya: The Road to Democracy', *Review of African Political Economy*, 53, pp. 79-87.

Ake, C., 1990, 'Academic Freedom and Material Base', CODESRIA Symposium on Academic Freedom, Research and the Social Responsibility of the Intellectual in Africa', Kampala, Uganda, November.

Ake, C., 1993, 'The Unique Case of African Democracy', *International Affairs*, Vol. 62, pp. 239-244.

Akinifesi, E. O., 1988, 'Balance of Payments, Debt Management and Nigeria's Economic Recovery', KEA/FES Conference on 'The North-South Dialogue: A Strategy for Africa', Nairobi, November.

Alagiah, G. and Bourke, G., 1988, 'Time for Decisions in Paris', *South*, April.

Alagiah, G., 1985a, 'Carry on Regardless', *South*, July.

Alagiah, G., 1985b, 'Spurning the IMF's shilling', *South*, July.

Alagiah, G., 1988, 'The Risks of Bread Politics', *South*, July.

Alatas, S. H., 1977, *The Myth of the Lazy Native: A Study of the Image of the Malays, Filipinos and Javanese from the Sixteenth Century to the Twentieth Century and Its Function in the Ideology of Colonial Capitalism*, London, Frank Cass.

Alavi, H., 1972, 'The State in Post-Colonial Societies: Pakistan and Bangladesh', *New Left Review*, 74, pp. 59-81.

Alavi, H., 1975, 'India and the Colonial Mode of Production', *Socialist Register*, London, Merlin Press.

Albert, I. O., 1995, 'University Students in the Politics of Structural Adjustment in Nigeria', in Mkandawire and Olukoshi, eds.

Alderman, H. and J. Sorenson, eds., 1994, *African Refugees: Development Aid and Repatriation*, Boulder, Col., Westview.

Alderman, H., 1994, 'Ghana: Adjustment Star Pupil?', in Sahn, ed.

Ali, A. G., 1990, 'Donors Wisdom vs. African Folly: What Academic Freedom and which High Moral Standing', CODESRIA Symposium on Academic Freedom, Research and the Social Responsibility of the Intellectual in Africa', Kampala, Uganda, November.

Allan, T. J., 1991, 'Afterword', in Aidoo.

Allen, V. L., 1969, 'The Study of African Trade Unionism', *Journal of Modern African Studies*, 7, pp. 289-307.

Allen, V. L., 1972, 'The Meaning of the Working Class in Africa', Journal of Modern African Studies, 10, pp. 169-189.

Allport, C., 1993, 'Integrating Theory into Women's History', Paper presented at the Conference on Teaching Women's History: Challenges and Solutions Organised by the Canadian Committee on Women's History, Trent University, August 20-22.

Alpers, E. A., 1984, '"Ordinary Household Chores": Ritual Power in a Nineteenth Century Swahili Women's Spirit Possession Cult', International Journal of African Historical Studies, Vol. 17, pp. 677-702.

Alpers, E. A., 1984, 'State, Merchant Capital, and Gender Relations in Southern Mozambique to the End of the Nineteenth Century: Some Tentative Hypotheses', African Economic History, Vol. 13, pp. 23-55.

Alpers, E. A., 1986, 'The Somali Community at Aden in the Nineteenth Century', Northeastern African Studies, 8, pp. 143-186.

Altbach, P. G., 1987, The Knowledge Context: Comparative Perspectives on the Distribution of Knowledge, Albany, N. Y., State University of New York.

Amadi, A. O., 1981, African Libraries: Western Tradition and Colonial Brainwashing, Metuchen, N.J. and London, Scarecrow Press.

Amadiume, I., 1987, Male Daughters, Female Husbands: Gender and Sex in an African Society, London, Zed Books.

Amin, S., 1972, 'Underdevelopment and Dependence in Black Africa: Historical Origins', Journal of Peace Research, 2, pp. 105-120.

Amin, S., 1973, Neo-Colonialism in West Africa, New York, Monthly Review Press.

Amin, S., 1974, Accumulation on a World Scale, Vols. 1 and 2, New York, Monthly Review Press.

Amin, S., 1975, 'Underdevelopment in Black Africa: Historical Origin', Journal of Modern African Studies, Vol. 10, pp. 503-524.

Amin, S., 1976, Unequal Development, New York, Monthly Review Press.

Amin, S., 1977, Imperialism and Unequal Development, New York, Harvester Press.

Amin, S., 1980a, Neo-Colonialism in West Africa, New York, Monthly Review Press.

Amin, S., 1980b, Class and Nation, Historically and in the Current Crisis, New York, Monthly Review Press.

Amin, S., 1990, 'The Issue of Democracy in the Contemporary Third World', CODESRIA Symposium on Academic Freedom, Research and the Social Responsibility of the Intellectual in Africa', Kampala, Uganda, November.

Amin, S., 1994, 'Ideology and Social Thought', CODESRIA Bulletin, n° 3, pp. 15-22.

Amin, S., 1995, 'The Causes of Africa's Economic Disaster', CODESRIA Eighth General Assembly, Dakar, June.

Amnesty International, 1990, 'Africa: Appeals For Academics and Students Who Are Prisoners of Conscience in 1990', CODESRIA Symposium on Academic Freedom, Research and the Social Responsibility of the Intellectual in Africa', Kampala, Uganda, November.

Amsden, A. M., 1971, International Firms and Labour in Kenya, 1945-70, London, Frank Cass.

Anderson, D. and R. Grove, eds., 1987, Conservation in Africa, Cambridge, Cambridge University Press.

Andrade, S. Z., 1990, 'Rewriting History, Motherhood, and Rebellion: Naming an African Women's Literary Tradition', Research in African Literatures, Vol. 21, pp. 91-110.

Andrae, G. and B. Beckman, 1986, 'The Nigerian Wheat Trap', in Lawrence, ed.

Angerman, A., et al., eds., 1989, Current Issues in Women's History, London, Routledge.

Ankomah, B., 1987, 'Debt Rescheduling Brightens the Outlook', African Business, February.

Antonius, G., 1969, The Arab Awakening: The Story of the Arab National Movement, Beirut, Librairie du Liban, [1938].

Anyandike, N., 1988, 'Planners Count the Cost after the SAP', African Business, September.

Appiah, K. A. and H. L. Gates eds., 1995, Identities, Chicago and London, University of Chicago Press.

Appiah, K. A., 1992a, In My Father's House, Africa in the Philosophy of Culture, New York, Oxford University Press.

Appiah, K. A., 1992b, 'Ofeimun's Misconception', West Africa, Vol. 10-16 August, pp. 1336.

Ardener, S. ed., 1975, Perceiving Women, London, J. M. Dent and Sons.

Argarwala, A. N. and S. P. Singh eds., 1958, The Economies of Underdevelopment, London, Oxford University Press.

Armah, A. K., 1968, The Beautiful Ones Are Not Yet Born, London, Heinemann.

Armah, A. K., 1970, Fragments, London, Heinemann.

Armah, A. K., 1972, Why Are We So Blest, London, Heinemann.

Armah, A. K., 1973, Two Thousand Seasons, London, Heinemann.

Armah, A. K., 1978, The Healers, London, Heineman

Armah, A. K., 1995, Osiris Rising a novel of Africa past, present and future, Popenguine, West Africa.

Arndt, H. W., 1978, The Rise and Fall of Economic Growth, Chicago, University of Chicago Press.

Arndt, H. W., 1981, 'Economic Development: A Semantic History', Economic Development and Cultural Change, Vol. 29, pp. 457-466.

Arrighi, G. and J. Saul, eds., 1973, Essays on the Political Economy of Africa, New York, Monthly Review Press.

Arrighi, G., 1973, 'Labor Supplies in Historical Perspective: A Study of the Proletarianization of the African Peasantry in Rhodesia', in Arrighi and Saul, eds.

ASA News, 22, 1989, January/March.

Asad, T. and H. Wolpe, 1976, 'Concepts of Modes of Production', Economy and Society, 5, pp. 471-505.

Atieno-Odhiambo, E. S., 1974, 'The Rise and Decline of the Kenya Peasant, 1881-1922', in The Paradox of Collaboration and Other Essays, Nairobi, East African Publishing House.

Atkins, K., 1993, The Moon Is Dead! Give Us Our Money! The Cultural Origins of an African Work Ethic, Natal, South Africa, 1843-1900, Portsmouth, NH, Heinemann.

Atmore, A. and N. Westlake, 1972, 'A Liberal Dilemma: A Critique of the Oxford History of South Africa', Race, 14, pp. 107-136.

Atwood, D. A., 1990, 'Land Registration in Africa: The Impact of Agricultural Production', World Development, Vol. 18, pp. 659-671.

Aubrey, L., 1995, 'Who is Leading Development in Africa into the Year 2000? A Look at the Role of African American Women in Non-Governmental Organizations', (Unpublished).

Aubrey, L., 1996, 'Toward a Pan African View of Women and (Co)Development: Considering Race in the Era of Development NGOs', (Unpublished).

Austen, J., 1966, *Mansfield Park*, edited by Tony Tanner, Harmondsworth, Penguin. [1814].

Austen, R. A., 1987, *African Economic History*, London, James Currey.

Austin, D., 1980, 'The Transfer of Power Why and How?', in Morris-Jones and Fisher, eds.

Austin, D., 1982, 'The British Point of No Return?', in Gifford and Louis, eds.

Austin, D., 1993, 'Reflections on African Politics: Prospero, Ariel and Caliban', *International Affairs*, Vol. 69, pp. 203-221.

Avineri, S., 1969, *Karl Marx on Colonialism and Modernization, New York, Doubleday*.

Awe, B., 1977, 'The Iyalode in the Traditional Yoruba Political System', in Schlegel, ed.

Awe, B., 1991, 'Writing Women into History: The Nigerian Experience', in Offen, et al., eds.

Awiti, A. and O. Ong'wen, 1990, 'Academic Freedom, State and Opportunism in Kenya', CODESRIA Symposium on Academic Freedom, Research and the Social Responsibility of the Intellectual in Africa, Kampala, Uganda, November.

Awoonor, K., 1971, *This Earth, My Brother*, New York, Doubleday.

Ayittey, G. B. N., 1991, *Indigenous African Institutions*, New York, Hudson-on-Ardsley.

Ayittey, G. B. N., 1992, 'Africa in the Postcommunist World', Problems of Communism, Vol. 41, pp. 207-17.

Ayittey, G., 1987, 'African Freedom of Speech', Index on Censorship, Vol. 16, n° 1, pp. 16-18.

Ayoade, J. A. A., 1988, 'States Without Citizen: An Emerging African Phenomenon', in Rothchild and Chazan, eds.

Azoulay, K. G., 1996, 'Outside Our Parents' House: Race, Culture, and Identity', Research in African Literatures, Vol. 27, pp. 129-142.

Ba, M., 1981, *So Long a Letter*, London, Heinemann.

Badran, M., 1989, 'The Origins of Feminism in Egypt', in Angerman, et al.

Bailey, R., 1987, 'Would a Freeport Benefit Your Country?' African Business, April.

Baker, P. H. et al., eds., 1993, *South Africa and the World Economy in the 1990s*, Cape Town, David Philip.

Bako, S., 1990, 'Education and Adjustment in Africa: The Conditionality and Resistance Against the World Bank Loan for Nigerian Universities', CODESRIA Symposium on Academic Freedom, Research and the Social Responsibility of the Intellectual in Africa, Kampala, Uganda, November.

Balaam, D. N. and M. J. Carey, eds., 1981, *Food Politics: The Regional Conflict*, London, Croom Helm.

Balibar, E., 1970, *Reading Capital*, London.

Balogh, T., 1949, *The Dollar Crisis: Cause and Cure*, London, Oxford University Press.

Balogun, J. and G. Mutahaba, 1990, 'The Dilemma of the Brain Drain', in Grey-Johnson, ed.

Bangura, Y., 1986, 'The Nigerian Economic Crisis', in Lawrence, ed.

Bangura, Y., 1989, 'Crisis and Adjustment: The Experience of Nigerian Workers', in Onimode, ed.

Bangura, Y., 1995, 'Perspectives on the Politics of Structural Adjustment, Informalization and Political Change in Africa', in Mkandawire and Olukoshi, eds.

Bangura, Y., 1995, 'The Pitfalls of Recolonization: A Comment on the Mazrui-Mafeje Exchange', CODESRIA Bulletin, n° 4, pp. 18-23.

Baran, P. A., 1956, The Political Economy of Growth, New York, Monthly Review Press.

Baran, P.A. and Paul M., 1966, Sweezy, Monopoly Capital, New York, Monthly Review Press.

Barber, K., 1987, 'Popular Arts in Africa', African Studies Review, Vol. 30, pp. 1-78.

Bardolph, J. ed., 1989, Short Fiction in the New Literatures in English, Nice, Charlet.

Barker, F., P., Hulme, and M., Iversen, eds., 1994, Colonial Discourse/Political Theory, Manchester and New York, Manchester University Press.

Barker, J., 1989, Rural Communities Under Stress: Peasant Farmers and the State in Africa, Cambridge, Cambridge University Press.

Barkin, D., et.al., 1990, Food vs. Feed Crops: Global Substitution of Grains in Production, Boulder, Lynne Rienner.

Barnett, D. L. and K. Njama, 1966, Mau Mau From Within, New York, Monthly Review Press.

Barraclough, G., 1978, Main Trends in History, New York and London, Holmes and Meier.

Barret, J., 1986, South African Women on the Move, London, Between the Lines.

Barrows, R. and M. Roth, 1990, 'Land Tenure and Investments in African Agriculture: Theory and Evidence', The Journal of Modern African Studies, Vol 28, pp. 265-297.

Barzun, J., 1987, 'Doing Research - Should the Sport be Regulated?' Columbia, February, 18-22.

Bassett, T. and D. Crummey, eds., 1993, Land Reform in African Agrarian Systems, Madison, The University of Wisconsin Press.

Bates, R. H. and M. I. Lofchie, eds., 1980, Agricultural Development in Africa, New York, Praeger.

Bates, R. H., 1971, Unions, Parties and Political Development: A Study of Mineworkers in Zambia, New Haven, Yale University Press.

Bates, R., H., 1981, Markets and States in Tropical Africa: The Political Basis of Agricultural Politics, Berkeley, University of California Press.

Bates, R. H., 1983, Essays on the Political Economy of Rural Africa, Berkeley, University of California Press.

Bates, R. H., V. Y. Mudimbe, and J. O'Barr, eds., 1993, Africa and the Disciplines: The Contributions of Research in Africa to the Social Sciences and Humanities, Chicago, University of Chicago Press.

Bathily, A., 1990, 'An Historical Perspective of the West African State (from the Earliest Times to the Post-Colonial Era), What State for Development? Lessons From History', CODESRIA/Rockefeller Foundation, Reflections on Development Symposium, Bellagio, Italy.

Bathily, A., 1994, 'The West African State in Historical Perspective', in Osaghae, ed.

Bay, E., ed., 1982, Women and Work in Africa, Boulder, Col., Westview.

Bayart, J-F., 1993, *The State in Africa: The Politics of the Belly*, London and New York, Longman.

Beck, I. and N. Keddie, eds., 1978, *Women in the Muslim World*, Cambridge, Cambridge University Press.

Becker, C. M., A. M. Hamer, and A. R, Morrison, eds., 1994, *Beyond Urban Bias in Africa: Urbanization in an Era of Structural Adjustment*, Portsmouth, NH, Heinemann.

Beckman, B, 1995, 'The Politics of Labour and Adjustment: the Experience of the Nigerian Labour Congress', in Mkandawire and Olukoshi, eds.

Beckman, B., 1980, 'Imperialism and Capitalist Transformation: Critique of a Kenyan Debate', Review of Africa Political Economy, n 19, pp. 48-62.

Bedran, S. and N. Bourenane, 1987 'State Policies, Agricultural Development and Food Production: The Algerian Experience', in Mkandawire and Bourenane, ed.

Beinart, W. and C. Bundy, 1987, *Hidden Struggles in Rural South Africa*, Johannesburg, Ravan Press.

Beinart, W., 1985, 'Chieftaincy and the Concept of Articulation: South Africa ca, 1900-1950', Canadian Journal of African Studies, 19, pp. 91-98.

Beinart, W., et.al., 1986, *Putting a Plough to the Ground*, Johannesburg, Ravan.

Bell, D., 1979, *The Cultural Contradictions of Capitalism*, London, Heinemann.

Bell, R. H. V., 1987, 'Conservation with a Human Face: Conflict and Reconciliation in African Land Use Planning', in Anderson, and Grove, eds.

Belling W. A., ed., 1968, *The Role of Labour in African Nation Building*, New York, Praeger.

Ben-Yehuda, N., 1985, *Deviance and Moral Boundaries*, Chicago, University of Chicago Press.

Bender, G. J., 1978, *Angola Under the Portuguese: The Myth and the Reality*, London, Heinemann.

Berardo, F. M., 1981, 'The Publication Process: An Editor's Perspective', *Journal of Marriage and the Family*, Vol. 43, pp. 771-779.

Berardo, F. M., 1989, 'Scientific Norms and Research, Publication Issues and Professional Ethics', Sociological Inquiry, Vol. 59, pp. 249-266.

Berardo, F. M., 1993, 'Scholarly Publication: A Scholarly Retrospective', *Marriage and Family Review*, Vol. 18, pp. 59-73.

Berg, E. J. and J. Butler, 1966, 'Trade Unions', in Coleman and Rosberg, eds.

Berger, R. A., 1988, 'Ngugi's Comic Vision', Research in African Literatures, Vol. 19, pp 1-25.

Berger, R. A., 1990, 'The Politics of Madness in Bessie Head', in Cecil Abrahams, ed.

Bergesen, A., 1980, 'Cycles of Formal Colonial Rule', in Hopkins and Wallerstein, eds.

Bergesen, A., 1983a, 'Modeling Long waves of Crisis in the World System', in Bergesen, ed.

Bergesen, A., ed., 1983b, *Crises in the World System*, Beverly Hills, Sage Publications.

Berman, B. and J. Lonsdale, 1980, 'The Development of the Labour Control System in Kenya, 1919-1929', *Canadian Journal of African Studies*, 14, pp. 37-54.

Berman, B. and J. Lonsdale, 1992, *Unhappy Valley: Conflict in Kenya and Africa*, London, James Currey.

Berman, B. J. and C. Leys, eds., 1994, *African Capitalists in African Development*, Boulder, Col., Lynne Rienner.

Berman, B. J., 1994, 'African Capitalism and the Paradigm of Modernity', in Berman and C. Leys, eds.

Berman, B. J., 1994, 'African Capitalists in African Development', in Berman and Leys, eds.

Berman, B., 1990, *Control and Crisis in Colonial Kenya*, The Dialectic of Domination, London, James Currey.

Bernal, M., 1987-1991, *Black Athena: The Afro-Asiatic Roots of Classical Civilization*, Volumes I and II. New Brunswick, NJ, Rutgers University Press, 1987.

Bernstein, H, 1979, 'African Peasantries: A Theoretical Framework', *Journal of Peasant Studies*, 6, pp. 421-443.

Bernstein, H. and B. Campbell, eds., 1985, *Contradictions of Accumulation in Africa*, Beverly Hills, Sage Publications.

Bernstein, H. and J. Depelchin, 1978-79, 'The Object of African History: A Materialist Perspective', History in Africa, [Part 1], 5, pp. 1-19, 1978; [Part 2], 6, pp. 17-43, 1979.

Bernstein, H., 1976b, 'Underdevelopment and the Law of Value: A Critique of Kay', *Review of African Political Economy*, 6, pp. 51-64.

Bernstein, H., 1977, 'Notes on Capital and Peasantry', *Review of African Political Economy*, 10, pp. 60-73.

Bernstein, H., 1985, *For Their Triumphs and For Their Tears*, London, International Defence and Aid.

Bernstein, H., ed., 1976a, *Underdevelopment and Development*, Harmondsworth, Penguin.

Berry, S., 1975, *Cocoa, Custom, and Socio-Economic Change in Rural Western Nigeria*, Oxford, Clarendon.

Berry, S., 1985, *Fathers Work for Their Sons*, Berkeley, University of California Press.

Bettelheim, C., 1972, 'Theoretical Comments', in Emmanuel.

Bgoya, W., 1994, 'Bgoya Reports on African Books Collective', Bellagio Publishing Network Newsletter, Vol. 10, pp. 5-6.

Bhabha, H., 1990, 'The Other Question: Difference, Discrimination, and the Discourse of Colonialism', in Ferguson, et al., ed.

Bhabha, H., 1994, *The Location of Culture*, London and New York, Routledge.

Bhagwati, J., 1991, *The World Trading System at Risk*, Hemel Hempstead, Harvester Wheatsheaf.

Bienefeld, M., 1975, 'The Informal Sector and Peripheral Capitalism: The Case of Tanzania', Bulletin of the Institute of Development Studies, 6.

Bienefeld, M., 1988, 'Dependency Theory and the Political Economy of Africa's Crisis', Review of African Political Economy, 43, pp. 68-87.

Bienen H. S. and M. Gersovits, 1986, 'Economic Stabilization, Conditionality, and Political Stability', Comparative Politics, Vol. 19, pp. 25-44.

Bienfeld, M. and M. Godfrey, 1975, 'Measuring Unemployment and the Informal Sector: Some Conceptual and Statistical Problems', *Bulletin of the Institute of Development Studies*, 7, pp. 4-10.

Biermann, W. and J. Wagao, 1986, 'The IMF and Tanzania - a solution to the Crisis?', in Lawrence, ed.

Biermann, W. and R. Kössler, 1985, 'The Settler Mode of Production: The Rhodesian Case', *Review of African Political Economy*, 18, pp. 106-116.

Bird, J., et al., eds., 1993, *Mapping the Futures. Local Cultures*, Global Change, London and New York, Routledge.

Birenbaum, R., 1995, 'Scholarly Communication Under Siege' *University Affairs*, Association of Universities and Colleges of Canada, August-September, p. 6.

Birmingham, D. and P. Martin, eds., 1985, *History of Central Africa*, 2 Volumes, London, Longman.

Black Historians' Response, 1995, 'The Significance of Race in African Studies', Chronicle of Higher Education, April 7, pp. B8.

Blaug, M., P. R. G. Layard and M. Woodhall, 1969, *The Causes of Graduate Unemployment in India*, London, Allen Lane.

Blier, S. P., 1993, 'Truth and Seeing: Magic, Custom, and Fetish in Art History', in Bates, Mudimbe, and O'Barr, eds.

Blomström, M. and B Hettne, 1984, *Development Theory in Transition*, London, Zed Books.

Bloom, A., 1988, *The Closing of the American Mind*, New York, Touchstone Books.

Blyden, E. W., 1857, *A Vindication of the African Race*, Monrovia.

Blyden, E. W., 1869, *The Negro in Ancient History*, Washington.

Blyden, E. W., 1994, *Christianity, Islam and the Negro Race*, Baltimore, MD: Black Classic Press [1887].

Boahen, A., 1989, 'Production of History Textbooks', paper presented to the Meeting of Experts on Textbooks and Teaching of History in African Schools, UNESCO, Nairobi, March.

Bock, G., 1991, 'Challenging Dichotomies: Perspectives on Women's History', in Offen, et al.

Boesen, J., 1979, 'On Peasantry and the 'Modes of Production' Debate', *Review of African Political Economy*, 15/16, pp. 154-161.

Bohannan, P. and P. Curtin, 1988, *Africa and Africans*, Prospect Heights, IL, Waveland Press.

Boone, S. A., 1986, *Radiance from the Waters: Ideals of Feminine Beauty in Mende Art*, New Haven, Yale University Press.

Booth, M., 1989, 'Mahfouz and the Arab Voice', *Index on Censorship*, Vol. 18, n° 1, pp. 14-16.

Bordo, S., 1990, 'Feminism, Postmodernism, and Gender-Scepticism', in Nicholson, ed.

Borgstrom, G., 1969, *Too Many*, New York, Macmillan.

Boserup, E., 1965, *The Conditions of Agricultural Growth: The Economics of Agrarian Changes Under Economic Pressure*, London, Allen and Unwin.

Boserup, E., 1970, *Woman's Role in Economic Development*, New York, St. Martin's Press.

Bouami, A. and M. Raki, 1987, 'Agricultural Policy and the Limits of Agricultural Development in Morocco', in Mkandawire and Bourenane, ed.

Bourke, G., 1987, 'A Model Debtor Reaches Its Moment of Truth', South, November.

Bourke, G., 1988, 'Cote d'Ivoire: Privatization Pace Slackens', African Business, February.

Bourne, L. S., 1988, 'Different solitudes and the restructuring of academic publishing: on barriers to communication in research', Environment and Planning A, Vol. 20, pp. 1423-4-5.

Bovill, E. W., 1920, 'The Encroachment of the Sahara on the Sudan', Journal of the African Society, Vol. 20.

Bowen, M. L., 1993, 'Socialist Transitions: Policy Reforms and Peasant Producers in Mozambique', in Bassett, and Crummey, eds.

Boyd, J. and M. Last, 1985, 'The Role of Women as 'Agents Religieux' in Sokoto', Canadian Journal of African Studies, 19, pp. 283-300.

Boyd, J., 1986, 'The Fulani Women Poets', in Kirk-Greene and Adamu, eds.

Bozzoli, B., 1983, 'Marxism, Feminism and South African Studies', Journal of Southern African Studies, Vol. 9, pp. 139-171.

Bradburd, D., 1984, 'Marxism and the Study of Pastoralists', Nomadic Peoples, Vol. 16.

Bradby, B., 1975, 'The Destruction of the Natural Economy', Economy and Society, 4, pp. 127-161.

Brain, J. L., 1976, 'Less Than Second Class: Women in Rural Settlement Schemes', in Hafkin and Bay.

Brandt, W., ed., 1980, North-South: A Programme for Survival, London, Pan Books.

Brantley, C., 1986, 'Mekatili and the Role of Women in Giriama Resistance', in Crummey, ed.

Bratton, M. and N. van de Walle, 1994, 'Neopatrimonial Regimes and Political Transitions in Africa', World Politics, Vol. 46, pp. 453-89.

Brennan, T., 1989, 'Cosmopolitans and Celebrities', Race and Class, Vol. 31, pp. 1-19.

Brenner, R., 1976, 'Agrarian Class Structure and Economic Development in Pre-Industrial Europe', Past and Present, Vol. 70, pp. 30-75.

Brenner, R., 1977, 'The Origins of Capitalist Development: A Critique of Neo-Smithian Marxism', New Left Review, 104, pp. 25-92.

Brett, E. A., 1973, Colonialism and Underdevelopment in East Africa, London, Heinemann.

Brickhill, P., 1994, 'APNET Report', Bellagio Publishing Network Newsletter, n° 11, pp. 8-9.

Brister, J., 1988, 'The Cooking Pots Are Broken', African Recovery, June.

Brokensha, D.W. et al., eds., 1980, Indigenous Knowledge Systems and Development, Lanham, Maryland, University Press of America.

Bromley, R., 1978, 'The Urban Informal Sector: Why is it Worth Discussing?' World Development, 6, pp. 1033-1039.

Brooks, G. E., 1976, 'The Signares of Saint-Louis and Goree: Women Entrepreneurs in Eighteenth Century Senegal', in Hafkin and Bay, eds.

Brown, D. M., 1991, 'Matigari and the Rehabilitation of Religion', Research in African Literatures, Vol. 22, pp. 172-180.

Brown, L. W., 1981, Women Writers in Black Africa, Westport, Conn., Greenwood Press.

Browne, R. S., 1985, 'The IMF in Africa: Inappropriate Technology', African Business, January.

Bruner, K., ed., 1981, The Great Depression Revisited, Hingham, Mass, Martinus Nijhoff.

Brzezinski, Z., 1989, The Grand Failure - The Birth and Death of Communism in the Twentieth Century, New York, Scribner's.

Buckoke, A., 1985, 'Tanzania: There may yet be an IMF Deal - Eventually', *African Business*, January.

Bujra, J., 1975, 'Women 'Entrepreneurs' of Early Nairobi', *Canadian Journal of African Studies*, Vol. 9, pp. 213-34.

Bukharin, N., 1972, *Imperialism and the Accumulation of Capital*, New York, Monthly Review Press.

Bundy, C., 1979, *The Rise and Fall of the South African Peasantry*, London, Heinemann.

Bunker, S., 1987, *Peasants Against the State: The Politics of Market Control in Bugisu, Uganda*, Urbana, University of Illinois Press.

Burawoy, M., 1972, *The Colour of Class in the Copper Mines*, Manchester, Manchester University Press.

Burawoy, M., 1982, 'The Hidden Abode of Underdevelopment: Labour Process and the State in Zambia', Politics and Society, 123-166.

Burawoy, M., 1985, *The Politics of Production*, London, New Left Books.

Burbaker, W. R., ed., 1989, *Immigration and Politics of Citizenship in Europe and North America*, Lanham, Maryland, University Press of America.

Buren, L. V., 1986, 'Zaire Claps Ceiling on Debt Payments', African Business, December.

Burr, W. R., 1993, 'Generativity in a Professional Sense', Marriage and Family Review, Vol. 18, pp. 75-80.

Buschman, J., 1992, 'A Response', Progressive Librarian, Vol. 5, pp. 51-3.

Byrne, E., 1993, 'Dialogue Suspended', *Index on Censorship*, Vol. 22, n° 2, pp. 21-23.

Cabral, A., 1969, *Revolution in Guinea*, London and New York, Love and Malcomson, Monthly Review Press.

Callaghy, T. M., 1993, 'Vision and Politics in the Transformation of the Global Political Economy: Lessons from the Second and Third Worlds', in Slater, Schutz, and Dorr, eds.

Callaghy, T. M., 1995, 'Africa and the World Economy: Still Caught Between a Rock and a Hard Place', in Harbeson and Rothchild, ed.

Callaghy, T., 1987, 'The State as Lame Leviathan: The Patrimonial-Administrative State in Africa', in Ergas, ed.

Callaghy, T., 1990, 'Lost Between State and Market: The Politics of Economic Adjustment in Ghana, Zambia and Nigeria', in Nelson, ed.

Callari, A., S., Cullenberg and C. Biewener, eds., 1995, *Marxism in the Postmodern Age: Confronting the New World Order*, New York, Guilford Press.

Callaway, B., 1987, *Muslim Hausa Women in Nigeria: Tradition and Change*, Syracuse, NY, Syracuse University Press.

Callinicos, A., 1989, *Against Postmodernism*, Oxford, Polity Press.

Campbell, B. K. and J. Loxley, eds., 1989, *Structural Adjustment in Africa*, London, Macmillan.

Campbell, H. and H. Stein, eds., 1991, *The IMF and Tanzania*, Harare, Sapes Books.

Canadian Journal of African Studies 22, n° 3, 1988, Special Issue on Current Research on African Women.

Cannon, B. D., 1985, 'Nineteenth Century Arabic Writings on Women and Society: The Interim Role of the Masonic Press in Cairo - (Al-Latif 1865-1895)', *International Journal of Middle East Studies*, Vol. 17, pp. 463-484.

Cardoso, F. H. and E. Faletto, 1979, *Dependency and Development in Latin America*, Berkeley University of California Press.

Cardoso, F. H., 1972, 'Dependency and Development in Latin America', *New Left Review*, 74, pp. 83-95.

Cardoso, F. H., 1977, 'The Consumption of Dependency Theory in the U.S', Latin American Research Review, 12, pp. 7-24.

Carney, J. and M. Watts, 1991, 'Disciplining Women? Rice, Mechanization, and the Evolution of Mandinka Gender Relations in Senegambia', Signs, Vol. 16, pp. 651-681.

Carrington, C. E., 1961, *The Liquidation of the British Empire*, London, Harrap.

Carroll, B. A., ed., 1976, *Liberating Women's History: Theoretical and Critical Essays*, Chicago, University of Illinois Press.

Carroll, B. B., 1990, 'The Politics of 'Originality': Women and the Class System of the Intellect', *Journal of Women's History*, Vol. 2, pp. 136-163.

Carver, R., 1992, 'A Licence to Kill', *Index on Censorship*, Vol. 21, n° 5, pp. 14,.

Castro, F., 1984, *The World Crisis*, London, Zed Books.

Catholic Bishops, 1992, 'The Catholic Bishops Speak Out - Pastoral Letter from the Catholic Bishops in Malawi', *Index on Censorship*, Vol. 21, n° 5, pp. 15-17.

Chakava, H., 1994, 'Publishing Ngugi: The Challenge, The Risk, and The Reward' *Bellagio Publishing Network Newsletter*, Vol. 10, pp. 3-5.

Chakrabarty, D., 1996, 'Postcoloniality and the Artifice of History: Who Speaks for 'Indian' Pasts? in Mongia, ed.

Chambers, R., 1983, *Rural Development: Putting the Last First*, Harlow, Longman.

Chambers, R., 1986, 'The Crisis of Africa's Rural Poor: Perceptions and Priorities', in JASPA.

Chanda, D., 1995, 'The Movement for Multi-Party Democracy in Zambia: Some Lessons in Democratic Transition', in Sachikonye, ed.

Chanock, M., 1985, *Law, Custom and Social Order: The Colonial Experience in Malawi and Zambia*, Cambridge, Cambridge University Press.

Chase-Dunn, C., 1978, 'Core-Periphery Relations: The Effects of Core Competition', in Kaplan, ed.

Chase-Dunn, C., 1979, 'Comparative Research on World System Characteristics', *International Studies Quarterly*, 23.

Chauncey, G., 1981, 'The Locus of Reproduction: Women's Labour in the Zambian Copperbelt, 1927-1953', *Journal of Southern African Studies*, Vol. 7, pp. 135-164.

Chazan, N., 1988, 'Ghana: Problems of Governance and the Emergence of Civil Society', in Diamond, Linz, and Lipset, eds.

Chazan, N., 1988, 'State and Society in Africa: Images and Challenges', in Rothchild and Chazan, eds.

Chazan, N., 1989, 'Gender Perspectives on African States', in Parpart and Staudt, eds.

Cheney, L. V., 1988, *Humanities in America: Reports to the President the Congress and the American People*, Washington, DC, National Endowment for the Humanities.

Cheney-Coker, S., 1991, *The Last Harmattan of Alusine Dunba*, Oxford, Heinemann.

Chinweizu, O. J. and I. Madubuike, 1980, *Towards the Decolonization of African Literature*, Enugu, Fourth Dimension Publishing.

Cherif, M. H., 1989, 'Report on History Textbooks in the Arabic-Speaking Countries of Africa', paper presented to the Meeting of Experts on Textbooks and Teaching of History in African Schools, UNESCO, Nairobi, March.

Chikhi, S., 1995, 'The Working Class, The Social Nexus and Democracy in Algeria', in Mamdani and Wamba-dia-Wamba, eds.

Chilcote, R. H., 1984, *Theories of Development and Underdevelopment*, Boulder, Col., Westview Press.

Chiposa, S., 1986, 'Zambia Withdraws Namboard Monopoly', *African Business*, March.

Chiposa, S., 1986, 'Zambian Unions Fight For Job Survival', *African Business*, October.

Chiposa, S., 1987a, 'Foreign Investors are Buying in Zambia', *African Business*, April.

Chiposa, S., 1987b, 'Heinz Re-thinks Zambian Investment', *African Business*, September.

Chiposa, S., 1987c, 'Zambia: Economy Enters Post-IMF Era', *African Business*, July.

Chiposa, S., 1988, 'Zambia: IMF Austerity Looms Once Again', *African Business*, November.

Chirimuuta, R., 1989, *Aids, Africa, and Racism*, London, Free Association Books.

Chirwa, C. W., 1994, 'The Politics of Ethnicity and Regionalism in Contemporary Malawi', *African Rural and Urban Studies*, Vol. 1, pp. 93-118.

Chirwa, C., 1994, 'Publishing in Zambia - Towards Privatization', *Bellagio Publishing Network Newsletter*, Vol. 10, pp. 9-10.

Chodak, S., 1973, 'Social Stratification in Sub-Saharan Africa', *Canadian Journal of African Studies*, 7, pp. 401-419.

Chole, E. and J. Ibrahim, eds., 1995, *Democratization Processes in Africa: Problems and Prospects*, Dakar, CODESRIA.

Chomsky, N., 1994, 'Sweet Home of Liberty', *Index on Censorship*, Vol. 3, pp. 9-18.

Chow, R., 1993, *Writing Diaspora: Tactics of Intervention in Contemporary Cultural Studies*, Bloomington and Indianapolis, Indiana University Press.

Clarence-Smith, G., 1985, 'Though Shall Not Articulate Modes of Production', *Canadian Journal of African Studies*, 19, pp. 19-22.

Clark, C. M., 1980, 'Land and Food, Women and Power, in Nineteenth Century Kikuyu', *Africa*, Vol. 50, pp. 357-369.

Clark, G., ed., 1988, *Traders Versus the State. Anthropological Approaches to Unofficial Economies*, Boulder, Col., Westview.

Clayton, A. and D. C. Savage, 1974, *Government and Labour in Kenya 1895-1963*, London, Frank Cass.

Cleaver, T. and M. Wallace, 1990, *Namibia: Women in War*, London, Zed Books.

Clegg, E. M., 1960, *Race and Politics: Partnership in the Federation of Rhodesia and Nyasaland*, London, Oxford University Press.

Cliffe, L. and R. Moorson, 1979, 'Rural Class Formation and Ecological Collapse in Botswana', Review of African Political Economy, n° 15/16, pp. 35-52.

Cliffe, L., 1977, 'Rural Class Formation in East Africa', *Journal of Peasant Studies*, 4, pp. 195-324.

Clow, D., 1986, 'Aid and Development the Context of Library-Related Aid', Libri, Vol. 36/2, pp. 85-97.

Clute, R. E., 1982, 'The Role of Agriculture in African Development', *African Studies Review*, Vol. 25, pp. 1-20.

Cobham, R., 1991, 'Boundaries of the Nation: Boundaries of the Self: African Nationalist Fictions in Nurrudin Farah's Maps', *Research in African Literatures*, Vol. 22, pp. 83-97.

Cobham, R., 1992, 'Misgendering the Nation: African Nationalist Fictions and Nurrudin Farah's Maps', in Parker, et.al. eds.

CODESRIA Bulletin, n° 1, 1991.

CODESRIA Bulletin, n° 4, 1990.

CODESRIA, 1990a, *Press Cuttings on the Abuses of Academic Freedom in Africa*, Dakar, Senegal.

CODESRIA, 1990b, 'CODESRIA Workshop on: Academic Freedom, Research and the Social Responsibility of the Intellectual in Africa', UDASA Newsletter, 11.

CODESRIA, 1990c, *The Kampala Declaration on Intellectual Freedom and Social Responsibility*, Dakar, Senegal.

CODESRIA, 1993, *Report of the Executive Secretary 20th Anniversary of CODESRIA*, Dakar, CODESRIA.

Cohen, D. W., 1994. *The Combing of History*, Chicago, University of Chicago Press.

Cohen, R., 1972, 'Classes in Africa: Analytical Problems and Perspectives', Socialist Register, London, Merlin Press.

Cohen, R., 1976, 'From Peasants to Workers in Africa', in Gutkind and Wallerstein, eds.

Colclough, C., 1991, 'Wage flexibility in sub-Saharan Africa: Trends and explanations', in Standing and Tokman, eds.

Cole, J. R., 1981, 'Feminism, Class, and Islam in Turn-of-the Century Egypt', *International Journal of Middle East Studies*, Vol. 13, pp. 387-407.

Cole, S. et al., 1988, 'Do Journal Rejection Rates Index Consensus?' *American Sociological Review*, Vol. 53, pp. 152-56.

Coleman, J. S. and C. G. Rosberg, eds., 1966, Political Parties and National Integration in Africa, Berkeley, University of California Press.

Coleman, J. S., 1958, Nigeria Background to Nationalism, Berkeley, University of California Press.

Collier, P. and D. Lal, 1986, *Labor and Poverty in Kenya*, Oxford, Clarendon.

Collier, P., 1993, 'Africa and the Study of Economics', in Bates, Mudimbe, and O'Barr, eds.

Collins, C., 1993, 'Famine Defeated: Southern Africa, UN Win Battle Against Drought', *Africa Recovery*, Vol. 9, pp. 1-12.

Comaroff, J., 1980, *The Meaning of Marriage Payments*, London, Academic Press.

Comaroff, J., 1985, Body of Power, Spirit of Resistance: The Culture and History of a South African People, Chicago, University of Chicago Press.

Commins, S. K. et al., eds., 1986, *Africa's Agrarian Crisis: The Roots of Famine*, Boulder, Lynne Rienner.

Compton, A., 1993, 'CD-ROM Technology: Hardware and Software', in AAAS, *CD-ROM for African Research Needs*, Washington, American Association for the Advancement of Science.

Conan, A. R., 1966, *The Problem of Sterling*, London, Macmillan.

Conkin, P. K., 1989, *Heritage and Challenge: The History and Theory of History*, Arlington Heights, Ill, Forum Press.

Connolly, P., 1985, 'The Politics of the Informal Sector: A Critique', in Redclift and Mingione, eds.

Conrad, J., 1925, 'Heart of Darkness', in Youth and Two Other Stories, Garden City, NY, Doubleday.

Cooper, F. et.al., 1993, *Confronting Historical Paradigms: Peasants, Labor, and the Capitalist World System in Africa and Latin America*, Madison, University of Wisconsin Press.

Cooper, F., 1977, *Plantation Slavery on the East Coast of Africa*, New Haven, Yale University Press.

Cooper, F., 1979, 'The problem of Slavery in African Studies: Review Article', *Journal of African History*, 20, pp. 103-125.

Cooper, F., 1980, 'Peasants, Capitalists, and Historians: A Review Article', *Journal of Southern African Studies*, 7, pp. 284-314.

Cooper, F., 1981, *From Slaves to Squatters*, New Haven, Yale University Press.

Cooper, F., 1981, 'Africa and the World Economy', *African Studies Review*, Vol. 24, pp. 1-86.

Cooper, F., 1987, *On the African Waterfront*, New Haven, Yale University Press.

Cooper, F., 1993a, 'Africa and the World Economy', in Cooper, et al. [1981].

Cooper, F., 1993b, 'Postscript: Africa and the World Economy', in Cooper, et al.

Cooper, F., 1994, 'Conflict and Connection: Rethinking Colonial African History', *American History Review*, Vol. 99, pp. 1516-1545.

Cooper, F., ed., 1983, *Struggle For the City: Migrant Labour, Capital and the State in Urban Africa*, Beverly Hills, Sage Publications.

Copans, J., 1977, 'African Studies: A Periodization', in Gutkind and Waterman, eds.

Coquery-Vidrovitch, C. and P. E. Lovejoy, eds., 1985, The Workers of African Trade, Beverly Hills, Sage Publications.

Coquery-Vidrovitch, C., 1976, 'The Political Economy of the African Peasantry and Modes of Production', in Gutkind and Wallerstein.

Coquery Vidrovitch, C., 1977, 'Research on an African Mode of Production', in Gutkind and Waterman, eds.

Cordell, D. D., et al., eds., 1987, *African Population and Capitalism*, Boulder, Col., Westview.

Cordell, D., 1985, 'The Pursuit of the Real: Modes of Production and History', *Canadian Journal of African Studies*, 19, pp. 58-63.

Cornia, G., R., van der Hoeven and T. Mkandawire, 1992, *Africa's Recovery in the 1990s: from Stagnation to Human Development*, New York, St. Martin's Press.

Coughlin, P. and G. K. Ikiara eds., 1991, *Kenya's Industrialization Dilemma*, Nairobi, Heinemann Kenya.

Coughlin, P. and G. K. Ikiara, 1988, *Industrialization in Kenya: In Search of a Strategy*, Nairobi Heinemann Kenya,.

Court, D., 1990, 'Universities and Academic Freedom in East Africa 1963-1983: Random Reflections From a Donor Perspective', CODESRIA Symposium on Academic Freedom, Research and the Social Responsibility of the Intellectual in Africa', Kampala, Uganda, November.

Cowen, M. P., 1976, 'Notes on Capital, Class and Household Production', Nairobi, mimeo.

Cowen, M. P., 1979, 'Capital and Household Production: The Case of Wattle in Kenya's Central Province, 1903-1964', Ph.D. dissertation, University of Cambridge.

Cowen, M. P., 1982, 'The British State and Agrarian Accumulation in Kenya', in Fransman, ed.

Creighton, T. R. M., 1960, *The Anatomy of Partnership: The Central African Federation*, London, Faber and Faber.

Crevey, L. E., 1986, *Women Farmers in Africa: Rural Development in Mali and the Sahel*, Syracuse, Syracuse University Press.

Crisp, J., 1984, 'Nigeria Workers Bear the Brunt of Austerity', *African Business*, November.

Crisp, J., 1984, *The Story of an African Working Class: Ghanaian Miners' Struggles, 1870-1980*, London, Zed Books.

Crisp, J., 1985, 'Zimbabwe's New Labour Act Caught in Crossfire', *African Business*, July.

Crowder, M., 1978, 'The 1939-45 War and West Africa', in Ajayi and Crowder, eds.

Crowder, M., 1981, *West Africa Under Colonial Rule*, London, Hutchinson.

Crummey, D. and C. Stewart, 1981, *Modes of Production in Africa The Precolonial Era*, Beverly Hills, Sage.

Crummey, D., 1981, 'Women and Landed Property in Gondarine Ethiopia', *International Journal of African Historical Studies*, Vol. 14, pp. 444-265.

Crummey, D., 1982, 'Women Property and Litigation Among the Bagemder Amhara, 1750s to 1850s', in Hay and Wright, eds.

Crummey, D., ed., 1986, *Banditry: Rebellion and Social Protest in Africa*, London, Heinemann.

Curtin, P. D. et al., 1978, *African History*, London, Longman.

Curtin, P. D., 1964, *The Image of Africa*, Madison, University of Wisconsin Press.

Curtin, P. D., 1975, *Economic Change in Precolonial Africa: Senegambia in the Era of the Slave Trade*, 2 Vols., Madison, University of Wisconsin Press.

Curtin, P. D., 1995, 'Ghettoizing African History', Chronicle of Higher Education, March 3, pp. A44.

Czerniewicz, L., 1993, 'Learning Lessons from African Publishing', in Steve Kromberg, et al.

D'Souza, D., 1992, *Illiberal Education: The Politics of Race and Sex on Campus*. New York, Vintage Books.

Daily Nation, 1984-1988, Nairobi, Kenya.

Dalton, G., 1964, 'The Development of Subsistence and Peasant Economies in Africa', *International Social Science Journal*, 16, pp. 378-89.

Daly, M. W., ed., 1985, *Modernization in the Sudan*, New York, Barbar Press.

Danning, J. ed., 1972, *The Multinational Enterprise*, London, Allen and Unwin.

Darch, C., 1993, 'The Western Cape Library Cooperative Project Leveling the Playing Field in the 1990s', in Patrikios and Levey, eds.

Darkoh, M. B. K., 1989, *Combating Desertification in the Southern African Region*, Nairobi, UNEP.

Darwish, A., 1992, 'The Hydra Grows Another Head', *Index on Censorship*, Vol. 21, n° 6, pp. 27-33.

Darwish, A., 1993, 'Creeping Intolerance and a State at Risk', *Index on Censorship*, Vol. 22, n 7, pp. 33.

Davidson, B. and B. Munslow, 1990, 'The Crisis of the Nation-State in Africa', *Review of African Political Economy*, n° 49, pp. 9-21.

Davidson, B., 1961a, *Old Africa Rediscovered*, London, Gollancz.

Davidson, B., 1961b, *Black Mother*, London, Gollancz.

Davidson, B., 1967, *The African Past: Chronicles From Antiquity to Modern Times*, New York, Grosset and Dunlap.

Davidson, B., 1973, *Black Star: A View of the Life and Times of Kwame Nkrumah*, London, Allen Lane.

Davidson, B., 1974, *Africa in History*, New York, Macmillan.

Davidson, B., 1978, *Let Freedom Come: Africa in Modern History*, Boston, Little Brown.

Davidson, B., 1991, *Africa in History: Themes and Outlines*, New York, Collier Books.

Davidson, B., 1992, *The Black Man's Burden: Africa and the Curse of the Nation-State*, New York, Times Books.

Davidson, B., 1993, 'Comments on Mamdani', *Monthly Review*, July-August, pp. 49-57.

Davies, C. B. and A. A. Graves, 1986, *Ngambika: Studies of Women in African Literature*, Trenton, NJ, Africa World Press.

Davies, C. B. and E. S. Fido, 1993, 'African Women Writers: Toward a Literary History', in Owomoyela, ed..

Davies, C. B., 1994, 'Uprooting the "Posts": From Post-Coloniality to Uprising Discourses', 21st Annual Spring Symposium: 'Reconstructing the Study and Meaning of Africa', Center for African Studies, University of Illinois at Urbana-Champaign, April.

Davies, I., 1966, *African Trade Unions*, Harmondsworth, Penguin.

Davies, R. et al., 1976, 'Class Struggle and the Periodization of the State in South Africa', *Review of African Political Economy*, No. 7, pp.4-30.

Davis, H. B., 1967, *Nationalism and Socialism: Marxist and Labour Theories of Nationalism to 1917*, New York.

Davison, J., 1989, *Voices from Muthira: Lives of Rural Gikuyu Women*, Boulder, Col, Lynne Rionner.

Davison, J., ed., 1988, *Agriculture, Women and Land: The African Experience*, Boulder, Col., Westview.

Dawe, A., 1942, *Memorandum on a Federal Solution for East Africa*, Colonial Office 967/57 27 - 7.

de Braganca, A. and I. Wallerstein, eds., 1982, *The African Liberation Reader*, 3 Volumes, London: Zed Press.

de Kiewet, C. W., 1956, *The Anatomy of South African Misery*, London.

de Vos, A., *Africa the Devastated Continent?* The Hague, 1975.

de Vroey, M., 1984, 'A Regulation Approach Interpretation of Contemporary Crisis', *Capital and Class*, 23.

de Waal, A., 1990, 'A Re-Assessment of Entitlement Theory in the Light of Recent Famines in Africa', *Development and Change*, Vol. 21, pp. 469-490.

Dean, P., 1982, *The Evolution of Economic Ideas*, Cambridge, Cambridge University Press.

Deaux, K., and J. Taynor, 1973, 'Evaluation of Male and Female Ability: Bias Works Both Ways', *Psychological Review*, Vol. 32, pp. 261-2.

Deger, S., 1986, *Military Expenditures in Third World Countries: The Economic Effects*, London, Routledge and Kegan Paul.

Dei, G. J. S., 1990, 'Indigenous Knowledge and Economic Production: The Food Crop Cultivation, Preservation and Storage Methods of a West African Community', Ecology of Food and Nutrition, Vol. 24, pp. 1-20.

Dembele, K., 1987, 'State Policies on Agriculture and Food Production in Mali', in Mkandawire and Bourenane, ed.

Demery, P and T. Addison, 1987, 'Stabilization Policy and Income Distribution in Developing Countries', World Development, 15, pp. 1483-1493.

Dennis, C., 1987, 'Women and the State in Nigeria: The Case of the Federal Military Government, 1984-5', in Afshar, ed.

Denoon, D. and A. Kuper, 1970, 'Nationalist Historians in Search of a Nation. The New Historiography', *African Affairs*, 69, pp. 329-349.

Denoon, D. and B. Nyeko, 1972, *Southern Africa Since 1800*, 1st edition, London, Longman.

Denoon, D. and B. Nyeko, 1984, *Southern Africa Since 1800*, 2nd edition, London, Longman.

Denoon, D., 1983, *Settler Capitalism: The Dynamics of Dependent Development in the Southern Hemisphere*, Oxford, Clarendon.

Denzer, L., 1976, 'Towards a Study of the History of West African Women's Participation in Nationalist Politics: The Early Phase, 1935-1950', *Africana Research Bulletin*, Vol. 6, pp. 65-85.

Denzer, L., 1981, 'Constance A. Cummings-John of Sierra Leone: Her Early Political Career', Tarikh, Vol. 7, pp. 20-32.

Denzer, L., 1987, 'The Influence of Pan-Africanism in the Political Career of Constance A. Cummings-John', in Hill.

Depelchin, J., 1976, 'Toward a Problematic History of Africa', Tanzania Zamani, 18, pp. 2-9.

Depelchin, J., 1977, 'African History and the Ideological Reproduction of Exploitative Relations of Production', *African Development*, 2, pp. 43-61.

Depelchin, J., 1992, *From Congo Free State to Zaire (1885-1974)*, Dakar, CODESRIA Book Series.

Dhlamini, Z., 1982, 'Women's Liberation', in de Braganca and Wallerstein, eds.

Diamond, L., 1987, 'Class Formation in the Swollen African State', *Journal of Modern African Studies*, Vol. 25, pp. 567-97.

Diamond, L., 1993a, 'Ex Africa ... A New Democratic Spirit Has Loosened the Grip of African Dictatorial Rule', Times Literary Supplement, 2 July.

Diamond, L., 1993b, 'The Globalization of Democracy', in Slater, Schutz, and Dorr, eds.

Diamond, L., 1995, 'Promoting Democracy in Africa: U.S. and International Policies in Transition', in J.W. Harbeson and D. Rothchild, eds.

Diamond, L., J. J. Linz, and S. M. Lipset, eds., 1988, *Democracy in Developing Countries*, Volume 2: Africa, Boulder, Col., Lynne Rienner.

Diejomaoh, V. P., 1987, 'Welcome Address', in ILO/JASPA.

Dieng, A. A., 1995, 'The Political Context of Structural Adjustment in Africa', in Mkandawire and Olukoshi, eds.

Dietz, M., 1992, 'Context Is All: Feminism and Theories of Citizenship', in Mouffe, ed.

Dinham, B. and C. Hines, 1983, Agribusiness in Africa, London, Earthscan.

Diop, C. A., 1974, The African Origin of Civilization: Myth or Reality, New York, Hill, [1944].

Diop, C. A., 1987a, Precolonial Black Africa, Trenton, NJ, Africa World Press.

Diop, C. A., 1987b, The Cultural Unity of Black Africa, Chicago, Third World Press, [1959].

Diouf, M. and M. Mamdani eds., 1994, Academic Freedom in Africa, Dakar, CODESRIA Book Series.

Dobb, M., 1979, Theories of Value and Distribution Since Adam Smith, New Delhi, Vikas Publishing House.

Dolphyne, F. A., 1991, Emancipation of Women, Oxford, ABC.

Dow, J.C.R., 1964, The Management of the British Economy, 1945-1960, Cambridge, Cambridge University Press.

Doyle, M., 1993, 'A Dangerous Place', Africa Report, Vol. 38, pp. 38-40.

Drachler, J., 1975, Black Homeland/Black Diaspora: Cross-Currents of the African Relationship, New York, Kennikat Press.

Drazen, A., 1980, 'Recent Development in Macroeconomic Disequilibrium Theory', Econometrica, 48, pp. 283-306.

Drazen, A., 1982, 'Unemployment in LDCs: Worker Heterogeneity, Screening and Quantity Restraints', World Development, 10, pp. 1039-1047.

Dubois, W. E. B., 1947, The World and Africa, New York, Viking.

Duignan, P. and L. H. Gann, 1967, Burden of Empire, New York, Praeger.

Duignan, P. and L. H. Gann, 1969-1975, Colonialism in Africa 1870-1960, 5 Volumes, Cambridge, Cambridge University Press.

Dumont, R., 1980, The Growth of Hunger, London, Marion Boyars.

Duncan, J. and Ley, D., 1993, Place, Culture, Representation, London and New York.

Duodu, C., 1987, 'How to Have your Cake and Eat it', South, June.

During, S., ed., 1993, The Cultural Studies Reader, London and New York, Routledge.

Durufle, G., 1989, 'Structural Disequilibria and Adjustment Policies in the Ivory Coast', in Campbell and Loxley, eds.

Eaton, J., 1966, Political Economy, New York, International Publishers.

ECA, 1987, Abuja Statement, The International Conference on Africa: The Challenge of Economic Recovery and Accelerated Development, Abuja, Nigeria, ECA/CERAD/87/75.

ECA, 1988, The Khartoum Declaration on the Human Dimension of Africa's Economic Recovery and Development, Khartoum, ECA.

ECA, 1988a, 'The Impact of the Economic Crisis on Vulnerable Groups in African Societies: Women', Khartoum Conference on the Human Dimension of Africa's Economic Recovery and Development, ECA/ICHD/88/46, March.

ECA, 1988b, 'Long-term Development and Structural Change: Manpower Planning and Utilization', Khartoum Conference on the Human Dimension of Africa's Economic Recovery and Development, ECA/ICHD/88/32, March.

ECA, 1988c, 'An Enabling Environment to Retain Africa's High-level Manpower', Khartoum Conference on the Human Dimension of Africa's Economic Recovery and Development, ECA/ICHD/88/33, March.

ECA, 1989, *African Alternative Framework to Structural Adjustment Programmes for Socio-Economic Recovery and Transformation*, E/ECA/CM.15/6/Rev.3.

ECA/ILO, 1988, 'Long term Development and Structural Change in Africa: Manpower Planning and Utilization', Khartoum Conference on the Human Dimension of Africa's Economic Recovery and Development sponsored by the UNECA, March.

Echeruo, M. J. C., 1996, 'Modernism, Blackface, and the Postcolonial Condition', Research in African Literatures, Vol. 27, pp. 172-186.

Egejuru, P. A., 1980, *Towards African Literary Independence: A Dialogue with Contemporary African Writers*, Westport, Conn., Greenwood Press.

Eicher, J., 1986, 'The Paper Qualification Syndrome and Unemployment of School-Leavers in West Africa', in JASPA.

Eke, P., 1986, 'Development Theory and the African Predicament', *Africa Development*, Vol. 11, pp. 1-40.

Ekejiuba, F., 1967, 'Omu Okwei, The Merchant Queen of Ossomari: A Biographical Sketch', *Journal of the Historical Society of Nigeria*, Vol. 3, pp. 633-46.

Elbl, I., 1992, 'Cross-Cultural Trade and Diplomacy: Portuguese Relations with West Africa, 1441-1521', *Journal of World History*, Vol. 3, pp. 165-204.

Eldridge, E. A., 1991, 'Women in Production: The Economic Role of Women in Nineteenth Century Lesotho', Signs, Vol. 16, pp. 707-731.

Elkan, W., 1960, *Migrants and Proletarians*, London, Oxford University Press.

Emecheta, B., 1979, *The Joys of Motherhood*, London, Allison and Busby.

Emecheta, B., 1979, *The Joys of Motherhood*, London, Heinemann.

Emecheta, B., 1982, *Destination Biafra*, London, Heinemann.

Emecheta, B., 1982, *Double Yoke*, London, Heinemann.

Emecheta, B., 1994, *Kehinde*, Oxford, Heinemann.

Emmanuel, A., 1972, *Unequal Exchange: A Study of the Imperialism of Trade*, New York, Monthly Review Press.

Emmanuel. A., 1974, 'Myths of Development and Myths of Underdevelopment', New Left Review, 85.

Engels, F., 1972, *The Origin of the Family, Private Property, and the State*, New York, Pathfinder Press.

Enloe, C., 1973, *Ethnic Conflict and Political Development*, Boston, Little Brown,

Entelis, J. P., 1992, 'The Crisis of Authoritarianism in North Africa: The Case of Algeria', Problems of Communism, Vol. 41, pp. 71-81.

Epp-Tieseen, E., 1990, *Missiles and Malnutrition: The Links Between Militarization and Underdevelopment*, Waterloo, ON, Project Ploughshares.

Ergas, Z., ed., 1987, *African State in Transition*, London, Macmillan.

Eriksen, T. L., 1979, *Modern African History: Some Historiographical Observations*, Uppsala.

Escobar, A., 1995, *Encountering Development: The Making and Unmaking of the Third World*, Princeton, NJ, Princeton University Press.

Esedebe, P., 1982, *Pan-Africanism: The idea and the movement, 1776-1963*, Washington, Howard University Press.

Eshiwani, G. S., 1985, 'Women's Access to Higher Education in Kenya: A Study of Opportunities and Attainment in Science and Mathematics Education', *Journal of Eastern African Research and Development*, 15, 91-110.

Essuman-Johnson, A., 1988, 'Ghana's Experience with Foreign Capital and the Debt Problem Since 1983', KEA/FES Conference on 'The North-South Dialogue: A Strategy for Africa', Nairobi, November.

Etherton, M., 1982, *The Development of African Drama*, London, Hutchinson.

Evasdaughter, E. N., 1989, 'Bessie Head's *A Question of Power* Read as a Mariner's Guide to Paranoia', *Research in African Literatures*, Vol. 20, pp. 72-83.

Eyoh, D., 1995, 'From Economic Crisis to Political Liberalization: Pitfalls of the New Political Sociology for Africa'.

Eyoh, D., 1995, 'From the Colonial Belly to the Ballot Box: Ethnicity and Politics in Post-Independence Africa', *Queen's Quarterly*, Vol. 102, pp. 39-51.

Eze O. C., 1984, *Human Rights in Africa*, Lagos, Macmillan

Fage, J. D., 1970, *Africa Remembered*, London, Oxford University Press.

Fage, J. D., 1978, *A History of Africa*, London, Hutchinson.

Fage, J. D., 1989, 'British African Studies Since the Second World War: A Personal Account', *African Affairs*, Vol. 88, pp. 397-413.

Fage, J. D., 1993, 'Reflections on the Genesis of Anglophone African History After World War II', *History in Africa*, Vol. 20, pp. 15-26.

Fallers, L., 1961, 'Are African Cultivators to be Called Peasants', *Current Anthropology*, 2, pp. 108-110.

Fanon, F., 1961, *The Wretched of the Earth*, New York, Grove and Wedenfeld.

Fanon, F., 1963, *The Wretched of the Earth*, London, Andre Deutsch.

FAO, 1983, *Production Yearbook*, Rome, FAO.

FAO, 1984, *World Food Report*, Rome, FAO.

FAO, 1985a, *Food and African Countries Affected by Calamities in 1983-85*, Situation Report n° 7, Rome, FAO.

FAO, 1985b, *Changes in Shifting Cultivation in Tropical Africa*, Rome, FAO.

Farah, N. R., 1990, 'Civil Society and Freedom of Research', CODESRIA Symposium on Academic Freedom, Research and the Social Responsibility of the Intellectual in Africa, Kampala, Uganda, November.

Farah, N., 1979, *Sweet and Sour Milk*, London, Allison and Busby.

Farah, N., 1981, *Sardines*, London, Allison and Busby.

Farah, N., 1983, *Close Sesame*, London, Allison and Busby.

Farah, N., 1986, *Maps*, London, Picador.

Farnsworth, et al., 1975, 'Potential Value of Plants as Sources of New Antifertility Agents: Part I', *Journal of Pharmaceutical Sciences*, Vol. 64, pp. 535-598 and Part II, 717-754.

Fatton, R., 1989, 'Gender, Class, and State in Africa', in Parpart and Staudt, eds.

Fatton, R., 1990, 'Liberal Democracy in Africa', *Political Science Quarterly*, Vol. 105, pp. 455-473.

Fatton, R., 1992, *Predatory Rule: State and Civil Society in Africa*, Boulder, Col, Lynne Rienner.

Faye, D., 1994, 'Publishing in Senegal: Current Developments', *Bellagio Publishing Network Newsleter*, Vol. 10, pp. 11.

Fearon, P., 1979, *The Origins and Nature of the Great Slump*, 1929-1932, London.

Featherstone, M., 1990, *Global Culture: Nationalism, Globalization and Modernity*, London, Sage.

Feder, E., 1977, *Strawberry Imperialism*, London, America Latina.

Feder, G. and R. Norohna, 1987, 'Land Rights Systems and Agricultural Development in Sub-Saharan Africa', World Bank Research Observer, Vol. 2, pp. 143-169.

Fei, J. H. C. and G. Ranis, 1964, *Development of the Labour Surplus Economy: Theory and Policy*, London, Irwin.

Feierman, S. and J. M. Janzen, eds., 1992, *The Social Basis of Health and Healing in Africa*, Berkeley, University of California Press.

Feierman, S., 1990, *Peasant Intellectuals: Anthropology and History in Tanzania*, Madison, University of Wisconsin Press.

Feierman, S., 1993, 'African History and the Dissolution of World History', in Bates, Mudimbe, and O'Barr, eds.

Ferber, M. A., 1986, 'Citations: Are They an Objective Measure of Scholarly Merit?' Signs, Vol. 11, pp. 381-389.

Ferguson, R. ed., et al., 1990, *Out There: Marginalization and Contemporary Cultures*, New York, New Museum of Contemporary Art.

Fernández-Armesto, F., 1995, 'Rewriting History', Index on Censorship, Vol. 24, n° 3, pp. 25-32.

Fieldhouse, D. K., 1982, 'Decolonization, Development and Dependence: A Survey of Changing Attitudes', in Gifford and Louis, eds.

Fieldhouse, D. K., 1988, 'Arrested Development in Anglophone Black Africa?', in Gifford and Louis, eds.

Fieldhouse, D., 1986, *Black Africa, 1945-1980*, London.

Fieldhouse, D.K., 1971, 'The Economic Exploitation of Africa: Some British and French Comparisons', in Gifford and Louis, eds.

Financial Review, 1988, Nairobi, Kenya.

Fines, R., 1991, with Dennis Daved, *Beyond Apartheid: Labor and Liberation in South Africa*, Johannesburg, Ravan Press.

Fitt, Y., et. al., 1976, *The World Economic Crisis. US Imperialism at Bay*, London, Zed Press.

Fleischman, J., 1993, 'An Uncivil War', *Africa Report*, Vol. 38, pp. 59-59.

Flint, J. E., 1982, 'African Historiography - A Subjective View', Center for African Studies, Dalhousie University.

Fong, G. C. M. et al., 1993, *A Bibliography of African, Middle Eastern and Indian History, 1970-1993*, Department of History, Concordia University, Montreal.

Fonow, M. M. and J. A. Cook, 1991a, 'Back to the Future: A Look at the Second Wave of Feminist Epistemology and Methodology', in Fonow and Cook, eds.

Fonow, M. M. and J. A. Cook, eds., 1991b, 'Beyond Methodology: Feminist Scholarship as Lived Research', Bloomington, Indiana University Press.

Ford, J., 1971, 'The Role of Trypanosomiasis in African Ecology: A Study of Tsetse-Fly Problem', Oxford, Clarendon Press.

Fortman, L., 1982, 'Women and Work in a Communal Setting: The Tanzanian Policy of Ujamaa', in Bay and Stichter.

Foster, E. B., 1992, 'The Construct of a Postmodernist Feminist Theory for Caribbean Social Science Research', Social and Economic Studies, 41, pp. 1-43.

Foster-Carter, A., 1976, 'The Modes of Production Controversy', New Left Review, 107, pp. 7-33.

Fran, K., 1984, 'Feminist Criticism and the African Novel', African Literature Today, Vol. 14, pp. 34-45.

Franck, T. M., 1960, Race and Nationalism: The Struggle for Power in Rhodesia - Nyasaland, London, Allen and Unwin.

Frank, A. G., 1967, Capitalism and Underdevelopment in Latin America: Historical Studies of Chile and Brazil, New York, Monthly Review Press.

Frank, A. G., 1969, Latin America: Underdevelopment or Revolution? New York, Monthly Review Press.

Frank, A. G., 1978, Dependent Accumulation and Underdevelopment, London, Macmillan.

Frank, A. G., 1980, Crisis: in the World Economy, London, Heinemann.

Frank, A. G., 1981, Crisis in the Third World, New York, Monthly Review Press.

Frank, A. G., 1984, Critique and Anti-critique: Essays on Dependence and Reformism, New York, Praeger.

Frank, A.G., 1981, In the Third World, London, Heinemann.

Franke, R. and B. Chasin, eds., 1980, Seeds of Famine, Montclair, NJ, Allanheld, Osmun and Co.

Fransman, M. ed., 1982, Industry and Accumulation in Africa, London, Heinemann.

Fraser, R., 1968, The Novels of Ayi Kwei Armah, London, Heinemann.

Freedman, D. H., 1986, 'Youth Unemployment Problems and Programmes in Africa', in JASPA.

Freund, B., 1981, Capital and Labour in the Nigerian Tin Mines, London, Longman.

Freund, B., 1984, 'Labor and Labor History in Africa: A Review of the Literature', African Studies Review, Vol. 27, n° 2, pp. 1-58.

Freund, B., 1984, The Making of Contemporary Africa: The Development of African Society Since 1800, London, Macmillan.

Freund, B., 1985 'The Modes of Production Debate in African Studies', Canadian Journal of African Studies, 19, pp. 23-29.

Freund, B., 1988, The African Worker, Cambridge, Cambridge University Press.

Freund, B., 1995, Insiders and Outsiders: The Indian Working Class in Durban 1910-1990, Portsmouth, NH, Heinemann.

Friedland, J. and M. Westlake, 1986, 'The Incredible Shrinking Lenders', South, September.

Friedland, W. H., 1969, Vuta Kamba: The Development of Trade Unions in Tanganyika, Stanford, Hoover Institution Press.

Friedland, W. H., 1974, 'African Trade Union Studies: Analysis of Two Decades', Cahiers D'Etudes Africaines, 14, pp. 575-589.

Friedman, M., 1953, *Essays In Positive Economics*, Chicago, University of Chicago Press.

Fukuyama, F., 1992, *The End of History and the Last Man*, London, Hamish.

Funkel, V. R., 1993, 'Savimbi's Sour Grapes', *Africa Report*, Vol. 38, pp. 25-28.

Furedi, F., 1989, *The Mau Mau War, in Perspective*, London, James Currey.

Fyfe, C. and D. McMaster, eds., 1981, *African Historical Demography*, Edinburgh: University of Edinburgh, Centre for African Studies.

Fyfe, C., 1994, 'The Emergence and Evolution of African Studies in the United Kingdom', 21st Annual Spring Symposium, Center for African Studies, University of Illinois at Urbana-Champaign.

Fyfe, C., ed., 1976, *African Studies Since 1945*, London, Longman.

Gailey, H. A., 1972, *History of Africa from 1800 to Present*, London, Holt.

Gaitskell, D., et al., 1983, 'Class, Race and Gender: Domestic Workers in South Africa', *Review of African Political Economy*, 27/28, pp. 86-106.

Galaty, J., 1992, 'The Land is Yours: Social and Economic Factors in the Privatization, Subdivision and Sale of Maasai Ranches', *Nomadic Peoples*, 30, pp. 26-40.

Galbraith, J. K., 1966, *The Affluent Society*, London, Penguin.

Galbraith, J. K., 1972, *The New Industrial State*, Harmondswoth, Penguin.

Galbraith, J. K., 1992, *The Culture of Contentment*, New York, Houghton Miflin.

Gallagher, J. and Robinson, R., 1953, 'The Imperialism of Free Trade', Economic History Review, 2nd ser., Vol. 6, pp. 1-15.

Gamble, A. and Walton, P., 1976, *Capitalism in Crisis*, London, Macmillan.

Gardinier, D. E., 1982, 'Decolonization in French, Belgian, and Portuguese Africa', in Gifford and Louis, eds.

Gardinier, D. E., 1988, 'Decolonization in French, Belgian, and Portuguese Africa, Bibliography', in Gifford and Louis, eds.

Gardner, R., 1967, *Sterling-Dollar Diplomacy*, New York.

Gavlak, D., 1995, 'Egyptian Couple Battling Their Beliefs - And Marriage', The Herald, Harare, July 27, p. 10.

Gayi, S. K., 1991, 'Adjustment and "safety-netting" Ghana's Programme of Actions to Mitigate the Social Costs of Adjustment (PAMSCAD)', *Journal of International Development*, 3, pp. 557-564.

Geertz, C., 1963, *Peddlers and Princes*, Chicago, University of Chicago Press.

Geiger, S., 1987, 'Women in Nationalist Struggle: TANU Activists in Dar es Salaam', *International Journal of African Historical Studies*, Vol. 20, pp. 1-26.

Geiger, S., 1990, 'What's Feminist About Women's Oral History?', *Journal of Women's History*, Vol. 2, pp. 169-182.

Geiss, I., 1974, *The Pan-African Movement*, Methuen, London.

Gelb, S., 1991, 'South Africa's economic crisis: an overview', in Gelb, ed.

Gelb, S. ed., 1991, *South Africa's Economic Crisis*, Cape Town, David Philip.

Gelles, R. J., 1993, 'From yellow pads, to Typewriters, to Wordprocessors: Confessions of a... Writer? Author? Scholar?' *Marriage and Family Review*, Vol. 18, pp. 381-92.

Gellner, E., 1993, 'The Mightier Pen? Edward Said and the Double Standards of Inside-out Colonialism', *The Times Literary Supplement*, 19 February, pp. 3-4.

George, S., 1976, *How the Other Half Dies*, Harmondsworth, Penguin.

George, S., 1988, *A Fate Worse Than Debt*, London.

Gerald, A., 1981, *African Language Literatures*, Washington, DC, Three Continents Press.

Gerry, C., 1974, *Petty Producers and the Urban Economy: A Case Study of Dakar*, Geneva, ILO/WEP Working Paper.

Gerry, C., 1978, 'Petty Production and Capitalist Production in Dakar: The Case of Self-Employed', *World Development*, 6, pp. 1147-1160.

Gerschenkron, A., 1962, *Economic Backwardness in Historical Perspective*, Cambridge, Mass, Harvard University Press.

Geschiere, P., 1985, 'Applications of the Lineage Mode of Production in African Studies', *Canadian Journal of African Studies*, 19, pp. 58-63.

Ghosh, P. K., ed., 1984, *Disarmament and Development: A Global Perspective*, Westport, Conn., Greenwood Press.

Gibbon, P., 1995, 'Towards a Political Economy of the World Bank, 1970-90.' In T.Mkandawire and A. Olukoshi, eds.

Gibbon, P., Y. Bangura, and A. Ofstad, 1992, *Authoritarianism Democracy and Adjustment: The Politics of Economic Reform in Africa*.Uppsala: Scandinavian Institute of African Studies.

Gibbs, J., 1988, 'Singing in the Dark Rain', Index on Censorship, Vol. 17, n° 2, pp. 18-22.

Gidney, M., 1994, 'APNET and VSO Launch a New Program to Support African Publishing', *Bellagio Publishing Network Newsletter*, Vol. 11, pp. 2-3.

Gifford, P. and W. R. Louis eds., 1971, *France and Britain in Africa*, New Haven Yale University Press.

Gifford, P. and W. R. Louis eds., 1982, *The Transfer of Power in Africa, Decolonization, 1940-1960*, New Haven and London, Yale University Press.

Gifford, P. and W. R. Louis, eds., 1988, *Decolonization and African Independence: The Transfers of Power, 1960-1980*, New Haven and London, Yale University Press.

Gikandi, S., 1991, 'The Epistemology of Translation: Ngugi, *Matigari*, and the Politics of Language', *Research in African Literatures*, Vol. 22, pp. 161-7.

Gilroy, P., 1993a, 'Travelling Theorist', New Statesman and Society, 12 February, p. 46.

Gilroy, P., 1993b, *The Black Atlantic: Modernity and Double Consciousness*, Cambridge, MA, Harvard University Press.

Glantz, M., 1987, 'Drought and Hunger in Africa', Cambridge, Cambridge University Press.

Glantz, M., ed., 1976, *The Politics of Natural Disaster: The Case of the Sahel Drought*, New York, Praeger.

Glick, P., 1993, 'Publish as Soon as you Can', Marriage and Family Review, Vol. 18, pp. 93-98.

Glover, D. and D. Tussie, 1993, 'Developing Countries in World Trade: Implications', in Tussie and Glover, eds.

Gluck, S., 1979, 'What's So Special about Women? Women's Oral History', Frontiers, Vol. 2, pp. 3-11.

Goddard, D., 1969, 'The Limits of British Anthropology', *New Left Review*, 58, pp. 79-89.

Godlewska, A. and Smith, N., eds., 1994, *Geography and Empire*, Cambridge, Ma., Blackwell.

Gold, A., 1985, 'Women in Agricultural Change: The Nandi in the 19th Century', in Ogot, ed.

Goldfrank, W.L., 1983, 'The Limits of Analogy: Hegemonic Decline in Great Britain and the United States', in Bergesen, ed.

Goldsworthy, D., 1971, *Colonial Issues in British Politics 1945-61*, Oxford, Clarendon Press.

Goliber, T. J., 1989, 'Africa's Expanding Population: Old Problems, New Policies', *Population Bulletin*, Vol. 44, pp. 1-51.

Goodwin, J., 1984, *Cry Amandla! South African Women and the Question of Power*, New York, African Publishing Co.

Goody, J., 1971, *Technology, Tradition and the State in Africa*, London, Oxford University Press.

Gorden, D. C., 1972, *Women of Algeria: An Essay of Change*, Cambridge, Cambridge University Press.

Gordimer, N., 1994, 'Soyinka the Tiger', in Maja-Pearce, ed.

Gottdiener, M. and N. Komninos, eds., 1989, *Capitalist Development and Crisis Theory*, New York, St. Martin's Press.

Gough, K., 1969, 'Anthropology: Child of Imperialism', *Monthly Review*, 10.

Goulbourne, H., 1979, 'Some Problems of Analysis of the Political in Backward Capitalist Social Formations', in H. Goulbourne, ed.

Goulbourne, H., ed., 1979, *Politics and State in the Third World*, London, Macmillan.

Graf, W. D., 1988, *The Nigerian State*, London, James Currey and Heinemann.

Graham, G., 1993, 'Book Publishing and Freedom: An Essay', *Media, Culture and Society*, Vol. 15, pp. 245-251.

Graham, M., et al., 1986, *Disarmament and World Development*, 2nd ed. Oxford, Pergamon Press.

Gray, R. F., 1963, *The Sonjo of Tanganyika: An Anthropological Study of an Irrigation-Based Society*, London, Oxford University Press.

Green, R. H., 1978, 'Basic Human Needs: Concept of Slogan, Synthesis or Smokescreen?' *IDS Bulletin*, June.

Greenland and Lal, eds., 1977, *Soil Conservation and Management in the Humid Tropics*, Chichester, John Wiley.

Grever, M., 1991, ''Pivoting the Center': Women's History as a Compulsory Examination Subject in All Dutch Secondary Schools in 1990 and 1991', *Gender and History*, 3, pp. 65-80.

Grey-Johnson, C., ed., 1990, *The Employment Crisis in Africa*, Harare, Sapes Trust.

Grigg, D. B., 1974, *The Agricultural Systems of the World*, Cambridge, Cambridge University Press.

Gugelberger, G. M. ed., 1985, *Marxism and African Literature*, London, James Currey.

Guha, R., 1963, *A Rule of Property for Bengal: An Essay on the Idea of Permanent Settlement*, Paris and The Hague, Mouton.

Gulliver, P. H., 1955, *Labour Migration in a Rural Economy*, Kampala, East African Institute of Social Research.

Gunner, E., 1979, 'Songs of Innocence and Experience: Women as Composers and Performers of Izibongo, Zulu Praise Poetry', *Research in African Literatures*, Vol. 10, pp. 239-267.

Gupta, P. S., 1975, *Imperialism and the British Labour Movement*, 1914-1964, London, Macmillan.

Gurnah, A., 1991, '*Matigari*: A Tract of Resistance', *Research in African Literatures*, Vol. 22, pp. 169-172.

Gutkind, P. C. W. and I. Wallerstein, eds., 1976, *The Political Economy of Contemporary Africa*, London, Sage Publications.

Gutkind, P. C. W. and P. Waterman eds., 1977, *African Social Studies: A Radical Reader*, New York, Monthly Review Press.

Gutkind, P. C., R. Cohen, and J. Copans, eds., 1978, *African Labour History*, London, Sage Publications.

Guy, J., 1980, *Society and Economy in Pre-Industrial South Africa*, London, Longman.

Guyer, J. I., 1995, *A Perspective on African Studies in the United States*, Report Submitted to the Ford Foundation.

Guyer, J., 1981, 'Household and Community in African Studies', *african Studies Review*, vol. 24, pp. 87-137.

Guyer, J., 1984, 'Women in the Rural Economy: Contemporary Variations', in Hay and Stichter, eds.

Haaland, R. and P. Shinnie, eds., 1985, *African Iron Working - Ancient and Traditional*, Oslo, Norwegian University Press.

Hafkin, N. and E. Bay, eds., 1976, *Women in Africa: Studies in Social and Economic Change*, Stanford, Stanford University Press.

Hagan, G. P., 1990, 'Academic Freedom and National Responsibility in an African State', CODESRIA Symposium on Academic Freedom, Research and the Social Responsibility of the Intellectual in Africa, Kampala, Uganda, November.

Haile, D., 1980, *Law and the Status of Women in Ethiopia*, Addis Ababa, UN/ECA.

Halim, R. A., 1990, 'Academic Freedom', CODESRIA Symposium on Academic Freedom, Research and the Social Responsibility of the Intellectual in Africa, Kampala, Uganda, November.

Hall, A.V. ed., 1984, *Conservation of Threatened Habitats*, Pretoria, Council for Scientific and Industrial Research.

Hall, C., 1991, 'Politics, Post-Structuralism and Feminist History', *Gender and History*, Vol. 3, pp. 204-210.

Hamalengwa, M., 1983, 'The Political Economy of Human Rights in Africa: Historical and Contemporary Perspectives', *Philosophy and Social Action*, n° 9, pp. 15-26.

Hamalengwa, M., 1991, *Thoughts Are Free*, Toronto, Africa in Canada Press.

Hamalengwa, M., 1992, *Class Struggles in Zambia, 1889-1989 & The Fall of Kenneth Kaunda*, Lanham, Mass., University Press of America.

Hamermesh, D., 1992, *Unemployment Insurance for Developing Countries, Population and Human Resources Department*, Working Paper 897, World Bank, May.

Hamid, M. B., 1992, 'Democratization in Africa: Some Exploratory Views', CODESRIA Seventh General Assembly, Dakar.

Hamill, J., 1994, 'Angola's Road From Under the Rubble', *The World Today*, 50, pp. 6-11.

Hammersmesh, D. S. et al., 1982, 'Scholarship, Citations and Salaries: Economic Rewards in Economics', *Southern Economic Journal*, Vol. 49, pp. 472-481.

Hammond, P. E., 1983, 'Power Changes and Civil Religion: The American Case', in A. Bergesen, ed.

Handlin, O., 1987, 'Libraries and Learning', *The American Scholar*, Vol. 56, pp. 205-218.

Hanks, R. S., 1993, '"Editors" Tips to Authors: "Guidelines for Novitiates" and "How to Avoid Fatal Mistakes"', *Marriage and Family Review*, Vol. 18, pp. 13-23.

Hansberry, W. L., 1977, *Africa and Africans as Seen by Classical Writers*, Washington, DC, Howard University Press.

Hansen, E. and K. A. Ninsin, eds., 1989, *The State Development and Politics in Ghana*, Dakar, CODESRIA Book Series.

Hansen, K. T., 1989, 'The Black Market and Women Traders in Lusaka, Zambia', in Parpart and Staudt, eds.

Hansen, W. and B. Schulz, 1981, 'Imperialism, Dependency and Social Class', *African Today*, 3rd Quarter 5-36.

Hansohm, D., 1986, 'The "success" of IMF/World Bank policies in Sudan', in Lawrence, ed.

Harbemas, J., 1976, *Legitimation Crisis*, London, Heinemann.

Harbeson, J. W. and D. Rothchild, eds., 1995, *Africa in World Politics: Post-Cold War Challenges*, Boulder, Col., Westview.

Harbeson, J. W., et al., eds., 1994, *Civil Society and the State in Africa*, Boulder, Col, Lynne Rienner.

Hargen, L. H., 1988, 'Scholarly Consensus and Journal Rejection Rates', *American Sociological Review*, Vol. 53, pp. 139-51.

Hargreaves, J. D., 1980, 'Assumptions, Expectations, and Plans: Approaches to Decolonization in Sierra Leone', in Morris-Jones and Fisher, eds.

Hargreaves, J. D., 1982, 'Towards the Transfer of Power in British West Africa', in Gifford and Louis, eds.

Hargreaves, J. D., 1988, *Decolonization in Africa*, London and New York, Longman,

Harlan, J. R., 1975, *Crops and Man*, Madison, American Society of Agronomy and Crop Science of America.

Harlow, V. and E. M. Chilver, eds., 1965, *History of East Africa*, Volume 2, London, Oxford University Press.

Harper, P. B., 1995, 'Nationalism and Social Division in Black Arts Poetry of the 1960s', in Appiah and Gates, eds.

Harries, P., 1985, 'Modes of Production and Modes of Analysis: The South African Case', *Canadian Journal of African Studies*, 19, pp. 30-37.

Harries, P., 1994, *Work, Culture, and Identity: Migrant Laborers in Mozambique and South Africa c. 1860-1910*, Portsmouth, NH, Heinemann.

Harrington, H., 1986, 'The World Bank's Gendarme', *African Business*, March.

Harris, B., 1978, 'Quasi-Formal Employment Structures and Behaviour in the Unorganized Urban Economy, and the Reverse: Some Evidence from South India', *World Development*, 6.

Harris, J. E., ed., 1982, *Global Dimensions of the African Diaspora*, Washington, D.C., Howard University Press.

Harris, L., 1986, 'Conceptions of the IMF's Role in Africa', in Lawrence, ed.

Harris, N., 1986, *The End of the Third World Newly Industrialising Countries and the Decline of an Ideology*, Harmondsworth, Penguin.

Harris, R. ed., 1976, *The Political Economy of Africa*, Cambridge, Cambridge University Press.

Harrison, P., 1982, *Inside the Third World*, Penguin, Harmondsworth.

Harrison, P., 1987, *The Greening of Africa*, London, Paladin.

Harsch, E., 1988, 'Privatization: No Simple Panacea', *Africa Recovery*, August.

Harsch, E., 1993, 'Accumulators and Democrats: Challenging State Corruption in Africa', *Journal of Modern African Studies*, Vol. 31, pp. 31-48.

Hart, K., 1973, 'Informal Income Opportunities and Urban Employment in Ghana', *Journal of Modern African Studies*, 11.

Harvey, D., 1982, *The Limits To Capital*, Oxford, Basil Blackwell.

Harvey, D., 1990, *The Condition of Postmodernity*, Cambridge, MA, Blackwell.

Haugerud, A., 1989, 'Land Tenure and Agrarian Change in Kenya', *Africa*, 59, pp. 61-90.

Haugerud, A., 1995, *The Culture of Politics in Modern Kenya*, Cambridge, Cambridge University Press.

Havel, V., 1988, 'Stories and Totalitarianism', *Index on Censorship*, 3.

Hawk, B. G., ed., 1993, *Africa's Media Image*, New York, Praeger.

Hawk, B., ed., 1992, *Africa's Media Image*, New York, Praeger.

Hawkins, T., 1987, 'How Debt Chokes Development', *African Recovery*, August.

Hawley, J. C., 1995, 'Ben Okri's Spirit-Child: Abiku Migration and Postmodernity', *Research in African Literatures*, Vol. 26, pp. 31-39.

Hay, M. J. and M. Wright, eds., 1982, *African Women and the Law: Historical Perspectives*, Boston, Boston University.

Hay, M. J. and S. Stichter, eds., 1984, *African Women South of the Sahara*, Harlow, Longman.

Hay, M. J., 1976, 'Luo Women and Economic Change During the Colonial Period', in Hafkin and Bay.

Hayes, C. J. H., 1941, *A Generation of Materialism*, New York, Harper and Row.

Hayter, T. and Watson, C., 1985, *Aid: Rhetoric and Reality*, London, Pluto Press.

Hazlewood, A., 1985, 'Kenya Land Transfer Programmes and their Relevance for Zimbabwe', *Journal of Modern African Studies*, 23, pp. 445-461.

Head, B., 1969, *When Rain Clouds Gather*, New York, Simon and Schuster.

Head, B., 1971, *Maru*, London, Victor Gollancz.

Head, B., 1974, *A Question of Power*, London, Heinemann.

Head, B., 1977, *The Collector of Treasures*, London, Heinemann.

Head, B., 1981, *Serowe: Village of the Rain Wind*, London, Heinemann.

Head, B., 1984, *A Bewitched Crossroad*, Johannesburg, Ad. Donker.

Head, B., 1990, *A Woman Alone. Autobiographical Writings*, Oxford, Heinemann.

Head, B., 1991, *A Gesture of Belonging: Letters from Bessie Head, 1965-1979*, Edited by Randolph Vigne, Portsmouth, NH, Heinemann.

Hear, N. V., 1984, 'Nigerian Layoff Rate May Be Slowing Down', *African Business*, October.

Hear, N. V., 1985, 'South Africa's Expulsion Threat Could Backfire', *African Business*, September.

Hear, N. V., 1985a, 'Nigeria: Profits Boom But The Screw Still Tightens', *African Business*, August.

Hear, N. V., 1985b, 'Nigeria: Manufacturers Present Their Case', *African Business*, November.

Hear, N. V., 1986, 'Migrant Workers had a Bleak Year', *African Business*, January.

Hear, N. V., 1986a, 'Nigeria and the IMF: The Shouting Match Rolls On', *African Business*, February.

Hear, N. V., 1986b, 'Privatization: The Big sell-off is Here', *African Business*, April.

Hedlund, H., 1979, 'Contradictions in the Peripherisation of a Pastoral Society: The Maasai', *Review of African Political Economy*, 15/16, 15-34.

Hegel, G. F. W., 1944, *The Philosophy of History*, New York.

Heilbroner, R. L., 1976, *Business Civilization in Decline*, New York, Norton.

Heillener, G. K., 1966, *Peasant Agriculture, Government and Economic Growth in Nigeria*, Homewood, IL, Irwin.

Heisler, H., 1970, 'A Class of Target Proletarians', *Journal of Asian and African Studies*, 5, pp. 161-175.

Heller, P. and A. Tait, 1984, 'Government, Employment and Pay: Some International Comparisons', Washington, *IMF Occasional Paper* n° 24.

Henige, D., 1982, *Oral Historiography*, London, Longman.

Henze, P. B., 1992, 'After Communism', *Problems of Communism*, Vol. 41, pp. 207-17.

Herbert, E. W., 1993, *Iron, Gender, and Power: Rituals of Transformation in African Societies*, Bloomington, Indiana University Press.

Hermida, A., 1992, 'Democracy Derailed', *Africa Report*, Vol. 37, pp. 13-17.

Hertzberg, H. and H. L. Gates, 1996, 'The African American Century', *The New Yorker*, April 29 and May, Vol. 6, pp. 9-10.

Hess, B. H. and M. M. Ferree, 1987, *Analysing Gender: A Handbook of Social Science Research*, London, Sage Publications.

Hetata, S., 1990, 'Censoring the Mind', *Index on Censorship*, Vol. 19, n° 6, pp. 18-19.

Hettne, B., 1983, 'The Development of Development Theory', *Acta Sociologica*, Vol. 26, pp. 247-6.

Hettne, B., 1990, *Development Theory and the Theory Worlds*, London, Longman.

Hiatt, R., 1994, 'The Case against double-blind reviewing', *CAUT Bulletin*, 41.

Higginbotham, E., 1990, 'Designing an Inclusive Curriculum: Bringing all Women into the Core', *Women's Studies Quarterly*, Vol. 18, pp. 7-23.

Hilferding, R., 1978, *Finance Capital*, London, Routledge and Kegan Paul.

Hill, F. W. G., 1993, 'Strategic Planning for University Libraries Welcome to Workshop Participants', in Patrikios and Levey, eds.

Hill, M. R, 1990, 'Creative Journals and Destructive Decisions: A Comment on Singer's "Academic Crisis"', *Sociological Inquiry*, Vol. 60, p. 299.

Hill, P., 1963, *The Migrant Cocoa Farmers of Southern Ghana*, Cambridge Cambridge University Press.

Hill, P., 1970, *Studies in Rural Capitalism in West Africa*, Cambridge, Cambridge University Press.

Hill, P., 1977, *Population, Prosperity and Poverty, Rural Kano 1900 and 1970*, Cambridge, Cambridge University Press.

Hill, R. A., 1987, *Pan-African Biography*, Los Angeles, Crossroads Press.

Himbara, D., 1994, *Kenyan Capitalists, the State and Development*, Boulder, Lynne Rienner.

Himmelfarb, G., 1987, *The New History and the Old*, Cambridge, Cambridge University Press.

Hinden, R., 1949, 'Dilemma in Colonial Policy' (Editorial), *Socialist Commentary* April.

Hindess, B, and P. Q. Hirst, 1975, *Pre-Capitalist Modes of Production*, London, Routledge and Kegan Paul.

Hindess, B, and P. Q. Hirst, 1977, *Mode of Production and Social Formation: An Auto-Critique of 'Pre-Capitalist Modes of Production*, London, Macmillan.

Hirji, K. F., 1990, 'Academic Pursuits Under the Link', *CODESRIA Bulletin*, n° 1.

Hirsch, E. D., 1988, *Cultural Literacy: What Every American Needs to Know*, New York, Vantage Books.

Hirsch, M. and E. F. Keller, eds., 1990, *Conflicts in Feminism*, New York, Routledge.

Hirschman, A., 1970, *Exit, Voice and Loyalty: Responses to Decline in Firms, Organizations and States*, Boston, Harvard University Press.

Hobsbawm, E. and T. Ranger, eds., 1983, *The Invention of Tradition*, Cambridge, Cambridge University Press.

Hobsbawm, E., 1970, 'Lenin and the 'Aristocracy of Labour', *Monthly Review*, April, pp. 47-55.

Hobson, J. A., 1972, *Imperialism: A Study*, London, Allen and Unwin.

Hodges, T., 1988, 'Ghana's Strategy For Adjustment With Growth', *Africa Recovery*, August.

Hodgkin, T., 1956, *Nationalism in Colonial Africa*, London, Muller.

Hodgkin, T., 1976, 'Where the Paths Began', in Fyfe, ed.

Hoffer, C. P., 1972, 'Mende and Sherbo Women in High Office', *Canadian Journal of African Studies* Vol. 4, pp. 151-64.

Holbrook, W. P., 1978, *The Impact of the Second World War on the Gold Coast, 1939-45*, Ph.D. dissertation, Princeton University.

Holm, J. and P. Molutsi, 1989, *Democracy in Botswana*, Gaborone, Macmillan.

Holston, J. and A. Appadurai, 1996, 'Cities and Citizenship', *Public Culture*, Vol. 8, pp. 187-204.

Hoogvelt, A. M. M., 1982, *The Third World in Global Development*, London, Macmillan.

Hooks, B. and C. West, 1991, *Breaking Bread. Insurgent Black Intellectual Life*, Boston, South End Press.

Hooks, B., 1981, *Ain't I a Woman: Black Women and Feminism*, Boston, South End Press.

Hooks, B., 1984, *Feminist Theory: From Margin to Center*, Boston, South End Press.

Hooks, B., 1988, *Talking Back: Thinking Feminist, Thinking Black*, Toronto, Between the Lines.

Hooks, B., 1989, 'Critical Interrogation: Talking Race, Resisting Racism', *Inscriptions*, Vol. 7-8, pp. 159-164.

Hooks, B., 1991, 'Black Women Intellectuals', in Hooks B. and West C.

Hooven, R. Van Der and F. Van Der Kraaij, eds., 1994, *Structuiral Adjustment and Beyond in Sub-Saharan Africa*, London, James Currey.

Hopkins, A. G., 1973, *An Economic History of West Africa*, London, Longman.

Hopkins, A. G., 1975, 'On Importing Gunder Frank into Africa', *African Economic History Review*, 2, pp. 13-21.

Hopkins, A. G., 1976, 'Clio-Antics: A Horoscope of African Economic History', in Fyfe, ed.

Hopkins, A. G., 1986, 'The Victorians and Africa: A Reconsideration of the Occupation of Egypt, 1882', *Journal of African History*, 27, pp. 363-391.

Hopkins, T. K. and I. Wallerstein, 1977, 'Patterns of Development in the Modern World-System', *Review* 1.

Hopkins, T. K. and I. Wallerstein, 1979, 'Cyclical Rhythms and Secular Trends of the Capitalist World Economy', *Review* 2.

Hopkins, T. K. and I. Wallerstein, eds., 1980, *Processes of the World-System*, Beverly Hills, Sage Publications.

Horton, S., et al., 1991, *Labour Markets in an Era of Adjustment: An Overview*, World Bank Working Paper Series 694, Economic Development Institute, World Bank, May.

Houessou, A. T., 1995, 'The Big Con: Europe Upside Down', *Journal of Black Studies*, Vol. 26, pp. 185-200.

Hountondji, P., 1990, 'Scientific Dependency in African Today', *Research in African Literatures*, Vol. 21, pp. 5-15.

House, W. J., 1981, 'Nairobi's Informal Sector: An Exploratory Study', in Killick, ed.

Hove, C., 1989, *Bones*, Oxford, Heinemann.

Howard, R. E., 1986, *Human Rights in Commonwealth Africa*, New Jersey, Rowman and Littlefield.

Huband, M., 1993, 'Targeting Taylor', *Africa Report*, Vol. 38, pp. 29-32.

Hughes, R., 1993, *Culture of Complaint: The Fraying of America*, New York, Oxford University Press.

Hunt, D., 1984, *The Impending Crisis in Kenya: the Case for Land Reform*, Gower, Aldershot.

Huntington, S. P., 1991, *The Third Wave: Democratization in the Late Twentieth Century*, Norman, University of Oklahoma Press.

Hutchful, E., 'Gender and African Marxism', in Imam, Mama, and Sow, eds., *Engendering African Social Science*, Dakar, CODESRIA.

Hutchful, E., 1988a, 'The Violence of Periphery States', *African Journal of Political Economy Vol???????*, pp. 48-74.

Hutchful, E., 1992, 'The International Dimensions of the Democratization Process in Africa', CODESRIA Seventh General Assembly, Democratization in Africa: Problems and Prospects.

Hutchful, E., 1995, 'Adjustment, Regimes and Politics in Africa', in Mkandawire and Olukoshi, eds.

Hutchful, E., ed., 1988b, *The IMF and Ghana, The Confidential Record*, London, Zed Press.

Hyden, G., 1980, *Beyond Ujamaa in Tanzania, Underdevelopment and the Uncaptured Peasantry*, London, Heinemann.

Hyden, G., 1980, *Beyond Ujamaa in Tanzania: Underdevelopment and an Uncaptured Peasantry*, Berkeley, University of California Press.

Hyden, G., 1983, *No Shortcuts To Progress*, London, Heinemann.

Hyden, G., 1990, 'Mamdani's One-Eyed Glimpse', *CODESRIA Bulletin*, n° 4, pp. 7-8.

Ibrahim, J., 1995, 'The Narcissism of Minor Differences and the Rise of Genocidal Tendencies in Africa: Lessons From Rwanda and Burundi', CODESRIA Eighth General Assembly.

ICFTU, 1985, *Trade Union Rights Survey of Violations*, Geneva, ICFTU.

IFAA., 1987, *The IMF, World Bank and Africa*, London, IFAA.

Ifeka-Moller, C., 1975, 'Female Militancy and Colonial Revolt: The Women's War of 1929, Eastern Nigeria', in Ardener, ed.

IFLA, 1995, Seminar on Information Provision to Rural Communities in Africa, Uppsala University Library: International Federation of Library Associations and Institutions.

Igbozurike, U. M., 1977, *Agriculture at the Crossroads: A Comment on Agricultural Ecology*, Ile-Ife, University of Ife Press.

Ikiara, G. K., 1987, 'Kenya, Manufacturing Under Bond Begins', *African Business*, October.

Illife, J., 1983, *The Emergence of African Capitalism*, London, Macmillan.

Illife, J., 1987, *The African Poor: A History*, Cambridge, Cambridge University Press.

ILO, 1972, *Employment, Incomes and Equality, A Strategy for Increasing Productive Employment in Kenya*, Geneva, ILO.

ILO, 1984, *World Labor Report 1*, Geneva, ILO.

ILO, 1985a, *World Labor Report 2*, Geneva ILO.

ILO, 1985b, *Equal Opportunities and Equal Treatment for Men and Women in Employment*, Geneva, ILO.

ILO, 1986, *Yearbook of Labour Statistics*, Geneva, ILO.

ILO, 1987, *World Labor Report 3*, Geneva, ILO.

ILO, 1989, *Tripartite Symposium on Structural Adjustment and Employment in Africa*, Geneva, ILO, 16-19 October.

ILO, 1995, *World Labour Report 8*, Geneva, ILO.

ILO/JASPA, 1982, *Paper Qualification Syndrome (PQS) and Unemployment of School Leavers*, Addis Ababa, JASPA.

ILO/JASPA, 1984a, *Rural-Urban Gap and Income Distribution: A Comparative Sub-Regional Study*, Addis Ababa, JASPA.

ILO/JASPA, 1984b, *Review of Studies of Manpower Planning - Nine English Speaking Countries and Nine French Speaking African Countries*, Addis Ababa, JASPA.

ILO/JASPA, 1985a, *Patterns of Rural Development and Impact on Employment and Incomes: The Case of Malawi*, Addis Ababa, JASPA.

ILO/JASPA, 1985b, *Informal Sector in Africa*, Addis Ababa, JASPA.

ILO/JASPA, 1985c, *Patterns of Rural Development and Impact on Employment and Incomes: The Case of Kenya*, Addis Ababa, JASPA.

ILO/JASPA, 1985d, *Patterns of Rural Development and Impact on Employment and Incomes: The Case of Somalia*, Addis Ababa, JASPA.

ILO/JASPA, 1985e, *Patterns of Rural Development and Impact on Employment and Incomes: The Case of Liberia*, Addis Ababa, JASPA.

ILO/JASPA, 1985f, *Patterns of Rural Development and Impact on Employment and Incomes: The Case of Sierra Leone*, Addis Ababa, JASPA.

ILO/JASPA, 1985g, *Patterns of Rural Development and Impact on Employment and Incomes: The Case of Sudan*, Addis Ababa, JASPA.

ILO/JASPA, 1985h, *Patterns of Rural Development and Impact on Employment and Incomes: The Case of Tanzania*, Addis Ababa, JASPA.

ILO/JASPA, 1985i, *Patterns of Industrialization: Impact on Employment and Incomes in Six Selected African Countries*, Addis Ababa, JASPA.

ILO/JASPA, 1986a, *Women's Employment: Patterns of Discrimination and Promotion of Equality in Africa: The Case of Ethiopia*, Addis Ababa, JASPA.

ILO/JASPA, 1986b, *Women's Employment: Patterns of Discrimination and Promotion of Equality in Africa: The Case of Tanzania*, Addis Ababa, JASPA.

ILO/JASPA, 1986c, *Women's Employment: Patterns of Discrimination and Promotion of Equality in Africa: The Case of Zimbabwe*, Addis Ababa, JASPA.

ILO/JASPA, 1986d, *Women's Employment: Patterns of Discrimination and Promotion of Equality in Africa: The Case of Kenya*, Addis Ababa, JASPA.

ILO/JASPA, 1986e, *Women's Employment: Patterns of Discrimination and Promotion of Equality in Africa: The Case of Sierra Leone*, Addis Ababa, JASPA.

ILO/JASPA, 1986f, *Youth Employment and Youth Programmes in Africa, A Comparative Sub-Regional Study: The Case of Zambia*, Addis Ababa, JASPA.

ILO/JASPA, 1986g, *Youth Employment and Youth Programmes in Africa, A Comparative Sub-Regional, Study: The Case of Malawi*, Addis Ababa, JASPA.

ILO/JASPA, 1986h, *Youth Employment and Youth Programmes in Africa, A Comparative Sub-Regional Study: The Case of Mauritius*, Addis Ababa, JASPA.

ILO/JASPA, 1986i, *Youth Employment and Youth Programmes in Africa, A Comparative Sub-Regional Study: The Case of Kenya*, Addis Ababa, JASPA.

ILO/JASPA, 1986j, *Youth Employment and Youth Programmes in Africa, A Comparative Sub-Regional Study: The Case of Botswana*, Addis Ababa, JASPA.

ILO/JASPA, 1986k, *Informal Sector Employment in Kenya*, Addis Ababa, JASPA.

ILO/JASPA, 1987, *Youth Employment Promotion in Africa*, Addis Ababa, JASPA.

ILO/JASPA, 1988, 'Recent Trends in Employment Equity and Poverty in African Countries', Khartoum Conference on the Human Dimension of Africa's Economic Recovery and Development, ECA/ICHD/88/25, March.

ILO/JASPA, 1989, *African Employment Report 1988*, Addis Ababa, JASPA.

Imam, A. M. and A. Mama, 1994, 'The Role of Academics in Limiting and Expanding Academic Freedom', in Diouf and Mamdani, eds.

Imam, A., 1988, 'The Presentation of African Women in Historical Writing', in Kleinberg, ed.

Imam, A., 1991, 'The Development of Women's Seclusion in Hausaland, Northern Nigeria', *Dossier*, Vol. 9/10, pp. 4-18.

Imam, A., A. Mama, and F. Sow, eds., 1997, *Gendering the Social Sciences in Africa*, Dakar, CODESRIA Book Series.

Index on Censorship, 1994, 'Immigration', Vol. 3, pp. 27-58.

Index on Censorship, London, United Kingdom.

Ingham, K., 1965, *A History of East Africa*, London, Longman.

Inglis, F., 1993, 'A Peregrine Spirit with an Eye for Eagles', *The Times Higher Education Supplement*, Vol. 5 March, pp. 25, 27.

Inikori, J. E. ed., 1982, *Forced Migration: The Impact of the Export Slave Trade on West African Societies*, New York, Africana Publishing Co.

Inikori, J. E., 1991, 'Ideology Versus the Tyranny of Paradigm: Walter Rodney and the Debate on African Development and Underdevelopment', mimeo.

Innes, D. et al., 1992, *Power and Profit: Politics, Labor, and Business in South Africa*, Cape Town, Oxford University Press.

Irele, A., 1990a, *The African Experience in Literature and Ideology*, Bloomington, Indiana University Press.

Irele, A., 1990b, 'The African Imagination', *Research.in African Literatures*, Vol. 21, pp. 49-67.

Irele, A., 1992, 'The Crisis of Legitimacy in Africa', *Dissent*, Vol. 39, pp. 93-111.

Irele, A., 1995, 'Dimensions of African Discourse', in Mrysiades and McCullough.

Isaacman, A. and B. Isaacman, 1977, 'Resistance and Collaboration in Southern and Central Africa, Ca. 1850-1920', *International Journal of African Historical Studies*, 10, pp. 31-62.

Isaacman, A. E., 1976, *The Tradition of Resistance in Mozambique*, London, Heinemann.

Isaacman, A. F., 1990, 'Peasants and Rural Social Protest in Africa', *African Studies Review*, Vol. 33, pp. 1-120.

Isaacman, A. F., 1993, 'Peasants and Rural Social Protest in Africa', in Cooper, et al.

Isaacman, A., 1984, 'State, Merchant Capital, and Gender Relations in Southern Mozambique to the End of the Nineteenth Century: Some Tentative Hypotheses', *African Economic History*, Vol. 13, pp. 25-55.

Isamah, A. N., 1995, 'Labour Response to Structural Adjustment in Nigeria and Zambia', in Mkandawire and Olukoshi, eds.

Islam, S., 1985, 'The Emperor's New Clothes', *South*, July.

Islam, S., 1987, 'Revolt of the Star Pupil', *South*, February.

Issawi, C., 1982, *An Economic History of the Middle East and North Africa*, New York, Columbia University Press.

Issue: A Journal of Opinion, African Studies, 1995, Vol. 23, n 1.

Issue: A Journal of Opinion, 1976, 6, nos. 2 and 3.

Iweriebor, I., 1990, 'Restriction to Research on Gender Issues in Africa with Particular Reference to Nigeria', CODESRIA Symposium on Academic Freedom, Research and the Social Responsibility of the Intellectual in Africa, Kampala, Uganda, November.

Iyayi, F., 1986, 'The Primitive Accumulation of Capital in a Neo-Colony, Nigeria', *Review of African Political Economy*, Vol. 35, pp. 27-39.

Izevbaye, D. S., 1977, 'The State of Criticism in African Literature', *African Literature Today*, Vol. 7, pp. 1-19.

Jackson, L. A., 1996, '"Stray Women' and 'girls on the move": Women, Spaces and Disease in Colonial and Post-Colonial Zimbabwe', 23rd Annual Spring Symposium of the Center for African Studies, University of Illinois at Urbana-Champaign, March.

Jackson, R. H., 1973, 'Political Stratification in Tropical Africa', *Canadian Journal of African Studies*, 7, pp. 381-400.

Jacobs, S., 1989, 'Zimbabwe: State, Class, and Gendered Models of Land Resettlement', in Parpart and Staudt, eds.

Jacoby, E. H., 1973, 'Transnational Corporations and Third World Agriculture', *Development and Change*, Vol. 6.

Jacoby, R., 1991, 'The Greening of the University: From Ivory Tower to Industrial Park', *Dissent*, 38, pp. 286-292.

Jaggar, A. M. and P. S. Rothenberg, 1984, *Feminist Frameworks*, New York, McGraw-Hill.

Jamal, V. and J. Weeks, 1973, 'The Vanishing Rural-Urban Gap in Sub-Saharan Africa', *International Labour Review*, 127.

Jamal, V., 1986, 'Rural-Urban Gap, Income Distribution and Poverty in Africa', in JASPA.

James, A., 1990, *In Their Own Voices: African Women Writers Talk*, London, James Currey.

James, C. L. R., 1962, *Nkrumah and the Ghana Revolution*, London, Allison and Busby.

James, C. L. R., 1963, *The Black Jacobins: Toussaint L'Ouverture and the San Domingo Revolution*, New York, Vintage Books, [1938].

James, V.U., 1991, *Resource Management in Developing Countries: Africa's Ecological and Economic Problems*, New York, Bergin and Garvey.

James, V.U., 1993, *Africa's Ecology: Sustaining the Biological and Environmental Diversity of a Continent*, Jefferson, N.C., McFarland and Co.

Jameson, F., 1991, *Postmodernism. Or. The Cultural Logic of Late Capitalism*, Durham, NC, Duke University Press.

Jansen, W., 1989, 'Ethnocentricism in the Study of Algerian Women', in Angerman, et al., eds.

Jara, A., 1993, 'Bargaining Strategies of Developing Countries in the Uruguay Round', in Tussie and Glover, eds.

Jason, P. and N. V. Hear, 1986, 'Nigeria's Crisis Sets Grim Options', *African Business*, July.

Jason, P., 1986, 'Lagos Union Seminar Halted by Arrests', *African Business*, July.

Jason, P., 1988, 'Nigeria Embarks On a New Debt Solution', *African Business*, November.

Jason, P.,1986, 'Lagos Union Seminar Halted by Arrests', *African Business*, July.

JASPA, 1986, The Challenge of Employment and Basic Needs in Africa, Nairobi, Oxford University Press.

JASPA/ILO, 1986b, *Women's Employment Patterns: Discrimination and Promotion of Equality in Africa: The Case of Tanzania*, JASPA, Addis Ababa.

JASPA/ILO, 1986c, *Women's Employment Patterns: Discrimination and Promotion of Equality in Africa: The Case of Zimbabwe*, JASPA, Addis Ababa.

JASPA/ILO, 1986d, *Women's Employment Patterns: Discrimination and Promotion of Equality in Africa: The Case of Ethiopia*, JASPA, Addis Ababa.

JASPA/ILO, 1986e, *Women's Employment Patterns: Discrimination and Promotion of Equality in Africa: The Case of Ethiopia*, JASPA, Addis Ababa.

JASPA/ILO, 1986f, *Women's Employment Patterns: Discrimination and Promotion of Equality in Africa: The Case of Sierra Leone*, JASPA, Addis Ababa.

JASPA/ILO, 1988, 'Recent Trends in Employment, Equity and Poverty in African Countries', Khartoum Conference on the Human Dimension of Africa's Economic Recovery and Development, March.

JASPA/ILO, 1988, *Women's Employment Patterns: Discrimination and Promotion of Equality in Africa: The Case of Kenya*, JASPA, Addis Ababa.

Jay, M., 1994, 'African Book Collective', *Africa Bibliography 1992*. Works on Africa Published in 1992, Edinburgh University Press (for International African Institute).

Jean-Germain, G., ed., *Coping With Uncertainty. Transitional Multi-Party Elections in Sub-Saharan Africa* (forthcoming).

Jeffries, R. D., 1975, 'Populist Tendencies in the Ghanaian Trade Union Movement', in Sandbrook and Cohen, eds.

Jeffries, R., 1978, *Class, Ideology and Power in Africa: The Railwaymen of Sekondi*, Cambridge, Cambridge University Press.

Jeffries, R., 1978, *Class, Power and Ideology in Ghana: The Railwaymen of Sekondi* Cambridge, Cambridge University Press.

Jenkins, K., 1995, *On 'What is History?': From Carr and Elton to Rorty and White*, New York, Routledge.

Jessop, B., 1989, 'Conservative Regimes and the Transition to Post-Fordism: The Case of Great Britain and West Germany', in Gottdiener and Komninos, eds..

Jewsiewicki, B. and D. Newbury, eds., 1986, *African Historiographies: What History for Which Africa*, Beverly Hills, Sage.

Jewsiewicki, B., 1989, 'African Historical Studies: Academic Knowledge as 'Usable Past' and Radical Scholarship', *African Studies Review*, Vol. 32, n° 3, pp. 1-76.

Jeyifo, B., 1990, 'The Nature of Things: Arrested Decolonization and Critical Theory', *Research in African Literatures*, Vol. 21, pp. 33-48.

Jeyifo, B., 1992, 'Literature in Postcolonial Africa', *Dissent*, Vol. 39, pp. 353-60.

Johnson, C., 1978, 'Madamu Alimotu Pelewera and the Lagos Market Women', *Tarikh*, Vol. 7, pp. 1-10.

Johnson, R. W., 1977, 'Sekou Toure: the Man and His Ideas', in Gutkind and Waterman, eds.

Johnson-Odim, C. and M. Strobel, 1990, *Restoring Women to History Teaching Packets For Integrating Women's History Into Courses on Africa, Asia, Latin America, the Caribbean, and the Middle East*, Bloomington, Organization of American Historians.

Johnson-Odim, C., 1993, 'The Debate Over Black Athena', *Journal of Women's History*, Vol. 4, pp. 83-89.

Johnston, R. J. and P. J. Taylor, eds., 1986, *A World in Crisis?*, Oxford, Basil Blackwell.

Johnston, R. J., 1986, 'Individual Freedom and the World-Economy', in Johnston and Taylor, eds.

Johnstone, F., 1970, 'White Prosperity and White Supremacy in South Africa Today', *African Affairs*, 69, pp. 124-139.

Johnstone, F., 1982, 'Most Painful to our Hearts: South Africa Through the Eyes of the New School', *Canadian Journal of African Studies*, 16, pp. 5-26.

Jolly, R., 1973a, 'Introduction', in Jolly, et al., eds.

Jolly, R., 1986, 'JASPA-Evolution of an Approach from Employment to Basic Needs in Economies Under Press', in JASPA.

Jones, E. D. ed., 1991, *African Literature Today*, Special issue on language.

Jones, E. D., ed. *African Literature Today*, Special Issue on Insiders and Outsiders.

Jones, E. D., ed., 1978, *African Literature Today*, Special Issue on Drama in Africa.

Jones, E. D., et al., eds., 1987, *African Literature Today*, Special Issue on Women, Vol. 15.

Jones, G., 1990, 'Africa's New Islamic Fundamentalists', *Index on Censorship*, Vol. 19, n° 8, pp. 20.

Jordan, G. and Weedon, C., eds., 1995, *Cultural Politics: Class, Gender, Race and the Postmodern World*, Cambridge, Ma., Blackwell.

Joseph, G. I. and J. Lewis, 1986, *Common Differences: Conflicts in Black and White Feminist Perspectives*, Boston, *South End Press*.

Joseph, R., 1984, 'State and Prebendal Politics in Nigeria', in Kasfir, ed.

Joseph, R., 1985, 'Cameroon Since Independence: Towards a New Conceptual Framework', Conference on African Independence and Consequences of the Transfer of Power, 1956-1980, University of Zimbabwe, Harare, January.

Joseph, R., 1987, *Democracy and Prebendal Politics in Nigeria*, Cambridge, Cambridge University Press.

Journal of African History, 1979, Vol. 20.

Jowitt, K., 1993, 'A World Without Leninism', in Slater, Schutz, and Dorr, eds.

Julien, E., 1992, *African Novels and the Questions of Orality*, Bloomington, Indiana University Press.

July, R. W., 1992, *A History of the African People*, 4th ed. Prospect Height, IL., Waveland.

Kaba, L., 1985, 'From Colonialism to Autocracy: Guinea Under Sekou Toure', Conference on African Independence and Consequences of the Transfer of Power, 1956-1980, University of Zimbabwe, Harare, January.

Kadenge, P., et al., 1992, 'Zimbabwe's Structural Adjustment Programme: The First Year Experience', in Mwanza, ed.

Kader, A. S., 1987, *Egyptian Women in a Changing Society, 1899-1984*, Boulder, Col., Lynne Rienner.

Kagan, A., 1992a, 'Review of Experiences with CD-ROM for International Development and Prospects for the Future', *IFLA Journal*, Vol. 18/1, pp. 70-71.

Kagan, A., 1992b, 'Liberation Technology', *Progressive Librarian*, Vol. 5, pp. 47-9.

Kagan, A., 1994, 'Access to Appropriate Electronic Documents', IFLA Anglophone Africa Seminar on Government Information and Official Publications, University of Zimbabwe, Harare, 15-18 December.

Kahler, M., 1984, *Decolonization in Britain and France: The Domestic Consequences of International Relations*, Princeton, NJ, Princeton University Press.

Kalipeni, E., 1992, 'Political Development and Prospects for Democracy in Malawi', *Transafrica Forum*, Vol. 9, pp. 27-40.

Kalipeni, E., 1995, 'Malawi-USA Relations and Political Change in Malawi: The Post-Independence Experience, 1964-1993', in Munene, et al., eds.

Kalipeni, E., 1995, 'The Fertility Transition in Africa', *Geographical Review*, Vol. 85, pp. 287-301.

Kalipeni, E., 1996, 'Democratic Transition and the Political Landscape in Malawi: The Politics of Regional Polarization', paper presented at the 27th Annual Meeting of the National Conference of Black Political Scientists, Savanna, Georgia, March.

Kalipeni, E., 1996, 'Demographic Response to Environmental Pressure in Malawi', Population and Environment, Vol. 17, pp. 285-308.

Kalua, B., et al., 1992, 'The Structural Adjustment Programme in Malawi: A Case of Successful Adjustment?' in Mwanza, ed.

Kamarck, A. M., 1976, The Tropics and Economic Development, Baltimore, Johns Hopkins University Press.

Kamumchuluh, J.T., 1975, 'Meru Participation in Mau Mau', Kenya Historical Review, 3.

Kanogo, T., 1987, 'Kikuyu Women and the Politics of Protest: Mau', in Macdonald, et al., eds.

Kanogo, T., 1987, Squatters and the Roots of Mau Mau, James Currey, London.

Kantor, B. S. and H. F. Kenny, 1976, 'The Poverty of Neo-Marxism: The Case of South Africa', Journal of Southern African Studies, 3, pp. 20-30.

Kaplan, B. H., ed., 1978, Social Change in the Capitalist World Economy, Beverly Hills, Sage.

Kaplan, P., 1982, 'Gender, Ideology and Modes of Production on the Coast of East Africa', Paideuma, Vol. 28, pp. 29-43.

Kaplinsky, R., 1980, 'Capitalist Accumulation in the Periphery - The Kenyan Case Re-Examined', Review of African Political Economy, n° 17, pp. 83-104.

Kaplinsky, R., ed., 1978, Readings on the Multinational Corporation in Kenya, Nairobi, Oxford University Press.

Kapteijns, L., 1977, African Historiography Written by Africans 1955-1973: The Nigerian Case, Leiden.

Kapteijns, L., 1985, 'Islamic Rationales for the Changing Social Roles of Women in the Western Sudan', in Daly, ed.

Karakatounian, M., 1988, 'Gabon: Oil Pulls Economy out of a Trough', African Business, June.

Kasfir, N., ed., 1984, State and Class in Africa, London, Frank Cass.

Kasozi, A. B. K., 1994, The Social Origins of Violence in Uganda 1964-1985, Kingston, McGill-Queen's University Press.

Katrak, K. H., 1985, 'From Pauline to Dikeledi: The Philosophical and Political Vision of Bessie Head's Protagonists', Ba Shiru, Vol. 12, pp. 26-35.

Katsouris, C., 1995, 'Naples Debt Deal Falls Short of Needs', Africa Recovery, 11-12, June.

Katz, S., 1980, Marxism, Africa and Social Class: A Critique of Relevant Theories, Occasional Monograph Series 14, Center for Developing Area Studies, McGill University, Montreal.

Kaunda, K., 1967, Humanism in Zambia and a Guide to its Implementation Part 1, Lusaka, Zambia Information Service.

Kay, G., 1975, Development and Underdevelopment: A Marxist Analysis, London, Macmillan.

Keatley, P., 1963, The Politics of Partnership: The Federation of Rhodesia and Nyasaland, Harmondsworth, Penguin.

Keddie, N. and B. Baron, eds., 1991, *Women in Middle Eastern History*, New Haven, Yale University Press.

Keddie, N. R., 1979, 'Problems in the Study of Middle Eastern Women', *International Journal of Middle East Studies*, Vol. 10, pp. 225-240.

Keegan, T., 1987, *Rural Transformations in Industrializing South Africa*, London, Macmillan.

Keen, M., 1988, 'Caution Sounded Over Untimely Privatization', *African Business*, August.

Keith, M. and Pile, S., eds., 1993, *Place and the Politics of Identity*, London and New York, Routledge.

Kelly, J., 1984, *Women, History and Theory: The Essays of Joan Kelly*, Chicago, University of Chicago Press.

Kelly, M., 1993, 'Views of Malawi - review of *Smouldering Charcoal* by Tiyambe Zeleza, Heinemann African Writers Series', *London Magazine*, Vol. 32, pp. 143-146.

Kemp, T., 1983, *Industrialization in the Non-Western World*, London, Longman.

Kennedy, E. C., 1989, *The Negritude Poets*, New York, Thunder's Mouth Press.

Kennedy, P., 1988, *African Capitalism: The Struggle for Ascendancy*, New York, Cambridge University Press.

Kenwood, A. G. and A. L. Lougheed, 1983, *The Growth of the International Economy, 1820-1980*, London, Unwin Hyman.

Kenya Times, 1984-1985, Nairobi, Kenya.

Kenyatta, J., 1938, *Facing Mount Kenya*, London, Seeker and Warburg.

Kesteloot, L. 'Senghor, 1990, Negritude and Francophonie on the Threshold of the Twenty-First Century', *Research in African Literatures*, Vol. 21, pp. 51-57.

Kettel, B., 1986, 'The Commoditization of Women in Tugen (Kenya) Social Organization', in Robertson and Berger, eds.

Keylard, M., 1992, 'CD-ROM Databases: Subjects and Formats', in CD-ROM for African Research Needs: 23-6.

Keynes, J. M., 1961, *The General Theory of Interest and Money*, London, Macmillan.

Khan, A. R., 1993, *Structural Adjustment and Income Distribution: Issues and Experience*, Geneva International Labour Office.

Khennas, S., ed., 1992, *Industrialization Mineral Resources and Energy in Africa*, Dakar, CODESRIA Book Series.

Ki-Zerbo, J., 1989, 'The Teaching of the History of Africa', paper presented at the Meeting of Experts on Textbooks and Teaching of History in African Schools, UNESCO, Nairobi, March.

Ki-Zerbo, J., ed., 1981, *UNESCO General History of Africa*, Vol. 1, London, Heinemann.

Kiernan, V. G., 1974, *Marxism and Imperialism*, London, Edward Arnold.

Kilby, P., 1983, 'Review of the 'Berg Report' (World Bank 1981)', *Labour, Capital and Society*, Vol. 16, p. 2.

Killick, T., 1978, *Development Economics in Action: A Study of Economic Policies of Ghana*, London, Heinemann.

Killick, T., ed., 1981, *Papers on the Kenyan Economy*, Nairobi, Heinemann.

Killingray, D., 1982, 'Military and Labour Recruitment in the Gold Coast during the 2nd World War', *Journal of African History*, 23, pp. 83-95.

Killingray, D., ed., 1986, *Africa and the Second World War*, Basingstoke, Macmillan.

Kimball, R., 1990, *Tenured Radicals: How Politics Has Corrupted Higher Education*, New York, Harper and Row.

Kimble, D. A., 1963, *Political History of Ghana*, Oxford, Clarendon.

Kimble, D. H. T., 1960, *Tropical Africa: Society and Politics*, Volume 2, New York.

Kimble, J., 1985, 'A Case for the Defence', *Canadian Journal of African Studies*, 19, pp. 64-72.

Kindleberger, C. P., 1973, *The World in Depression, 1929-1939*, Berkeley, University of California Press.

King, G., 1986, 'Tanzania Wins IMF Deal Once Thought Impossible', *African Business*, July.

King, R. H., 1993, 'Appiah's Humanistic Africanism',·*Mississippi Quarterly*, Vol. 46, pp. 267-271.

Kinsman, M., 1983, '"Beasts of Burden": The Subordination of Southern Tswana, ca.1800-1840', *Journal of Southern African Studies*, Vol. 9, pp. 39-54.

Kipling, R., 1941, *Kim*, Garden City, Doubleday, Doran, [1901].

Kirk-Greene, A. H. M. and M. Adamu, eds., 1986, *Pastoralists of the West African Savannah*, Manchester, Manchester University Press.

Kirk-Greene, A. H. M., 1982, 'A Historiographical Perspective on the Transfer of Power in British Colonial Africa', Gifford and Louis, eds.

Kirk-Greene, A. H. M., 1988, 'Anglophone Africa and the Transfers of Power: Bibliography', in Gifford and Louis, eds.

Kirk-Greene, A. H. M., ed., 1979, *The transfer of Power: The Colonial Administration in the Age of Decolonization*, London, Oxford University Press.

Kirkman, W. P., 1966, *Unscrambling an Empire*, London, Chatto and Windus.

Kitching, G., 1977, 'Modes of Production and Kenyan Dependency', *Review of African Political Economy*, 8, pp. 56-74.

Kitching, G., 1980, *Class and Economic Change in Kenya: The Making of an African Petite Bourgeoisie*, New Haven and London, Yale University Press.

Kitching, G., 1985, 'Politics, Method, and Evidence in the "Kenya Debate"', in Bernstein and Campbell, eds.

Kitching, G., 1985, 'Suggestions for a Fresh Start on an Exhausted Debate', *Canadian Journal of African Studies*, 19, pp. 116-126.

Kitching, G., 1989, *Development and Underdevelopment in Historical Perspective* London, Routledge.

Kjekshus, H., 1977, *Ecology Control and Economic Development in East African History*, London, Heinemann.

Klein, M. A., ed., 1980, *Peasants in Africa: Historical and Contemporary Perspectives*, Beverly Hills: Sage Publications.

Klein, M., 1985, 'The Use of Mode of Production in Historical Analysis', *Canadian Journal of African Studies*, 19, pp. 9-12.

Kleinberg, S. J. ed., 1988, *Retrieving Women's History*, New York, Berg/UNESCO.

Koepp, S., 1987, 'in The Shadows of The Twin Towers', *Time*, November 2.

Komisar, L., 1992, 'The Claws of Dictatorship in Zaire', *Dissent* 39, pp. 326-330.

Kondratieff, N.D., 1935, 'The Long Waves in Economic Life', *The Review of Economic Statistics*, Cambridge, Mass.

Kowet, D. K., 1978, *Land, Labor Migration and Politics in Southern Africa*, Uppsala, Scandinavian Institute of African Studies.

Kromberg, S. et al., 1993, *Publishing for Democratic Education*, Johannesburg, SACHED Books.

Kuper, L. and M. G. Smith, eds., 1969, *Pluralism in Africa*, Berkeley, University of California Press.

Kuzwayo, E., 1985, *Call Me Woman*, London, Women's Press.

Kydd, J. G. and R. E. Christiansen, 1982, 'Structural Change in Malawi Since Independence: Consequences of a Development Strategy Based on Large-Scale Agriculture', *World Development*, 10, pp. 355-376.

La Guma, A., 1972, *In the Fog of the Season's End*, London, Heinemann.

Lachaud, J., 1994, *The Labour Market in Africa*, Geneva, ILO.

Laclau, E., 1971, 'Feudalism and Capitalism in Latin America', *New Left Review*, 67, pp. 19-38.

Laishley, R., 1993, 'Africa Faces Crisis Over Aid as World Bank Lending Falls', *Africa Recovery*, October.

Lake, O., 1995, 'Toward a Pan-African Identity: Diaspora African Repatriates in Ghana', *Anthropological Quarterly*, Vol. 68, pp. 21-36.

Lall, S., 1975, 'Is Dependence a Useful Concept in Analyzing Underdevelopment?' *World Development*, 3, pp. 799-810.

Lamb, G., 1974, 'Peasant Politics: Conflict and Development', in Murang'a, Friedman.

Lamb, G., 1975, 'Marxism, Access and the State', *Development and Change*, Vol. 6, pp. 119-135.

Lamb, R. and Seifulaziz, M., 1983, 'Soil Erosion, Real Cause of Ethiopian Famine', *Environmental Conservation*, Vol. 10, pp.157-9.

Lambert, S., H. Schneider and A. Suwa, 1991 'Adjustment and Equity in Côte d'Ivoire: 1980-86', *World Development*, Vol. 19, pp. 1563-1576.

Lamont, M., 1987, 'How to Become a Dominant French Philosopher: The Case of Jacques Derrida', *American Journal of Sociology*, Vol. 93, pp. 584-622.

Lancaster, F. W., 1978, *Toward Paperless Information System*, New York, Academic.

Lancaster, F. W., 1982, *Libraries and Librarians in an Age of Electronics*, Arlington, VA: Information Resources.

Landell-Mills, P., 1992, 'Governance, Cultural Change, and Empowerment', *Journal of Modern African Studies*, Vol. 30, pp. 543-567.

Lane, A. J., 1976, 'Women in Society: A Critique of Frederick Engels', in Carroll, ed.

Langdon, S., 1976, 'Partners in Underdevelopment? The Transinationalization Thesis in a Kenyan Context', *Journal of Commonwealth and Comparative Politics*, Vol. 14, pp. 42-63.

Langdon, S., 1976, 'The State and Capitalism in Kenya', *Review of African Political Economy*, 8, pp. 90-98.

Langdon, S., 1980, 'Industry and Capitalism in Kenya - Contributions to a Debate'. Conference on the African Bourgeoisie, Dakar, Dec.

Langley, J. A., 1973, *Pan-Africanism and Nationalism in West Africa: A Study in Ideology and Social Classes*, Oxford, Clarendon Press.

Lanning and L. Mueller, 1979, *Africa Undermined: Mining Companies and Economic Growth in Nigeria*, Harmondsworth, Penguin.

Lappe, F. M. and J. Collins, 1977, *Food First: The Myth of Scarcity*, London, Souvenir Press.

Lapper, R., 1987a, 'Severe with a Snarl from Debtors At Bay', *South*, July.

Lapper, R., 1987b, 'Debt-Equity: My Kingdom for a Loan', *South*, June.

Lash, S. and Urry, J., 1994, *Economies of Signs and Space*, London, Sage Publications.

Last, M., 1985, 'Reform in West Africa: the Jihad Movements of the Nineteenth Century' in Ajayi and Crowder, eds. Volume 2.

Law, R., 1978, 'In Search of A Marxist Perspective on Pre-Colonial Tropical Africa', *Journal of African History*, Vol. 19, pp. 441-452.

Lawrence, P. ed., 1986, *World Recession and the Food Crisis in Africa*, London, Review of African Political Economy and James Currey.

Laws of Malawi. '"Defamation", Cap. 7, p. 01, Chapter XVIII, Volume 2.

Laws of Malawi. 'Censorship and Control of Entertainments', Cap. 21.01, Volume IV.

Lazarus, N., 1990, *Resistance in Postcolonial African Fiction*, New Haven and London, Yale University Press.

Lazarus, N., 1994, 'National Consciousness and the Specificity of the (post) Colonial Intellectualism', in Barker, Hulme, and Iversen, eds.

Lazarus, N., 1994, 'National consciousness and the specificity of the (post) colonial intellectualism', in Barker, Hulme, and Iversen, eds.

Lazreg, M., 1990, 'Women, Work and Social Change in Algeria', in S. Stichter and Parpart, eds.

Leacock, E. B., 1981, *Myths of Male Dominance: Collected Articles on Women Cross-Culturally*, New York, Monthly Review.

Lebowitz, M. A., 1995, 'Situating the Capitalist State', in Callari, Cullenberg and Biewener, eds.

Lee, B., 1995, 'Critical Internationalism', *Public Culture*, Vol. 7, pp. 559-592.

Lee, J. M., 1967, *Colonial Development and Good Government*, London, Oxford University Press.

Legassick, M., 1979, 'Review Article:Records of Protest and Challenge', *Journal of African History*, 20, pp. 451-455.

Legum, C., 1965, *Pan-Africanism: A Short Political Guide*, New York, Praeger.

Legum, C., 1992, 'The Postcommunist Third World: Focus on Africa', *Problems of Communism*, Vol. 41, pp. 195-206.

Lemarchand, R., 1992, 'Uncivil States and Civil Societies: How Illusion Became Reality', *Journal of Modern African Studies*, Vol. 30, pp. 177-191.

Lemelle, S. and Kelley, R. D. G., eds., 1994, *Imagining Home: Class, Culture and Nationalism in the African Diaspora*, London and New York, Verso.

Lenin, V.I., 1978, *Imperialism: The Highest Stage of Capitalism*, Moscow, Progress Publishers.

Leo, C., 1984, *Land and Class in Kenya*, Toronto, Toronto University Press.

Lerner, G., 1979, *The Majority Finds Its Past: Placing Women in History*, New York, Oxford University Press.

Lerner, G., 1990, 'Reconceptualizing Differences Among Women', *Journal of Women's History*, 2.

Leslie, C. M., 1990, 'Scientific Racism: Reflections on Peer Review, Science and Ideology', *Social Science and Medicine*, Vol. 31, pp. 891-912.

Leubuscher, C., 1956, *Bulk Buying in the Colonies*, London, Oxford University Press.

Lever, H. and C. Huhne, 1985, *Debt and Danger. The World Financial Crisis*, London, Penguin.

Levey, L.A. ed., 1993, *A Profile of Research Libraries in Sub-Saharan Africa: Acquisitions, Outreach, and Infrastructure*, Washington, American Association for the Advancement of Science.

Levey, L.A., 1992, 'CD-ROM Costs and Implementation Issues', in CD-ROM for African Research Needs: 13-22.

Levey, L.A.,1991, *Computer and CD-ROM Capability in Sub-Saharan African University and Research Libraries*, Washington, American Association for the Advancement of Science.

Levi, J. and M. Havinden, 1982, *The Economics of African Agriculture*, Harlow, Longman.

Levins, R., 1995, 'Beyond Democracy: The Politics of Empowerment', in Callari, Cullenberg and Biewener, eds.

Levitzon, N., 1973, *Ancient Ghana and Mali*, New York, Africana Publishing Company.

Lewis, A., 1966, *Development Planning: the Essentials of Economic Policy*, London, Allen and Unwin.

Lewis, B., 1984, 'The Impact of Development Policies on Women', in Hay and Stichter, eds.

Lewis, E., 1995, 'To Turn as on a Pivot: Writing African Americans into a History of Overlapping Diasporas', *The American Historical Review*, Vol. 100, pp. 765-787.

Lewis, P., 1992, 'Africa Writes Back', *Stand Magazine*, Vol. 33, pp. 74-83.

Lewis, W. A., 1954, 'Economic Development with Unlimited Supplies of Labour', *Manchester School*, 22.

Lewis, W. A., 1958, 'Unlimited Labour: Further Notes', *Manchester School*, 26.

Leys, C. 1977, 'Underdevelopment and Dependency: Critical Notes', *Journal of Contemporary Asia*, Vol. 7, pp. 92-107.

Leys, C., 1971, 'Politics in Kenya: The Development of a Peasant Society', *British Journal of Political Science*, 1, pp. 307-337.

Leys, C., 1973, 'Interpreting African Underdevelopment: Reflections on the ILO Report in Kenya', *African Affairs*, 72, pp. 419-429.

Leys, C., 1974, *Underdevelopment in Kenya: The Development of a Peasant Society*, Berkeley, University of California Press.

Leys, C., 1976, 'The 'Overdeveloped' Post-Colonial State: A Re-Evaluation', *Review of African Political Economy*, Vol. 5, pp. 39-48.

Leys, C., 1977, 'Underdevelopment and Dependency: Critical Notes', *Journal of Contemporary Asia*, 7, 92-107.

Leys, C., 1980, 'Kenya: What Does 'Dependency' Explain?' *Review of African Political Economy*, 17, pp. 108-113.

Leys, C., 1982, 'Accumulation, Class Formation and Dependency: Kenya', in Fransman, ed.

Leys, C., 1982, 'African Economic Development in Theory and Practice', 3, pp. 99-124.

Leys, N., 1926, Kenya, London, Hogarth.

Leys, R., 1986, 'Drought and Drought Relief in Southern Zimbabwe', in Lawrence, ed.

Li, P. S., ed., 1990, Race and Ethnic Relations in Canada, Toronto, Oxford University Press.

Liabes, D., 1995, 'Entrepreneurs, Privatisation and Liberalisation: The Pro-Democracy Movement in Algeria', in Mamdani and Wamba-dia-Wamba, eds.

Liepetz, A., 1987, Mirages and Miracles: The Crises of Global Fordism, London, Verso.

Likimani, M., 1985, Passbook Number F.47927: Women and Mau Mau in Kenya, London, Macmillan.

Lipsey, R. G., 1971, An Introduction to Positive Economics, 4th ed., London, Widenfeld and Nicholson.

Lipton, M., 1976, 'The Debate About South Africa: Neo-Marxists and Neo-Liberals', African Affairs, 78, pp. 57-80.

Lipton, M., 1977, Why Poor People Stay Poor: Urban Bias in World Development, London, Temple Smith.

Lipton, M., 1985, Capitalism and Apartheid, Cape Town, David Philip.

Little, A., 1986, 'Examinations, Work and Quality of Learning: Evidence From Africa', in JASPA.

Little, I. et al., 1970, Industry and Trade In Some Developing Countries: A Comparative Study, London, Oxford University Press.

Little, K., 1973, West African Women in Towns: An Aspect of Africa's Social Revolution, Cambridge, Cambridge University Press.

Livingstone, I. and H. W. Ord, 1970, An Introduction to Economics For East Africa, London, Heinemann.

Livingstone, I., 1981, Rural Development and Incomes in Kenya, Addis Ababa, JASPA.

Livingstone, I., 1986, 'Alternative Strategies for Development South of the Sahara: Contrast Between the ILO/JASPA and the World Bank Approaches', in JASPA.

Livingstone, I., 1987, 'Youth Employment Policies and Programmes in Anglophone African Countries: Past Experience and Future Action', in ILO/JASPA.

Lloyd, M. E., 1990, 'Gender Factors in Reviewer Recommendations for Manuscript Publication', Journal of Applied Behavior Analysis, Vol. 23, pp. 539-43.

Loeb, K., 1992, White Man's Burden, Toronto, Lugus.

Lofchie, M. and S. K. Commins, 1982, 'Food Deficits and Agricultural Policies', Journal of Modern African Studies, Vol. 20, pp. 1-25.

Lofchie, M. F., 1975, 'Political and Economic Origins of African Hunger', Journal of Modern African Studies, 13.

Lofchie, M. F., 1989, The Policy Factor: Agricultural Performance in Kenya and Tanzania, Boulder, Col., Lynne Rienner.

Lonsdale, J. and B. Berman, 1979, 'Coping with the Contradictions: The Development of the Colonial State, 1895-1914', *Journal of African History*, Vol. 20, pp. 487-505.

Lonsdale, J. M., 1989, 'The Conquest State, 1895-1904', in Ochieng' ed.

Lonsdale, J. M., 1990, 'Mau Mau of the Mind: Making Mau Mau and Remaking Kenya', *Journal of African History*, Vol. 31, pp. 393-422.

Lonsdale, J., 1981, 'States and Social Processes in Africa: A Historical Survey', *African Studies Review*, Vol. 24, n° 2 and 3, pp. 139-225.

Lorimer, R., 1993, 'The socioeconomy of scholarly and cultural book publishing', *Media, Culture and Society*, Vol. 15, pp. 203-216.

Louis, W. R. and R. Robinson, 1982, 'The United States and the Liquidation of British Empire in Tropical Africa, 1941-1951', in Gifford and Louis, eds.

Louis, W. R., 1977, *Imperialism at Bay: The Role of the United States in the Decolonization of the British Empire 1941-45*, New York, Oxford University Press.

Louis, W. R., ed., 1976, *Imperialism: The Robinson and Gallagher Controversy*, New York, New Viewpoints.

Lovejoy, P. E., 1993, *Slow Death for Slavery: The Course of Abolition in Northern Nigeria, 1897-1936*, Cambridge, Cambridge University Press.

Lovejoy, P. E., ed., 1981, *Ideology of Slavery in Africa*, Beverly Hills, Sage.

Lovejoy, P. E., ed., 1983, *Transformations in Slavery: A History of Slavery in Africa*, Cambridge, Cambridge University Press.

Lovejoy, P. E., ed., 1986, *Africans in Bondage: Studies in Slavery and the Slave Trade*, Madison, University of Wisconsin Press.

Lovett, M., 1989, 'Gender relations, Class Formation, and the Colonial State in Africa', in Parpart and Staudt, eds.

Low, D. A. and Allison Smith, eds., 1976, *History of East Africa*, London, Oxford University Press.

Low, D. A. and J. M. Lonsdale, 1976, 'Introduction: Towards the New Order 1945-1963', in D.A. Low and Allison Smith, eds.

Low, D. A., 1973, *Lion Rampant*, London, Blackwell.

Low, D. A., 1982, 'The Asian Mirror to Tropical Africa's Independence', in Gifford and Louis, eds.

Low, D. A., 1988, 'The End of the British Empire in Africa', in Gifford and Louis, eds.

Lowe, C., 1988, 'African Debt: North and South Have a "Meeting of Minds"', *African Business*, November.

Loxley, J., 1984, 'The Berg Report and the Model of Accumulation in Sub-Saharan Africa', *Review of African Political Economy*, 27/28, pp. 197-204.

Luckham, R., 1980, 'Armaments and Underdevelopment in Africa', *Alternatives*, Vol. 5.

Lutz, C., 1990, 'The Erasure of Women's Writing in Sociocultural Anthropology', *American Ethnologist*, Vol. 17, pp. 611-627.

Luxemburg, R., 1968, *The Accumulation of Capital*, New York, Monthly Review Press.

Lycett, A., 1986, 'What Baker is Cooking Up For Africa', *African Business*, April.

Lynd, G. E., 1967, *The Politics of African Trade Unionism*, New York, Praeger.

Lyotard, J., 1984, *The Postmodern Condition*, Minneapolis, Minnesota University Press.

Maack, M., 1986, 'The Role of External Aid in West African Library Development', *Library Quarterly*, Vol. 56, pp. 1-16.

Macdonald, S., et al., eds., 1987, *Images of Change and War: Cross-Cultural and Historical Perspectives*, Madison, University of Wisconsin Press.

MacGaffey, J., 1987, *Entrepreneurs and Parasites: The Struggle for Indigenous Capitalism in Zaire*, Cambridge, Cambridge University Press.

Machua, W., 1988, 'Woes that Beset Kenya's Manufacturing Sector', *Daily Nation*, Nairobi, 10-29-88.

Mackenzie, F., 1989, 'Land and Territory: The Interface Between Two Systems of Land Tenure, Murang'a District, Kenya', *Africa*, Vol. 59, pp. 91-109.

MacKenzie, J. M., 1993, 'Occidentalism: Counterpoint and Counterpolemic', *Journal of Historical Geography*, Vol. 19, pp. 339-44.

Macmillan, H., 1967, *The Blast of War 1939-1945*, London, Macmillan.

Macmillan, W. M., 1930, *Complex South Africa*, London.

Madly, M., 1993, 'Resource-Sharing, Outreach, and Community Participation to Promote Library Services in Natal', in Patrikios and Levey, eds.

Mafeje, A. and S. Radwan, eds., 1995, *Economic and Demographic Change in Africa*, Oxford, Clarendon Press.

Mafeje, A., 1971, 'The Ideology of Tribalism', *Journal of Modern African Studies*, 9, pp. 253-261.

Mafeje, A., 1976, 'The Problem of Anthropology in Historical Perspective: An Inquiry into the Growth of the Social Sciences', *Canadian Journal of African Studies*, 10, pp. 307-333.

Mafeje, A., 1981, 'On the Articulation of Modes of Production', *Journal of Southern African Studies*, 8, pp. 123-138.

Mafeje, A., 1990, 'The African Intellectuals: An Inquiry into their Genesis and Social Options', CODESRIA Symposium on Academic Freedom, Research and the Social Responsibility of the Intellectual in Africa, Kampala, Uganda, November.

Mafeje, A., 1991, *African Households and Prospects for Agricultural Revival in Sub-Saharan Africa*, Dakar, CODESRIA.

Mafeje, A., 1992, 'Theory of Democracy and the African Discourse: Breaking Bread with my Fellow-Travellers', CODESRIA Seventh General Assembly, Democratization in Africa: Problems and Prospects.

Mafeje, A., 1995, '"Benign" Recolonization and Malignant Minds in the Service of Imperialism', *CODESRIA Bulletin*, n° 2, pp. 17-20.

Mafeje, A., 1995, 'Demographic and Ethnic Variations: A Source of Instability in Modern African States', CODESRIA Eigth General Assembly, Dakar.

Magdoff, H., 1978, *Imperialism: From the Colonial Age to the Present*, New York, Monthly Review Press,.

Magubane, B. M., 1987, *The Ties That Bind: African-American Consciousness of Africa*, Trenton, NJ, Africa World Press.

Magubane, B., 1971, 'A Critical Look at Indices Used in the Study of Social Change in Colonial Africa', *Current Anthropology*, 12, pp. 4-5.

Magubane, B., 1976, 'The Evolution of the Class Structure in Africa', in Gutkind and Wallerstein, eds.

Magubane, B., 1979, *The Political Economy of Race and Class in South Africa*, New York, Monthly Review Press.

Maja-Pearce, A., 1991, 'Africa in the Land of the Zombies', *Index on Censorship*, Vol. 20, n° 9, pp. 9-12.

Maja-Pearce, A., 1992, 'The Press in Central and Southern Africa - Malawi', *Index on Censorship*, Vol. 21, n° 4, pp. 42-73.

Maja-Pearce, A., ed., 1994, *Wole Soyinka: An Appraisal*, London, Heinemann.

Makaranga, M., 1987, 'Tanzania: Liberalized Trade Hits Manufacturers', *African Business*, August.

Makinda, S. 1993, 'Somalia: From Humanitarian Intervention to Military Offensive?' *The World Today*, Vol. 49, pp. 184-6.

Malima, K., 1986, 'The IMF and World Bank conditionality: The Tanzanian case', in Lawrence, ed.

Mallon, F. E., 1993, 'Dialogues Among the Fragments: Retrospect and Prospect', in Cooper, et. al.

Maloba, W., 1993, *Mau Mau and Kenya: An Analysis of a Peasant Revolt*, Bloomington, Indiana University Press.

Maloka, T., 1994, '"Faction Fights" or "Fixed Bayonets" Against Sticks and Stones? Basotho Migrants and Violence on the Mines 1886-1939', *Afrika Zamani*, 2 (New Series), pp.193-210.

Maloka, T., 1994, 'Faction Fights' or 'Fixed Bayonets' Against Sticks and Stones? Basotho Migrants and Violence on the Mines 1886-1939', *Afrika zamani*, 2 (New Series), pp. 193-210.

Mamdani, M, T. Mkandawire and E. Wamba-dia-Wamba, eds., 1988, *Social Transformation and Democratization in Africa*, CODESRIA, Working Paper 1.

Mamdani, M. and E. Wamba-dia-Wamba, eds., 1995, *African Studies in Social Movement and Democracy*, Dakar, CODESRIA Book Series.

Mamdani, M., 1972, *The Myth of Population Control*, New York, Monthly Review Press.

Mamdani, M., 1976, *Politics and Class Formation in Uganda*, New York, Monthly Review Press.

Mamdani, M., 1983, *Imperialism and Fascism in Uganda*, London, Heinemann.

Mamdani, M., 1990a, 'The Intelligentsia, the State and Social Movements: Some Reflections on Experiences in Africa', CODESRIA Symposium on Academic Freedom, Research and the Social Responsibility of the Intellectual in Africa, Kampala, Uganda, November.

Mamdani, M., 1990b, 'A Glimpse at African Studies, Made in USA', *CODESRIA Bulletin*.

Mamdani, M., 1992, 'Africa: Democratic Theory and Democratic Struggles', *Dissent*, Vol. 39, pp. 312-18.

Mamdani, M., 1993, 'The Sun is not Always Dead at Midnight', *Monthly Review*, July-August, pp. 27-48.

Mamdani, M., 1995, 'A Critique of the State and Civil Society Paradigm in Africanist Studies', in Mamdani and Wamba-dia-Wamba, eds.

Mamdani, M., 1996, *Citizen and Subject. Contemporary Africa and the Legacy of Late Colonialism*, Princeton, NJ, Princeton University Press.

Mandala, E. C., 1990, *Work and Control in a Peasant Economy: A History of the Lower Tchiri Valley in Malawi, 1886-1945*, Wisconsin, University of Wisconsin Press.

Mandala, E., 1984, 'Capitalism, Kinship and Gender in the Lower Tchiri (Shire) Valley of Malawi, 1860-1960: An Alternative Theoretical Framework', *African Economic History*, Vol. 13, pp. 137-169.

Mandaza, I., 1990, 'Democracy in the African Reality', *Southern Africa Political & Economic Monthly*, Vol. 11, pp. 8-9.

Mandaza, I., ed., 1986, *Zimbabwe: The Political Economy of Transition*, Dakar, CODESRIA Book Series.

Mandel, E., 1972, *Decline of the Dollar*, New York, Monad Press.

Mandel, E., 1975, *Late Capitalism*, London, New Left Books.

Mandel, E., 1977, *Marxist Economic Theory*, London, Marlin Press.

Mandela, W. (ed. A. Benjamin), 1984, *Winnie Mandela: Part of My Soul*, New York, Viking Press.

Mangum, S., et al., 1992, *Strategies for Creating Transitional Jobs During Structural Adjustment*, Working Paper 947, August.

Mann, K., 1985, *Marrying Well: Marriage, Status and Social Change Among the Educated Elite in Colonial Lagos*, Cambridge, Cambridge University Press.

Mann, K., 1991, 'Women, Landed Property, and the Accumulation of Wealth in Early Colonial Lagos', *Signs*, Vol. 16, pp. 682-706.

Manning, P., 1974, 'Notes Towards a Theory of Ideology in Historical Writing on Modern Africa', *Canadian Journal of African Studies*, 8, pp. 235-253.

Manning, P., 1990, *Slavery and African Life: Occidental, Oriental, and African Slave Trades*, Cambridge, Cambridge University Press.

Mansergh, N., 1949, *The Coming of the First World War: A Study in the European Balance, 1878-1914*, London, Longman.

Mapanje, J., 1981, *Of Chameleons and Gods*, London, Heinemann.

Mapanje, J., 1989, 'Censoring the African Poem', *Index on Censorship*, Vol. 18, n° 9, pp. 7-11.

Mapanje, J., 1995, 'Letter From Malawi: Bitter Sweet Tears', *Index on Censorship*, Vol. 24, n° 2, pp. 81-87.

Marable, M. and L. Mullings, 1994, 'The Divided Mind of Black America: Race, Ideology and Politics in the Post Civil Rights Era', *Race & Class*, Vol. 36, pp. 61-72.

Marchand, M. and J. Parpart, eds., 1995, *Feminism/Postmodernism/Development*, London, Routledge Press.

Marei, S. A., 1976, *The World Food Crisis*, London, Longman.

Marglin, S. A. and F. A, Marglin, eds., 1990, *Dominating Knowledge: Development, Culture and Resistance*, Oxford, Clarendon Press.

Marglin, S., 1990, 'Towards the Decolonization of the Mind', in Marglin and Marglin, eds.

Markovits, I. L., 1977, *Power and Class in Africa*, Englewood Cliffs, NJ, Prentice-Hall.

Marks, S., 1972, 'Liberalism, Social Realities, and South African History', *Journal of Commonwealth and Political Studies*, 10, pp. 243-249.

Martin, J. and M. Westlake, 1967, 'Baker's Dough is Failing to Rise', *South*, January.

Martin, R., 1992, 'Building Independent Mass Media in Africa', *Journal of Modern African Studies*, Vol. 30, pp. 331-40.

Marx, K. and F. Engels, 1976, *The Communist Manifesto*, New York, International Publishers.

Marx, K., 1967, *Capita*, New York, International Publishers.

Marx, K., 1980, *Pre-Capitalist Economic Formations*, New York, International Publishers.

Mashinini, E., 1991, *Strikes Have Followed Me All My Life: A South African Autobiography*, London, Routledge.

Matocha, L., 1993, 'Do Editors Have Obligations to Authors?' *Marriage and Family Review*, Vol. 18, pp. 31-39.

Maxwell, S., 1992, 'Food Security in Africa' *Africa Recovery*, Vol. 6, pp. 1-12.

Mazrui, A. A. and M. Tidy, 1984, *Nationalism and New States in Africa*, London, Heinemann.

Mazrui, A. A., 1978, *Political Values and the Educated Class in Africa*, London, Heinemann.

Mazrui, A. A., 1984, *Nationalism and New States in Africa*, London, Heinemann.

Mazrui, A. A., 1986, *The Africans: A Triple Heritage*, Boston, Little Brown.

Mazrui, A. A., 1990, 'The Impact of Global Changes on Academic Freedom in Africa: A Preliminary Assessment', CODESRIA Symposium on Academic Freedom, Research and the Social Responsibility of the Intellectual in Africa, Kampala, Uganda.

Mazrui, A. A., 1994, 'Decaying Parts of Africa Need Benign Colonization', *International Herald Tribune*, August 4.

Mazrui, A. A., 1995, 'Africa and Other Civilizations: Conquest and Counterconquest', in Harbeson and Rothchild, eds.

Mazrui, A. A., 1995, 'Self-Colonization and the Search For Pax Africana: A Rejoinder', *CODESRIA Bulletin*, Vol. 2, pp. 20-22.

Mazrui, A. A., ed., 1993, *Unesco General History of Africa*, Vol. 7, Africa Since 1935, Oxford, Heinemann.

Mazrui, A. and L. Mphande, 1993, 'Orality and the Literature of Combat: The Legacy of Fanon', *Paintbrush*, Vol. 39&40, pp. 159-183.

Mazrui, A. and L. Mphande, Forthcoming, *Beyond Decolonization: The Politics of Language in Africa*.

Mazundar, D., 1975, *The Urban Informal Sector*, Washington, DC, World Bank Staff Working Paper No.211.

Mazundar, D., 1976, 'The Urban Informal Sector', *World Development*, 4, pp. 655-679.

Mba, N., 1982, *Nigerian Women Mobilised: Women's Political Activity in Southern Nigeria, 1900-1945*, Berkeley, Institute of International Studies, University of California.

Mba, N., 1989, 'Kaba and Khakhi: Women and the Militarized State in Nigeria', in Parpart and Staudt, eds.

Mbanga, T., 1994, 'Zimbabwe Book Fair '94 Breaks Records', *Bellagio Publishing Network Newsletter*, Vol. 11, pp. 1-2.

Mbembe, A., 1992a, 'Provisional Notes on the Postcolony', *Africa*, Vol. 62, pp. 3-37.

Mbembe, A., 1992b, 'The Banality of Power and the Aesthetics of Vulgarity in the Postcolony', *Public Culture*, 4, pp. 1-30.

Mbilinyi, M., 1982, 'Wife, Slave and Subject of the King: The Oppression of Women in the Shambala Kingdom', *Tanzania Notes and Records*, 88/89, pp. 1-13.

Mbilinyi, M., 1984, 'Women in Development Ideology: The Promotion of Competition and Exploitation', *The African Review*, Vol. 2, pp. 14-33.

Mbilinyi, M., 1989, 'This is an Uncomfortable Business': Colonial State Intervention in Urban Tanzania', in Parpart and Staudt, eds.

Mboya, T., 1963, *Freedom and After*, London, Andre Deutsch.

McClintock, A., 1992, 'The Angel of Progress: Pitfalls of the Term "Post-Colonialism"', *Social Text*, Vol. 10, pp. 84-98.

McCracken, P., 1977, *Towards Full Employment and Price Stability*, Paris, OECD.

McFadden, P., 1987, 'The State and Agri-business in Swazi Economy', in Mkandawire and Bourenane, ed.

McGee, T. G., 1973, *Hawkers in Hong Kong: A Study of Planning Policy in a Third World City*, Hong Kong, University of Hong Kong.

McGee, T. G., 1974, *The Persistence of Proto-Proletariat: Occupational Structures and Planning for the Future of Third World Cities*, Los Angeles, University of California, School of Architecture and Urban Planning.

Mcharazo, A.A.S., 1995, 'Summary of S. Arunachalam's 'Accessing Information Published in the Third World: Should Spreading the Word from the Third World Always be Like Swimming Against the Current', Workshop on Access to Third World Journals, *The African Book Publishing Record*, Vol. 20/4, p. 245.

Mchombo, S. A., Forthcoming,'The Democratic Transition in Malawi: Its Roots and Prospects', in Jean-Germain, ed.

Mchombu, K. J., 1982, 'On the Librarianship of Poverty', *Libri*, Vol. 32/.3, pp. 241-50.

Meadows, D., et al., 1972, *The Limits of Growth*, New York, New American Library.

Mehretu, A., 1989, *Regional Disparity in Sub-Saharan Africa: Structural Readjustment of Uneven Development*, Boulder, Colorado, Westview.

Meier, G. E. and Seers, D., eds., 1985, *Pioneers in Development*, New York, Oxford University Press.

Meier, G.M. ed., 1976, *Leading Issues in Development Economics*, New York, Oxford University Press.

Meier, G.M., 1982, *Problems of a World Monetary Order*, 2nd ed., Oxford, Oxford University Press.

Meillassoux, 1974, 'Development or Exploitation: Is the Sahel Good Business?' *Review of African Political Economy*, 1, pp. 27-33.

Meillassoux, C., 1981, *Maidens, Meal and Money: Capitalism and the Domestic Economy*, Cambridge, Cambridge University Press.

Melotti, U., 1981, *Marx and the Third World*, London, Macmillan.

Mende, T., 1973, *From Aid To Recolonization*, London, Harrap.

Menshikov, S., 1975, *The Economic Cycle: Postwar Development*, Moscow, Progress Publishers.

Mersha, G., 1990, 'The State and Civil Society with Special Reference to Ethiopia', CODESRIA Symposium on Academic Freedom, Research and the Social Responsibility of the Intellectual in Africa, Kampala, Uganda, November.

Meynaud, J. and A. S. Bey, 1967, *Trade Unionism in Africa*, London, Methuen.

Mhone, G. C. Z., 1987, 'Agriculture and Food Policy in Malawi: A Review of the Evidence', in Mkandawire and Bourenane, eds.

Mhone, G. C. Z., 1994, *Labour Market Policy and Structural Adjustment in Zimbabwe*, Cape Town, Labour Law Unit, University of Cape Town.

Mhone, G. C., Forthcoming, 'Gender Biases in Economics', in Imam, Mama, and Sow, eds.

Michael, P., J. Petras, and R. Rhodes, 1975, 'Imperialism and the Contradictions of Development', *New Left Review*, 85.

Michaels, W. B., 1995, 'Race into Culture: A Critical Genealogy of Cultural Identity', in Appiah and Gates, eds.

Middleton, J., 1966, *The Effects of Economic Development on Traditional Political Systems in Africa South of the Sahara*, The Hague.

Miliband, R., 1995, 'Reclaiming the Alternative', in Callari, Cullenberg and Biewener, eds.

Miller, C. L., 1985, *Blank Darkness: Africanist Discourse in French*, Chicago, University of Chicago Press.

Miller, C. L., 1990, *Theories of Africans: Francophone Literature and Anthropology in Africa*, Chicago, University of Chicago Press.

Miller, C. L., 1993, 'Literary Studies and African Literature: The Challenge of Intercultural Literacy', in Bates, Mudimbe, and O'Barr, eds.

Miller, J. C., 1982, 'The Significance of Drought, Disease and Famine in the Agriculturally Marginal Zones of West Central Africa', *Journal of African History*, Vol. 23, pp. 17-61.

Miller, J., 1994, 'The case for double-blind reviewing', *CAUT Bulletin*, 41.

MilLer, N. and R. Yeager, 1994, *Kenya: The Quest for Prosperity*, Boulder, Westview.

Minkley, G., 1986, 'Re-examining Experience: The New South African Historiography', *History in Africa*, 13, pp. 269-281.

Miracle, M., 1967, *Maize in Tropical Africa*, Madison, Wisconsin University Press.

Misser, F., 1987, 'Zaire: New Lines of Credit Follow Debt-Repayment Move', *African Business*, January.

Mistry, P. S., 1988, *African Debt: The Case for Relief in Sub-Saharan Africa*, Oxford, International Association Oxford.

Mitchell, J. C., 1956, *The Kalela Dance*, Rhodes-Livingstone Paper, 27.

Mitchell, J. C., 1961, 'The Causes of Labour Migration', Commission for Technical Co-Operation in Africa South of the Sahara, Abidjan, Sixth International Labour Conference Publication, 79.

Mkandawire and N. Bourenane, eds., 1987, *The State and Agriculture in Africa*, Dakar, CODESRIA Book Series.

Mkandawire, T. and A. Olukoshi, eds., 1995, *Beyond Liberalism and Oppression: The Politics of Structural Adjustment in Africa*, Dakar, CODESRIA Book Series.

Mkandawire, T., 1987, 'The State and Agriculture in Africa: Introductory Remarks', in Mkandawire, and Bourenane, ed.

Mkandawire, T., 1989, 'Problems and Prospects of the Social Sciences in Africa', *Eastern African Social Science Review*, Vol. 5, n° 1.

Mkandawire, T., 1989, 'Problems and Prospects of the Social Sciences in Africa', *Eastern Africa Social Science Research Review*, Vol. 5, pp. 1-17.

Mkandawire, T., 1989, 'The Road to Crisis, Adjustment and De-industrialization', *Africa Development*, Vol. 14.

Mkandawire, T., 1992, 'Adjustment, Political Conditionality and Democratization in Africa', CODESRIA Seventh General Assembly, Democratization in Africa: Problems and Prospects.

Mkandawire, T., 1992, 'Adjustment, Political Conditionality and Democratization in Africa', CODESRIA Seventh General Assembly, Democratization in Africa: Problems and Prospects.

Mkandawire, T., 1993, *Report of the Executive Secretary to the 20th Anniversary of CODESRIA*, Dakar, CODESRIA.

Mkandawire, T., 1995, 'Africa's Three Generations of Scholars', *CODESRIA Bulletin*, Vol. 3, pp. 1-3.

Mkandawire, T., 1995, 'Beyond Crisis: Towards Democratic Developmental States in Africa', CODESRIA Eighth General Assembly, Dakar, June.

Mkandawire, T., 1995, 'Fiscal Structure, State Contraction and Political Responses in Africa', in Mkandawire and Olukoshi, eds.

Mkandawire, T., 1996, 'African Studies: Paradigms, Problems, and Prospects', Center for African Studies, University of Illinois at Urbana-Champaign, February.

Mkandawire, T., 1996, 'Stylizing Accumulation in Africa: The Role of the State in Policy Making'.

Modelski, G., 1978, 'The Long Cycle of Global Politics and the Nation-State', *Comparative Studies in Society and History*, 20.

Modelski, G., 1983, 'Long Cycles of world Leadership', in Thompson ed.

Mohri, K., 1979, 'Marx and Underdevelopment', *Monthly Review*, April, 32-42.

Moll, P. G. et al., 1991, *Redistribution: How Can it Work?*, Cape Town, David Philip.

Moll, P. G., 1991, *The Great Economic Debate: The Radical's Guide to the South African Economy*, Johannesburg, Skotaville.

Mongia, P., ed., 1996, *Contemporary Postcolonial Theory: A Reader*, New York: Arnold.

Moore, S. F., 1986, *Social Facts and Fabrications: 'Customary' Law on Kilimanjaro, 1880-1980*, Cambridge, Cambridge University Press.

Moore, S. F., 1993, 'Changing Perspectives on a Changing Africa: The Work of Anthropology', in Bates, Mudimbe, and O'Barr, eds.

Morel, E. D., 1969a, *Red Rubber*, New York, Negro Universities Press.

Morel, E. D., 1969b, *The Black Man's Burden*, New York, Monthly Review Press.

Morgan, D. J., 1980, *The Official History of Colonial Development*, 5 Volumes, London, Humanities Press.

Morgan, D.J., 1980a, *The Official History of Colonial Development, Volume 5, Guidance Towards Self-Government in British Colonies 1941-1971*, Atlantic Heights, Humanities Press.

Morgan, D.J., 1980b, *The Official History of Colonial Development, Volume 2, Developing British Colonial Resources, 1945-51*, Atlantic Heights: Humanities Press.

Morgan, D.J., 1980c, *The Official History of Colonial Development, Volume 1, The Origins of British Aid Policy, 1924-1945*, Atlantic Heights Humanities Press.

Morgan, D.J., 1980d, *The Official History of Colonial Development, Volume 4, Changes in British Aid Policy, 1951-1970*, Atlantic Heights, Humanities Press.

Morgan, W. B., 1969, 'Peasant Agriculture in Tropical Africa', in Thomas and Whittington, eds.

Morna, C. L., 1988, 'How Does Ghana's Economic Miracles Look From the Inside?' *African Business*, November.

Morna, C. L., 1991, 'Cutting Back on Campus', *Africa Report*, March-April.

Morris, G., 1980, *The Official History of Colonial Development*, 5 Vols. Atlantic Highlands, Humanities Press.

Morris, W. H. et al., eds., 1980, *Decolonization and Africa: The British and French Experience*, London, Frank Cass.

Morris-Jones, W. H. and G. Fisher, eds., 1980, *Decolonization and After*, London, Frank Cass.

Mortimer, M., 1990, *Journeys Through the French African Novel*, Portsmouth, NH, Heinemann.

Moser, C. O., 1978, 'Informal Sector or Petty Commodity Production: Dualism or Dependence in Urban Development', *World Development*, 6, pp. 1041-1064.

Moss, R. P. and R. J. Rathbone, eds., 1975, *The Population Factor in African Studies*, London, University of London Press.

Mott, F. L. and S. H. Mott, 1980, 'Kenya's Record Population Growth: A Dilemma of Development', *Population Bulletin*, Vol. 35, pp. 1-43.

Mouffe, C., ed., 1992, *Dimensions of Radical Democracy: Pluralism, Citizenship, Community*, London, Verso.

Moyo, S., 1995, *The Land Question in Zimbabwe*, Harare, Sapes Books.

Mphande, L., 1994, 'The Malawi Writers Group: Before and After Structural Adjustment', Committee For Academic Freedom in Africa Newsletter, Vol. 6, pp. 6-14.

Mphande, L., 1996, 'Dr. Hastings Kamuzu Banda and the Malawi Writers Group: The (un)Making of a Cultural Tradition', *Research in African Literatures*, Vol. 27, pp. 80-101.

Mrysiades, and McCullough, eds., 1995, *Order and Partialities: Theory, Pedagogy and the 'Postcolonial'*, Albany, NY, SUNY Press.

Msambichaka, L. A., 1987, 'State Policies and Food Production in Tanzania', in Mkandawire and Bourenane, eds.

Mudimbe, V. Y. and K. A. Appiah, 1993, 'The Impact of African Studies on Philosophy', in Bates, Mudimbe, and O'Barr, eds.

Mudimbe, V. Y., 1988, *The Invention of Africa: Gnosis, Philosophy, and the Order of Knowledge*, London, James Currey.

Mudimbe, V. Y., 1994, *The Idea of Africa*, Bloomington and Indianapolis, Indiana University Press.

Mugomba, A. T. and M. Nyaggah, eds., 1977, *Independence without Freedom: The Political Economy of Colonial Education in Southern Africa*, Santa Barbara, Cal., ABC-Clio.

Munachonga, M. L., 1989, 'Women and the State, Class, and Gendered Models of Land Resettlement', in Parpart and Staudt, eds.

Munene, M. et al., eds., 1995, *The United States and Africa: From Independence to the Cold War*, Nairobi, East African Educational Publishers.

Munro, J. F., 1976, *Africa and the International Economy, 1800-1960*, London, J. M. Dent and Sons.

Muntemba, M. S., 1982, 'Women and Agricultural Change in the Railway Region of Zambia: Dispossession and Counter-strategies, 1930-1970', in Bay, ed.

Muriuki, G., 1974, *A History of the Kikuyu, 1500-1900*, Nairobi, Oxford University Press.

Murray, C. and R. J. Hernstein, 1994, *The Bell Curve*, New York, Free Press.

Musisi, N. B., 1991, 'Women: 'Elite Polygyny', and Buganda State Formation', in *Signs*, Vol. 16, pp. 757-786.

Mwanza, A. M. ed., 1992, *Structural Adjustment Programmes in SADC*, Harare, Sapes Books.

Mwanza, A., et al., 1992, 'The Structural Adjustment Programme in Zambia: Lessons From Experience', in Mwanza, ed.

Mwanza, J., 1988, 'External Debt Crisis: Zambia's Experience', KEA/FES Conference on North-South Dialogue: A Strategy for Africa, Nairobi, November.

Mwarania, K. M., 1988, 'Foreign Capital Flows and External Debt Crises in Sub-Saharan Africa', *KEA/FES Conference on North-South Dialogues: A Strategy for Africa*, Nairobi, November.

Myint, H., 1971, *The Economics of the Developing Countries*, London, Hutchinson.

N'jie, M. D., 1986, 'Gambia Vows To "Off-load Excess"', *African Business*, May.

N'jie, M. D., 1987, 'Gambia: For Better or Worse, Reforms Affect All', *Africa Business*, January.

N'jie, M. D., 1988, 'Gambia: Private Sector Buys up State Shares', *African Business*, May.

Nabudere, D. W., 1977, *The Political economy of Imperialism*, London, Zed Press.

Nabudere, D. W., 1980, *Imperialism and Revolution in Uganda*, London, Onyx Press.

Nabudere, D. W., 1989, *The Crash of International Capital and Its Implications for the Third World*, Harare, Sapes Trust.

Nafsziger, E. W., 1988, *Inequality in Africa. Political Elites, Proletariat, Peasants and the Poor*, Cambridge, Cambridge University Press.

Naipaul, V. S., 1967, *The Mimic Men*, London, Deutsch.

Napoli, J., 1992, 'Egyptian Sleight of Hand', *Index on Censorship*, Vol. 21, n° 2, pp. 21-23.

Nawal-el-Saadawi, 1983, *Woman at Point Zero*. London: Zed Press.

Ndiaye, R., 1988, 'Oral Culture and Libraries', *IFLA Journal*, Vol. 14/1, pp. 40-46.

Nduru, M., 1989, 'Reporting the Sudan', *Index on Censorship*, Vol. 18, n° 1.

Nelson, J., ed., 1990, *Economic Crisis and Policy Choice: The Politics of Economic Adjustment in the Third World*, Princeton: Princeton University Press.

Nelson, N. ed., 1981, *African Women in the Development Process*, London, Frank Cass.

Netting, R. M., et al., eds., 1984, *Households Comparative and Historical Studies of the Domestic Group*, Berkeley, University of California Press.

New African, 1984-1985, London, UK.

New African, 1993, 'Aids: The Epidemic That Never Was', December, pp. 8-13.

New African, 1994, 'Aids is Not African', October, pp. 10-17.

Newa, J. M., 1993, 'The Sustainability of Information Technology Innovations, CD-ROM at the University of Dar es Salaam', in Patrikios and Levey, eds.

Newbury, C. and B. G. Schoepf, 1989, 'State, Peasantry, and Agrarian Crisis in Zaire: Does Gender Make a Difference?', in Staudt and Parpart, eds.

Newbury, D., 1985, 'Mode of Production Analysis and Historical Production', *Canadian Journal of African Studies*, 19, pp. 38-45.

Newman, L., 1991, 'Critical Theory and the History of Women: What's at Stake in Deconstructing Women's History', *Journal of Women's History*, 2, pp. 58-68.

Newsweek, 1987, November 2.

Ngara, E., 1994, *The African University and Its Mission*, Lesotho, The Institute of Southern African Studies.

Ngubane, H., 1992, 'Clinical Practice and Organization of Indigenous Healers in South Africa', in Feierman and Janzen, eds.

Ngwira, M., 1993, 'Information Delivery in Malawi with Special Emphasis on the Bunda College of Agriculture Library', in Patrikios and Levey, eds.

Nicholson, L. J., ed., 1990, *Feminism/Postmodernism*, New York, Routledge.

Nicholson, M., 1986, 'Breaking out of the Debtors' Prison', *African Business*, May.

Nicol, D., 1993, 'Race, Ethnohistory and Other Matters: A Discussion of Kwame Anthony Appiah, 'In My Father's House: Africa in the Philosophy of Culture', *African Studies Review*, Vol. 36, pp. 109-116.

Nicolaus, M., 1970, 'The Theory of the Labour Aristocracy', *Monthly Review*, April.

Nielsen, J. M., ed., 1990, *Feminist Research Methods: Exemplary Readings in the Social Sciences*, Boulder, Co., Westview Press.

Nigam, S. B. L., 1979, 'Employment Market Information Programmes in English-Speaking African Countries', Workshop on Improvements of Employment Market Information Systems in English-Speaking African Countries, Addis Ababa, JASPA.

Niven, A. 1989, 'Achebe and Okri: Contrasts in the Response to Civil War', in Bardolph, ed.

Nixon, F., 1982, 'Import-Substituting Industrialization', in Fransman, ed.

Njondo, K., 1988, 'Anatomy of Privatization Scheme: The Togo Example', *African Business*, February.

Njururi, B., 1988, 'Kenya: State Begins Unloading Loss-Making Parastatals', *African Business*, July.

Nkinyangi, J. A., 1983, 'Who Conducts Research in Kenya?', in Shaeffer and Nkinyangi eds.

Nkrumah, K., 1963, *Neo-Colonialism: The Highest Stage of Imperialism*, London, Panaf Books.

Nkrumah, K., 1964, *Conciencism*, London, Heinemann.

Nkrumah, K., 1965, *Neo-Colonialism: The Last Stage of Imperialism*. London, Nelson.

Nnoli, O. ed., 1993, *Dead-End To Nigerian Development*, Dakar, CODESRIA Book Series.

Nnoli, O., 1995, 'Ethnic Conflicts and Democratization in Africa', CODESRIA Eighth General Assembly, Dakar.

Noel, A., 1987, 'Accumulation, Regulation and Social Change: An Essay on French Political Economy', *International Organization*, 41.

Nonneman, G., ed., 1996, *Political and Economic Liberalization*, Boulder, Col., Lynne Rienner.

Nwapa, F., 1966, *Efuru*, London, Heinemann.

Nyang'oro, J. E. and T. M. Shaw eds., 1989, *Corporatism in Africa: Comparative Analysis and Practice*, Boulder, Colorado, Westview.

Nyariki, L. and R. Makotsi, 1995, 'Problems of Book Marketing and Distribution in Kenya', *African Publishing Review*, Vol. 4/2, p. 11.

Nyerere, J., 1969, *Nyerere on Socialism*, Dar es Salaam, Oxford University Press.

Nyong'o', P. A. ed., 1987, *Popular Struggles for Democracy in Africa*, London, Zed Press.

Nyong'o, A., 1988, 'Political Instability and the Prospects for Democracy in Africa', *Africa Development*, 13.

Nyong'o, P. A., 1989, 'State and Society in Kenya: The Disintegration of Nationalist Coalitions and the Rise of Presidential Authoritarianism 1963-1978', *African Affairs*, Vol. 88.

Nyong'o, P. A., 1992, 'Discourses on Democracy in Africa', CODESRIA Seventh General Assembly, Democratization in Africa: Problems and Prospects.

Nzegwu, N., 1996, 'Questions of Identity and Inheritance: A Critical Review of Kwame Anthony Appiah's *In My Father's House*', *Hypatia*, Vol. 11, pp. 175-201.

Nzomo, M., 1993, 'The Gender Dimensions of Democratization in Kenya: Some International Linkages', *Alternatives*, Vol. 18, pp. 61-73.

Nzomo, M., 1995, 'The Political Economy of the African Crisis: Gender Impacts and Responses', *International Journal*, 51.

O'Brien, J., 1986, 'Sowing the Seeds of Famine: The Political Economy of Food Deficits in Sudan', in Lawrence, ed.

O'Connor, J., 1973, *The Fiscal Crisis of the State*, New York, St. Martin's Press.

O'Connor, J., 1984, *Accumulation Crisis*, Oxford Basil Blackwell.

O'Loughlin, J., 1986, 'World-Power Competition and Local Conflicts in the Third World', in Johnston and Taylor, eds.

Obbo, C., 1980, *African Women: Their Struggle for Economic Independence*, London, Zed Press.

Obbo, C., 1986, 'Stratification and the Lives of Women in Uganda', in Robertson and Berger, eds.

Ochieng', W. R., 1974, 'Undercivilization in Black Africa', *Kenya Historical Review*, 2, pp. 45-57.

Ochieng', W. R., ed., 1989, *A Modern History of Kenya, 1895-1980*, London, Evans Brother.

Odamtten, V. O., 1994 *The Art of Ama Ata Aidoo: Polylectics and Reading Against Neocolonialism*, Gainesville, University of Florida Press.

Ofeimun, O., 1992, 'Africa's Many Mansions', *West Africa*, 20-26 July, 1231-32.

Offen, K., et al., eds., 1991, *Writing Women's History: International Perspectives*, Bloomington, Indiana University Press.

Offen, K., R. R., 1991, Pierson and J. Rendall, 'Introduction', in Offen, et al., eds.

Ofuatey-Kodjoe, W., ed., 1986, *Pan-Africanism: New Directions in Strategy*, Lanham, Md., University Press of America.

Ogot, B. A. and T. Zeleza, 1988, 'Kenya: The Road to Independence', in Gifford and Louis, eds.

Ogot, B. A. ed., 1973, *Zamani: A Survey of East African History*, Nairobi, Longman/EAPH.

Ogot, B. A., 1967, *A History of the Southern Luo*, Nairobi, East African Publishing House.

Ogot, B. A., 1978, 'Three Decades of Historical Studies in East Africa, 1947-1977', *Kenya Historical Review*, 6, pp. 22-33.

Ogot, B. A., 1981, 'Description of the Project', in Ki-Zerbo, ed.

Ogot, B. A., 1984, 'Whose History? - The Dilemmas of Research in Early African History', Chairman's Address to the Historical Association of Kenya Annual Conference, Nairobi, August.

Ogot, B. A., ed., 1985, *Kenya in the Nineteenth Century*, Nairobi, Bookwise and Anyange.

Ogren, K. J., 1994, ''What is Africa to Me?': African Strategies in the Harlem Renaissance', in Lemelle and Kelley, eds.

Ogunsanwo, O., 1995, 'Intertextuality and Post-Colonial Literature in Ben Okri's The Famished Road', *Research in African Literatures*, Vol. 26, pp. 40-52.

Ojaide, T., 1992, 'African Poetry', *Stand Magazine*, Vol. 33, pp. 54-59.

Ojo-Ade, F., 1990, 'Of Human Trials and Triumphs: Bessie Head's Collection of Treasures', in Cecil Abrahams, ed.

Ojulu, E., 1987, 'Uganda: Government is Revising Investment Plan', *African Business*, March.

Okali, C., 1983, 'Kinship and Cocoa Farming in Ghana', in Oppong, ed.

Okara, G., 1991, 'Towards the Evolution of an African Language for African Literature', in Rutherford and Petersen, eds.

Okonkwo, R., 1986a, 'Adelaide Casely-Hayford', in Okonkwo, ed.

Okonkwo, R., ed., 1986b, West African Nationalists, Enugu, Delta Press.

Okoth-Ogendo, H. W. O., 1981, 'Land Ownership and Land Distribution in Kenya's Large-Farm Areas', in Killick, ed.

Okoth-Ogendo, H. W. O., 1993, 'Agrarian Reform in Sub-Saharan Africa: An Assessment of State Responses to the African Agrarian Crisis and Their Implications for Agricultural Development', in Bassett and Crummey, eds.

Okpewho, I., 1983, 'Myth and Modern Fiction: Armah's two Thousand Seasons', *African Literature Today*, Vol. 13, pp. 1-23.

Okpewho, I., 1988, 'African Poetry: The Modern Writer and Oral Tradition', *African Literature Today*, Vol. 18, pp. 3-25.

Okri, B., *Songs of Enchantment. Sequel to the Famished Road*, London, Vintage.

Okri, B., 1991, *The Famished Road*, London, Vintage.

Olagunju, T., et al., 1993, *Transition to Democracy in Nigeria* (1985-1993), Ibadan, Safari Books.

Olden, A., 1987, 'Sub-Saharan Africa and the Paperless Society', *Journal of the American Society for Information Science*, 38/4, pp. 298-304.

Olukoshi, A. O., 1995, 'Africa: Democratizing Under Conditions of Economic Stagnation', CODESRIA Eighth General Assembly, Dakar.

Olukoshi, A. O., 1995, 'Bourgeois Social Movements and the Struggle for Democracy in Nigeria: An Inquiry into the "Kaduna Mafia"', in Mamdani and Wamba-dia-Wamba, eds.

Olukoshi, A., 1995, 'The Politics of Structural Adjustment in Nigeria', in Mkandawire and Olukoshi, eds.

Olukoshi, A., ed., 1991, *The Politics of Structural Adjustment in Nigeria*, London, James Currey.

Olusanya, G. O., 1973, *The Second World War and Politics, in Nigeria, 1939-1953*, London, Evans Brothers.

Omaar, R., 1993, 'The Best Chance for Peace', *Africa Report*, Vol. 38, pp. 44-8.

Omer-Cooper, J. D., 1987, *History of Southern Africa*, London, James Currey.

Onimode, B., 1985, *An Introduction to Marxist Political Economy*, London, Zed Books.

Onimode, B., 1988, *A Political Economy of the African Crisis*, London, Zed Books.

Onimode, B. ed., 1989a, *The IMF, World Bank and the African Debt Crisis*, Vol. 1, *The Social and Political Impact*, London, IFAA and Zed Books.

Onimode, B., ed., 1989b, *The World Bank, the IMF and the African Debt*, Vol. 2, London, Zed.

Onimode, B., 1992, 'The Democratization Process and the Economy', Codersia Seventh General Assembly, Democratization in Africa: Problems and Prospects, Dakar.

Onoge, O. F., 1973, 'The Counter-Revolutionary Tradition in African Studies: The Case of Applied Anthropology', *The Nigerian Journal of Economic and Social Studies*, 15.

Oppong, C., 1983, *Female and Male in West Africa*, Cambridge, Cambridge University Press.

Oreidin, O., 1988, 'Nigeria Now Faces Brain Drain', *Daily Nation*, Nairobi, October 19.

Orr, C. A., 1966, 'Trade Unionism in Colonial Africa', *Journal of Modern African Studies*, 4, pp. 65-84.

Osaghae, E. ed., 1994, *Between State and Civil Society in Africa*, Dakar, CODESRIA Book Series.

Osaghae, E., 1994, 'Towards a Fuller Understanding of Ethnicity in Africa: Bringing Rural Ethnicity Back', in Osaghae, ed.

Osman, A.M., 1985, 'Large menu, Little Food', *South*, July.

Osoba, S., 1977, 'The Nigerian Power Elite, 1952-65', in Gutkind and Waterman eds.

Overall, C., 1992, 'What's Wrong with Prostitution? Evaluating Sex Work', *Signs*, Vol. 17.

Overholt, C., et al., eds., 1985, *Gender Roles in Development Projects*, Kumarian Press.

Owen, R. and B. Sutcliffe, eds., 1972, *Studies in the Theory of Imperialism*, London, Longman.

Owolabi, K. A., 1995, 'Cultural Nationalism and Western Hegemony: A Review Essay on Appiah's Universalism', *Africa Development*, Vol. 20, pp. 113-123.

Owomoyela, O., 1992, 'Socialist Realism or African Realism? A Choice of Ancestors', Research in African Literatures, Vol. 22, pp. 21-40.

Owomoyela, O. ed., 1993a, *A History of Twentieth-Century African Literatures*, Lincoln and London, University of Nebraska Press.

Owomoyela, O., 1993b, 'The Question of Language in African Literatures', in Owomoyela, ed.

Owomoyela, O., 1994, 'With Friends Like These ... A Critique of Pervasive Anti-Africanisms in Current African Studies Epistemology and Methodology', African Studies Review, Vol. 37, pp. 77-101.

Owusu, M., 1992, 'Democracy and Africa - a View From the Village', *Journal of Modern African Studies*, Vol. 30, pp. 369-396.

Oxaal, I., T. Barnett, and J. Booth, eds., 1975, *Beyond the Sociology of Development*, London, Routledge and Kegan Paul.

Oxenham, J., 1986, 'Possible Levers Against the Paper Qualification Syndrome', in JASPA.

Oyegoke, L., 1996, 'Leaky Mansion? Appiah's Theory of African Cultures', *Research in African Literatures*, Vol. 27, pp. 143-148.

Oyugi, W. O. and A. Gitonga, eds., 1987, *Democratic Theory and Practice in Africa*, Nairobi, Heinemann Kenya.

Padmore, G., 1936, *How Britain Rules Africa*, New York, Negro Universities Press.

Padmore, G., 1949, *Africa: Britain's Third Empire*, New York, Negro Universities Press.

Padmore, G., 1956, *Pan-Africanism or Communism? The Coming Struggle for Africa*, London, Dobson.

Palma, G., 1978, 'Dependency: A Formal Theory of Underdevelopment or a Methodology for the Analysis of Concrete Situations of Underdevelopment?', *World development*, 6, pp. 881-924.

Palmer R. and N. Parsons, eds., 1977, *The Roots of Rural Poverty in Central and Southern Africa*, London, Heinemann.

Palmer, I., 1973, *How Revolutionary is the Green Revolution?*, London.

Palmer, R., 1977, *Land and Racial Discrimination in Southern Rhodesia*, London, Heinemann.

Paludi, M. A. and L.A. Strayer, 1984, 'Differential evaluations of performance as a function of author's name', *Sex Roles*, Vol. 12, pp. 353 -361.

Paludi, M. A. and W.D. Bauer, 1983, 'Goldberg revisited: What's in an author's name?' *Sex Roles*, Vol. 9, pp. 387-390.

Panter-Brick, K., 1988, 'Independence, French Style', in Gifford and Louis, eds.

Parfit, T. W., 1990, 'Lies, Damned Lies and Statistics: The World Bank/ECA Structural Adjustment Controversy', *Review of African Political Economy*, 47, pp. 128-43.

Park, P., 1986, 'Tanzania: Static Economy Awaits Joint Action', *African Business*, November.

Parker, A., et al., eds., 1992, *Nationalisms and Sexualities*, New York and London, Routledge.

Parpart, J. L. and K. A. Staudt, eds., 1989, *Women and the State in Africa*, Boulder and London Lynne Rienner.

Parpart, J. L., 1986, 'The Household and the Mine Shaft: Gender and Class Struggles on the Zambian Copperbelt, 1926-64', *Journal of Southern African Studies*, Vol. 13, pp. 36-56.

Parpart, J. L., 1990, 'Wage Earning Women and the Double Day: the Nigerian Case', in Stichter and Parpart, eds.

Parpart, J. L., 1992, 'Listening to Women's Voices: The Retrieval and Construction of African Women's History', *Journal of Women's History*, Vol. 4, pp. 171-79.

Parpart, J. L., 1995, 'Is Africa a Postmodern Invention?' *Issue*: A Journal of Opinion, Vol. 23, pp. 16-18.

Parpart, J., 1983, *Capital and Labour on the African Copperbelt*, Philadelphia, Temple University Press.

Parry, B., 1994, 'Resistance Theory/Theorising Resistance, or two Cheers for Nativism', in Barker, Hulme and Iversen eds.

Parsons, J., 1977, *Population Fallacies*, London, Elek Pemberton.

Parsons, T., 1951, *The Social System*, Glencoe, Free Pres.

Patrikios, H. A., 1992, 'Medline in Zimbabwe', in *CD-ROM for African Research Needs*: 30-37.

Patrikios, H. A., 1993, 'A Minimal Acquisitions Policy for Journals at the University of Zimbabwe Medical Library' in Patrikios and Levey, eds.

Patrikios, H.A. and L.A. Levey eds., 1993, *Survival Strategies in African University Libraries: New Technologies in the Service of Information, Proceedings from a Workshop*, University of Zimbabwe, Harare.

Patterson's, O., 1991, *Freedom in the Making of Western Culture*, Vol. 1. New York, Basic Books.

Patterson, D. K. and G. W. Hartwig, eds., 1978, *Disease in African History*, Durham, Duke University Press.

Payer, C., 1982, *The World Bank*, New York, Monthly Review Press.

Peace, A., 1975, 'The Lagos Proletariat: Labour Aristocrats or Populist Militants', in Sandbrook and Cohen, eds.

Peace, A., 1975, 'The Lagos Proletariat: Labour Aristocrats or Populist Militants.' In R. Sandbrook and R. Cohen, eds.

Pearse, A., 1983, 'Apartheid and Madness: Bessie Head's *A Question of Power*', *Kunapipi*, 5, pp. 81-88.

Peattie, L., 1987, 'An Idea In Good Currency and How It Grew: The Informal Sector', *World Development*, 15, pp. 851-860.

Peet, R., 1986, 'The Destruction of Regional Cultures', in Johnston and Taylor, eds.

Peires, J. B., 1989, *The Dead Will Arise: Nongqaawusa and the Great Xhosa Cattle-Killing of 1856-7*, London, James Currey.

Pereira, A., 1993, 'Peace in the Third World? The Case of Angola', *Dissent*, 40, pp. 291-4.

Perelman, M., 1977, *Farming For a Profit in a Hungry World: Capital and the Crises in Agriculture*, Montclair, N.J., Allanheld.

Perham, M., 1963, *Colonial Reckoning*, London, Rex Collins.

Perold, H., ed., 1985, *Working Women*, Johannesburg, Ravan Press.

Personal Narratives Group, 1989, *Interpreting Women's Lives: Feminist Theory and Personal Narratives*, Bloomington, in Indiana University Press.

Peters, P., 1983, 'Gender, Developmental Cycles and Historical Process: A Critique of Recent Research on Botswana', *Journal of Southern African Studies*, 9, pp. 100-122.

Petras, J., 1989, 'Metamorphosis of Latin America's Intellectuals', *CODESRIA Bulletin*, n° 1, 1990, reprinted from *Economic and Political Weekly*, 14, April 8.

Philipp, T, 1978, 'Feminism and Nationalist Politics in Egypt', in Beck and Keddie.

Picard, L. A. and M. Garrity, eds., 1994, *Policy Reform for Sustainable Development in Africa: The Institutional Imperative*, Boulder, Lynne Rienner.

Pierson, R. R., 1991, 'Experience, Difference, Dominance and Voice in the Writing of Women's History', in Offen, et al., eds.

Pierson, R. R., 1992, 'Colonization and Canadian Women's History', *Journal of Women's History*, 4, pp. 134-156.

Pike, W., 1985a, 'The Sour Fruits of Success', *South*, July.

Pike, W., 1985b, 'Uganda: The Recovery That Never Was', *African Business*, June.

Pilling, G., 1973, 'Imperialism, Trade and Unequal Exchange: The Work of Arghiri Emmanuel', *Economy and Society*, 2, pp. 164-185.

Place, F. and P. Hazell, 1993, 'Productivity Effects of Indigenous Tenure Systems in Sub-Saharan Africa', *American Journal of Agricultural Economics*, 75, pp. 10-19.

Plange, N., 1979, 'Underdevelopment in Northern Ghana: Natural Causes or Colonial Capitalism?', *Review of African Political Economy*, 15/16, pp. 4-14.

Plant, R., 1994, *Labour Standards and Structural Adjustment*, Geneva, ILO.

Polychroniou, C., 1991, *Marxist Perspectives on Imperialism: A Theoretical Analysis*, New York, Praeger.

Porter, R. C., 1989, 'Recent Trends in LDC Military Expenditures', *World Development*, 17.

Portes, A., 1984 *Latin American Class Structure: Their Composition and Change During the Last Decades*, Washington, DC, The Johns Hopkins University of Advanced International Studies, Occasional Paper n° 3.

Post, K., 1972, 'Peasantization and Rural Political Movements in Western Africa', *European Journal of Sociology*, 13, pp. 223-254.

Pratt, C., 1982, 'Colonial Governments and the Transfer of Power in East Africa', in Gifford and Louis, eds.

Prebisch, R., 1971, *Change and Development in Latin America: The Great Task*, New York, Praeger.

Presley, C. A., 1991, *Kikuyu Women, the Mau Mau Rebellion and Social Change in Kenya*, Boulder, Col., Westview Press.

Quijano, A., 1974, 'The Marginal Role of the Economy and the Marginalized Labour Force', *Economy and Society*, 3.

Radice, H. ed., 1975, *International Firms and Modern Imperialism*, Harmondsworth, Penguin.

Rahmato, D., 1993, 'Land, Peasants, and the Drive for Collectivization in Ethiopia', in Bassett and Crummey, eds.

Raikes, P., 1988, *Modernizing Hunger. Famine Food Surplus and Farm Policy in the EEC and Africa*, Portsmouth, N.H., Heinemann.

Rake, A., 1986, 'Tanzania Strikes a Deal with the World Bank', *African Business*, June.

Rake, A., 1988, 'Breaking The Debt Cycle', *New Africa*, June.

Ranaivosoa, G., 1988a, 'Madagascar: Perestroika Comes to the Great Red Isle', *African Business*, August.

Ranaivosoa, G., 1988b, 'Madagascar is to Privatize Banks', *African Business*, October.

Ranger, T. O. ed., 1968, *Emerging Themes in African History*, Nairobi, Oxford University Press.

Ranger, T. O., 1968, 'Introduction', in Ranger, ed.

Ranger, T. O., 1971, 'The Historiography in Dar es Salaam', *African Affairs*, 70, pp. 50-61.

Ranger, T. O., 1976, 'Towards a Usable African Past', in Fyfe, ed.

Ranger, T. O., 1977, 'The People in African Resistance: A Review', *Journal of Southern African Studies*, 4.

Ranger, T. O., 1983, 'The Invention of Tradition in Colonial Africa', in Hobsbawm and Ranger, eds.

Ranger, T. O., 1969, *Revolt in Southern Rhodesia, 1896-7*, London Heinemann.

Ranger, T. O., 1979, 'White Presence and Power in Africa', *Journal of African History*, 20, pp. 463-469.

Ranger, T. O., 1983, 'The Invention of Tradition in Colonial Africa', in Hobsbawm and Ranger, eds.

Rapley, J., 1993, *Ivorien Capitalism*, Boulder, Lynne Rienner.

Read, M., 1942, 'Migrant Labour in Africa and its Effects on Tribal Life', *International Labour Review*, 14, pp. 605-631.

Redclift, N. and E. Mingione, eds., 1985, *Beyond Employment: Household, Gender and Subsistence*, Oxford: Basil Blackwell.

Reinharz, S., 1992, *Feminist Methods in Social Research*, New York, Oxford University Press.

Rendall, J., 1991, '"Uneven Developments": Women's History, Feminist History and Gender History in Great Britain', in Offen, et al.

Republic of Kenya. *Economic Survey 1988*, Nairobi, Government Printer, 1988.

Republic of Kenya. *Sessional Paper No. 10 of 1965: African Socialism and Its Application to Kenya*, Nairobi, Government Printer, 1965.

Review of African Political Economy, 'Editorial', 15/16, 1979.

Review of African Political Economy, 1975, 'Editorial', 3, pp. 1-9.

Review of African Political Economy, 1977, Special Issue: Capitalism in Africa 8.

Rhodes, R. I. ed., 1970, *Underdevelopment and Revolution*, New York, Monthly Review Press.

Rich, A., 1979, *On Lies, Secrets, and Silence*, New York, Norton.

Richards, P. ed., 1975, *African Environment: Perspectives and Prospects*, London.

Richards, P., 1983, 'Ecological Change and the Politics of African Land Use', *African Studies Review*, Vol. 26, n° 2, pp. 1-72.

Richter, L. E., 1986, 'Upgrading Employment-Market Reporting: Realities and Priorities in Africa', in JASPA.

Riddell, J. B., 1992, 'Things Fall Apart Again: Structural Adjustment Programmes in Sub-Saharan Africa.' *Journal of Modern African Studies*, Vol. 30, pp.53-68.

Rigby, P., 1992, *Cattle, Capitalism and Class*, Philadelphia, Temple University Press.

Ritsenthaler, R. E., 1960, 'Anlu: A Woman's Uprising in the British Cameroons', *African Studies*, 19, pp. 151-56.

Roberts, A. D., 1970, 'The Lumpa Church of Alice Lenshina', in Rotberg and Mazrui, eds.

Roberts, A. D., 1978, 'The Earlier Historiography of Colonial Africa', *History in Africa*, 5.

Roberts, B. C. and L. G. de Bellccombe, 1967, *Collective Bargaining in African Countries*, London, Macmillan.

Roberts, B. C., 1964, *Labour in the Tropical Territories of the Commonwealth*, London, Bell.

Roberts, D., 1975, *Capitalism in Crisis*, New York, Pathfinder Press.

Roberts, H., ed., 1981, *Doing Feminist Research*, London, Routledge and Kegan Paul.

Roberts, P. A., 1987, 'The State and the Regulation of Marriage: Sefwi Wiawso (Ghana) 1900-1940', in Afshar, ed.

Roberts, R., 1984, 'Women's Work and Women's Property: Household Social Relations in the Maraka Textile Industry of the Nineteenth Century', *Comparative Studies in Society and History*, 26, pp. 48-69.

Robertson, C. and I. Berger eds., 1986, *Women and Class in Africa*, New York, Africana Publishing Co.

Robertson, C. and M. A. Klein eds., 1983, *Women and Slavery in Africa*, Madison, The University of Wisconsin Press.

Robertson, C., 1984, *Sharing the Same Bowl: A Socioeconomic History of Women and Class in Accra, Ghana*, Bloomington, Indiana University Press.

Robertson, C., 1987, 'Developing Economic Awareness: Changing Perspectives in Studies of African Women 1976-1985', *Feminist Studies*, 13, pp. 97-135.

Robinson, K. E., 1965, *The Dilemmas of Trusteeship*, London, Oxford University Press.

Robinson, P. T., 1994a, 'The National Conference Phenomenon in Francophone Africa', *Comparative Studies in Society and History*, 36, pp. 575-610.

Robinson, P. T., 1994b, 'Democratization: Understanding the Relationship Between Regime Change and the Culture of Politics', *African Studies Review*, 37, pp. 39-67.

Robinson, P. T., 1995, 'Confronting the Challenge of Co-Development', *African Commentary*, August, pp. 4-5.

Robinson, R. and J. Gallagher with A. Denny, 1961, *Africa and the Victorians: The Official Mind of Imperialism*, London, Macmillan.

Robinson, R., 1980, 'Andrew Cohen and the Transfer of Power in Tropical Africa, 1940-1951', in Morris-Jones and Fisher, eds.

Rodney, W., 1969, *West Africa and the Atlantic Slave Trade*. Historical Association of Tanzania.

Rodney, W., 1981, *A History of the Guyanese Working People, 1881-1905*, Baltimore and London, The Johns Hopkins University Press.

Rodney, W., 1982, *How Europe Underdeveloped Africa*, Washington, DC, Howard University Press, [1972].

Roelker, J. R., 1976, *Mathu of Kenya*, Stanford, Stanford University Press.

Rogers [Geiger], S., 1980, 'Anti-Colonial Protest in Africa: A Female Strategy Reconsidered', *Heresies*, 3, pp. 22-25.

Rogers [Geiger], S., 1990, 'Women and African Nationalism', *Journal of Women's History*, 2, pp. 227-244.

Rogers, B., 1980, *The Domestication of Womem*, London, Tavistock.

Romdhane, M. B., 1995a, 'Secular Political Opposition Groups in Tunisia', in Mamdani and Wamba-dia-Wamba, eds.

Romdhane, M. B., 1995b, 'The Politics of Structural Adjustment: The Case of Tunisia', in Mkandawire and Olukoshi, eds.

Romero, P. W. ed., 1988, *Life Histories of African Women*, London, The Ashfield Press.

Roodkowsky, M. and L. Leghorn, 1977, *Who Really Starves? Women and World Hunger*, New York, Friendship Press.

Rosaldo, R. 1994, 'Social justice and the crisis of national communities', in Barker, Hulme, and Iversen, eds.

Rosberg, C. G. and J. Nottingham, 1966, *The Myth of 'Mau Mau'*, Nairobi, Transafrica.

Roseberry, W., 1993, 'Beyond the Agrarian Question in Latin America', in Cooper, et.al.

Rosenfeld, C. P., 1986, *Empress Taytu and Menelik II: Ethiopia, 1883-1910*, Trent, NJ, The Red Sea Press.

Rostow, W. W., 1971, *Stages of Economic Growth, A Non-Communist Manifesto*, Cambridge, Cambridge University Press.

Roszak, T., 1993, 'Politics of Information and the Fate of the Earth', *Progressive Librarian*, 6/7, pp. 3-14.

Rotberg, R. I. and A. A. Mazrui eds., 1970, *Protest and Power in Black Africa*, New York, Oxford University Press.

Rotberg, R. I., 1972, *The Rise of Nationalism in Central Africa*, Cambridge, Harvard University Press.

Roth, M., 1993, 'Somalia Land Policies and Tenure Impacts: The Case of the Lower Shebelle', in Bassett and Crummey, eds.

Rothchild, D. and N. Chazan eds., 1988, *The Precarious Balance: The State and Society in Africa*, Boulder, Colorado, Westview.

Rothchild, D., 1987, 'Hegemony and State Softness: Some Variations in Elite Responses', in Ergas, ed.

Rousseas, S., 1979, *Capitalism and Catastrophe*, Cambridge, Cambridge University Press.

Roxborough, I., 1984, *Theories of Underdevelopment*, London, Macmillan.

Rushton, J. P. and A.F. Bagaert, 1989, 'Population Differences in Susceptibility to AIDS: An Evolutionary Analysis', *Social Science and Medicine*, 28, pp. 1211-1220.

Ruthenberg, H., 1971, *Farming Systems in the Tropics*, London, Oxford University Press.

Rutherford, A. and K. H. Petersen eds., 1991, *Chinua Achebe: A Celebration*, Portsmouth, N.H., Heinemann.

Sachikonye, L. ed, 1995, *Democracy, Civil Society and the State: Social Movements in Southern Africa*, Harare, Sapes Books,.

Sacks, K., 1982, *Sisters and Wives*, London, Greenwood.

Sahn, D. E. and J. Arulpragasam, 1994, 'Adjustment Without Structural Change: The Case of Malawi', in Sahn, ed.

Sahn, D., ed., 1994, *Adjusting to Policy Failure in African Economies*, Ithaca and London: Cornell University Press.

Said, E. W., 1979, *Orientalism*, New York, Vantage Books.

Said, E. W., 1993, *Culture and Imperialism*, New York, Alfred Knopf.

Salim, A. I. ed., 1984, *State Formation in Eastern Africa*, Nairobi, Heinemann Educational Books.

Samatar, A. I., 1992, 'Destruction of State and Society in Somalia, Beyond the Tribal Convention'.

Samuelson, P. A. and W. D. Nordhaus, 1985, *Economics*, 12th ed., New York, McGraw-Hill.

Sandbrook, R. and R. Cohen, eds., 1975, *The Development of an African Working Class: Studies in Class Formation and Action*, London, Longman.

Sandbrook, R., 1975, *Proletarians and African Capitalism: The Kenyan Case, 1960-1972*, London, Cambridge University Press.

Sandbrook, R., 1985, *The Politics of Africa's Economic Stagnation*, Cambridge, Cambridge University Press.

Sandbrook, R., 1990, 'Taming the African Leviathan', *World Policy Journal*, ,pp. 675-701.

Sanger, C., 1982, *Safe and Sound: Disarmament and Development in the Eighties*, Ottawa, Deneu Publishers.

Sangster, J. and P. T. Zeleza, 1994, 'Academic Freedom in Context', *Journal of Canadian Studies*, 29, pp. 39-144.

Sangster, J., 1995, 'Beyond Dichotomies: Re-Assessing Gender and Women's History in Canada', *Left History*, 3, pp. 109-121.

Santos, M., 1979, *The Shared Space*, London, Methuen.

Sarris, A. H. and R. V. den Brink, 1994, 'From Forced Modernization to Perestroika: Crisis and Adjustment in Tanzania', in Sahn, ed.

Sassen, S., 1996, 'Whose City Is It? Globalization and the Formation of New Claims', *Public Culture*, 8, pp. 205-233.

Saul, J. S. and G. Arrighi, 1968, 'Socialism and Economic Development in Tropical Africa', *Journal of Modern African Studies*, 6, pp. 141-168.

Saul, J. S. and G. Arrighi, 1969, 'Nationalism and Revolution in Sub-Saharan Africa', *Socialist Register*.

Saul, J. S. and R. Woods, 1979, 'African Peasantries', in Shanin, ed. *Peasants and Peasant Societies*, Harmondsworth, Penguin.

Saul, J. S., 1974, 'The State in Post-Colonial Societies - Tanzania', *Socialist Register*, London, Merlin.

Saul, J. S., 1975, 'The Labour Aristocracy Thesis Reconsidered', in Sandbrook and Cohen, eds.

Saul, J. S., 1976, 'Uganda: Unsteady State: Uganda, Obote and General Amin', *Review of African Political Economy*, 5, pp. 12-38.

Saul, J. S., 1977, 'Nationalist, Socialism, and Tanzanian History', in Gutkind and Waterman, eds.

Saul, J. S., 1979, 'The Dialectic of Class and Tribe', in *The State and Revolution in Eastern Africa*, London, Heinemann.

Scarlett, L., 1981, 'Tropical Africa: Food or Famine?', in Balaam and Carey, eds.

Schade, R., 1993, 'Towards an Integrated Curriculum: Suggestions for Balancing History Courses at the University Level', Paper presented at the Conference on Teaching Women's History: Challenges and Solutions Organised by the Canadian Committee on Women's History, Trent University, August 20-22.

Schapera, I., 1947, *Migrant Labor and Tribal Life*, London, Oxford University Press.

Scheub, H., 1985, 'A Review of African Oral Tradition and Literature', *African Studies Review*, 28, pp. 1-72.

Schipper, M., 1990, *Beyond Boundaries: Text and Context in African Literature*, Chicago, Ivan R. Dee.

Schissel, H., 1985, 'The Harder They Fall', *South*, July.

Schlegel, A. ed., 1977, *Sexual Stratification: A Cross-Cultural View*, New York, Columbia University Press.

Schmidt, E., 1991, 'Patriarchy, Capitalism, and the Colonial State in Zimbabwe', *Signs*, 16, pp. 732-756.

Schonfield, A., 1965, *Modern Capitalism*, London, Oxford University Press.

Schrank, B., 1994, 'Half and Half: The Case for Single-blind Reviewing the Other Way Round', *CAUT Bulletin*, 41.

Schumpeter, J., 1951, *Imperialism and Social Classes*, London, Blackwell.

Schuster, M. and S. van Dyne, 1984, 'Placing Women into the Liberal Arts: Stages of Curriculum Transformation', *Harvard Educational Review*, 54.

Schwegman, M. and M. Bosch, 1991, 'The Future of Women's History: A Dutch Perspective', *Gender and History*, 3, pp. 129-146.

Scott, J. C., 1976, *The Moral Economy of the Peasant*, New Haven, Yale University Press.

Scott, J. C., 1985, *Weapons of the Weak: Everyday Forms of Peasant Resistance*, New Haven, Yale University Press.

Scott, J. W., 1988, *Gender and the Politics of History*, New York, Columbia University Press.

Scott, R., 1966, *The Development of Trade Unions in Uganda*, Nairobi, East African Publishing House.

Seddon, D. ed., 1978, *Relations of Production: Marxist Approaches to Economic Anthropology*, London, Frank Cass.

Seeley, J., 1986, 'The Use of Bibliographic Databases in African Studies', *African Research and Documentation*, 41, pp. 7-12.

Seidman, G. W., 1984, 'Women in Zimbabwe: Post-independence Struggles', *Feminist Studies*, 10, pp. 419-440.

Selassie, B. H., 1985, 'Somali Independence, 1960: A Nation in Search of a State', Conference on African Independence and Consequences of the Transfer of Power, 1956-1980, University of Zimbabwe, Harare, January.

Selassie, S. G., 1986, 'Patterns of Women's Employment in Africa', in ILO/JASPA.

Seligson, M. A. and J. T. Passé-Smith eds., 1993, *Development and Underdevelopment: The Political Economy of Inequality*, Boulder and London, Lynne Rienner.

Sembene, O., 1986, *God's Bits of Wood*, Translated by Francis Price, Oxford, Heinemann.

Sen, A. K., 1970, *Growth Economics*, Harmondsworth, Penguin.

Sen, A., 1984, *Poverty and Famines: An Essay on Entitlement and Deprivation*, Oxford, Clarendon Press.

Sender, J. and S. Smith, *Development of African Capitalism in Africa*, London, Methuen.

Senghor, L. S., 1964, *On Socialism*, New York, Praeger.

Serequeberham, T., 1996, 'Reflection on *In My Father's House*', *Research in African Literatures*, 27, pp. 111-118.

Sethuraman, S. V., 1981, *The Urban Informal Sector in Developing Countries: Employment, Poverty and Environment*, Geneva, ILO.

Shaeffer S. and J. A. Nkinyangi eds., 1983, *Educational Research Environments in the Developing Countries*, Ottawa, International Development Research Centre, 213e.

Shao, I. et al., 1992, 'Structural Adjustment in a Socialist Country: The Case of Tanzania', in Mwanza, ed.

Shapiro, D., 1995, 'Population Growth, Changing Agricultural Practices, and Environmental Degradation in Zaire', *Population and Environment*, 16, pp. 221-236.

Sharawy, H., 1995, 'The Conflicts Between Islamic Fundamentalism and the National State in the Arab North Africa', CODESRIA Eighth General Assembly, Dakar.

Shaw, T. M., 1982, 'Beyond Underdevelopment: The Anarchic State in Africa', African Studies Association.

Shaw, T. M., 1993, *Reformism and Revisionism in Africa's Political Economy in the 1990s*, London and New York, Macmillan and St. Martin's Press.

Shenton, B. and M. Watts, 1979, 'Capitalism and Hunger in Northern Nigeria', *Review of African Political Economy*, 15/16, pp. 53-62.

Sherman, M. A., 1990, 'The State and the University in Africa: In Quest for Intellectual Freedom and Development', CODESRIA Symposium on Academic Freedom, Research and the Social Responsibility of the Intellectual in Africa, Kampala, Uganda, November.

Shiner, C., 1994, 'The World's Worst War', *Africa Report*, 39, pp. 13-16.

Shipton, P., 1992, 'Debts and Trespasses: Land, Mortgages, and the Ancestors in Western Kenya', *Africa*, 62, pp. 357-388.

Shiroya, O. J. E., 1968, 'The Impact of World War II on Kenya: The Role of Ex-servicemen in Kenyan Nationalism', Ph.D. dissertation, Michigan State University.

Shivji, I. G., 1987, *The Concept of Human Rights in Africa*, London: Codersia Book Series.

Shivji, I. G., 1989, 'The Pitfalls of the Debate on Democracy', *CODESRIA Bulletin*, n° 2.

Shivji, I. G., 1989, *The Concept of Human Rights in Africa*, Dakar: CODESRIA Book Series.

Shivji, I. G., 1990, 'The Jurisprudence of the Dar-es-Salaam Declaration on Academic Freedom', CODESRIA Symposium on Academic Freedom, Research and the Social Responsibility of the Intellectual in Africa, Kampala, Uganda, November.

Shivji, I. G., ed., 1986, *The State and the Working People in Tanzania*, Dakar, CODESRIA Book Series.

Shohat, B. 'Notes on the "Post-Colonial"', *Social Text*, 10, pp. 99-113, 1992.

Shopo, T. D., 1986, 'The Political Economy of Hunger', in Mandaza, ed.

Shuman, J. and D. Rosenau, 1972, *The Kondratieff Wave*, New York, World Publishing.

Sicherman, C. M., 1989, 'Ngugi wa Thiong'o and the Writing of Kenyan History', *Research in African Literatures*, 20, pp. 347-70.

Sievers, S., 1989, 'Dialogue: Six (or More) Feminists in Search of a Historian', in Angerman, et al., eds.

Signs, 1991, Special Issue. *Family, State, and Economy in Africa*, 16, 4 (Summer).

Simiyu, V. G., 1987, 'The Democratic Myth in the African Traditional Societies', in Oyugi and Gitonga, eds.

Simmons, E. B., 1988, *Economic Research on Women in Rural Development in Nigeria*, London, Overseas Liaison Committee.

Simpson, C., 1993, 'The Undemocratic Game', *Africa Report*, 38, pp. 49-51.

Sinclair, J., 1982, 'Improvement Within the System - A Delusion', in de Braganca and Wallerstein, eds.

Sindermann, C. J., 1982, *Winning the Games Scientists Play*, New York, Plenum.

Sindiga, I., 1985, 'The Persistence of High Fertility in Kenya', *Social Science and Medicine*, 20, pp. 71-84.

Singer, B. D., 1989, 'The Criterial Crisis of the Academic World', *Sociological Inquiry*, 59, pp. 127-143.

Singer, B. D., 1990, 'Reply to Hill', *Sociological Inquiry*, 60, pp. 301-2.

Singh, A, 1982, 'Industrialization in Africa: A Structuralist View', in Fransman, ed.

Siow, A., 1991, 'Are First Impressions Important in Academia?', *The Journal of Human Resources*, 26, pp. 236-255.

Skiff, A., 1980, 'Toward a Theory of Publishing or Perishing', *The American Sociologist*, 15, pp. 175-183.

Skinner, E. P., 1960, 'Labour Migration and its Relationship to Socio-economic Change in Mossi Society', *Africa*, 30, pp. 375-399.

Sklar, R. L., 1993, 'The African Frontier for Political Science', in Bates, Mudimbe, and O'Barr, eds.

Slater, R. O., B. M. Schutz, and S. R. Dorr eds., 1993, *Global Transformation and the Third World*, Boulder, Col., Lynne Rienner.

Smith, J. et al., 1984, *Households and the World Economy*, Beverly Hills, Sage Publications.

Smith, P., 1985, 'Nigeria: An IMF Deal Without IMF Money', *African Business*, October.

Smith, S., 1980, 'The Ideas of Samir Amin: Theory or Tautology?', *Journal of Development Studies*, 17.

Smith, T., 1982, 'Patterns in the Transfer of Power: A Comparative Study of French and British Decolonization', in Gifford and Louis, eds.

Smith, W., 1968, 'Industrial Sociology in Africa: Foundations and Prospects', *Journal of Modern African Studies*, 6.

Smock, D. R., 1969, *Conflict and Control in African Trade Unions: A Study of the Nigerian Coal Miners' Union*, Stanford, Hoover Institution Press.

Social and Economic Studies, 1980, Special Issue on Sir Arthur Lewis, Vol. 29, n° 4.

Soejarto, D. D. et al., 1978, 'Fertility Regulating Agents From Plants', *Bulletin of the World Health Organization*, 56, pp. 343-352.

Solway, J., 1994, 'Drought as a 'Revelatory Crisis': An Exploration of Shifting Entitlements and Hierarchies in the Kalahari, Botswana', *Development and Change*, 25, pp. 471-495.

South Magazine, London, 1984-1985.

South, London, South Publication, 1984-90.

Southall, A., 1970, 'The Illusion of Tribe', *Journal of Asian and African Studies*, 5.

Southall, R., 1982, *South Africa's Transkei*, London, Heinemann.

Souza, P. and V. Tokman, 1976, 'The Informal Urban Sector in Latin America', *International Labour Review*, 114.

Soyinka, W, 1965, *The Interpreters*, London, Heinemann.

Soyinka, W., 1978, *Myth, Literature and the African World*, Cambridge, Cambridge University Press.

Soyinka, W., 1993, 'Religion and Human Rights', *Index on Human Rights*, Vol. 17, n° 5, pp. 82-85.

Sparks, S., 1988, '"Camdessus" Quiet Revolution', *South*, September.

Spaulding, J., 1984, 'The Misfortunes of Some - The Advantages of Others: Land Sales by Women in Sinnar', in Hay and Stichter.

Spelman, E. V., 1988, *Inessential Woman: Problems of Exclusion in Feminist Thought*, Boston, Beacon Press.

Spender, D., 1981, 'The Gatekeepers: A Feminist Critique of Academic Publishing', in Roberts, ed.

Spivak, G. C., 1995, 'Acting Bits/Identity Talk', in Appiah and Gates, eds.

602 References

Stamp, P., 1986, 'Kikuyu Women's Self-Help Groups', in Robertson and Berger.
Standard, Nairobi, Kenya, 1984-1985.

Standing, G. and V. Tokman eds., 1991, *Towards Social Adjustment: Labour Market Issues in Structural Adjustment*, Geneva, ILO.

Stasiulis, D. K., 1990, 'Theorising Connections: Gender, Race, Ethnicity and Class,' in Li, ed.

Staudt, K. A. and J. L. Parpart eds., 1989, *Women and the State in Africa*, Boulder, Col., Lynne Rienner.

Steady, F. C., 1975, *Female Power in African Politics: The National Congress of Sierra Leone*, Los Angeles, California Institute of Technology.

Steady, F. M. ed., 1981, *The Black Woman Cross-Culturally*, Cambridge, Mass., Schenkman Publishing.

Stebbing, E. P., 1935, 'The Encroaching Sahara: The Threat to the West African Colonies', *Geographical Journal*, 85.

Stedman, S. J., 1993, *Botswana: The Political Economy of Democratic Development*, Boulder Lynne Rienner.

Stein, H. and E. W. Nafziger, 1991, 'Structural Adjustment, Human Needs, and the World Bank Agenda.' *Journal of Modern African Studies*, Vol. 29, pp.173-189.

Stern, S. J., 1993a, 'Africa, Latin America, and the Splintering of Knowledge: From Fragmentation to Reverberation', in Cooper, et al.

Stern, S. J., 1993b, 'Feudalism, Capitalism, and the World System in the Perspective of Latin America', in Cooper, et al.

Stevenson, R. F., 1968, *Population and Political Systems in Tropical Africa*, New York, Columbia Press.

Stichter, S. and J. L. Parpart eds., 1988, *Patriarchy and Class: African Women in the Home and the Workforce*, Boulder, Col., Westview Press.

Stichter, S. and J. L. Parpart eds., 1990, *Women, Employment and the Family in the International Division of Labour*, Philadelphia, Temple University Press.

Stichter, S., 1982, *Migrant Labour in Kenya. Capitalism and African Response, 1895-1975*, London, Longman.

Stichter, S., 1975-6, 'Women and the Labour Force in Kenya, 1895-1964', *Rural Africana*, 29, pp. 45-67.

Stichter, S., 1990, 'Women, Employment and the Family: Current Debates', in Stichter and Parpart, eds.

Stoneman, C. , 'The World Bank and the IMF in Zimbabwe', in Campbell and Loxley, eds., 1989.

Stratton, F., 1983, 'Narrative Method in the Novels of Ngugi', *African Literature Today*, 13, pp. 122-135.

Stratton, F., 1994, *Contemporary African Literature and the Politics of Gender*, New York, Routledge.

Streeten, P., 1977, *Basic Needs*, Washington, DC, World Bank.

Streeten, P. P., 1984, 'Structural Adjustment: A Study of the Issues and Options', in Meier and Seers, eds.

Strobel, M., 1979, *Muslim Women of Mombasa, 1890-1975*, New Haven, Yale University Press.

Stunt, R., 1992, 'CD-ROM Database Production', in *CD-ROM for African Research Needs.*

Sturges, P. and R. Neil, 1990, *The Quiet Struggle: Libraries and Information for Africa*, London, Mansell.

Sudarkasa, N., 1977, 'Women and Migration in Contemporary West Africa', in Wellesley Editorial Committee.

Summers, C., 1991, 'Intimate Colonialism: The Imperial Production of Reproduction in Uganda, 1907-1925', *Signs*, 16, pp. 787-807.

Sunday Nation, Nairobi, Kenya, 1984-1985.

Sunday Standard, Nairobi, Kenya, 1984-1985.

Suret-Canale, J., 1988, *Essays on African History: From the Slave Trade to Neocolonialism*, Trenton, NJ, Africa World Press.

Sussman, M. B., 1993a, 'Commentary on Publishing', *Marriage and Family Review, 18*, pp. 109-117.

Sussman, M. B. 1993b, 'The Charybdis of Publishing in Academia', *Marriage and Family Review, 18*, pp. *161-169.*

Sutcliffe, R. B., 1971, *Industry and Underdevelopment*, London, Addison-Wesley.

Swai, B., 1980, 'Crisis in Africanist History', *Tanzania Zamani*, 22.

Swai, B., 1979, 'Pragmatism, Opportunism, and Africanist History', *Utafiti*, 4.

Swainson, N., 1977, 'The Rise of a National Bourgeoisie in Kenya', *Review of African Political Economy*, 8, pp. 39-55.

Swainson, N., 1980, *The Development of Corporate Capitalism in Kenya, 1917-1977*, London, Heinemann.

Swantz, M., 1985, *Women in Development: A Creative Role Denied? The Case of Tanzania*, New York, St. Martin's Press.

Sweetman, D., 1984, *Women Leaders in African History*, London, Heinemann.

Sweezy, P. M., 1970, *Theory of Capitalist Development*, New York, Monthly Review Press.

Sy, M. A., A. Ba and N. Ndiaye, 1992, 'Demographic Implications of Development Policies in the Sahel: The Case of Senegal' in Toure and Fadayomi, eds.

Syzmanski, R. and J. Agnew, 1981, *Order and Scepticism: Human Geography and the Dialectic of Science*, Washington: Association of American Geographers.

Szereszewski, R., 1965, *Structural Changes in the Economy of Ghana, 1891-1911*, London.

Tabatabai, H., 1986, *Food Crisis and Development Policies in Sub-Saharan Africa*, Geneva, International Labour Office.

Tahi, M. S., 1992, 'The Arduous Democratization Process in Algeria', *Journal of Modern African Studies*, 30, pp. 397-419.

Taiwo, O., 1995, 'Appropriating Africa: An Essay on New Africanist Schools',*Issue: A Journal of Opinion*, 23, pp. 39-44.

Taiwo, O., 1984, *Female Novelists of Modern Africa*, London, Macmillan.

Talle, A., 1988, *Women at a Loss: Changes in Maasai Pastoralism and Their Effects on Gender Relations*, Stockholm, University of Stockholm.

Taylor, A. J. P., 1938, *Germany's Bid For Colonies, 1884-5*, London, Macmillan.

Taylor, P. J., 1986, 'The World Systems Project', in Johnston and Taylor.

Temu, A. J. and B. Swai, 1981, *Historians and Africanist History: A Critique*, London, Zed Press.

Terray, E., 1972, *Marxism and 'Primitive' Societies*, New York, Monthly Review Press.

The Cambridge History of Africa: Vol. 1: *From the Earliest Times to c.500 B. C.*, edited by J. D. Clark; Vol.2: *From c.500 B. C. to A. D. 1050*, edited by J. D. Fage; Vol. 3: *From c.1050 to c.1600*, edited by Roland Oliver; Vol. 4: *From c.1600 to c.1790*, edited by Richard Gray; Vol. 5: *From c.1790 to c.1870*, edited by J. E. Flint; Vol. 6: *From 1870 to 1920*, edited by G. N. Sanderson; Vol. 7: *From 1920 to 1942*, edited by A. D. Roberts; and Vol. 8: *From 1943 to the 1970s*, edited by Michael Crowder, Cambridge, Cambridge University Press, 1975-1986.

The Standard, Nairobi.

Thiombiano, T., 1987, 'State Policies on Agriculture and Food Production in Burkina Faso 1960-83', in Mkandawire and Bourenane, eds.

Thomas, C., 1984, *The Rise of the Authoritarian State in Peripheral Societies*, New York, Monthly Review Press.

Thomas, M. F. and G. W. Whittington eds., 1969, *Environment and Land Use in Africa*, London.

Thomas, N., 1990, 'Narrative Strategies in Bessie Head's Stories', in Abrahams, ed.

Thompson, J., 1988, 'Preparing For the Sale of the Decade', *African Business*, April.

Thompson, W. R. ed., 1983, *Contending Approaches to World System Analysis*, Beverly Hills, Sage.

Thornton, J., 1990, 'The Historian and the Precolonial African Economy', *African Economic History*, 19, pp. 45-54.

Thornycroft, P , 1985, 'Innovation is the Name of the Game', *African Business*, May.

Thrift, N., 1986, 'The Geography of International Economic Disorder', in Johnston and Taylor.

Throup, D., 1987, *Economic and Social Origins of Mau Mau 1945-1953*, London, James Currey.

Tiamiyu, M. A., 1989, 'Sub-Saharan Africa and the Paperless Society: A Comment and a Counterpoint', *Journal of the American Society for Information Science*, 40/5, pp. 325-28.

Tidy, M. with D. Leeming, 1981, *A History of Africa 1840-1914*, Vol.1: *1840-1880*, Volume 2: *1880-1914*, London, Hodder and Stoughton.

Timberlake, L, 1988, *Africa in Crisis: The Causes, the Cures of Environmental Bankruptcy*, London, Earthscan.

Time, 1987, November 2.

Titiola, S. T., 1987, 'The State and Food Policies in Nigeria', in Mkandawire and Bourenane, eds.

Todaro, M. P., 1971, 'Income Expectation, Rural-Urban Migration and Employment in Africa', *International Labour Review*, 104.

Todaro, M. P. and J. R. Harris, 1970, 'Migration, Unemployment and Development: A Two Sector Analysis', *American Economic Review*, 60.

Todaro, M. P. and J. R. Harris, 1968, 'Urban Unemployment in East Africa: An Economic Analysis of Policy Alternatives', *East African Economic Review*, 4.

Tokman, V. E., 1978, 'An Exploration into the Nature of Informal - Formal Sector Relationships', *World Development*, 6.

Toure, M. and T. O. Fadayomi eds., 1992, *Migrations, Development and Urbanization Policies in Sub-Saharan Africa*. Dakar, CODESRIA Book Series.

Toyo, E., 1986, 'Food and Hunger in a Petroleum Neocolony: A Study of the Food Crisis in Nigeria', in Lawrence, ed.

Traeger, L. , 1987, 'A Re-examination of the Urban Informal Sector in West Africa', *Canadian Journal of African Studies*, 21.

Trapido, S., 1972, 'South Africa and the Historians', *African Affairs*, 71, pp. 444-448.

Trebilock, C., 1981, *The Industrialization of the Continental Powers*, London: Longman.

Tshishimbi, B., 1994, 'Missed Opportunity for Adjustment in a Rent-Seeking Society: The Case of Zaire', in Sahn, ed.

Tsie, B., 1995, *The Political Economy of Botswana in SADCC*, Harare, Sapes Books.

Tucker, J. E., 1983, 'Problems in the Historiography of Women in the Middle East: The Case of Nineteenth Century Egypt', *International Journal of Middle East Studies*, 15, pp. 21-36.

Tucker, J. E., 1985, *Women in Nineteenth Century Egypt*, Cambridge, Cambridge University Press.

Tucker, M. E., 1988, 'A "Nice-Time Girl" Strikes Back: An Essay on Bessie Head's *A Question of Power*', *Research in African Literatures*, 19, pp. 170-181.

Turnham, D., 1970, *The Employment Problem in Less Developed Countries*, Paris, OECD.

Turok, B. et.al., 1993, *Development and Reconstruction in South Africa: A Reader*, Johannesburg, Institute for African Alternatives.

Tussie, D. and D. Glover eds. , *The Developing Countries in World Trade*, Boulder and Ottawa, Lynne Rienner and IDRC, 1993.

Twose, N., 1984, *Cultivating Hunger: Food For the Rich*, London, Oxfam.

Udall, A. T. and Sinclair, 1982, 'The "Luxury Unemployment" Hypothesis: A Review of Recent Evidence', *World Development*, 10.

UDASA, 1990, 'The Dar Es Salaam Declaration on Academic Freedom and the Social Responsibility of Academics', *UDASA Newsletter*, 11.

Uguru, 1988, 'Nigeria's Brain Drain', *New African*, October.

Uku, R., 1995, 'Mixed Results from CFA Devaluation', *Africa Recovery*, June.

UNCTAD, 1979, 'The Reverse Transfer of Technology: A Survey of its Main Features, Causes and Policy Implications', TD/B/C/47, New York.

UNCTAD, 1986, *Handbook of International Trade and Development Statistics*, New York, UNCTAD.

UNCTAD, 1988, *Trade and Development Report 1988*, New York, United Nations.

UNCTAD, 1989, *Trade and Development Report 1989*, New York, United Nations.

UNDP and World Bank, 1989, *African Economic and Financial Data*, Washington DC, World Bank.

UNESCO General History of Africa, 1981-1993, Vol.1: *Methodology and African Prehistory*, edited by J. Ki-Zerbo; Vol. 2: *Ancient Civilizations of Africa*, edited by G. Mokhtar; Vol. 3: *Africa from the Seventh to Eleventh Century*, edited by D. T. Niane; Vol. 5: *Africa from the Sixteenth to the Eighteenth Century*, edited by B. A. Ogot; Vol. 6: *The Nineteenth Century Until 1880*, edited by J. F. A. Ajayi; Vol. 7: *Africa Under Foreign Domination, 1880-1935*, edited by A. A. Boahen; and Vol. 8: *Africa Since 1935*, edited by A. A. Mazrui, London, Heinemann.

UNESCO, 1986, 'Final Report, Meeting of a Working Group to Determine the Procedure for the Pre-History of Africa', BREDA, Dakar, November.

UNESCO, 1981-1993, *General History of Africa*, 8 Volumes, various editors. London, Heinemann.

UNESCO, 1984, *Social Science Research and Women in the Arab World*, London, Francis Pinter

UNIDO, 1986a, *Industrial Development Review Series: Ghana*, Vienna, UNIDO.

UNIDO, 1986b, *Industrial Development Review Series: Egypt*, Vienna, UNIDO.

UNIDO, 1986c, *Industrial Development Review Series: Côte d'Ivoire*, Vienna, UNIDO.

UNIDO, 1986d, *Industrial Development Review Series: Tanzania*, Vienna, UNIDO.

UNIDO, 1986e, *Industrial Development Review Series: Mali*, Vienna, UNIDO.

UNIDO, 1986f, *Industrial Development Review Series: Central African Republic*, Vienna, UNIDO.

UNIDO, 1986g, *Industrial Development Review Series: Zaire*, Vienna, UNIDO.

UNIDO, 1987, *Industrial Development Review Series: Botswana*, Vienna, UNIDO.

UNIDO, 1988a, *Industrial Development Review Series: Kenya*, Vienna, UNIDO.

UNIDO, 1988b, *Industrial Development Review Series: Nigeria*, Vienna, UNIDO.

UNIDO, 1988c, *Industrial Development Review Series: Somalia*, Vienna, UNIDO.

UNIDO, 1989, *Industrial Development Review Series: Ghana*, Vienna, UNIDO.

United Nations, 1988, *Financing Africa's Recovery: Report and Recommendation of the Advisory Group on Financial Flow to Africa*, New York, United Nations.

United Nations, 1989, *Uruguay Round: Papers on Selected Issues*, New York, UNCTAD.

Urdang, S., 1983, 'The Last Transition? Women and Development in Mozambique', *Review of African Political Economy*, 27/28, pp. 8-23.

Urdang, S., 1984, 'Women in National Liberation Movements', in Hay and Stichter, eds.

Urdang, S., 1979, *Fighting Two Colonialisms: Women in Guinea-Bissau*, New York, Monthly Review Press.

Uzoigwe, G. N., 1974, *Britain and the Conquest of Africa*, Ann Arbor: University of Michigan Press.

Vail, L. and L. White, 1981, *Capitalism and Colonialism in Mozambique*, London, Heinemann.

van Adams, et al., 1992, *The World Bank's Treatment of Employment and Labour Market Issues*, World Bank Technical Paper No. 177.

van Allen, J., 1976, '"Aba Riots" or Igbo "Women's War"? Ideology, Stratification, and the Invisibility of Women', in Hafkin and Bay, eds.

van Onselen, C., 1976, *Chibaro: African Mine Labour in Southern Rhodesia*, Nottingham, Pluto.

van Onselen, C., 1982a, *New Babylon*, Vol. 1 of *Studies in the Social and Economic History of the Witwatersrand, 1886-1914*, London, Longman.

van Onselen, C., 1982b, *New Nineveh*, Vol. 2 of *Studies in the Social and Economic History of the Witwatersrand, 1886-1914*, London, Longman.

van Sertima, I., ed., 1985, *Black Women in Antiquity*, New Brunswick, NJ, Transaction Books.

Van Zwanenberg, R. M. A., 1975, *Colonial Capitalism and Labour in Kenya 1919-1939*, Nairobi, East African Publishing House.

Vance, C. S., ed., 1984, *Pleasure and Danger: Exploring Female Sexuality*, Boston, Routledge & Kegan Paul.

Vandemoortele, J., 1987, 'Youth Employment in Sub-Saharan Africa: Some Facts and Figures', in ILO/JASPA.

Vandemoortele, J., 1991, 'Labour Market Informalisation in Sub-Saharan Africa', in Standing and Tokman, eds.

Vansina, J., 1993, 'Unesco and African Historiography', History in Africa, 20, pp. 337-352.

Vansina, J., 1985, Oral Tradition as History, Nairobi, Heinemann Kenya.

Vansina, J., 1965, Oral Tradition: A Study in Historical Methodology, London, Routledge and Kegan Paul.

Vaughan, M. , 1994, 'Colonial Discourse Theory and African History, or has Postmodernism passed us by?', Social Dynamics, 20, pp. 1-23.

Vaughan, M, 1985, 'Household Units and Historical Process in Southern Malawi', Review of African Political Economy, 34, pp. 35-45.

Virashawmy, R., 1987, 'State Policies and Agriculture in Mauritius', in Mkandawire and Bourenane, eds.

Visel, Robin, 1990, '"We Bear the World and We Make It": Bessie Head and Olive Schreiner', Research in African Literatures, 21, pp. 114-124.

Volman, D., 1993, 'Africa and the New World Order', Journal of Modern African Studies, 31, pp. 1-30.

von Albertini, R., 1971, Decolonization: The Administration and Future of the Colonies, New York, Doubleday.

von Freyhold, M., 1977, 'The Post-Colonial State and its Tanzanian Version', Review of African Political Economy, 8, pp. 75-89.

Wa Thiong'o, N., 1967, A Grain of Wheat, London, Heinemann.

Wa Thiong'o, N., 1977, Petals of Blood, London, Heinemann.

Wa Thiong'o, N., 1981, Writers in Politics, London, Heinemann.

Wa Thiong'o, N., 1982, Devil on the Cross, London, Heinemann.

Wa Thiong'o, N., 1986, Decolonising the Mind.The Politics of Language in African Literature, Nairobi, Heinemann Kenya.

Wa Thiong'o, N. , 1987, Matigari, Oxford, Heinemann.

Wa Thiong'o, N., 1993, Moving the Centre: The Struggle for Cultural Freedoms, London, James Currey.

Walker, C., 1982, Women and Resistance in South Africa, London, Onyx Press.

Walker, C., ed., 1990, Women and Gender in Southern Africa to 1945, London: James Currey.

Wallerstein, I., 1973, 'Class and Class Conflict in Contemporary Africa', Canadian Journal of African Studies, 7, pp. 375-380.

Wallerstein, I., 1974, The Modern World System, New York, Academic Press.

Wallerstein, I., 1976, 'The Three Stages of African Involvement in the World Economy', in Gutkind and Wallerstein, eds.

Wallerstein, I. , 1979, The Capitalist World Economy, Cambridge, Cambridge University Press.

Wallerstein, I. , 1980, The Modern World System, Vol. 11., New York, Academic Press.

Wallerstein, I., 1983, Historical Capitalism, London, Verso.

Wallerstein, I., 1984, The Politics of the World Economy, Cambridge, Cambridge University Press.

Wallerstein, I., 1995, 'Revolution as Strategy and Tactics of Transformation', in Callari, Cullenberg and Biewener, eds.

Wamba-dia-Wamba, E., 1992, 'Africa in Search of a New Historical Mode of Politics', CODESRIA Seventh General Assembly, Democratization in Africa: Problems and Prospects.

Wamba-dia-Wamba, E., 1986, 'How is Historical Knowledge Recognized?', *History in Africa*, 13, PP. 331-344.

Wamba-dia-Wamba, E., 1987, *History of Neo-Colonialism or Neo-Colonialist History?* Africa Research and Publications Project. Working Paper n° 5.

Ward, B. and R. Dubois, 1972, *Only One Earth*, London, Andre Deutsch..

Ward, K. B. and L. Grant, 1985, 'The Feminist Critique and a Decade of Published Research in Sociology Journals', *Sociological Quarterly*, 26, pp. 139-157.

Warren, B., 1973, 'Imperialism and Capitalist Industrialization', *New Left Review*, 81.

Warren, B., 1980, *Imperialism: Pioneer of Capitalism*, London, Verso.

Warren, N., 1990, 'Pan-African Cultural Movements: From Baraka to Karenga', *The Journal of Negro History*, 75, pp. 16-27.

Wastberg, P., ed., 1968, *The Writer in Modern Africa*, Uppsala, The Scandinavian Institute of African Studies.

Waterman, P., 1975, 'The Labour Aristocracy Thesis in Africa: Introduction to a Debate', *Development and Change*, 6, pp. 57-73.

Waterman, P., 1977, 'On Radicalism in African Studies', in Gutkind and Waterman, eds.

Watson, W., 1958, *Tribal Cohesion in a Money Economy*, Manchester, Manchester University Press.

Watts, M., 1983, *Silent Violence: Food, Famine and Peasantry in Northern Nigeria*, Berkeley, University of California Press.

Weber, R. P., 1983, 'Cyclical Theories of Crises in the World System', in Bergesen, ed.

Webster, J. B. , 1982, 'Footnote to Flint - A View From the Periphery', Center for African Studies, Dalhousie University.

Webster, J. B. ed., 1979, *Chronology, Migration and Drought in Interlacustrine Region*, London, Longman.

Weeks, J., 1973, 'Does Employment Matter?', in Jolly, et.al., eds.

Weeks, J. , 1975, 'Policies for Expanding Employment in the Informal Urban Sector of Developing Countries', *International Labour Review*, 111.

Weeks, J., 1991, 'The Myth of Labour Market Clearing', in Standing and Tokman, eds.

Wehler, H-U. , 1969, *Bismarck and Imperialism*, Cologne, Koln.

Weil, R., 1993, 'Somalia in Perspective', *Review of African Political Economy*, 57, pp. 103-9.

Weir, A., 1986, 'IMF Has Role in Nigerian Debt Plans', *African Business*, May.

Weir, A., 1985, 'Togo Privatises National Steel Corporation', *African Business*, January.

Weisbord, R., 1973, *Ebony Kinship: Africa, Africans, and the Afro-American*, Westport, Conn. , Greenwood Press.

Weiss, R., 1986, *The Women of Zimbabwe*, London, Kesho Publications.

Wellesley Editorial Committee, 1977, *Women in National Development: The Complexities of Change*, Chicago, University of Chicago Press.

West, C., 1991, 'The Dilemma of the Black Intellectual', in Hooks and West.

West, C., 1993, *Race Matters*, New York, Vintage Books.

West, M. O., 1993, 'Pan-Africanism, Capitalism and Racial Uplift: The Rhetoric of African Business Formation in Colonial Zimbabwe', *African Affairs*, 92, pp. 263-283.

Westlake, M., 1985, 'The IMF's African Nightmare', *South*, July.

Westlake, M., 1986a, 'US Economy: Standby For the Crash', *South*, December.

Westlake, M., 1986b, 'The Price of an African Recovery', *South*, August.

Westlake, M., 1987, 'The Bulls Head South', *South*, June.

Westlake, M., 1988, 'Shape Up or Ship Out', *South*, September.

Westlake, M. and L. Jayawardena, 1985, 'Zones of Special Interest', *South*, November.

Westlake, M. with K. Watkins, 1987,'Farm Trade Wars', *South*, November.

Wheelwright, J., 1991, 'Pressed to Quit', *Index on Censorship*, 20, 6, pp. 26-27.

White, E. F., 1990, 'Africa on My Mind: Gender, Counter Discourse and African-American Nationalism', *Journal of Women's History*, 2, pp. 73-97.

White, E. F., 1987, *Sierra Leone's Settler Women Traders: Women on the Afro-European Frontier*, Ann Arbor, University of Michigan Press.

White, L., 1984, 'Women in the Changing African Family', in Hay and Stichter.

White, L., 1990, *The Comforts of the Home: Prostitution in Colonial Nairobi*, Chicago and London, University of Chicago Press.

Wickins, P., 1981, *An Economic History of Africa: From the Earliest Times to the Partition*, Cape Town, Oxford University Press.

Widstrand, C. ed., 1975, *Multinational Firms in Africa*, Uppsala: Scandinavian Institute of African Studies.

Wilkinson, J., ed., 1992, *Talking With African Writers*, London, James Currey.

Wilks, I., 1970, 'African Historiographical Traditions', in Fage, ed.

Wilks, I. , 1975, *Asante in the Nineteenth Century*, Cambridge, Cambridge University Press.

Wilks, I., 1976, in Fyfe, ed.

Wilks, I., 1988, 'She Who Blazed a Trail: Akyaawa Yikwan of Asante', in Romero, ed.

Williams, A., 1992, 'Towards a Theory of Cultural Production in Africa', *Research in African Literatures*, 22, pp. 5-20.

Williams, C. H., 1986, 'The Question of National Congruence', in Johnston and Taylor, eds.

Wilson, A., 1991, *The Challenge Road: Women and the Eritrean Revolution*, Trenton, NJ, The Red Sea Press.

Wilson, G., 1941, *An Essay on the Economics of Detribalization in Northern Rhodesia*, Lusaka, Rhodes-Livingstone Institute.

Wilson, G. and M. Wilson, 1945, *The Analysis of Social Change*, Cambridge, Cambridge University Press.

Wilson, H. S., 1994, *African Decolonization*, London and New York, Edward Arnold.

Wolf, E., 1982, *Europe and the People Without History*, Berkeley, University of California Press.

Wolpe, H., 1978, 'A Comment on "The Poverty of Neo-Marxism"', *Journal of Southern African Studies*, 4, pp. 240-256.

Wolpe, H., 1980, ed. , *The Articulation of Modes of Production*, London, Routledge and Kegan Paul.

Wolpe, H., 1990, *Race, Class and the Apartheid State*, Trenton, NJ, Africa World Press.

Women's History, 1991, *Journal of Women's History*, 2, pp. 58-68.

Wood, A. P. and E. C. W. Shula, 1987, 'The State and Agriculture in Zambia: A Review of the Evolution and Consequences of Food and Agricultural Policies in a Mining Economy', in Mkandawire and Bourenane, eds.

World Bank and UNDP, 1989, *Africa's Adjustment and Growth in the 1980s*, Washington, DC, World Bank.

World Bank, 1975, *The Assault on World Poverty*, Baltimore, Johns Hopkins University.

World Bank, 1981, *Accelerated Development in Sub-Saharan Africa: An Agenda for Action*, Washington, DC, World Bank.

World Bank, 1982, *World Development Report*, New York, Oxford.

World Bank, 1984, *Towards Sustainable Development in Sub-Saharan Africa: A Joint Programme of Action*, Washington, DC, World Bank.

World Bank, 1986, *World Development Report 1986*, New York, Oxford University Press.

World Bank, 1987, *World Development Report 1987*, New York, Oxford University Press.

World Bank, 1989, *Sub-Saharan Africa, From Crisis to Sustainable Growth: A Long-Term Perspective Study*, Washington, DC, The World Bank.

World Bank, 1990, *World Development Report 1990*, New York, Oxford University Press.

World Bank, 1992a, *World Development Report 1992*, New York, Oxford University Press.

World Bank, 1992b, *Developing Effective Employment Service*, Human Resources Sector Operations Division, World Bank, June.

World Bank, 1993, *World Development Report 1993, New York, Oxford University Press*.

World Bank, 1995, *A Continent in Transition: Sub-Saharan Africa in the Mid-1990s*, Washington, DC, World Bank.

World Resources Institute and IIED, 1986, *World Resources 1986*, New York, Basic Books.

World Resources Institute and IIED, 1988, *World Resources 1988-89*, New York, Basic Books.

Wright, D., 1990, 'Somali Powerscapes: Mapping Farah's Fiction', *Research in African Literatures*, 21, pp. 21-34.

Wright, H. M., 1977, *The Burden of the Present: Liberal-Radical Controversy Over Southern African History*, London, Rex Collins.

Wright, H. M., ed., 1976, *The 'New Imperialism': Analysis of Late Nineteenth-Century Expansion*, Lexington, MA, D. C. Heath and Company.

Wright, M., 1983, 'Technology, Marriage and Women's Work in the History of Maize Growers in Mazabuka, Zambia: A Reconnaissance', *Journal of Southern African Studies*, 10, pp. 71-85.

Wylie, D., 1989, 'The Changing Face of Hunger in Southern Africa', *Past and Present*, 122, pp. 159-199.

Yankaner, A., 1991, 'How Blind is Blind Review?', *American Journal of Public Health*, 81, pp. 843-845.

Young, C., 1988, 'The Colonial State and the Post-Colonial Crisis', in Gifford and Louis, ed.

Young, C., 1985, 'The Colonial State and Its Connection to Current Political Crises in Africa', Conference on African Independence and Consequences of the Transfer of Power, 1956-1980, University of Zimbabwe, Harare, January.

Young, C. and T. Turner, 1985, *The Rise and Decline of the Zairean State*, Madison, University of Wisconsin Press.

Young, S., 1977, 'Fertility and Famine: Women's Agricultural History in Southern Mozambique', in Palmer and Parsons, eds.

Zack-Williams, A., 1995, 'Development and Diaspora: Separate Concerns?', *Review of African Political Economy*, 65, pp 49-358.

Zartman, I. W. ed., 1995, *Collapsed States. The Disintegration and Restoration of Legitimate Authority*, Boulder, Col, Lynne Rienner.

Zeleza, P. T., 1982, *Dependent Capitalism and the Making of the Kenyan Working Class During the Colonial Period*, Unpublished Ph.D. dissertation, Dalhousie University.

Zeleza, P. T., 1984, *Record of Minutes of UNESCO-Sponsored Consultative Meeting on the Revision of History Textbooks in East and Central Africa*, Nairobi, December.

Zeleza, P. T. , 1985, *Record of Minutes of UNESCO-Sponsored Consultative Meeting on the Revision of History Textbooks in East and Central Africa*, Nairobi, March.

Zeleza, P. T. and I. Sindiga, 1985, 'The Western Expatriate Scholar and Africa', History Department Staff Seminar Paper, Kenyatta University.

Zeleza, P. T. , 1985a, 'The Problems of Teaching African Economic History', Paper presented to UNESCO Conference on Revision of History Textbooks in East and Central Africa, Nairobi, March.

Zeleza, P. T., 1985b, *Record of Minutes of UNESCO-Sponsored Consultative Meeting on the Revision of History Textbooks in East and Central Africa*, Nairobi, March.

Zeleza. P. T., 1987, *Imperialism and Labour: The International Relations of the Kenyan Labour Movement*, Kisumu, Anyange Publications.

Zeleza, P. T. and B. A. Ogot, 1988, 'Kenya's Road to Independence and After', in Gifford and Lewis, eds.

Zeleza, P. T., 1988a, 'Women and the Labour Process in Kenya Since Independence', *Transafrican Journal of History*, 17:69-107.

Zeleza, P. T., 1988b, *Labour, Unionization and Women's Participation in Kenya 1963-1987*, Nairobi, Friedrich Ebert Foundation.

Zeleza, P. T., 1988, 'African Sugar in the World Market', *Journal of African Research and Development*, 18, pp 1-23

Zeleza, P. T., 1989, 'Review of *Africa's Adjustment and Growth in the 1980s*, by World Bank and UNDP, Washington, The World Bank, 1989', *Journal of Eastern African Research and Development*, 19, pp. 187-91.

Zeleza, P. T. , 1989, 'The Problems of Writing History Textbooks: The Case of East Africa', Paper presented to Meeting of Experts on Textbooks and Teaching of History in African Schools, UNESCO, Nairobi, March.

Zeleza, T., M. Sharman, and M. Williams, 1989, *Revising for Government for K.C.S.E.*, Nairobi and London, Evans Brothers.

Zeleza, P. T., 1991, 'Economic Policy and Performance in Kenya Since Independence', *Journal of Eastern African Research and Development*, 21, pp. 35-76.

Zeleza, P. T., 1991, unpublished, 'Rethinking Development and Underdevelopment, Review Essay'.

Zeleza, P. T., 1992, *Smouldering Charcoal*, Oxford, Heinemann.

Zeleza, P. T., 1992, 'The Development of African Capitalism: Review Article', *Africa Development*, 15, pp. 129-136.

Zeleza, P. T. , 1993, 'The Strike Movement in Colonial Kenya', *Transafrican Journal of History*, 21, pp. 1-23.

Zeleza, P. T., 1993, *A Modern Economic History of Africa*, Vol 1 *The Nineteenth Century*, Dakar, CODESRIA Book Series.

Zeleza, P. T., 1994, *The Joys of Exile: Stories*, Toronto, House of Anansi Press.

Zeleza, P. T., 1994, 'Noma Award Acceptance Speech', *The African Book Publishing Record*, 20/4, pp. 238.

Zeleza, P. T., 1994a , *Maasai*, New York: Rosen Publishing Group.

Zeleza, P. T., 1994b, *Kamba*, New York, Rosen Publishing Group.

Zeleza, P. T., 1994c, *Mijikenda*, New York: Rosen Publishing Group.

Zeleza, P. T., 1995, 'The Moral Economy of Working Class Struggle: Strikers, the Community and the State in the Mombasa General Strike', *Africa Development*, 20, pp. 51-87.

Zeghidi, S., 1995, 'Tunisian Trade Unionism: A Central Pole of Social and Democratic Challenge', in Mamdani and Wamba-dia-Wamba, eds.

Zeleza, P. T., 1995, 'Bullies in Uniform: Military Misrule in Nigeria', *Africa Development*, 20, pp. 3-51.

Zeleza, P. T., 1995, 'Review of *Adjusting to Policy Failure in African Economies*', by D. Sahn, ed., *The International Journal of African Historical Studies*, 25, pp. 416-418.

Zeleza, P. T. and M. C. Diop, 1995, 'Report of CODESRIA Eighth General Assembly', *CODESRIA Bulletin*, n° 4, pp. 3-15.

Zeleza, P. T., 1995, 'Memories of Struggle. Review of *Nehanda*, by Yvonne Vera. Toronto: Tsar Publications, 1994', *The Toronto Review*, 13, pp. 87-89.

Zeleza, P. T., 1996, 'Review of *African Decolonization*', *Journal of African History*.

Zeleza, P. T., forthcoming, *A Modern Economic History of Africa*, Vol. 2: *The Twentieth Century* .

Zeleza, P. T. and M. C. Diop, eds., forthcoming, *Beyond Crisis in Africa: Struggles and Transformations*, Dakar, CODESRIA Book Series.

Zeleza, P. T., forthcoming, 'Resurrecting South Africa From the Ashes of Apartheid', *Afrika Zamani*.

Zeleza, P. T., 'Pan-African Trade Unionism: Unity and Discord', *Transafrican Journal of History*, Vol. 15, pp. 164-190.

Zelinsky, W., 1975, 'The Demigod's Dilemma', *Annals of the Association of American Geographers*, 55.

Zell, H. M., 1995, 'Effective Promotion and Marketing, and the Size of the Export Market for African Books', *African Publishing Review*, 4/2, pp. 16-18.

Zell, H. M., 1993, 'Publishing in Africa: The Crisis and the Challenge', in Owomoyela, ed.

Zghal, A., 1995, 'The 'Bread Riot' and the Crisis of the One-Party System in Tunisia', in Mamdani and Wamba-dia-Wamba, eds.

Ziemann, W. and Lanzendorfer, 1977, 'The State in Peripheral Societies', *Socialist Register*, London, Merlin.

Zimbabwe Association of University Teachers, 1990, 'Statement on Academic Freedom', *UDASA Newsletter*, 11.

Zolberg, A. R., 1992, 'The Spectre of Anarchy: African States Verging on Dissolution', *Dissent*, 39, pp. 303-311.

Zupnick, E. *Britain's Post-War Dollar Problem*, New York, Columbia University Press.

Index

A

academic freedom 10, 39-42
African historiography 148-150,
 156-157, 178-182
Afrocentrism 504
Afropessimism 127
agribusiness 263-266, 268
agriculture 248, 252-253,
 255-256, 269-271
 cash crops 256-259
 employment 338
 and environment 245-247
 and peasants 271-272
 and the state 249-251, 385-386
aid
 and libraries 74, 80
anthropology 90, 495, 498
apartheid 92-93, 415
articulation of modes of
 production (AMOP) 98-102

B

balance of payments 280, 301

C

capitalism 96
 and apartheid 92-93
 and dependency 97
 and the state 127
censorship 28, 32
 effects of 390-392, 398
 enforcement of 392
 operation of 393-398, 400, 402-403
 origins of 392, 400
 and publication 394-398
 self-censorship 399-401
citizenship 374-375
civil society 29-30, 426-427
 and intellectuals 28-31

and social sciences 28-31
and the state 30, 123-124, 126, 393,
 413-414, 517
class 91, 102-103, 179, 237-238
classification 152-153, 155
Colonial Development and
 Welfare (CD&W) 225, 230, 234,
 238-239, 257
colonialism 230
 and democracy 411-412
 and development 218-220
 and economy 227-232, 235-238
 and food production 254-259
 and health 254-255
 and history 145-146
 and nationalism 428-429
 resistance to 232-233
 See also colonization, imperialism,
 neo-colonialism, post-colonialism
colonization
 reasons for 120-121
 state formation 386-388
community 450-455

D

databases 77
debt 302-303, 305
debt relief 305-309
decolonization 91, 118-119, 121,
 259
 and nationalism 116-118, 121
deconstruction 494
democracy 373, 375-376, 388,
 411, 422-423, 427, 437-438,
 443-446, 461
 and colonialism 411-412
 and development 41-42
 and economy 386
 liberal 374-375, 410
 rejection of 421